W9-CZF-724

Teacher's Manual

Spaceship Earth

Earth Science

Teacher's Edition: Manual and Annotations

Printed in the U.S.A.

Rationale

For the first five years of a student's life, most learning is accomplished through touching, seeing, listening and experimentation. He or she can walk, run, comply with social norms, assemble toys, count, speak fluently one of the world's most difficult languages, and even rationalize away problems. All by the age of five and without formal instruction!

With this in mind, the greater part of *Spaceship Earth/Earth Science* involves discussions revolving around activities of various kinds. This gives the student a chance to continue learning by seeing, touching, listening and experimenting. In a few cases, for reasons of economy in time and money, the teacher may carry out the activity as a demonstration. In *all* cases, the student is given a reason for participating in the activity and a basis for discussing the results in a meaningful fashion.

The equipment for carrying out the activities is neither complicated nor expensive. Most of the materials called for can be bought in local stores. A complete list of materials for the year appears in this manual and individual chapter lists appear with the notes for each chapter.

Throughout the book an attempt has been made to incorporate the major conceptual schemes outlined in the National Science Teachers Association's publication, *Theory Into Action*. NSTA calls them "big ideas" that summarize areas of inquiry and provide an understanding of the means by which scientific knowledge is obtained.

Briefly, any course in science should emphasize major concepts that link it to all other courses in science. Such concepts are:

- I All matter is composed of units called fundamental particles; under certain conditions these particles can be transformed into energy and vice-versa.
- II Matter exists in the form of units which can be classified into hierarchies of organizational levels.
- III The behavior of matter in the universe can be described on a statistical basis.
- IV Units of matter interact. The bases of all ordinary interactions are electromagnetic, gravitational, and nuclear forces.
- V All interacting units of matter tend toward equilibrium states in which the energy content (enthalpy) is a minimum and the energy distribution (entropy) is most random. In the process of attaining equilibrium, energy transformations occur. Nevertheless, the sum of energy and matter in the universe remains constant.
- VI One of the forms of energy is the motion of units of matter. Such motion is responsible for heat and temperature and for the states of matter: solid, liquid, and gaseous.
- VII All matter exists in time and space and, since interactions occur among its units, matter is subject in some degree to changes with time. Such changes may occur at various rates and in various patterns.

In addition to the basic concepts, there are several basic processes stressed by the program that can be effectively used as threads for binding together any and all courses of science. The processes are observing, classifying, measuring, predicting, inferring, formulating hypotheses, experimenting, interpreting data, and using numbers. Marginal notes indicate some places where these processes or concepts appear in the body of the text. You and the students can find others.

The text itself has been written with the student in mind. Chapter openers have been designed to entertain the reader and develop interest in the study of what is to follow. The self check questions at the end of each topic deviate from many of the kinds of questions that are traditionally asked. Quite often the questions require involvement with other students or activities at home.

The end of chapter questions, another textbook tradition, have been replaced with games or puzzles of one kind or another. The games are both fun and instructive. In addition, the games and puzzles lend themselves to easy grading by the teacher and provide immediate learning reinforcement for the student. Since they summarize chapter content, they even serve as spot sources of material for designing quiz or test questions.

Feature articles have been added to the text to stand out from the regular flow. In some cases the feature takes the student beyond the minimum required subject depth. In other cases the feature outlines an optional activity, the work of a particular scientist, or a recent development in the subject area.

Photos and art have been chosen or designed to accent the reading. Each piece is specifically mentioned in the text.

Supplementary information, answers to questions, and materials lists are found in the margins of this Teacher's Edition. The manual, to which this is a preface, furnishes background material and descriptions of extra activities that are too long for marginal annotations. We hope that the manual notes and annotations will suggest other subjects for study and other ways of illustrating earth science. All of this because we believe that science students at all levels of ability are entitled to a course that is not watered down.

Schedule

This suggested schedule is based upon the following assumptions:

(1) The school year is 180 days.

(2) Each earth science class has one 45 minute period per day.

(3) 15 periods will be used for review at the end of the school year.

(4) 11 periods will be used for other school activities, movies, and field trips.

There is ample time in this schedule to allow for additions to the course that follow particular interests of teacher and students, or for pursuing any subjects in greater depth. The numbers of periods listed on the chart include time for tests and discussions of tests.

Chapter	Number of Periods	Comments
Introduction	2	Introduction to course and book, and any activities you choose to get the course going.
1	8	Two periods for an activity you might like to add to the text material. Begin to mix subject matter of Chapters 1, 2, and 3, if you wish.
2	6	One period for discussion of current articles and one additional activity. Mix Chapters 1, 2, and 3, if you wish.
3	7	One period for bringing Chapters 1, 2, and 3 together.
4	10	Two days for discussion of recent articles. One period for making class chart on "The Planets Compared" and one for in-depth discussion.
5	10	Two periods for discussion of recent articles and two for extra activities.
6	5	One period for discussion of recent articles or extra activities.
7	9	One period for extra activities or additional discussion of climates. Begin to mix Chapters 7 and 8, if you wish.
8	10	15 minutes each period to make weather forecast and one period for extras.
9	9	One period of work in class on recording of data and observations in notebook. One field trip around the school building and one extra activity.
10	7	One period for reading and discussion of accounts of volcano activity from other sources.
11	7	Allows for use of fault box and one extra epicenter location activity.

Chapter	Number of Periods	Comments
12	5	Allows for one field trip around the school.
13	11	One period of random work on stream tables at beginning, and one extra activity.
14	9	Allows for an extra activity and a lecture on the nature of the sea floor.
15	8	One period for arguments or repeating of activities. Another day for reading of current information.
16	8	Allows time to play detective with the Holmes mystery.
17	9	One period for field trip around the school for signs of the passage of time, or an additional activity.
18	7	Allows for discussion of controversial topics. Accept all arguments and give equal time.
19	12	Four periods for discussion of local issues. One period for an extra activity. One period for writing letters in class.

Visual Aids

A bibliography of films for classroom use, many of them suitable for junior high and high school even though the title indicates college level only: Potter, Noel Jr., Richard L. Batells, and George R. Rapp, Jr., *An Annotated Bibliography of 16-mm Films Useful in College-Level Geology and Earth Science Courses*. Council on Education in the Geological Sciences, American Geological Institute, 2201 M Street, NW, Washington, D. C. 20037.

A set of 25 overhead visuals covering numerous topics in earth science (including astronomy) can be obtained from the Houghton Mifflin Company. Title: *Earth Science Overhead Visuals*. Contact the nearest regional office:

Pennington-Hopewell Rd., Hopewell, N.J. 08525
666 Miami Circle, NE, Atlanta, Ga. 30324
1900 So. Batavia Ave., Geneva, Ill. 60134
6626 Oakbrook Blvd., Dallas, Texas 75235
777 California Ave., Palo Alto, Calif. 94304.

Three maps of the lunar surface in varying sizes: *Lunar Reference Mosaics,* Catalogue No. D301.49/4:LEM (LEM-1, Scale 1:5,000,000, 34″ × 35″, $1.00; LEM-1A, Scale 1:10,000,000, 18″ × 19″, $0.35; LEM-1B, Scale 1:2,500,000, 2 sheets 56″ × 70″, $2.00.) They are available from: Superintendent of Documents, U.S. Government Printing Office, Washington, D. C. 20402.

Photographs and 35-mm color slides from the Gemini and Apollo programs can be obtained. For a catalog write to: Technology Application Center, University of New Mexico, Box 181, Albuquerque, New Mexico 87106.

You can send for current satellite photographs showing the topography of the land in your area. Request that the photographs show cloud cover if you wish to use them for the weather unit. State your latitude and longitude to receive the 70 mm slides for the opaque projector or black-and-white prints. For a catalog and prices write to: EROS Data Center, 132 South Dakota Street, Sioux Falls, South Dakota 57198.

Student Preface

The photographs are close-ups of soap bubbles. For a demonstration, pour one part dishwashing liquid into two parts water. Mix gently, to avoid making bubbles. Pour the mixture into a shallow glass or clear plastic pan. A depth of half an inch will do. Blow through a straw to make the bubbles—the smaller the straw diameter the better. If the pan is on a white background, the setup for the photographs will be modelled.

Total Materials List

The following is a list of materials for the entire course. The quantity of each item specified is considered to be a reasonable maximum number required for a class of 30. You will probably need to adjust the amounts to your particular class arrangement. For example, you may prefer to have students work in pairs whenever possible. In that case, where 30 items are listed, you need only 15. On the other hand, you may be able to build or acquire a larger number of earth boxes than the number in the list.

If you modify aquariums for use as earth boxes by inserting opaque partitions, students will not be able to look through both sides of the box. For this reason, you may want to have more aquariums than the number suggested. An aquarium modified in this way can be conveniently used by three students. If the box is transparent all around, it can be used by five or six students. (Instructions for making an earth box are in the Background section for Chapter 7 of this manual.) The number of students per box may also be determined by the spatial arrangement of the room and the sizes and shapes of the tables.

CLASSROOM EQUIPMENT

Glassware

Beaker, 250 ml, or paper cups 20
Beaker, 500 ml, or pint jar 10
Beaker, 1000 ml, or 250 ml (optional) 2(6)
Beaker, 2000 ml or large pan 1
Bottle, 1 qt, with 2 hole stopper 6
Bottle, 1 gram 30
Eyedropper 30
Funnel, long stem 10
Glass plate, 4″ × 4″ 30
Graduated cylinder, 50 ml 30
Graduated cylinder, 1000 ml 10
Jar, gallon size, wide-mouth 10
Microscope slide 30
Test tube, large 30

Other Containers

Bottle, plastic, about 250 ml 30
Deep pan or dish, for "moonstuff" 6
Dishpan or equivalent 15
Frying pan, small, with cover 1
Gallon can with plastic liner, or plastic jug 1 or more
Pan, 12″ × 8″ × 4″ 15
Pie plate, or equivalent 10
Pill bottle, transparent, about 50 ml 30
Plastic cup 15
Salt shaker 10
Shoe box 6
Tin can, #10 or 1 qt 10
Tube, plastic, 2″ × 24″, with drain and clamp 10

Measuring Devices

Meterstick 15
Protractor 30
Ruler, 6 or 12 in, 15 or 30 mm 30
Small thermometer 24
Stop clock 5
Stop watch 5
Watch, with second hand 10

Instruments

Astrolabe, made by students 6–30
Barometer 1
Flashlight 6
Hot plate, or Bunsen burner 5
Magnetic compass 30
Magnifying glass 30
Microscope, binocular 1
Paper punch 1
Propane torch, or Bunsen burner 5
Record player (optional) 1
Slide or overhead projector 1
Sling psychrometer 1
Smoke generator (optional) 1
Stapler 1 or more
Wind-speed indicator 1

Tools

Drawing compass 30
Knife, plastic 6
Plastic spoon 30
Scissors 15
Screen sieve at least 2 sets of 4
Screen sieve, 2 sizes 30
Small hammer 10

Miscellaneous Equipment

Aluminum funnel and rubber or plastic tube, 6 in (optional) 6

Ball chain with eyelet 30
Bar magnet 30
Board, 1 m long, with cloth to cover half lengthwise 15
Board, 2′ × 3″, $\frac{1}{4}$–$\frac{1}{2}$ in thick 10
Dowels, or round pencils, 6 in long 20
Earth box or aquarium 10
Fault box (optional) 1 or more
Globe, 10–12 in 6
Light bulb, 300w photoflood 6
Paper or cardboard tube 11″ × 1$\frac{1}{2}$″ diam 30
Plastic or rubber tubing 12 ft
Ringstand with clamp 10
Rod for ringstand 5
Rope 50 ft
Socket and extension cord 6
Spring clip 30
Stream table 10
Trough, 3 ft long 6
Tubing and pail, for emptying stream table if necessary 10
Turntable 10
Wood or cardboard, 8″ × 10″ 30
Wood or clay base for punk sticks 12
Wooden block, 2″ × 4″ to fit across a modified aquarium or earthbox for 15.1 activity 20 or more

Supplies

Ball, rubber or plastic, 3–4 in diam 10
Ball, $\frac{7}{8}$–1 in diam 10
Ball, 2–3 in diam, for earth 5
Ball, 1 in diam, for moon 5
Candle stub 10
Clamshell 30
Colored pencil, 5 colors 30
Colored pencil, 2 colors 60
Cork, for test tube to be broken 30
Cork, for pop bottle 30
Cork, for 250 ml plastic bottle if necessary 30
Crushed stone, small 15 lb
Crushed stone, large 15 lb
Duplicated copies of world map 30
Felt-tip marker 30
Grease pencil, 2 colors 15
Iron filings 2 qt
Marbles, nuts, or washers 6 doz
Modeling clay, 2 colors 10 lb
Nail, to make hole & stopper in plastic jug used with stream table 10
Needle or pin 5
Paper clips 8 boxes
Phonograph record, old, 10–12 in 6–30
Portland cement 30 qt
Quarter dollar 15
Record, fast music (optional)
Rubber band, large 36
Sand 180 lb
Sandstone, porous 30 pieces
Sandstone with calcite cement 30 pieces
Small pebbles or crushed stone 3 qt
Soft wire 6 ft
Thumbtack 100
Towel 10
Washers, 3 in diam 15 doz
Weight, 1 kg, or rock of similar weight 30
World map, to have thumbtacks stuck in it 1

Optional rock samples 30 each

Basalt
Gabbro
Gneiss
Granite
Limestone
Marble
Obsidian
Porphyry
Quartzite
Rhyolite
Shale
Slate

Optional mineral samples

Mica
Orthoclase
Quartz

Optional fossils

Fossil molds
Fossil casts

EXPENDABLE MATERIALS

Baking soda 1 box
Calcium chloride 1 cup
Crushed limestone 15 lb
Fixative 1 bottle
Food coloring, in squeeze bottle 10
Glue, water soluble 1qt
Gold or silver paint, spray paint or model paint 1 oz
Hydrochloric acid, dilute 50 ml
Moth balls $\frac{1}{2}$ lb
Plaster of Paris 11 lb

Popping corn 1 cup
Portland cement 6 lb
Putty or equivalent to seal earth box partition
Salt, 26–oz box 3
Sugar cubes 1 box
Topsoil 20 lb
Turpentine 1 qt

Clear plastic cup 30
Paper cup 245
Bottle, plastic, gallon size, for use with stream table 10

Adding machine tape 4–20 rolls
Black construction paper 18 sheets
Cardboard, 16″ × 8″ 10
Cardboard, 8″ × 10″ 30
Cardboard, 2″ × 2″ slide frames 30
Graph paper, 5 squares/in 360 sheets

Masking tape 3 rolls
Newspaper 60
Paper, 8½″ × 11″ sheets 280
Paper, legal-size pad 30
Paper stars 6–12
Paper strip, for Sec. 4.2 30
Paper towel 30
Scrap paper, sheets blank on back, for asterisks 150–200
Tissues 1 box
Tracing paper 50 sheets
Transparent tape 3 rolls
Two-sided tape 1 roll
Unglazed paper 150–180 sheets

Balloons, small, round 30
Chalk 6 pieces
Cloth, cotton, colored 1 sq yd
Crayons 70
Ice
Match books 12
Modeling clay 10 lb
Nylon stocking 2

Pipe cleaner, or fine wire 15 pieces
Plastic sandwich bag with rubber band 100
Punk stick 45
String 50 ft
Test tube, 1 × 10 cm, to be broken 30
Thread 10 ft
Yarn 10–15 ft

TEACHER'S DEMONSTRATIONS

2.4 Glass or plastic bottle with cap, 4–6 in tall 1
Rice, cake decoration candies, or popcorn to fill bottle
Felt-tip marker 1

3.4 Large demonstration globe or world map 1

7.7 Candle 1
Paper bag 2
String 8–10 in
Yardstick, or meterstick 1

10.5 Pumice, or scoria several samples

11.1 Cellophane from cigarette pack 1 piece
Cigarette 1

12.7 Bunsen burner and ringstand 1
Evaporating dish 1
Hydrochloric acid, dilute 50 ml
Limestone fragment 1

13.2 Dirt, or sand
Snow
Stream table

Unit 1 Earth in the Universe

Chapter 1 Stars

OBJECTIVES

After completing this chapter, the student should be able to

(1) describe stars—their sizes, distances, numbers, colors and radiations.
(2) use constellations to locate stars.
(3) explain how to determine visual magnitude and distinguish it from luminosity.
(4) explain the relationship between surface temperature and luminosity of stars by making a scatter diagram.
(5) describe how stars are born, what gives them their energy, and the different stages in a star's life history.

BACKGROUND

1.3 Use of sky maps in Appendix C

Sky maps can be used in class, and for outdoor observing, with chapters in Unit 1. The maps show the main constellations for the north and south views of the sky in the Northern Hemisphere for the four seasons of the year. The brightest stars are named on the maps and magnitude is indicated by the star symbols. All of the constellations and stars named in Chapter 1 can be located on the sky maps; almost all of these stars are visible in the fall sky and can be located there with the help of the maps.

In Chapter 2, the positions of the Great Nebula of Orion and the Great Galaxy of Andromeda can be

1.3 Demonstration of the different distances to the stars in Ursa Major

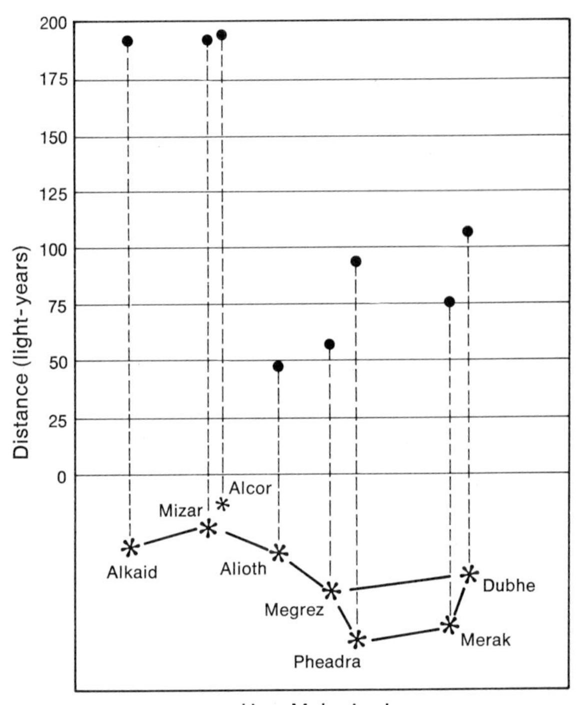

found from the sky maps. In Chapter 3, the celestial equator, right ascension, and declination are described in connection with the equatorial system of coordinates. This system enables you and the students to translate from any other maps of stars or other celestial objects to these sky maps, and vice versa. You can also translate directly from positions given in the *American Ephemeris and Nautical Almanac* for planets and stars, sun and moon, to these sky maps with this equatorial system.

For Chapter 4, the positions of the planets can be located along the ecliptic as shown on these sky maps, in relation to the celestial equator.

1.5 Magnitude measurement

About the year A.D. 150, the famous astronomer Ptolemy of Alexandria, Egypt, listed many of the brightest visible stars. Then he classified them by their brightness.

Ptolemy called the 20 brightest stars first magnitude. Stars he could barely see with the naked eye he placed in sixth magnitude, and all the rest of the stars in between, in second, third, fourth, and fifth magnitudes with decreasing brightness. He estimated the magnitudes of about 300 stars this way. In his book on astronomy, the *Almagest,* he described the magnitudes of the stars. He also explained how he thought all celestial objects, sun, stars, and planets, circle in great transparent spheres around the stationary earth. This Ptolemaic system was accepted by most people for some 1500 years.

Although Ptolemy's system has been replaced by the Copernican system, with the planets moving around the sun, Ptolemy's scale of steps of magnitude is still the basic yardstick for the measurement of the brightness of stars. The full magnitude scale as it has been expanded since Ptolemy's time is shown by the chart.

The unit of brightness or magnitude remains the same along the magnitude scale. Each step of magnitude is about 2.5 times the brightness of the next step, so our eyes can just distinguish the brightness of one magnitude from the next.

When astronomers wanted to show the brightness of the planets, moon, and sun, they extended the magnitude scale from 1 to 0 (zero), and then into negative magnitudes. In this way, they did not have to change the magnitudes already given to many of the stars. They also found that some of the 20 brightest stars that Ptolemy called magnitude 1 were much

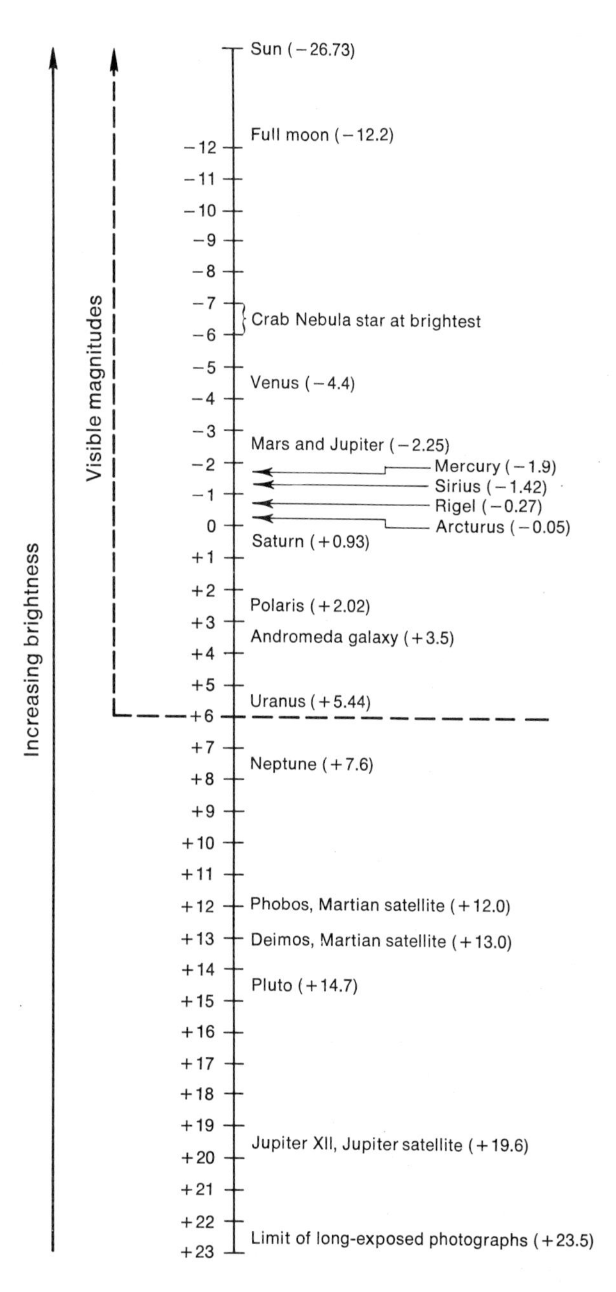

brighter than others. So four of these stars were given negative magnitudes.

With the invention and development of the telescope in the seventeenth century, many more dim stars could be observed. The magnitude scale had to be extended far up into the positive magnitudes. The scale kept getting longer and longer until now it covers about 50 magnitude steps from the brightest celestial magnitude, the sun (−27), to the present faintest magnitude of objects photographed with telescopes (+23).

The measurement of star brightness by eye gives apparent visual magnitude. Astronomers now use many other measures of brightness. Measuring the brightness of stars from photographs gives photographic magnitude. And the effect of light from stars on photoelectric cells, producing an electric current that increases with brightness, gives photoelectric magnitude.

All these apparent magnitudes refer to the brightness of stars as they appear in the sky. But astronomers need to compare the actual brightness of stars, no matter how far away they may be. So they have set up an absolute magnitude scale that gives the brightness of the stars as if they were all at the same distance from us, 10 parsecs or 32.6 light-years away. (One parsec is 3.26 light-years.) At this distance, the magnitude of the sun is +4.79 on the absolute scale. The sun would be a dim star, but still visible to us. With its apparent visual magnitude of −1.44, the brightest star in the sky, Sirius, is only 2.7 parsecs away. Its absolute magnitude is less, +1.41, the brightness it would have if it were 10 parsecs away. In terms of absolute brightness, Sirius is then many times brighter than the sun.

The luminosity of the stars, used in the H-R diagrams, is the total energy flow outward from the surface of a star, in this case expressed in units of the sun's luminosity. Luminosity is often derived roughly by measuring the absolute visual magnitude of stars and making a correction on this for the spectral class of the star.

What Happened to that Star?

During the summer of 1967, a large new telescope sweeping the skies at the Mullard Radio Astronomy Observatory of Cambridge University in England picked up some very puzzling radio signals from space. Early in 1968, five astronomers at this observatory announced that they had discovered four rapidly pulsating radio sources. At first they thought the radio waves from these sources were weak interference from some transmitter. Then they found that the positions of the sources among the stars did not change, indicating that the sources could not have been on the earth.

The signals were repeated faster than any ever observed. Those from one source repeated every 1.33730113 seconds. The signals were so extremely strong that it was finally decided they could not come from other intelligent beings in space, although this possibility was considered.

The discoverers suggested that these pulsing sources might be white dwarf stars, or stars consisting almost entirely of neutrons, in which gravity vibrations set up shocks in the stars' atmospheres. The shocks, in turn, created the radio signals. Astronomers all over the world were excited by these pulsating radio sources, or pulsars, as they were soon named.

The following list of some of the discoveries already made about pulsars shows how much effort was devoted to these mysterious sources:

(1) Over two dozen more pulsars have been discovered by British, American, and Australian observers.
(2) Pulsing light waves were found coming from the Crab Nebula and several other pulsars—the most rapid light pulses ever discovered.
(3) Pulsars were found to be sources of strong X-rays, and of vast outpourings of infrared (heat) rays. X-rays, light waves, and radio pulses were all found to be pulsing together at the same rate.
(4) Pulsars were found to lie within 1000 to 2000 parsecs of the earth, many of them within the disk of our Milky Way galaxy, of which they are probably a part.
(5) Rough measurements indicated that the pulsars are very small—probably smaller than the earth. These are the smallest stellar objects known.

The explanation of pulsars is still an unsolved problem. In one of the final stages of certain stars, when their nuclear fuel is gone, they contract to white dwarf stars, about the size of the earth. About 250 white dwarfs are known, although they are difficult to discover because their light is so weak. These white dwarfs may later contract further, under gravitational compression, until everything is squeezed out of them except neutrons, forming even smaller neutron stars, which might be the source of the rapid pulsations.

The identification of one pulsar, the Crab, with a remnant of a supernova (A.D. 1054) may indicate that pulsars are formed by a supernova explosion. Rotation may explain the extreme regularity of the radio and other pulses, emitted from an active spot on the neutron star as it rotates into view.

1.8 The H-R diagram used by astronomers

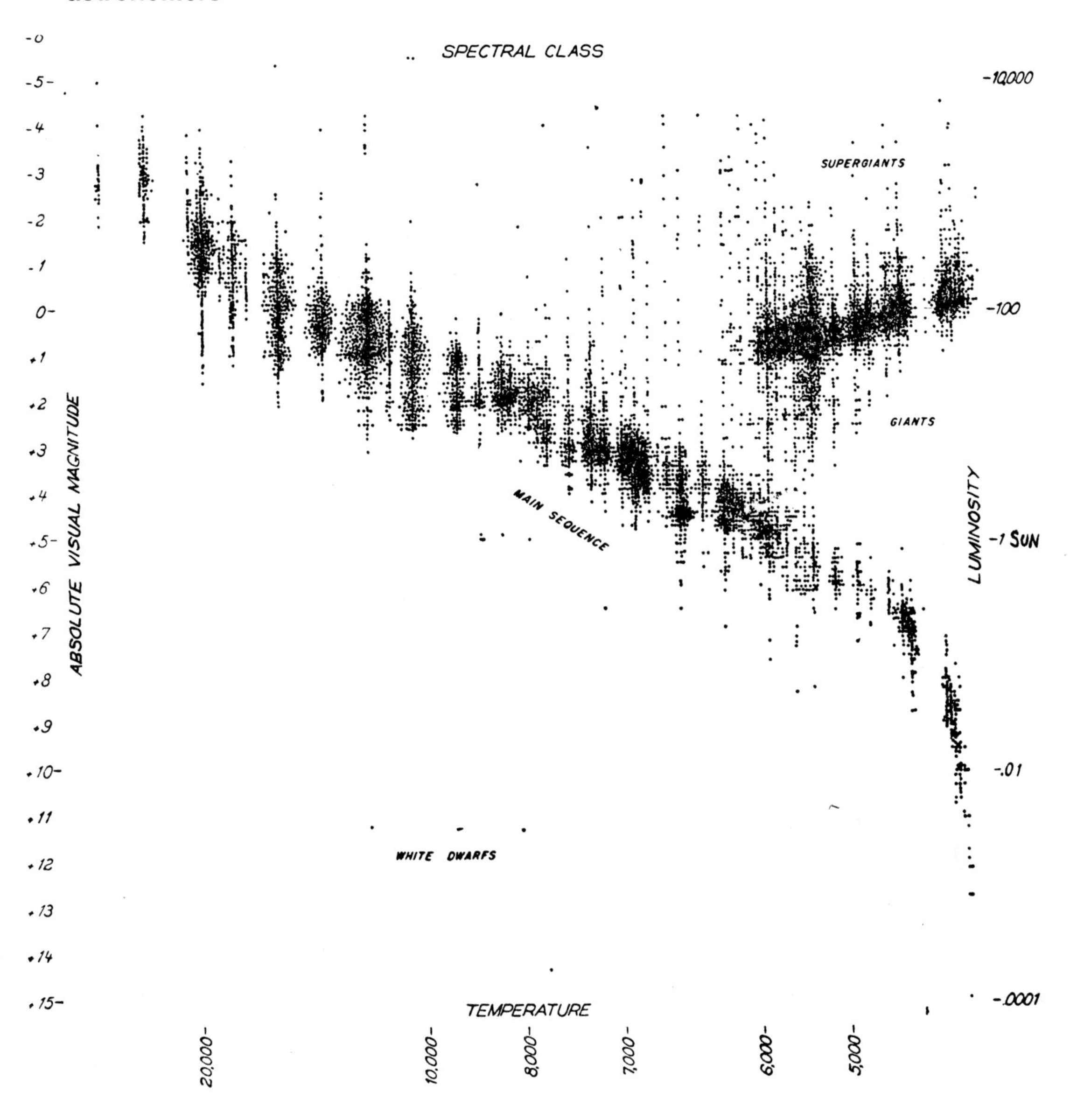

MATERIALS

1.4 How many stars can you see?

CLASS OF 30

paper tubes, 11 × 1.5 in diameter approximately 30

paper 30 sheets

pencils 30

1.5 Stars look bright or dim.

CLASS OF 30

sky map 30

paper 30 sheets
pencils 30

1.6 Stars have different real brightnesses.

CLASS OF 30
paper 30 sheets
pencils 30

1.8 Scatter stars on paper.

CLASS OF 30
paper clips (25 per student) 8 boxes
graph paper, 5 plus squares to inch 30 sheets
newspaper or pads of paper 30
pencils 30

REFERENCES

Corliss, William R. *Mysteries of the Universe*. Thomas Y. Crowell Company, New York, 1971. Clear and lively discussions of the great mysteries of astronomy. A good book to read with this one—many of the mysteries are introduced in this unit.

Howard, Neale E. *The Telescope Handbook and Star Atlas*. Thomas Y. Crowell Company, New York, 1967. A useful introduction to astronomy for the layman. It is of special interest because it includes chapters on telescopes and sky photography.

Menzel, Donald H. *A Field Guide to the Stars and Planets*. Houghton Mifflin Co., Boston, 1964. Excellent background, particularly for Chapter 1. Many photographs, descriptions, and maps.

Menzel, Donald H. *Astronomy*. Random House, New York, 1970. A fine, well illustrated source of elementary astronomical information, done with a minimum of details.

Ronan, Colin A. *Astronomers Royal*. Doubleday and Company, New York, 1969. A very human account of the lives and discoveries of astronomers, from Elizabethan to modern times.

Wood, Harley. *Unveiling the Universe: The Aims and Achievements of Astronomy*. American Elsevier Publishing Company, New York, 1968. A well written, relatively complete guide to astronomy and observing.

Scientific American articles:
Gorenstein, Paul and Wallace Tucker. *Supernova Remnants*. July, 1971.
Ostriker, Jeremiah. *The Nature of Pulsars*. January, 1971.
Ruderman, Malvin. *Solid Stars*. February, 1971.

See also the magazines *Sky and Telescope, Natural History,* and *Science and Children*.

Chapter 2 Galaxies

OBJECTIVES

After completing this chapter, the student should be able to
(1) describe nebulas, where they occur, and how they may form.
(2) describe the basic characteristics and groups of galaxies.
(3) describe the earth's place in the universe, including the sun, the Milky Way, the Local Group of galaxies, and more distant galaxies.

(4) describe electromagnetic radiation in terms of wavelength, frequency, amplitude, and speed.
(5) describe the principal kinds of radiational increases or decreases along the electromagnetic spectrum.
(6) describe how the redshift is used to indicate that galaxies are moving away from us.
(7) describe three current ideas about how the universe originated and developed.

BACKGROUND

2.1 Nebulas and galaxies

The fuzzy blurs of the Orion Nebula (text Figure 2/1) and the Andromeda galaxy (text Figure 2/2) are in the constellations of the same names. Originally all fuzzy spots were called nebulas and were believed to be simply clouds of gas and dust. Much later, using the largest telescopes, astronomers found that fuzzy spots were of two kinds: (1) nebulas of gas and dust nearby in the Milky Way, such as the Orion Nebula; and (2) galaxies with hundreds of billions of stars at great distances outside the Milky Way, such as the Andromeda galaxy. The Milky Way is such a galaxy itself.

2.4 Demonstration of density of stars in the Milky Way

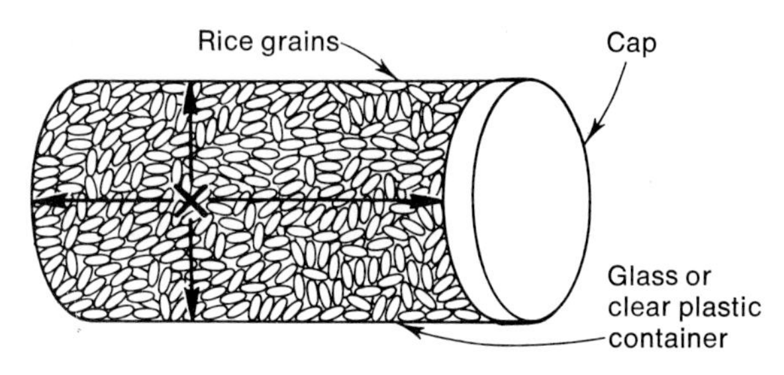

Fill a glass or plastic bottle with rice or candies and cap it. Lay the bottle on its side, as shown in the drawing, to stand for stars in the Milky Way. Mark an "X" with a felt-tip pen on the bottle for the earth's position with the sun in the Milky Way. Have students count the rice they could see if they stood at X and looked out toward both ends of the bottle. Then have them count the rice they could see looking up and down. Looking toward the ends gives a greater density like the greater density of star clouds in the disk of the Milky Way. Toward the center, stars are denser than toward the edge from X, or Earth. Vertically, only a few stars are seen above and below the disk of the Milky Way.

2.8 Demonstration of parallax and variable star methods for calculating distances

Astronomers use a parallax method to find the distance of nearer objects such as the moon, sun, planets, and nearby stars, and a variable star method for somewhat greater distances. But the redshift method must be used for estimating the distances to faraway galaxies.

To show parallax, have each student hold a finger about six inches in front of his or her eyes and close first one eye and then the other. Ask them to note how the position of the finger seems to change from side to side against some object in the background. In order to find the distance from the eyes to the finger, the distance between the eyes (the baseline) and the angle of a line from each eye to the finger must be known. Then, simple math gives the distance to the finger. Half the angle between the lines from the finger to the eyes is known as the parallax.

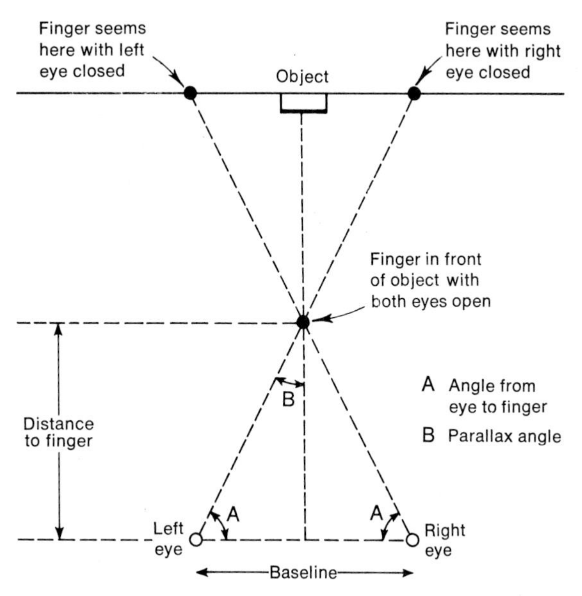

Substitute a planet or star for the finger and other stars for the background objects. For the distance between the eyes substitute the diameter of the earth or of the earth's orbit around the sun. Then the distance to the planet or star can be calculated by parallax.

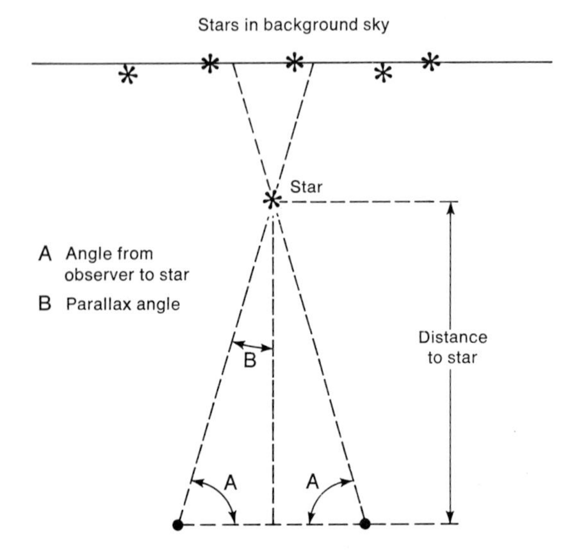

The parallax method can only be used for a few thousand very close stars. At greater distances, the angles get too small to measure precisely. As a project for school or homework tell students to hold a finger six inches and then 12 inches from their eyes and see how much smaller the angle becomes. Ask how they can prove that the angle does become smaller. Draw diagrams of the two positions of the finger and their eyes and measure the parallax angles with a protractor.

To measure greater distances in space, astronomers turned to a variable-star method. Variable stars brighten and dim in the sky over a period of time. The actual brightness, or luminosity, of certain variable stars is known from the period of time it takes them to brighten and dim. Certain stars with certain periods of variation have given luminosities. Once the period of variation of the stars has been observed, their apparent brightness in the sky tells their distance, because their real brightness is known.

Such variable stars, some of which are called Cepheid variables, have been used to measure the distances to the nearest galaxies beyond our own Milky Way. The variables were first found in the Large and Small Magellanic Clouds, at distances of about 160,000 and 180,000 light-years from us. These are the two galaxies closest to the Milky Way in the Local Group.

MATERIALS

2.2 Model nebulas and galaxies.

CLASS OF 30—nebulas

unglazed paper, 4 × 5 in (half of an $8\frac{1}{2} \times 11$ in sheet) 300–360 sheets
newspapers, full-width sheets folded or paper towels 30 sheets
small hammers 10 shared
food coloring in squeeze bottles 10
salt in shakers 10
or salt in bulk, 26 oz box 1
fixative 1 bottle

CLASS OF 30—galaxies

round dishes, pie plates, or deep paper plates 10
turntables (lazy Susans) 10
food coloring in squeeze bottles 10
cups or test tubes 10
water

2.3 Galaxies come in different shapes.

CLASS OF 30

galaxy photographs in textbook (Figure 2/11) 30
Hubble galaxy classification (Appendix H) 30
paper 30 sheets
pencils 30

2.4 We live inside the Milky Way.

TEACHER

glass or plastic bottle, 4 to 6 in tall with cap 1
rice, cake decoration candies, or popcorn to fill bottle
felt-tip marker 1

2.5 Model energy radiations.

CLASS OF 30

paper pads, legal size 30
spring clips 30
paper clips, large 30
eyelets to slide on paper clips 30
ball chains 30
rulers, 15 to 30 mm, or 6 to 12 in 30
clock or watches with second hands 1 clock visible to all or 10 watches
pencils 30

2.8 What is the redshift?

CLASS OF 30—redshift
paper pads, legal size 30
spring clips 30
paper clips, large 30
eyelets to slide on paper clips 30
ball chains 30
rulers 30
clock or watches with second hands 1 clock visible to all or 10 watches
pencils 30

CLASS OF 30—parallax (optional)
protractor 1 (or 30 if whole activity is done individually)

2.9 What kind of universe is this?

CLASS OF 30
balloons, small and round 30
felt-tip markers 10 shared

REFERENCES

Abetti, Giorgio and Margherita Hack, translated by V. Barocas, *Nebulae and Galaxies.* Thomas Y. Crowell Company, New York, 1965. An authoritative and relatively complete book, written for interested laymen and scientists in other fields. A fine, well illustrated source of background information.

Scientific American articles:

Iben, Icko, Jr. *Globular-Cluster Stars.* July, 1970.

Maran, Stephen. *The Gum Nebula.* December, 1971.

Rees, Martin and Joseph Silk. *The Origin of Galaxies.* June, 1970.

Schmidt, Maarten and Francis Bello. *The Evolution of Quasars.* May, 1971.

Chapter 3 Place and Time

OBJECTIVES

After completing this chapter, the student should be able to

(1) devise reference systems of points and/or lines for places or things in various two- and three-dimensional situations.
(2) state the direct evidence from photographs that the earth turns.
(3) explain how places are located on the earth with the geographic reference points of the poles and reference lines of the equator and Prime Meridian.
(4) find latitude and longitude of places from globes or maps and locate places on globes or maps from latitude and longitude data.
(5) describe the time zone system for setting time around the earth.
(6) locate objects on sky maps from celestial sphere data, or give the data from the places of objects on sky maps.

BACKGROUND

3.2 Locating points on a sphere

You may wish to have a contest to illustrate the value of latitude and longitude for locating points. If you can do so, arrange the seats in two rows, back to back. These will be row A and row B. Students will work in pairs, back to back, one in each row. Give each student a plastic foam ball. Have each one in row A mark a dot on his or her sphere. Then have each one in row A also mark an X.

Each student in row A will then tell the other member of the pair (over the shoulder) where to place the dot and the X. Then, have each one in row B mark a triangle on his or her sphere, and tell the other team member in row A how to locate it on his or her sphere.

The winners will be the two students with the most closely matched pair of spheres.

3.3 Appendix I enrichment activity: Earth's celestial pole and time of rotation

Activity Time: 20 to 25 minutes. Students can do this activity individually, sharing a protractor among several of them if necessary. The motion of all stars (measured from tracks) through the same angles (within limits of error) shows that they all move around the same point at the center of the arcs, the north celestial pole. This apparent motion of stars is actually caused by the turning of the earth on its axis of rotation.

Be sure that students draw long enough lines from the center through arc or curve endings to extend beyond the edge of the protractor. Students will be surprised to see that Polaris and all other stars in the same photograph moved through exactly the same number of degrees. The apparent motion of the stars is due to Earth's turning, so the stars pass through the same number of degrees no matter what their distance in the sky from the pole.

A typical calculation follows: If the student's average measure for the star tracks is 15 degrees, then dividing 360 degrees by 15 degrees equals 24, the number of times the part of the circle covered by the star tracks divides the whole circle. Multiply 24 by the time of the photograph, one hour, and this equals 24 hours for the time it would take the stars to complete the full circle, or the earth's period of rotation on its axis.

3.5 Positions of spacecraft in orbit

Position	Longitude	Latitude
1	81°W	28°N
2	75°W	26°N
3	60°W	23°N
4	45°W	17°N
5	30°W	10°N
6	15°W	2°N
7	0°	7°S
8	15°E	14°S
9	30°E	22°S
10	45°E	28°S
11	60°E	32°S
12	75°E	34°S
13	90°E	33°S
14	105°E	31°S
15	120°E	27°S
16	135°E	22°S
17	150°E	14°S
18	165°E	6°S
19	180°E	0°
20	165°W	8°N
21	150°W	16°N
22	135°W	22°N
23	120°W	26°N
24	105°W	28°N
25	90°W	29°N

3.8 Time zones adjust time to the sun

The time meridian on the other side of the earth, exactly opposite the Prime Meridian set by Greenwich, is called the International Date Line. This is 180 degrees both east and west of the Prime Meridian. By agreement between countries, you lose a whole day

(24 hours) when you cross the International Date Line going west. If you cross it going east, you gain a day. Thus, Wednesday noon to the east of the Date Line becomes Thursday noon to the west of it.

When the sun reaches 12 noon at Greenwich, it is 12 midnight at the Date Line. Then you are moving into another day, for the next day starts at midnight. But it is only 12 noon of the day before at Greenwich. It will be another 12 hours to 12 midnight at Greenwich, when a new day starts there. So you have two days going around the earth all the time. You go ahead one day by agreement when you cross the Date Line going west away from the sun, and go back one day when you cross it going east toward the sun.

3.8 Appendix K enrichment activity: Local sun time

Activity Time: 30 minutes. The activity can be done indoors or outdoors, at school or home, but a fairly sunny day is needed, so the sun will cast a good shadow. The activity must be done at noon. Do the activity with the whole class, if the time can be arranged.

Longitude (Section 3.4) and time zones (Section 3.8) are related to local sun time. Students see how predictions are made on the basis of facts at hand—your longitude, the longitude of the time-zone meridian, and the time for the sun to move 1 degree—verified, in this case, by the sun itself. Inference, then, is often used in science to check guesses or hypotheses.

You can ask the class to find the longitude from a large detailed map or atlas, and calculate the total time difference and prediction of when the sun will cross your local meridian.

Students can work in groups of four to six, each group with a compass or watch. All groups may use a compass and clock together, if necessary. Caution students to align the N-S line to the geographic, not magnetic pole, adjusting for the magnetic declination in your location, which may be obtained from the map in the manual notes for Chapter 6.

Principal sources of error: not pointing the line on the paper directly north and south; not having the exact time-zone time; not finding the local longitude correctly; and not calculating the time difference correctly. But students' results may be surprisingly accurate.

3.10 The celestial-sphere reference system

You can demonstrate the celestial-sphere reference system either indoors with a star-object for Polaris or outdoors with Polaris itself, using an astrolabe of type B. As shown in the drawing, point the fixed upright of the astrolabe toward Polaris, then hold astrolabe tube at 90° angle to the upright and sweep the tube back and forth. Where the tube points, along the walls of the room or along the sphere of the sky, stands for the celestial equator. Polaris stands for the north celestial pole. This shows the basic reference line (celestial equator) and reference point (north celestial pole) of the equatorial system.

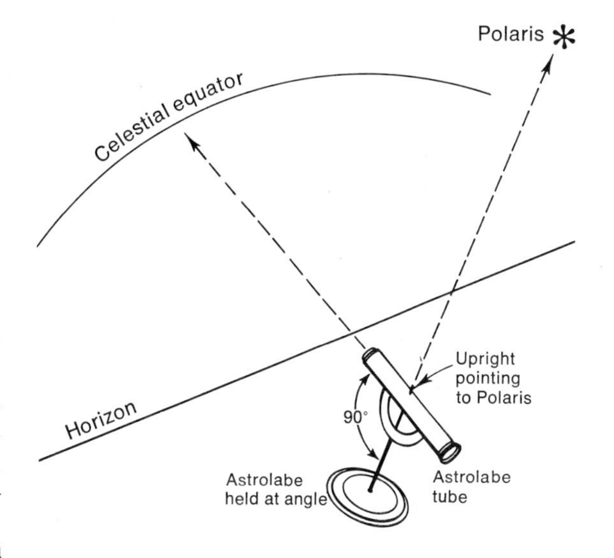

3.10 The use of declination and ascension

Familiarity with locating objects in the sky can be gained by a classroom activity involving sighting on a sphere suspended from the ceiling.

Hang a sphere six inches or so in diameter, big enough to sight on from across the room. Have each student determine its declination with an astrolabe, from his or her seat. The reading should be written on a piece of paper with a secret identifying mark instead of a name. Next have each student use a compass (if compasses will work in your room) and determine the ascension in degrees from the north. After the ascensions have been written, collect the papers and redistribute them so that each student has someone else's paper.

Have the students leave their seats and locate the seats from which the sightings they received were made. The correctness of the choice can be verified by the identifying mark on the paper.

MATERIALS

3.1 How are places located?

CLASS OF 30
rulers, 6 or 12 in 30
pencils 30

3.2 Locate places on a ball like the earth.

CLASS OF 30
balls, rubber or plastic 30
felt-tip pens or grease pencils 30

3.4 Latitude and longitude tell where you are.

TEACHER
globe, large demonstration model, or map 1

3.5 Make a space flight around the earth.

CLASS OF 30
globes, 10 to 12 in 6
rubber bands 6
graph paper, 5-plus squares to inch 30 sheets
pencils 30

3.6 How well can you judge time?

CLASS OF 30
clock or watches, with second hands 1 clock visible to all or 10 watches
paper 6 sheets
pencils 6

3.7 Time is measured and set.

CLASS OF 30
globes, 10 to 12 in 6
flashlights 6

3.9 Measuring the altitude of Polaris.

CLASS OF 30
astrolabes 6
stars, silver or gold, gummed, or cut-out paper stars 6–12
transparent tape 1 roll

Chapter 4 Solar System

OBJECTIVES

After completing this chapter, the student should be able to

(1) name the planets in their order from the sun, distinguish the rocky planets and gas giants, and give an approximate idea of their size.
(2) describe the ecliptic and zodiac and tell how planets move on the ecliptic and why.
(3) state the law of gravitation simply, in terms of how mass and distance affect the attraction between two bodies.
(4) describe some of the physical characteristics of each of the planets.
(5) explain the evidence that there may be other planetary systems like our solar system.

BACKGROUND

4.1 A model of the solar system

A spectacular scale model of the solar system can be made with paper, a pencil, compass, and some walking. Cut out a disk, 110 cm in diameter, to represent the sun. Cut out additional disks, scaled to the proper size, for the planets. Distances and diameters are given in the following table:

Object	Average distance from sun (Earth's distance = 1)	Distance converted to miles	Scale model diameter (Earth's diameter = 1)
Sun	—	—	110.0 cm
Mercury	0.4	0.025	0.4
Venus	0.7	0.044	1.0
Earth	1.0	0.062	1.0
Mars	1.5	0.093	0.5
Jupiter	5.2	0.32	10.1
Saturn	9.5	0.59	9.4
Uranus	19.2	1.19	4.0
Neptune	30.1	1.87	4.3
Pluto	39.5	2.45	1.0

Transfer the distances to a local map, and then tack the representative disks along a route at the correct distances. Have the students follow the route.

4.3 The 1969 retrograde motion of Mars

Because of the ever changing relative positions of Earth and Mars, the paths of retrograde motions are not all the same. The 1969 motion shown on page T20 and that of 1971 show this clearly.

The diagram shows how the changing line of sight from Earth to Mars causes the curves in the apparent path of Mars.

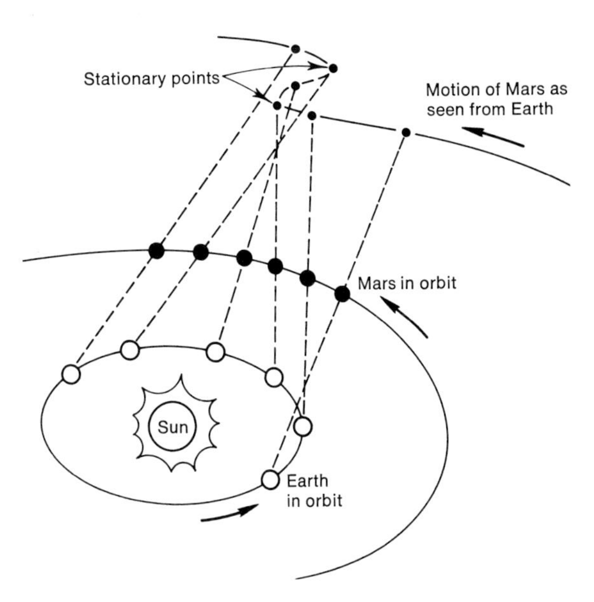

4.7 What are the planets like?

Encourage students to imagine a trip into space and to go as far beyond the text as they wish in exploring the topics in this section. Have as many books and articles available as you can. Special topics include: French discovery of Saturn's tenth moon; radio signal (radar) study of Mercury; strong radio signals from Jupiter; flyby and orbiting of Mars by Mariner spacecraft; American and Soviet spacecraft studies of Venus's atmosphere and surface; Jupiter's Red Spot; and the question whether or not Pluto is an escaped moon of Neptune.

MATERIALS

Introduction

CLASS OF 30
paper 30 sheets
pencils 30

4.1 Model the solar system yourself.

CLASS OF 30
balls, $\frac{7}{8}$ to 1 in 10
yardsticks or meter sticks 10
adding machine tape (optional) 20 rolls
paper 10 sheets

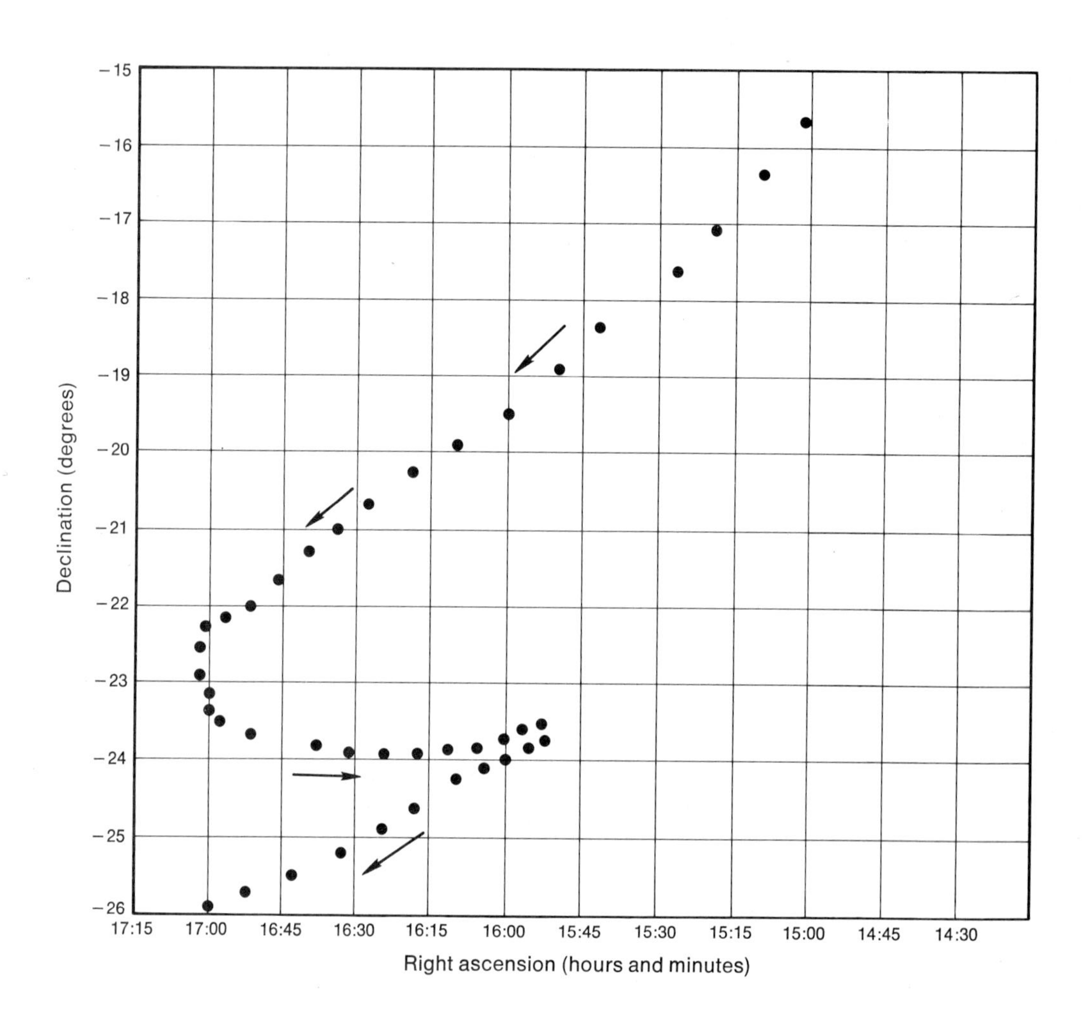

4.2 Where do the planets appear in the sky?

CLASS OF 30—visible sky objects

paper strips 30

CLASS OF 30—ecliptic model

phonograph records, 10 or 12 in 6

pencils 6

paper, $8\frac{1}{2} \times 11$ in 6

chalk 6 pieces

4.3 Planets wander in the sky.

CLASS OF 30

graph paper, 5-plus squares to inch 30

pencils 30

4.4 How and why do the planets move?

CLASS OF 30

thumbtacks 60 plus

string 30 ft plus

wood or cardboard bases, 8×10 in 30

4.5 Measure the size of the sun.

CLASS OF 30

yardsticks or meter sticks 5

cardboard, 6×8 in squares 10

needles or pins 5

scissors 5

rulers, millimeter 5

4.9 What is happening to the sun?

CLASS OF 30

graph paper, 5-plus squares to inch 30 sheets
pencils 30

REFERENCES

Bova, Benjamin. *Planets, Life and LGM.* Addison-Wesley, Reading, Mass. 1970. Discusses the possibility of life outside our solar system.

Jastrow, Robert. *Red Giants and White Dwarfs: Man's Descent from the Stars.* Harper and Row, New York, 1971. Covers much of the material in Unit 1. Excellent photographs of planets, stars, and moon. Stresses origins of the universe, stars, the solar system, and life on earth.

Knight, David. *First Book of Meteors and Meteorites: An Introduction to Meteoritics.* Franklin Watts, Inc., New York, 1969. Historical information, facts and figures, and descriptions of meteorite falls in an unusually well illustrated book.

Lawless, James, Clair Folsome and Keith Kvenvolden. *Organic Matter in Meteorites. Scientific American,* June, 1972.

Masursky, Harold et al. *Mariner 9 Television Reconnaissance of Mars and Its Satellites: Preliminary Results. Science,* Vol. 175, 21 January, 1972.

Wilson, Curtis. *How Did Kepler Discover His First Two Laws? Scientific American,* March, 1972.

Chapter 5 Moon and Earth

OBJECTIVES

After completing the activities in this chapter, the student should be able to

(1) use models of natural events and processes effectively and understand how models can be used to answer many questions.
(2) describe the positions of the earth, moon, and sun during the lunar month as the moon passes through its phases and as solar and lunar eclipses occur.
(3) describe the actual and relative sizes of earth and moon, and the paths they follow around the sun.
(4) describe various gravitational reactions between the earth and moon that make tides on earth, bands on seashells, and quakes on the moon.
(5) describe some of the differences between terrestrial and lunar landscapes and explain the meteoritic and volcanic processes that shaped lunar features.

BACKGROUND

5.1 Systems

The concept of systems is vividly displayed in this chapter and can be emphasized with the students, particularly in connection with Sections 5.1 and 5.5. A system is an organized group of bodies or parts (such as the two-member system of the earth and moon, now joined by other members, the artificial satellites) interacting with each other. They interact through many forms of energy (such as motion, gravitation, magnetism, and radiation), and tend toward a balance, or equilibrium, of these energies in their relations with each other.

Natural systems, such as the earth-moon combination, occur in nature. They in turn form parts of larger systems, like the solar system, and in turn the Milky Way galaxy. Artificial systems, such as man-made satellites, automobiles, and works of art, are produced by people. They also are formed of many parts, which act and interact with each other to produce desired effects. Artificial systems have a human purpose. But, presumably, purpose can be disregarded by the scientist who is trying to explain the functioning of a natural system.

The parts of systems can be treated as variables. Investigations can be made of the changes in and relations between these variables. Motions of satellites around earth and moon in our system have revealed variations in the gravitational fields of both bodies that help explain the small changes in their orbits.

5.1 The partial solar eclipse

The completed graph on the next page shows the relative movements of the sun and the moon during the period under consideration. If each pair of observations is plotted on a separate graph, as shown in the drawings, the eclipse may be easier for the students to visualize. This could be done on the chalkboard.

5.1 The difference between sidereal and synodic months

The sidereal month is the period required for the moon to orbit the earth once and return to a given starting position in relation to the stars. However, as the diagram shows, the moon must travel farther to return to its starting position in relation to the sun. This period, based on the phases of the moon, is the synodic month. (See diagram on page T23.)

5.2 The path of the moon around the sun

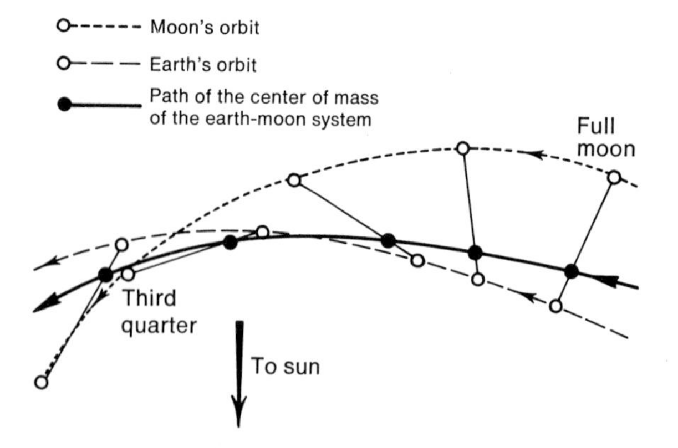

The earth and the moon rotate around a common center of mass, as illustrated in the accompanying drawing. The center of mass of the earth-moon system is inside the earth, of course, but it is actually this center of mass that follows the elliptical orbit around the sun. If you project the three orbits farther, you will see that the orbit of the moon is always concave to the sun.

5.7 Model of the Aristarchus and Cobrahead area of the moon, made in cement powder

Joseph Jackson

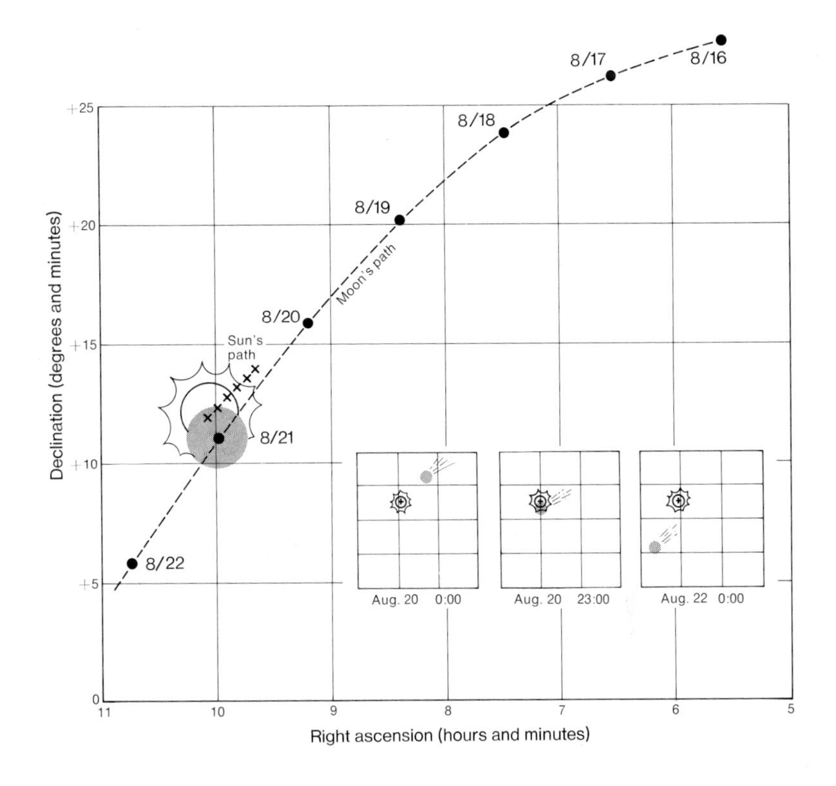
8/16
8/17
8/18
8/19
8/20
8/21
8/22
Moon's path
Sun's path
Declination (degrees and minutes)
Right ascension (hours and minutes)
+25
+20
+15
+10
+5
0
11
10
9
8
7
6
5
Aug. 20 0:00
Aug. 20 23:00
Aug. 22 0:00

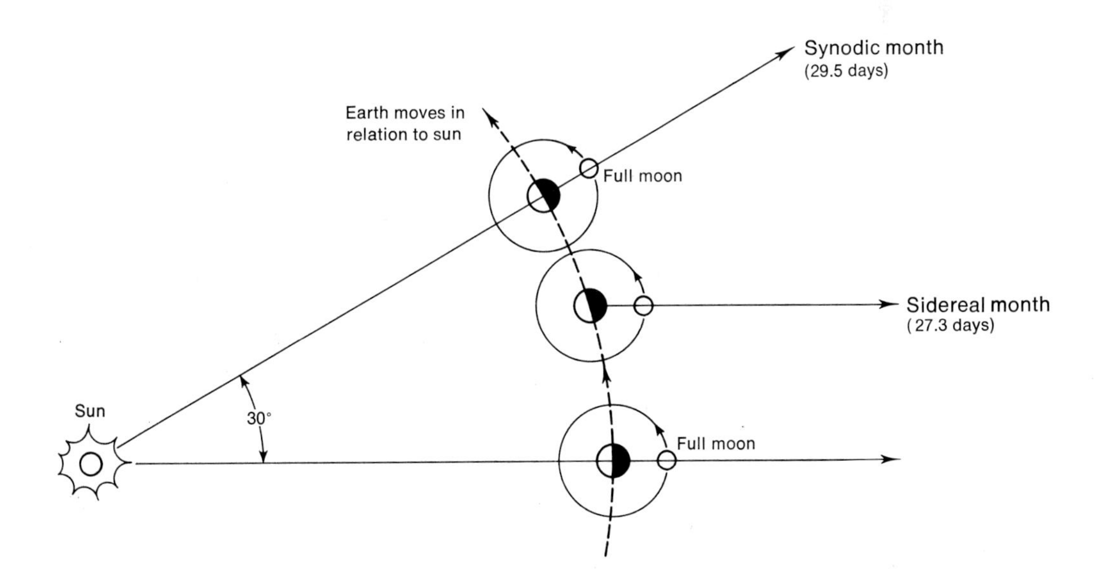
Synodic month
(29.5 days)
Earth moves in relation to sun
Full moon
Sidereal month
(27.3 days)
Sun
30°
Full moon

5.5 Variations in the length of the synodic month through geologic time, with standard error bars

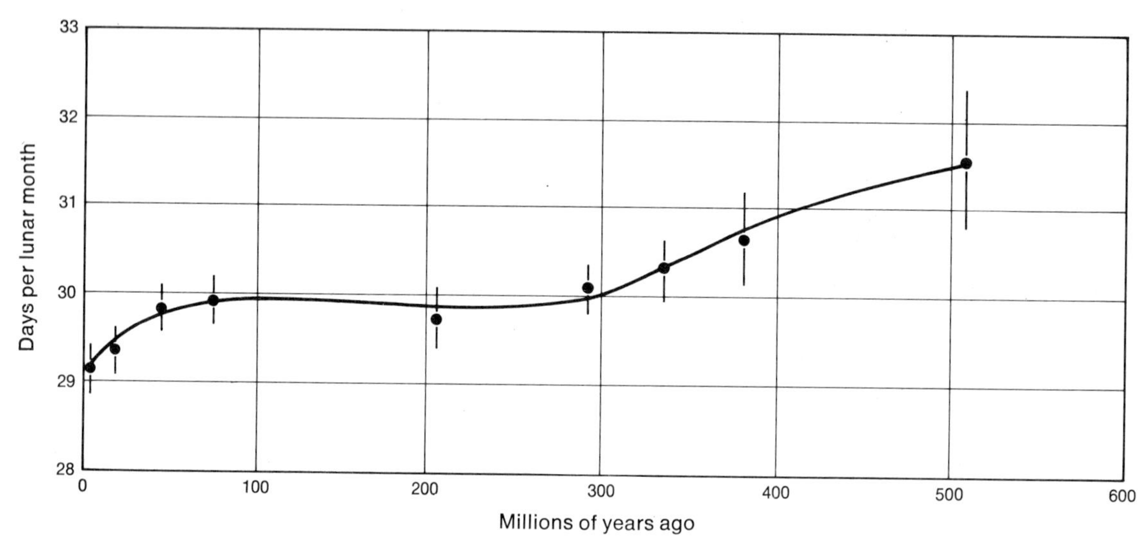

From G. Panella, *Science,* Vol. 162, 11/15/68, p. 795.

MATERIALS

5.1 Make the moon's shapes.

CLASS OF 30
iron stands 5
clamps and rods 5
earth ball, plastic or rubber (2 to 3 in diameter) 5
moon ball, clay or dough ($\frac{1}{4}$ in diameter) 5
thread, black 10 ft

5.2 What paths do earth and moon follow?

CLASS OF 30
boards, plywood or cardboard (2 ft $\times$ 3 in $\times$ $\frac{1}{4}$ to $\frac{1}{2}$ in) 10
yardsticks or meter sticks 10–15
compasses, drawing 30

5.3 Earth and moon are called a double planet.

CLASS OF 30
half dollars or poker chips 15
thumbtacks 15
yardsticks or meter sticks 10–15

5.5 What do lines on seashells tell?

CLASS OF 30
graph paper, 5-plus squares to inch 30 sheets
pencils 30

5.7 Make some lunar landscapes.

CLASS OF 30—moon model with meteorites
cement powder 90–120 cups or 25–30 quarts
deep pans or dishes 6
marbles, nuts, or washers 6 doz
pencils or sticks 12–15

CLASS OF 30—moon model with moonstuff
glass bottles, quart 6
stoppers for bottles, 2-hole 6
plastic tubing 12 ft
cement powder 12–16 cups or 3–4 quarts

REFERENCES

Aerospace Bibliography. National Aeronautics and Space Administration, Washington, D. C., 1970.

Apollo 15 at Hadley Base. NASA EP-94, National Aeronautics and Space Administration, Washington, D. C., 1971.

Apollo 15 Preliminary Examination Team. *The Apollo 15 Lunar Samples: A Preliminary Description. Science,* Vol. 175, 28 January, 1972.

Dyal, Palmer and Curtis Parkin. *The Magnetism of the Moon. Scientific American,* August, 1971.

Journal of Geophysical Research, November 15, 1969. A series of articles on the results from the Surveyor 7 soft-landing near crater Tycho. Also included is a sequence of photographs of the earth taken by Surveyor 7 showing the waning and waxing earth in phases as seen from the moon.

Mason, Brian. *The Lunar Rocks. Scientific American,* October, 1971.

Moore, Patrick and Barbara Middlehurst. *Lunar Transient Phenomena; Topographical Distribution. Science,* Vol. 155, 27 January, 1967. Historical survey and discussion of pink spots, red patches, and hazes observed on the moon.

Musgrove, Robert, compiler. *Lunar Photographs from Apollos 8, 10, and 11.* NASA SP-246, National Aeronautics and Space Administration, Washington, D. C., 1971.

Nature, International Journal of Science, 19 July, 1969, pages 243 to 267 and elsewhere in issue. A collection of articles on the moon to mark the first (Apollo 11) manned landing on the moon.

Science, Vol. 167, 30 January, 1970. A whole issue devoted to reports on the analysis of lunar soil, rocks, and minerals brought back by the Apollo 11 mission.

Wood, John. *The Lunar Soil. Scientific American,* August, 1970.

Unit 2 The Earth's Atmosphere

Chapter 6 Atmosphere

OBJECTIVES

After completing the activities in this chapter, the student should be able to

(1) describe and give his own example of fields, and describe the magnetic and gravitational fields of the earth.

(2) describe the magnetosphere formed as the earth moves through the solar wind.

(3) explain how temperature, pressure, and the composition of the atmosphere vary with increasing height above the earth's surface.

(4) describe the principal heavy gases mixed in the air and the layers of gases at greater heights.

(5) describe air pressure and water vapor in the air at and near the earth's surface.

BACKGROUND

6.2 A map of the variations in magnetic declination as of 1970

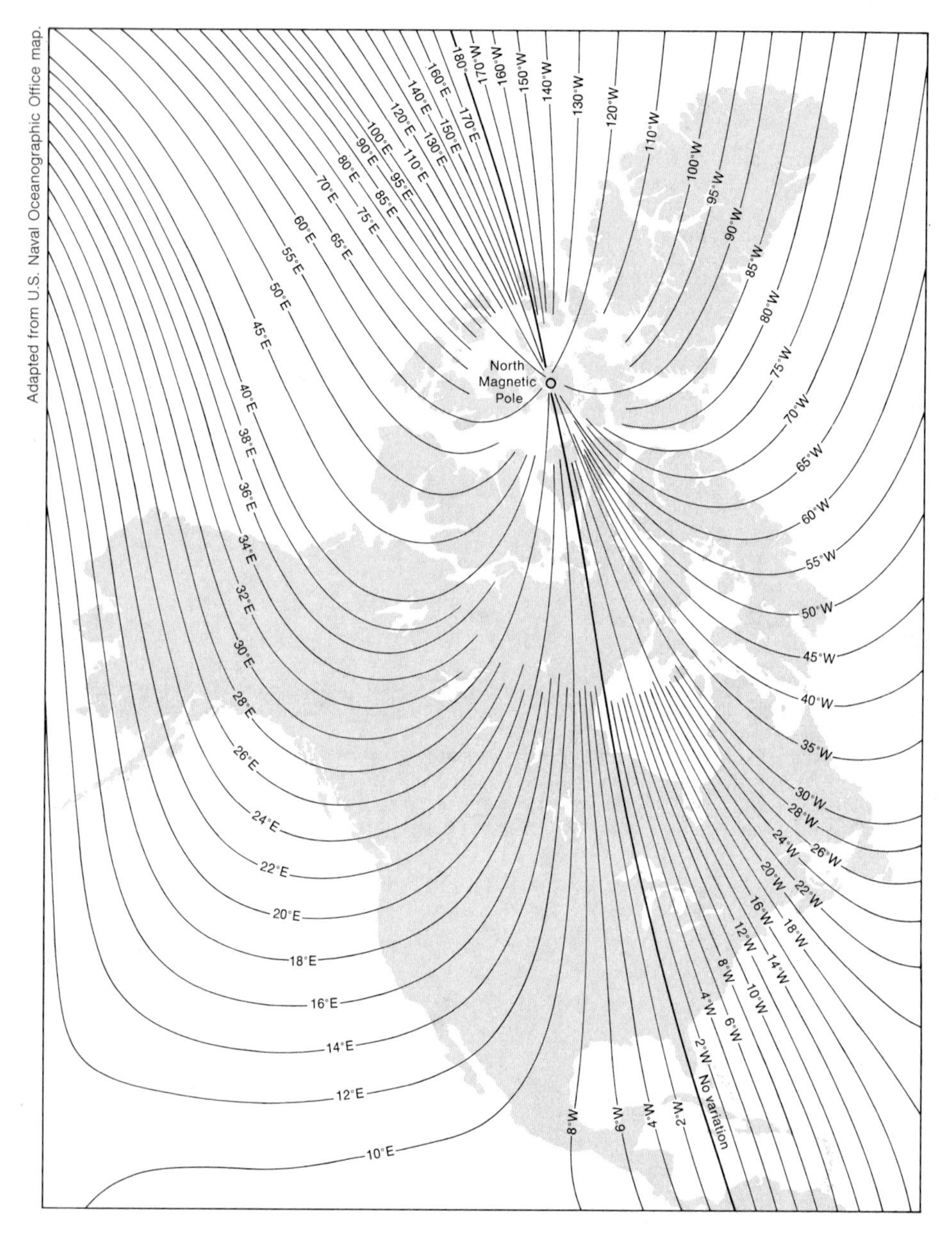

The lines on this map show the amount and direction of compass "declination" or deviation of magnetic north from true north.

Are There Tides in the Atmosphere?

A map of the lunar air tide strengths around the earth. The units are microbars (a microbar is one thousandth of a millibar).

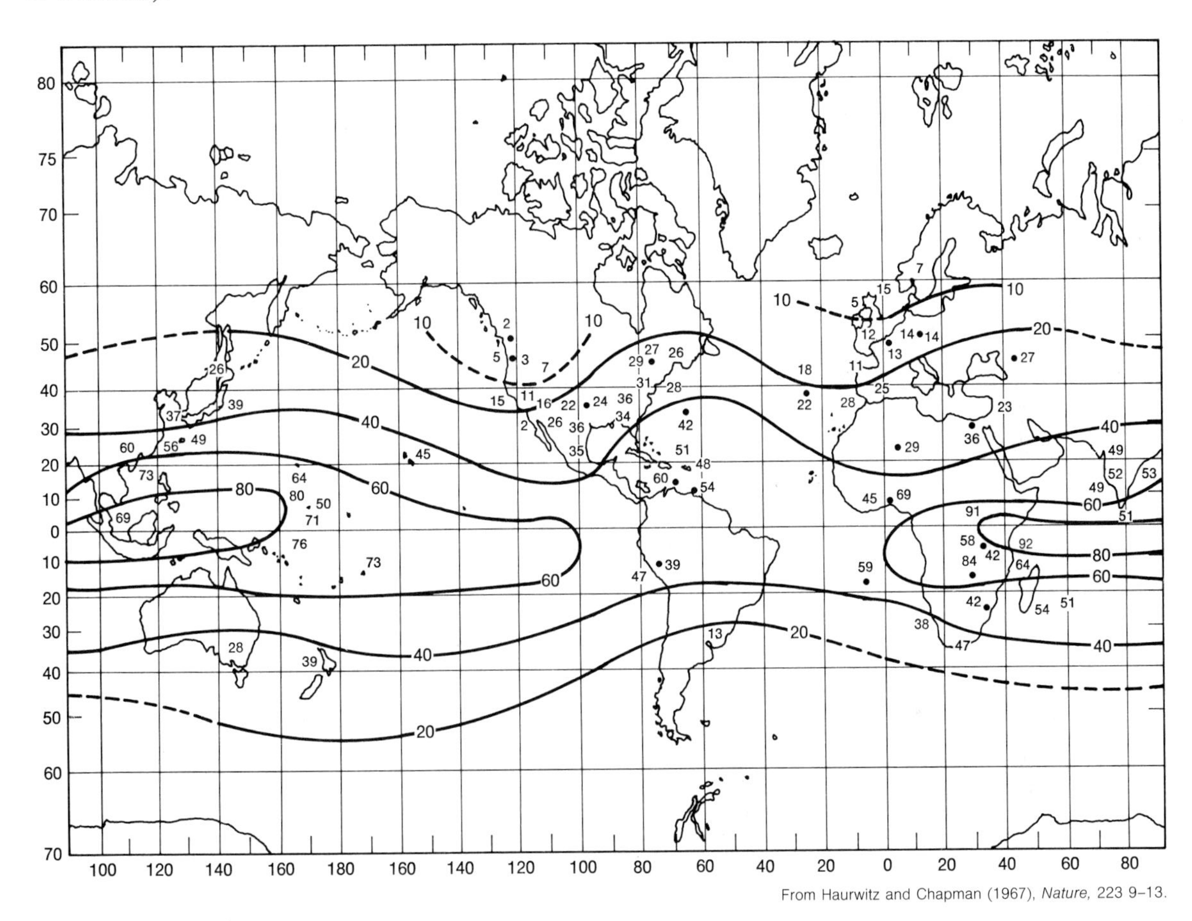

From Haurwitz and Chapman (1967), *Nature,* 223 9-13.

MATERIALS

6.1 A magnet makes its own field.

CLASS OF 30

bar magnets, 1 to 1.5 in long 30
cardboard sheets, 8 × 10 in 30
iron filings or iron powder 2 quarts
paper 30 sheets
pencils 30

6.2 Model the earth's magnetic field.

CLASS OF 30

modeling clay or dough to make 30 balls, 1.5 to 2 in diameter 15 cans
scissors 15
compasses, magnetic 30
cardboard sheets, 8 × 10 in 30
paper 30 sheets
pencils 30

6.3 The solar wind shapes the magnetic field.

CLASS OF 30

large trays or dishes, 2 to 3 in deep 15
pencils 30
water

6.5 What goes on in the atmosphere?

CLASS OF 30

graph paper, 5-plus squares to inch 30 sheets
pencils 30

6.6 How much oxygen is in the air?

CLASS OF 30

pint or quart jars or 500-ml beakers 10
millimeter rulers, 15 cm 10
transparent tape 1 roll
large trays or dishes, 2 to 3 in deep 10
candle stubs 10
water

6.8 Question 3, page 180.

CLASS OF 30

graph paper, 5-plus squares to inch 30 sheets
pencils 30

REFERENCES

Chapman, Sydney. *IGY: Year of Discovery*. University of Michigan Press, Ann Arbor, Michigan, 1959. This book received a prize as the best science book of the year. In it, the president of the International Geophysical Year describes discoveries about the earth's atmosphere made as a result of the IGY.

Chapter 7 Sun, Water, and Wind

OBJECTIVES

After completing this chapter, the student should be able to:

(1) describe and give examples of the ways in which heat energy is transferred.

(2) explain why the heat energy coming to the earth varies with the seasons.

(3) describe the relationship between solar radiation and broad climate zones.

(4) explain the paths followed by water in its various forms through the water cycle.

(5) explain the formation of high and low pressure eddies in the atmosphere.

BACKGROUND

7.1 Instructions for making an earth box

To get the best results with many activities, you need a relatively high and narrow box with transparent sides. It will contain the convection currents and be useful for limiting the volumes of smoke or water needed for various activities or demonstrations.

Type 1 Make a wooden base from a board 20 in long and 6 in wide. This will be somewhat larger than the box, for ease of handling. The box itself can be fashioned with sides of glass or plastic, measuring

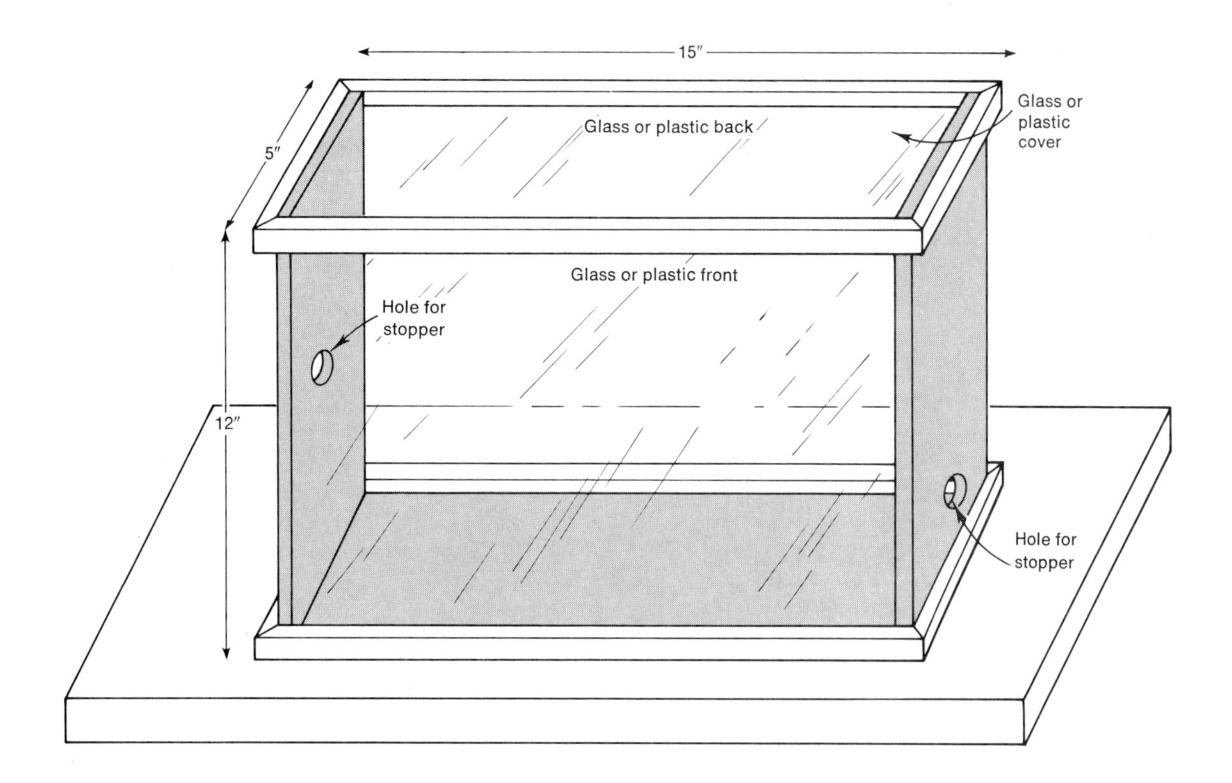

12″ × 15″. The ends can be of wood, glass, or plastic, $4\frac{3}{4}$″ × 12″. There should be a hole 1 in from the bottom in one end (for draining out water) and a hole 6 in from the bottom in the other end. The holes should be large enough to permit the insertion of rubber stoppers. The top can be of glass or plastic 5″ × 15″. Plan to have the sides overlap the ends. Make frames to go around the top and the bottom from $\frac{3}{8}$″ × $\frac{3}{4}$″ strips, as shown in the illustration. You will need a piece of cardboard the same size as the top for an opaque shield. Caulk all the seams on the inside, and paint or varnish the wooden base to make it waterproof. The sides and ends can be cemented together with a cement appropriate to the material used.

If you use plastic, make sure the plastic is of a type that will not melt in the heat from a 300 watt bulb.

Type 2 You can make a box entirely of glass or plastic, with the dimensions and holes the same as those of the Type 1 box. This will be without the wooden base or the wood strip frames. Make the top a double layer of glass or plastic, with the bottom layer fitting inside and the top layer projecting so it will rest on the upper edges of the sides and ends.

Type 3 If you do not need an earth box that will hold water, you can make one from a cardboard box with low sides such as a canned soft drink case. Cut out the top of the cardboard box, leaving a half inch margin all around for the support and attachment of a glass or plastic panel. Fasten the panel onto the cardboard with package sealing tape, inside and out. Repeat the operation with the bottom of the cardboard box. Cut out one side of the cardboard box, leaving a half inch margin. This will form the open top of the earth box. Reinforce all the corners with tape. Fit a glass or plastic top.

Type 4 All the earth box activities have also been performed in a modified medium size aquarium (20″ × $10\frac{1}{2}$″ × 12″), including the fault demonstration in Chapter 11.

Modification 1: Make a longitudinal partition from plywood or plastic. Be sure that it will fit tightly against the glass or plastic top that will be needed. Insert it about 5 in from one side of the aquarium. The partition can be held in place with modeling clay, and imperfect seams can be filled with the clay. This arrangement is satisfactory for the activities that do not require water—the convection currents, weather and climate activities, the smog activity, and the fault demonstration.

Modification 2: Use the longitudinal partition, waterproofed, and add the transverse partition required for the water cycle activity. The latter partition need not be more than 2 in high. Caulk all the seams below the projected water level with glazing compound. Modeling clay does not seem to work as well for this purpose.

Modification 3: Use the longitudinal partition as in Modification 1 and add a transverse partition as high as the aquarium. Sturdy cardboard will do. The transverse partition can be held in place and caulked with modeling clay. This modification is for the supplementary activity in Chapter 19, which compares radiation in smoky air to radiation in clear air.

7.1 Sample data for the sun and atmosphere model activity

	Thermometer position				
Minutes	**A**	**B**	**C**	**D**	**E**
Start	21	21	21	21	21
1	25	21	21	22	22
2	29	21.5	21	23	24
3	33	22	21	24	26
4	37	23	21	24	26.5
5	38	23	21	25	27

Notes:

1. Light bulb is placed $4\frac{1}{4}$ inches from the top.
2. Difference in temperature for *A* caused by radiation, *B* and *C* by convection, *D* by radiation, and *E* by conduction and radiation.
3. The diameter of wire in *E* is a factor.

7.5 Air moves by convection

Punk or incense sticks come in such varied odors as pine, strawberry, and lemon, as well as the usual ones. Suit yourself.

7.8 and 7.9 Prevailing winds and the Coriolis effect

Large-scale air movements are often presented as convection cells and air flows being deflected with the Coriolis effect to yield latitudinal zones of "prevailing winds." Although this appears to be good theoretical science and forms a neat logical package, it is not correct. This theory is also difficult to relate to the newspaper and TV weather maps that the students see. The material in Chapter 7 introduces large-scale weather patterns in such a manner that students can move easily to the weather maps in Chapter 8.

Many descriptions of air flow assert that air rises at the equator, travels northward in the northern hemisphere, and then sinks and moves southwestward as the (northeast) trade winds. This circulation, according to the descriptions, results from the formation of a large convection cell (the Hadley cell). Actual observations show this explanation to be basically in error. A Hadley convection cell is driven by a strong difference in temperature between two latitude belts, with the heat source at one latitude belt causing rising motion, and the heat sink at the cooler latitude belt associated with descending motion. The observed zone of strongest temperature contrast on the earth runs from about 20° to 50° north latitude. The strongest convection currents would be expected in this zone, and the currents should yield winds flowing from northeast to southwest. However, the wind direction in this range of latitude, based upon data collected over a period of many years, is on the average more or less from the southwest. This average wind flow might be interpreted as a Hadley cell, but the air flow does not follow the usual explanation.

The Coriolis effect, or deflecting influence of the earth's rotation on moving air, has been used to explain "prevailing winds"—the trade winds, the westerlies, and the polar easterlies. The rotation of the earth certainly does have an effect on air flow, but the rotation is only one of several factors that result in the actual wind pattern near the surface. Other factors are temperature differences in air such as those caused by differing radiation on cloudy and clear areas, or different temperatures of contiguous land and sea areas. Another obvious factor is topography.

An examination of a series of weather maps shows that there may be on any day many heat sources and heat sinks, and that these sources and sinks typically change positions daily. As text Figure 7/15 shows, the major air currents near the surface at a given time (which are the air streams around highs and lows) travel in a variety of directions in the temperate and polar zones. Thus, a "prevailing wind" is an average over a long period. At a particular north-temperate-zone locality, the wind may blow from the southwest on the average over a period of a month. However, the students will readily observe that the local winds often blow from other directions. For instance, the summary of hourly observations at Albany, N.Y. from 1951 through 1960 shows that the wind was recorded from the south, the west, or in between an average of 36.3% of the time. The most common wind in the quadrant was from the south (average 14.8%). It would be difficult to call these winds "prevailing westerlies."

The actual origins of highs and lows are complicated and not completely known. It may be best to emphasize this lack of knowledge in class, and cite it as a good example of the fact that many everyday phenomena will not be understood without more research.

You can demonstrate and describe the Coriolis effect in terms of the activities given here, or have the students do these themselves. Make it clear, however, that this effect is only one among many factors that yield the actual large-scale air circulation. You can emphasize the Coriolis effect in presenting this material as much, or as little, as you choose.

If you are interested in a review of modern concepts, refer to the publications of Lorenz, Starr, and Fultz in the bibliography.

7.9 Demonstrations of the Coriolis effect

The Coriolis effect lends itself to illustration in various ways. The simplest way is to use the setup shown in the drawing. A pin or thumbtack serves as a center around which the paper will rotate. Moving the pencil along the ruler will prove that the pencil is moving in a straight line. If the paper is rotated counterclockwise, you have a model of the Northern Hemisphere; the resulting line curves to the right. If the paper is rotated clockwise, you have a model of the Southern Hemisphere.

Another way to illustrate the Coriolis effect is to rotate a globe of the earth from west to east slowly with one hand. At the same time, hold a pencil vertically and move it slowly straight down in front of the globe from the north pole toward the equator. Look at the globe behind the pencil. The pencil seems to be following a path that curves clockwise.

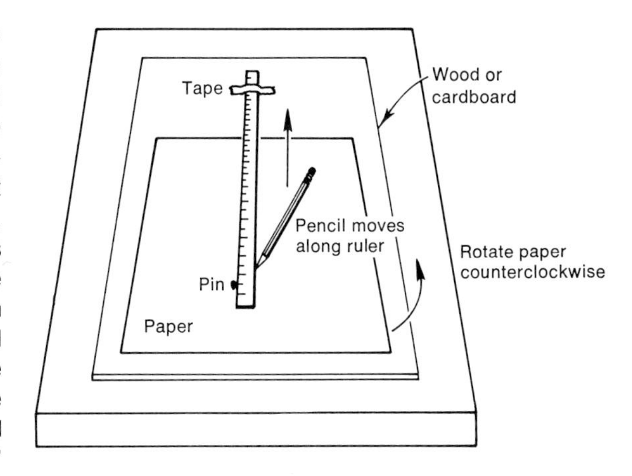

A third way to demonstrate the effect is with a record player. This method requires a piece of writing paper, a piece of carbon paper, a one inch steel ball, and a record. The writing paper will model a flat earth. The upper surface will represent the Southern Hemisphere, since the turntable rotates clockwise, and the lower surface will represent the Northern Hemisphere. The record player should be propped up so that the turntable is at an angle of 15–20°, and the speed should be $33\frac{1}{3}$ or 45 rpm.

To illustrate the effect on the Southern Hemisphere, put the carbon paper on top of the other sheet and both on the record. Drop the ball from three or four inches above the turntable. It will leave a carbon trail on the paper.

To illustrate the effect on the Northern Hemisphere, put the carbon paper below the other sheet. The curved path is due to the fact that the ball is rolling from a zone of the record that is moving at one speed to a zone of the record that is moving at another speed.

7.10 Climate activity

If you can, do this activity when the air in the room is the driest. If it is impossible to avoid moist air, dry out the air in the box with calcium chloride for a half hour or so before the activity is performed. For the best results, raise up the mountain so that its top is close to the lid of the box. It will thus form an effective barrier to the convection current. Look for the first evidence of change on the mountain; the most obvious results will show up early. As the moisture builds up in the box, it will condense everywhere.

7.10 Data for answer to question 3, page 212

Station	Annual average high	Days of rain	Climate
Seattle	59°F	150	cool & moist
Phoenix	85°F	39	warm & dry
Denver	64°F	86	cool & dry
Memphis	85°F	105	warm & moist

MATERIALS

7.1 Model the sun and atmosphere.

CLASS OF 30

aquariums or earth boxes with lids and opaque shields 6
light bulbs, 300-watt, reflector 6
sockets and extension cords 6
ring stands 6
clamps for sockets 6
thermometers, small 24
wire, soft 6 ft
plastic foam or stopper 6
graph paper, 5-plus squares to inch 30 sheets
clock or watches, with second hands 1 clock visible to all or 10 watches
black construction paper 6 sheets
pencils, 5 colors 6 sets

PREPARATION

Note: Activities in Sections 7.1 and 7.3 can be done conveniently at the same time by different groups.

7.3 Heating of the earth varies.

CLASS OF 30—angles vs. heat energy

construction paper, strips 2 × 12 in 6
candle stubs 6
light bulbs, 150 to 300-watt 6
extension cords 6
ring stands and clamps 6

CLASS OF 30—heating the earth (optional)

globe of earth, 10 in to 12 in diameter 1
ring stand and clamp 1
light bulb, 150 to 300-watt 1
extension cord 1
thermometers 2

7.5 Air moves by convection.

CLASS OF 30—heat sources

punk sticks 12
matches 12 books
wood or clay bases 12

CLASS OF 30—movement of air

aquariums or earth boxes with lids and opaque shields 6
punk sticks 12
wood or clay bases 12
ring stands and clamps 6
light bulbs 6
sockets and cord 6
funnel and rubber or plastic tubing (optional)

PREPARATION

Note: Activities in Sections 7.5 and 7.6 can be done conveniently at the same time by different groups.

7.6 You can move water through air.

CLASS OF 30

aquariums or earth boxes with lids and opaque shields 6
punk sticks 12
wooden partitions 6
putty or plastic to seal partition 1 box
dry, cold sand, refrigerated overnight 6 quarts
water 6 quarts
light bulbs on extension cords 6
ring stands and clamps 6

7.7 What changes happen in the water cycle?

TEACHER—question 4, page 205

yardstick or meter stick 1
paper bags, 5 × 3 × 10 in 2
candle 1
string, 8 to 10 in 3 pieces

7.8 Warm and cool air mix around the earth.

TEACHER
globe of earth, any size 1

7.10 There is more to climate than temperature.

CLASS OF 30
aquariums or earth boxes with lids and equipment 6
dishes or cups 6
blocks 12
calcium chloride, powdered 1 cup
black construction paper 6 sheets

REFERENCES

Lorenz, Edward N. *The Nature and Theory of the General Circulation of the Atmosphere*. World Meteorological Organization, Unipub, New York, 1967.

Starr, Victor P. *Physics of Negative Viscosity Phenomena*. McGraw-Hill, New York, 1968.

Fultz, Dave, et al. *Studies of Thermal Convection in a Rotating Cylinder with Some Implications for Large-Scale Atmospheric Motions*. Meteorological Monographs, American Meteorological Society, *4*:21, Boston, 1959.

The titles may seem forbidding, and the books are technical, but the generalities and many of the specifics can be readily understood by a reader who is not an expert meteorologist. These publications contain relevant information and concepts that have not been included in many of the elementary texts.

Chapter 8 Weather

OBJECTIVES

After completing the work in this chapter, the student should be able to

(1) describe the main components of the weather and explain a number of the varying conditions that determine it.

(2) explain how the varying conditions related to weather are measured.

(3) describe the conditions that are most reliable for a weatherman in predicting the weather in his local area.

(4) read and explain weather maps as they appear in local newspapers and on TV.

(5) explain how air masses, fronts, and cyclones are related to the weather.

(6) describe the character of some of the more extreme forms of the weather, such as thunderstorms, hurricanes, and tornadoes.

BACKGROUND

8.1 Weather maps and forecasting

Examine maps of forecasts and actual weather, noting differences indicating poor forecasts and similarities indicating good forecasts. For both Sections 8.1 and 8.4 you might assign students to see if radio and TV broadcasts agree or disagree with class predictions in Section 8.4.

8.4 Conclusions to be drawn from the weather chart

The best general way to predict the weather is to know where the fronts are, which way they are moving, and where you are in relation to the highs and lows. Thus, the conditions upwind will be important. Significant observations may be summarized as follows:

8.5 Average surface winds in January and July

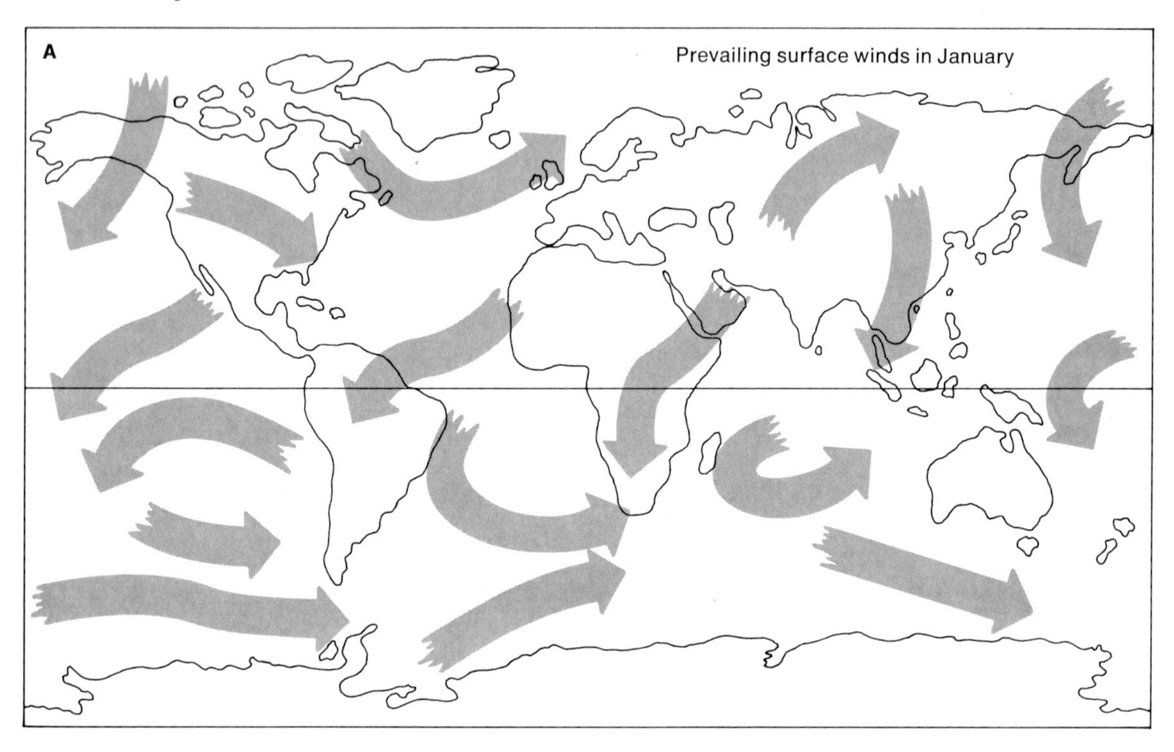

Variable	Observations favor precipitation? No	Yes
Pressure change	steady, rising	falling
Cloud type	cumulus (fair weather type)	cirrus-warm front on the way; stratus-warm front near; nimbus-rain or snow
Cloud cover	absent, sparse	over 50%
Difference between dewpoint and temperature	more than five degrees	less than five degrees
Conditions upwind	clear	precipitation

The significance of wind speed and direction depends on your location and particular conditions. The students will probably conclude that the present temperature and pressure and the temperature change are not very helpful in predicting precipitation.

8.5 The air mass concept

The term "air mass" has been avoided in this book because it can be misleading. As mappable structures of air currents, highs and lows do move. But a "polar air mass" does not move from Canada to the United States as a discrete entity. You can demonstrate this by calculating the outflow from a "typical" polar "air mass", a high pressure area, by using surface winds on a daily weather map. You will see that the original supply of polar air contained within the "air mass" would be depleted long before the "air mass" itself seems to die out. This calculation will serve to point out that the "air mass" is continuously being replenished from above.

For the sake of simplicity, upper air currents are not mentioned in the text. You may wish to discuss them,

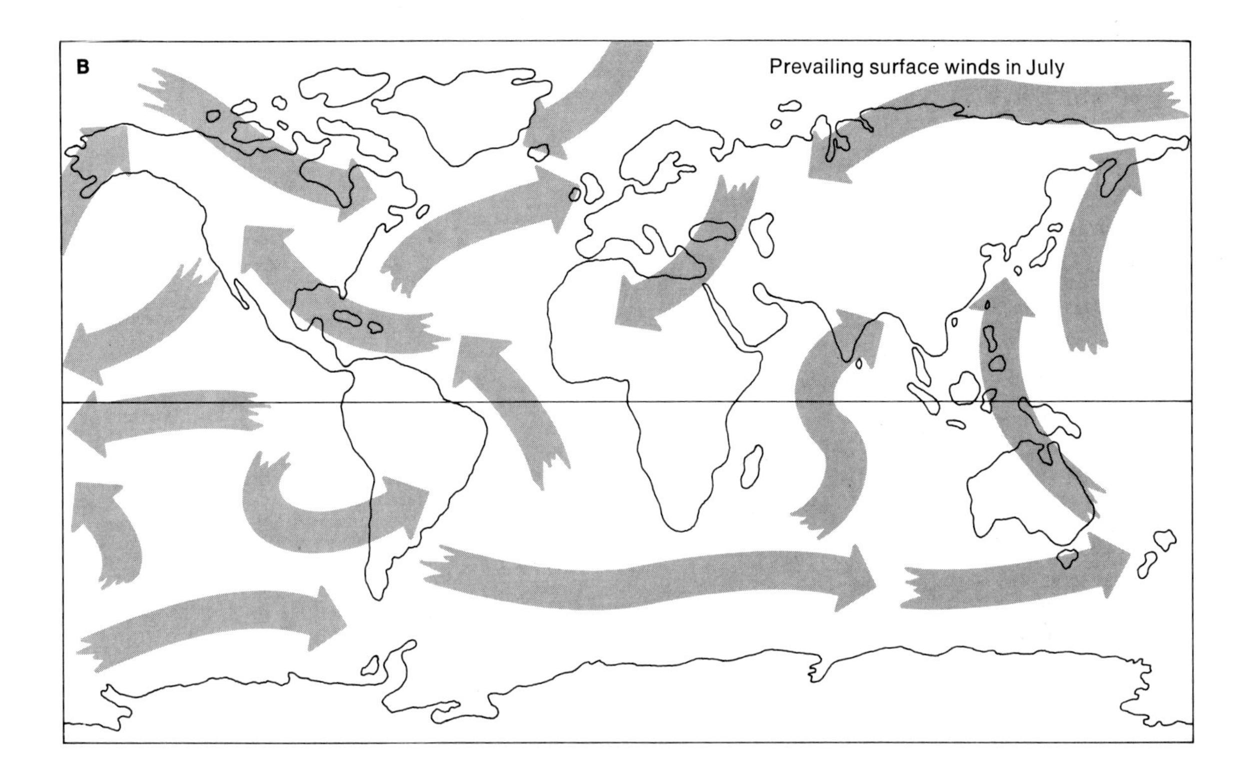

since they are occasionally referred to by TV weathermen. Highs and lows generally move in the flow of the upper air currents, but the cause and effect relationship is not known. There may be a common unknown cause.

8.5 Major ocean currents of the earth

Discuss with the class the effects that warm currents have on the weather of high latitude countries, such as Great Britain. (Why is the Emerald Isle green?) Conversely, the cool currents can affect land areas; an example is the effect of the California Current on the warm moist air flowing across the Pacific Ocean. (Why is it foggy in San Francisco?) Map, page T36.

8.6 Weather fronts

The map on page T37 of a cold front and a warm front was designed to show how the fronts sectioned in text Figures 8/10 and 8/12 are related in a larger scheme. The map shows a common arrangement of the two kinds of fronts. Often, the cold front moves up behind the warm front, and a zone of occlusion spreads along the warm front from the low pressure center. Such an occlusion was predicted in the prediction map of Figure 8/1. Note the relationships of clouds and rain to the fronts, and the large size of the whole system.

You may have to explain to the students the vertical distortion of clouds in Figures 8/10 and 8/12. This method of illustration permits an accurate depiction of the cloud types at various distances from the surface fronts. It also avoids the unnatural sharp line of demarcation between warm and cold air that is found in many diagrams of frontal systems.

MATERIALS

8.4 Predict the weather from changing conditions.

CLASS OF 30

thermometer, Fahrenheit 1
barometer 1
anemometer or some type of wind-speed indicator 1
sling psychrometer or wet-and-dry bulb thermometer 1
wind direction indicator 1

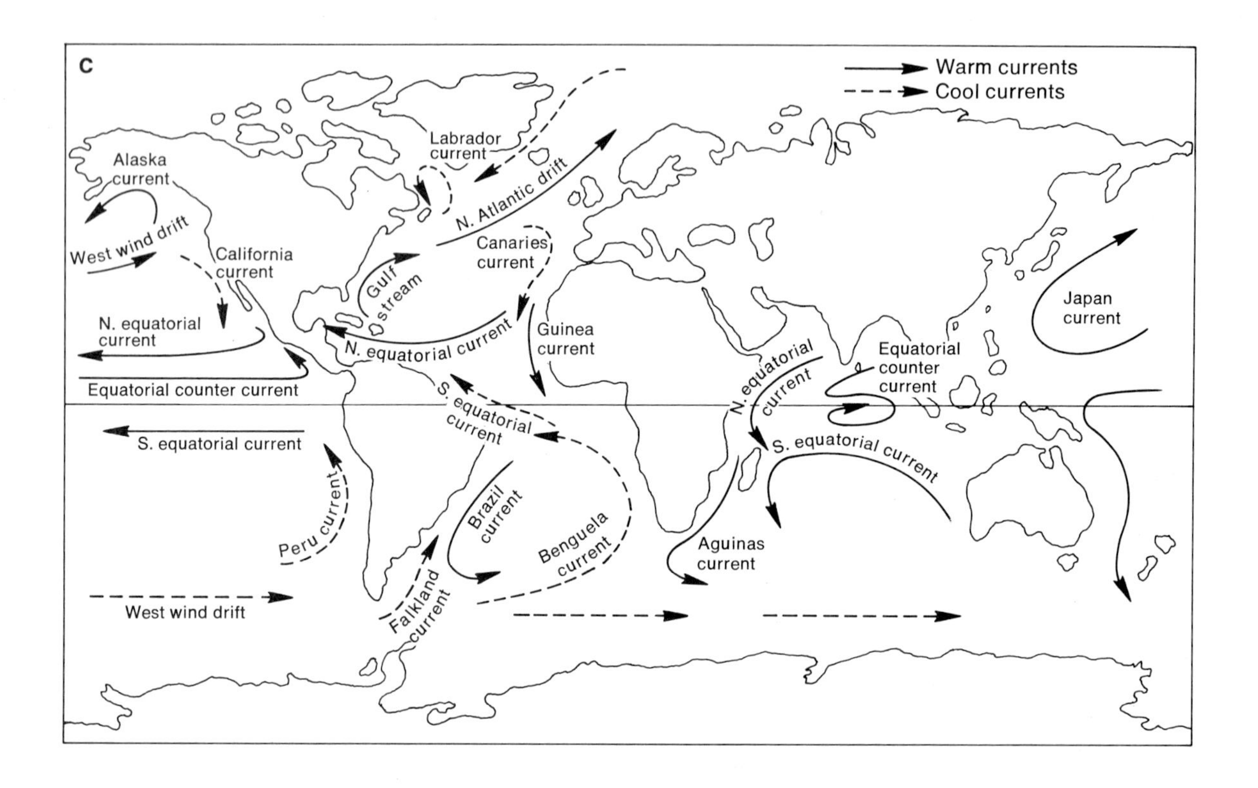

8.6 Make your own weather fronts.

CLASS OF 30

earth boxes with lids 6
light bulbs, 100-watt, sockets and extensions 6
ring stands and clamps 6
cups or tins for hot water 6
small plastic bags, such as sandwich bags, for cold sand 25–30
paper clips or rubber bands to close sand-bags 25–30
cold sand, refrigerated 1 gallon
source of hot water

PREPARATION

Note: Have boxes and materials on hand. Refrigerate sand thoroughly overnight. Let students fill sandbags and hot water cups.

Note: Review drawings of earth box setups for fronts, inversions, and cyclones.

REFERENCES

Aviation Weather. Federal Aviation Administration and U.S. Department of Commerce, 1965. U.S. Government Printing Office. An elementary meteorology text for pilots and flight operations personnel.

Nicks, Oran W., editor. *This Island Earth*. NASA SP-250. Superintendent of Documents, U.S. Government Printing Office, Washington, D.C., 1971. Photographs taken from space. Excellent views of cloud formations as well as land forms.

Space—Environmental Vantage Point. National Oceanic and Atmospheric Administration, NOAA PI 70033, U.S. Government Printing Office, Washington, D.C., 1971. Descriptions of weather satellites and programs, and many satellite photos.

Strahler, Arthur N. *The Earth Sciences*. Harper & Row, Publishers, New York, 1971. Chapter on, "Air Masses,

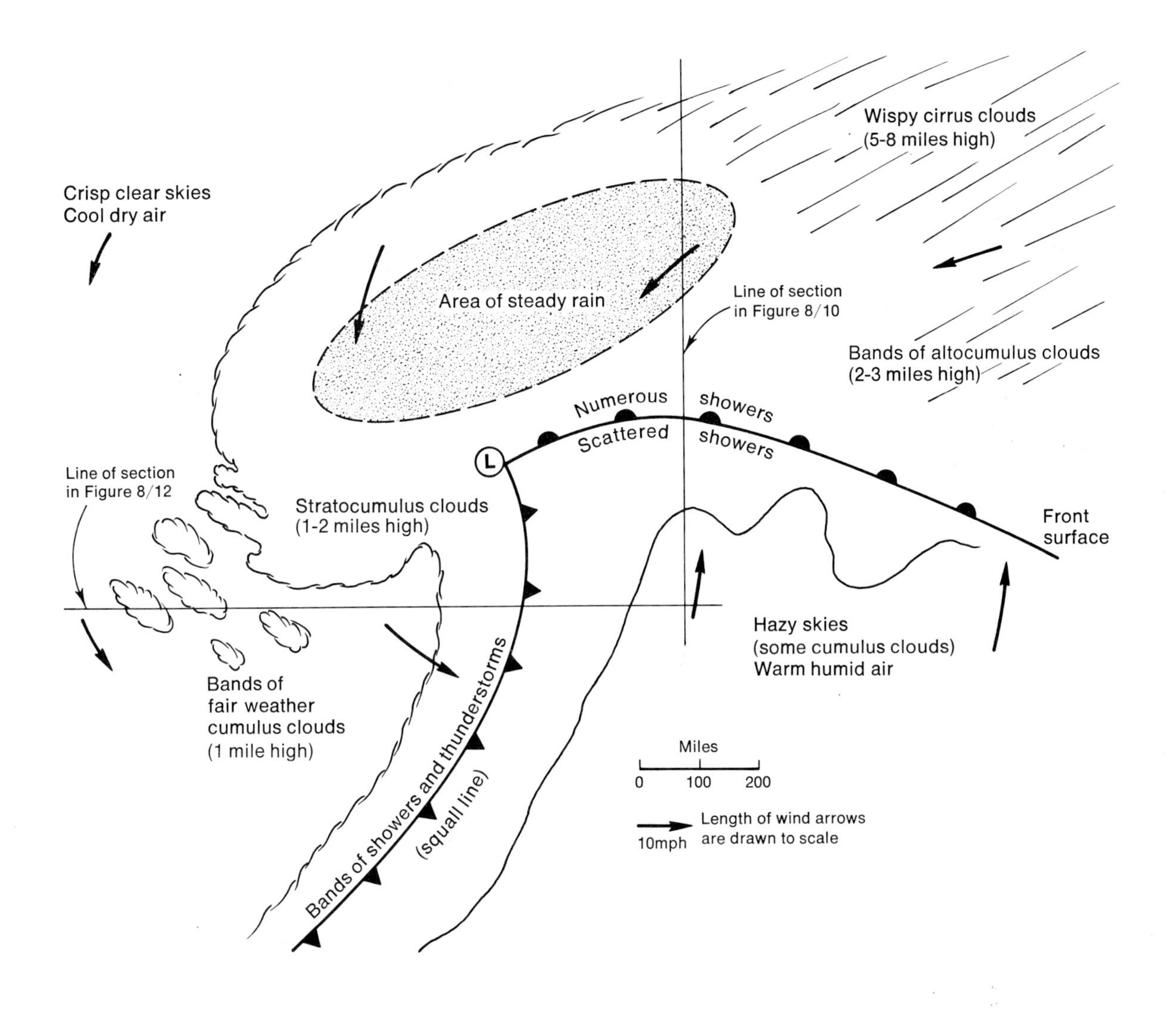

Fronts, and Storms," gives additional review and coverage, with many interesting details.

Weather in Motion. National Aeronautics and Space Administration, Washington, D.C. 2-546, Publication EP-79. U.S. Government Printing Office, Washington, D.C., 20402. A pamphlet describing the ATS (Applications Technology Satellite)–III satellite and its weather photography mission. On the cover comes a removable photographic animation ("movable effect card") which is exceptionally effective, showing clouds moving underneath the "stationary" satellite.

You can subscribe to a weekly set of weather maps produced by the Weather Service. Each set contains daily maps for the previous week. The cost is $7.50 per year.

Write for "Daily Weather Maps—Weekly Series," to the Superintendent of Documents, U.S. Government Printing Office, Washington, D.C., 20402.

Unit 3 The Earth's Crust

Chapter 9 Rocks

OBJECTIVES

After completing this chapter, the student should be able to

(1) explain how igneous, sedimentary, and metamorphic rocks are formed.
(2) classify common rock samples as sedimentary, igneous, or metamorphic.
(3) explain and demonstrate how crystal size relates to cooling rate in igneous rocks.
(4) explain and demonstrate how sediment can be cemented together to form sedimentary rocks.
(5) describe how fossil molds and casts are formed.
(6) explain or demonstrate how banding and crystal orientation become a distinguishing characteristic of some metamorphic rocks.

BACKGROUND

9.1 Examining rocks and minerals

If it is at all possible, have actual rock and mineral samples for the students to examine and compare with Figures 9/1 and 9/2. The rock types in Figure 9/1 are recommended. Suggested minerals are quartz, orthoclase feldspar, and mica (either biotite or muscovite). These are the common minerals of granite.

You might have the students discover the ways in which the three minerals differ from each other—a lesson in mineral identification. Quartz is recognized by its glassy shine (vitreous luster) and its irregular broken surface. Orthoclase breaks in such a way that some surfaces are irregular, but there are also flat surfaces (cleavage planes) at right angles to each other. The mineral is said to cleave if it breaks along planes. Mica cleaves along only one plane, causing it to split into flakes. Under magnification, these characteristics can be seen in the minerals in a piece of coarse granite.

If you wish to study minerals in more detail, refer to the texts and charts in the sources listed in the references.

9.10 Making model fossils

Students can make model fossil molds by pressing leaves, sticks, snail shells, or other objects into modeling clay. You might have a guessing game. Students could bring in objects which might become fossilized, and make molds. Then, those on one side of the room could try and guess what the "fossils" made on the other side represent, and vice versa.

The students can then pour plaster of Paris into the molds and make model fossil casts.

SKULLDUGGERY

Directions to Question Baseball Game

Write the numbers 1 through 17 on separate slips of paper and dump them into a container. Mix them well. Allow the class to choose teams as they would in sandlot baseball.

Announce some kind of reward for the team winning the game. Bonus points to be added to next test score work well. Emphasize that they as individuals can hurt chances for their team by not doing homework.

Each member of the team "up to bat" first is to jot down the answer to the question number you draw from the container. The answers are collected and four of the answers are drawn, one at a time, from the pile.

9.13 Answer to question 2, page 270

Characteristics	Rock type: Sedimentary	Metamorphic	Igneous
Arrangement of crystals	random and cemented together*	intergrown, may be banded	intergrown, random but no cement
Pore spaces	may have many, of different sizes	none	none**
Layering	almost always in outcrop	sometimes in appearance	none, except for lava flows
Banding	often	sometimes	none
Color	many possible	many possible	many possible

*except when formed by evaporation
**except in cases like pumice

(6) At the surface.
(7) Other dissolved minerals could act as cement; temperature changes cause evaporation of water; cement of calcite could be from calcite dissolved in ground water.
(8) Plant or animal is trapped in sediment; living parts are left as empty spaces (mold) or mold is filled with sediment (cast).
(9) The high heat would destroy the animal or plant before a fossil could form.
(10) There are empty spaces between the particles.
(11) The one with cement should be less homogeneous.
(12) Shells of sea animals, ripple marks or mud cracks might be found.
(13) Spaces between particles in sedimentary rock fill in; cement and particles blend; more crystals are formed; color may change; minerals may combine to form others.
(14) Crystals may form bands in the rehardened rock; crystals are turned.
(15) Fossils could be deformed or destroyed.
(16) The continual change from one form to another.
(17) The rock cycle does not have to take place in any special order; seasons and moon phases are in orderly succession.

If the first answer drawn is correct, the team has "a man on first." If the second answer drawn is correct, the man moves to second base and so on. If all four answers are correct, they have a home run. The runner stops with a wrong answer and the team has an out. Draw another number and repeat the process. A runner scores by being forced around the bases by a runner behind him. Three outs retires the side.

Keep track of the base running on the chalkboard. Let the team in "the field" help judge the correctness of the answers. Continue playing for the number of innings desired.

Answers:
(1) Concrete is made by man.
(2) Rocks provide information about past events.
(3) Radioactivity alone wouldn't be sufficient because surface heat is readily lost; added to heat due to deep burial, it could be a factor.
(4) Radioactivity is continually contributing heat.
(5) By the size of the crystals; deepest (cooled slowest) produces largest crystals.

MATERIALS

9.1 Rocks are made of minerals.

CLASS OF 30

rock samples: granite, rhyolite, gabbro, porphyry 30 of each
mineral samples: quartz (not crystals), orthoclase, mica (biotite or muscovite) 30 of each
magnifying glasses 30

9.4 Make rocks from melted minerals.

CLASS OF 30

moth balls $\frac{1}{2}$ lb
hammers 10
crayons, any color 10
test tubes, 1 × 10 cm, to be broken 30
corks 30
hot plate or Bunsen burner with ring stand 1
beaker, 1000 ml (2) or 250 ml (6) or large trays (2)

magnifying glasses 10
pan or large cake pan, 12 × 8 × 4 in 1
towels 10
ice
sand 50 lb

TEACHER
bottle or shallow dish, transparent 1

9.5 Cooling rate affects the size of crystals in a rock.

CLASS OF 30
obsidian 30 pieces
basalt 30 pieces
granite 30 pieces

9.6 Rocks are broken down and carried away.

CLASS OF 30
quartz sand

9.7 Cemented sediment makes sedimentary rocks.

CLASS OF 30
gravel (pebbles or crushed stone) 3 quarts
paper cups, unwaxed if possible 45
glue, water-soluble 1 quart
pipe cleaners or fine wire 15 pieces
sandstone, porous 15 pieces
eyedroppers 30

9.8 Dissolved animal shells can cement sediment.

CLASS OF 30
hydrochloric acid, dilute (use vinegar if it works on your samples)
clamshells 30
sandstone with calcite cement (not hard sandstone cemented with quartz)

9.9 Some sedimentary rocks need no cement.

CLASS OF 30
salt 1 box
sugar 1 lb
food coloring, squeeze bottle 1
baking soda 1 box
shallow pie tins 6

9.10 Sedimentary rocks contain records of the past.

CLASS OF 30
cups, clear plastic 15
sand
sugar cubes 1 box
tissues 1 box
samples of fossil molds and casts, if possible

9.12 Sedimentary rock may become another kind of rock.

CLASS OF 30
sandstone 30 pieces
limestone 30 pieces
shale 30 pieces
quartzite 30 pieces
marble 30 pieces
slate 30 pieces

9.13 Igneous rock can also be changed.

CLASS OF 30
clay 5 lb
gneiss 30 pieces
washers, 3 in diameter 15 doz
magnifying glasses 15
microscope 1

REFERENCES

Gilluly, James, Aaron C. Waters, and A. O. Woodford. *Principles of Geology.* W. H. Freeman and Company, San Francisco, 1968, pp. 595–619.

Leet, L. Don and Sheldon Judson. *Physical Geology.* Prentice-Hall, Inc., Englewood Cliffs, New Jersey, 1971, pp. 617–637.

Rapp, George, Jr. *Color of Minerals.* Houghton Mifflin Company, Boston, 1971.

Romey, William D. *Field Guide to Plutonic and Metamorphic Rocks.* Houghton Mifflin Company, Boston, 1971.

Chapter 10 Volcanoes

OBJECTIVES

After completing this chapter, the student should be able to

(1) explain and demonstrate that minerals expand when they melt.
(2) explain and demonstrate how expanding minerals can force melted rock to the surface of the earth.
(3) show with the aid of a map and data that volcanic activity takes place in certain areas forming a pattern around the earth.
(4) demonstrate that volcanic eruptions cannot be accurately predicted through the use of past records of activity.
(5) explain modern methods of attempting to predict future volcanic eruptions.
(6) identify volcanic rock samples by sorting a few of the more common types from a mixture of volcanic and nonvolcanic samples.

BACKGROUND

10.2 The model volcano

Crayons or paraffin wax melted in a closed space will exert pressure as they expand. Moth balls may be used instead of crayons, but the activity will take longer. The activity can be performed as a classroom demonstration if time, space, and materials are a problem. Otherwise, let students mix their own plaster.

The volcano may be modified by sealing a thermometer within the plaster model. The bulb should be in the area of the crayon. Crayon or paraffin is a mixture of hydrocarbons and therefore will begin to melt in the range of 45° to 50° C. (Naphthalene flakes or moth balls melt at 80.2° C.) In this way the eruption can be predicted by the temperature shown on the thermometer. This modification could be worked into your discussion of Section 10.9, on modern methods of predicting eruptions.

Pitfalls and cautions in making a model volcano:

(1) Plaster mixture must be thick enough to stick to an overturned spoon. If the mixture is too soupy, the entire chunk may crack when it is heated the next day.
(2) Be sure that the plaster is completely set before heating. Otherwise it will crack and allow the model mineral to spurt out the sides.
(3) Do not place crayons against the walls of the cup. This mistake provides an escape other than the yarn vent.
(4) **Do not heat the model volcano without the yarn vent.** The results are extremely unpredictable. Chunks of hot plaster could shoot out and cause injury.

10.8 The Vesuvius activity

It is impossible to pick out a perfect list of Vesuvius' eruptions. Various published lists differ because each compiler has his own definition of what constitutes an eruption.

The early records are no doubt incomplete, but several eruptions after that of A.D. 79 were significant. In 472, ashes drifted all over Europe. The 1036 eruption included the first lava flows in recorded history.

The eruptions with asterisks, listed below, from 1631 to 1944, actually represent climactic eruptions in presumed 10–35 year cycles. Each one was preceded by lesser ones and followed by a relatively quiet period. Cycles cannot be detected before 1631.

This activity is intended to be open-ended and to promote class discussion. Students' examination of text Figure 10/14 should indicate the possibility of some kind of a cycle but lead to the conclusion that

Years when Mt. Vesuvius erupted (all dates A.D.)		
1631*	1804	1895
1682*	1805	1900
1689	1822*	1903
1694*	1838*	1904
1707*	1850*	1906*
1737*	1858	1913
1760	1861	1926
1767*	1871–72*	1929
1779*	1875	1944*
1794*	1891	

accurate prediction requires more information. Some may sense the 10–35 year cycle and predict an eruption before 1975. The discussion can be led to the suggestion that according to the modern cycle concept a major eruption may be expected at any time.

MATERIALS

10.1 Melted minerals take up space.

CLASS OF 30
crayons 30
graduated cylinders 30
paper towels 30
beaker, 2000 ml, **or large pan** 1
hot plate 1
water

10.2 Melted minerals build up pressure.

CLASS OF 30
paper cups, 3 oz size, unwaxed if possible 30
plastic spoons 30
yarn 10–15 ft
plaster of Paris 5 lb
crayons 30
propane torches or gas burners 5
ringstand and clamp 5
scissors 5

10.5 Gases in magma leave holes.

TEACHER
volcanic rock, especially pumice and scoria several samples

PREPARATION
Pumice can be obtained beforehand at drug stores. It is sold for removing calluses. Compare with cinders from school track, or slag, if there are steel mills in your area.

10.6 The dissolved materials can kill.

CLASS OF 30—question 4, page 289
corks to fit pop bottles 30

10.7 The "where" for an eruption is no secret.

CLASS OF 30
world map that can have thumbtacks stuck into it 1
thumbtacks 100
mimeographed copies of world map 30

10.8 The "when" for an eruption is difficult to predict.

CLASS OF 30
adding machine tape, at least 200 feet 1 roll
tape, any kind 1 roll
meter stick 10
colored pencils 15
scissors 5

REFERENCES

Bullard, Fred M. *Volcanoes: In History, in Theory, in Eruption*. University of Texas Press, Austin, 1962.

Macdonald, Gordon A., and Douglass H. Hubbard. *Volcanoes of the National Parks in Hawaii* (2nd ed.). Hawaii Natural History Assn., Hawaii Volcanoes National Park, Hawaii, 1961.

Thorarinsson, Sigurdur. *Surtsey: The New Island in the North Atlantic*. The Viking Press, New York, 1967.

Chapter 11 Earthquakes

OBJECTIVES

After completing this chapter, the student should be able to

(1) relate earthquakes to movements beneath the surface of the earth.
(2) demonstrate, using data and a map, that earthquake epicenter patterns match patterns of volcanic activity.
(3) explain and demonstrate how energy results and travels from the breaking of a rock layer.
(4) explain the difference between two types of earthquake waves.
(5) explain and demonstrate how to locate epicenters of earthquakes by using seismograph tracings and information of wave-travel speed.

BACKGROUND

11.2 The fault box

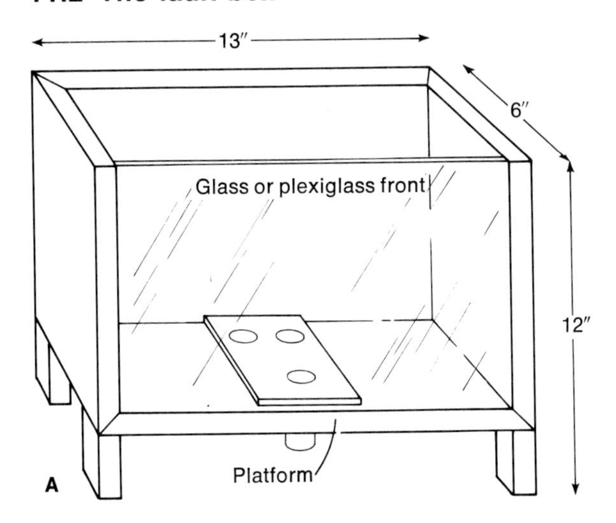

Type 1 A wooden box of any size will do. Plans can be modified to fit the materials available. It is important to have the box at least 8 inches deep. (Microscope cases work well.)

Remove one side of the box and replace it with transparent plastic or window glass. Cut a $\frac{1}{4}$-inch board, piece of plastic, or heavy sheet metal to form a platform that has one edge no longer than $\frac{1}{3}$ the width of transparent side. The other dimension is not important.

Cut three legs, 5 inches long, from a $\frac{3}{4}$-inch dowel rod. Fasten them to the platform and drill holes in the bottom of the box to accommodate them. One edge of the platform should be against the transparent side of the box.

Place the box in a position so that the platform legs can hang freely beneath the box until some force pushes them up. Cover the bottom of the box with a sheet of strong pliable plastic or rubber, and add the sand.

Raise the platform slowly until a desired fault is produced. The platform can be held at several successive heights as it is raised, for examination and discussion of the progress of the faulting.

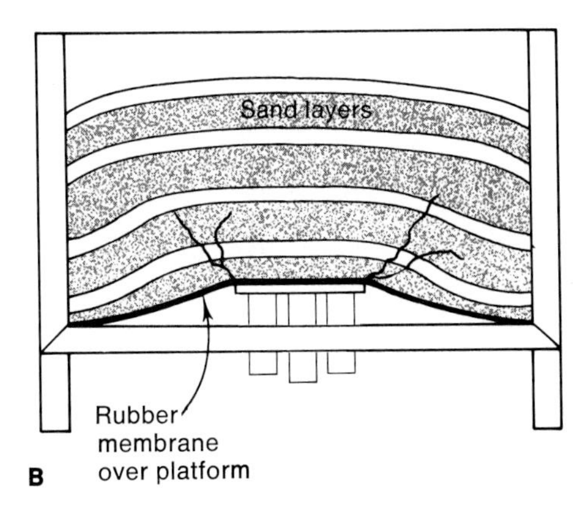

Type 2 An earth box or modified aquarium can become a fault box with the addition of a platform that can be raised from above.

Make a wooden platform the width of the box and about four inches long. Attach to it on one side a vertical piece (about ½″ × 2″ × the box height) to serve as a handle for lifting. An L-shaped metal strengthener will help at the joint.

Cover the bottom of the box with a sheet of plastic or rubber before you put in the sand.

Type 3 You can make your own version of the apparatus described in the text if you have a bicycle tire pump. A sturdy plastic bag will do to lift the sand. Seal the bag onto one end of a six inch length of glass tubing. Cut a piece of rubber tubing long enough to run from the glass tubing along the bottom of the box, up over the end, and to the pump. A tubing of surgical rubber will stretch enough to fit over the valve at the end of the pump's air hose.

This arrangement works as described here in a modified aquarium. If you use an earth box with holes in its ends, the rubber tubing can go through the lower hole.

11.4 Answer to question 1, page 310

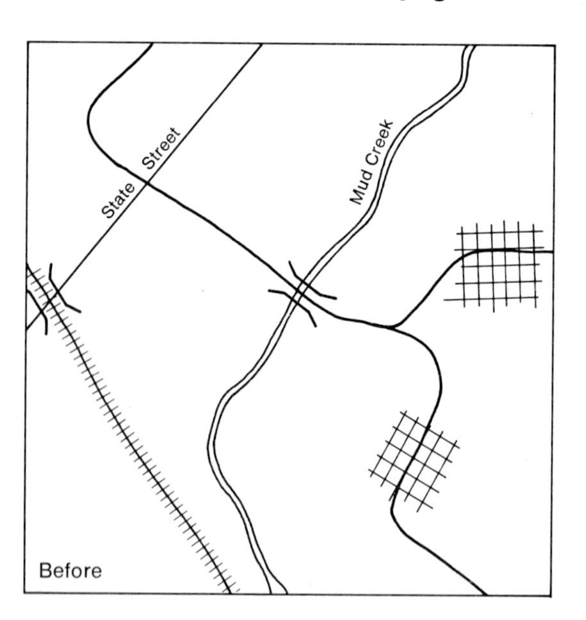

MATERIALS

11.1 Earthquakes change the earth's surface.

CLASS OF 30

mimeographed copies of world map 30

TEACHER

cigarette

cellophane from cigarette pack

11.2 Land in earthquake areas moves up and down.

CLASS OF 30

graph paper, 5-plus squares to inch 30

fault box (optional) 1 or more

protractors 5

11.5 Sensitive instruments detect earthquakes.

CLASS OF 30

rubber bands, large 30

weights, 1 kg, **or rocks** of similar weight 30

11.7 Energy is released when a rock layer breaks.

CLASS OF 30

wooden laths, 2 ft long 30

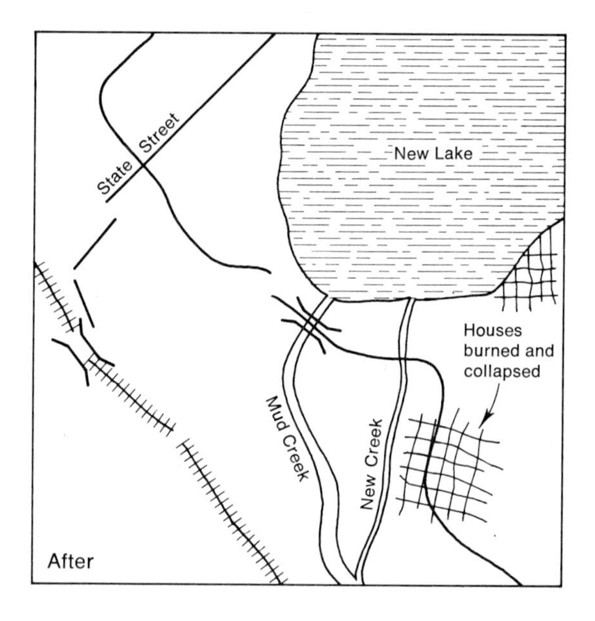

11.9 Suggested diagrams for illustrating how energy is transferred in P and S waves.

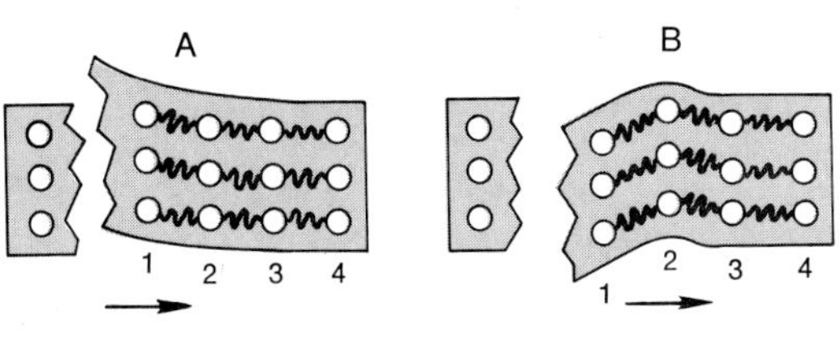

Rock breaks. First row of particles moves up.

Second row moves up, pulled by first row.

C

D

Third row moves up, pulled by second row.

Fourth row moves up, pulled by third row.

A

B

Rock breaks. First row of particles crashes into next row.

Second row crashes into third row.

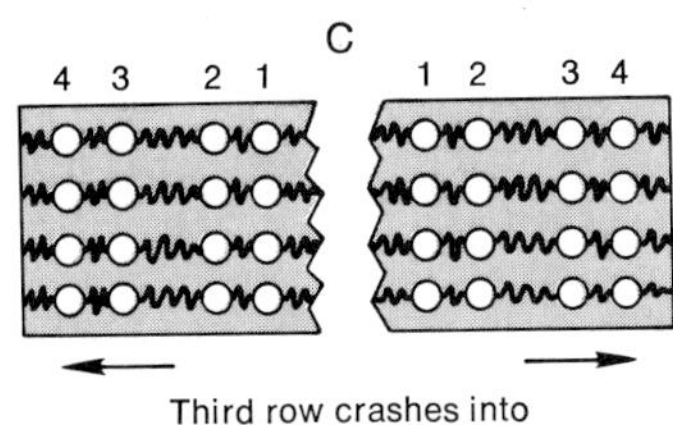

Third row crashes into fourth row and so on.

11.8 The released energy shows up elsewhere.

CLASS OF 30

wooden laths, 2 ft long 30

salt shakers, filled (6) **or box of salt** (1 lb)

11.10 Energy can be passed along by rows of particles.

CLASS OF 30

rope 50 ft

11.13 Three seismographs can determine the epicenter of an earthquake.

CLASS OF 30

paper, can be pages of students' notebooks 30 sheets

compasses, drawing 30

REFERENCES

Coffman, Jerry L. *Earthquake Investigation in the United States.* U.S. Department of Commerce, C&GS Special Publication 282, Revised Edition, 1969. A pamphlet with maps and descriptions of earthquakes and information on seismic waves and seismographs. It also contains a list of publications of the National Earthquake Information Center.

Engle, Eloise. *Earthquake! The Story of Alaska's Good Friday Disaster.* The John Day Company, Inc., New York, 1966.

Preliminary Determination of Epicenters Monthly Listing. Superintendent of Documents, U.S. Government Printing Office, Washington, D.C. 20402. You can have these mailed to you, if you want to keep a record of earth activity.

The California Earthquake of April 18, 1906—Report of the State Earthquake Investigation Commission. Carnegie Institution of Washington, Publication No. 87, 2 volumes and atlas, 1908, reprinted 1969. The first detailed study of an earthquake, and an all-time classic.

Unit 4 The Changing Crust

Chapter 12 Weathering

OBJECTIVES

After completing this chapter the student should be able to

(1) illustrate the natural tendency for things to become disordered.
(2) explain the difference between mechanical and chemical weathering, listing agents and examples of each type.
(3) associate mechanical and chemical weathering with the deterioration of materials around him and his home.
(4) explain why water can be called the most versatile of the weathering agents.
(5) explain how a soil is formed from hard rock.
(6) explain differences in soils and reasons for the differences.

BACKGROUND

12.4 Mechanical weathering activity

The best results can be obtained from medium hard rock types, such as limestone or hard shale. Softer shale chips may be soft enough to "melt" or weather too fast. Some shale will leave a deposit on the inside of the bottle that is very difficult to wash off. Crushed limestone is recommended; chip size about 1–2 cm. Limestone fragments show abrasion of edges and corners well.

Soak the chips in water before class, at least a half hour, so they absorb all the water they can. This facilitates the weathering process. Whatever size of bottle is used, it is best to have the water level just a bit higher than the top level of the rock chips. Excess water merely dilutes the resulting suspension.

If you wish to have more weathered material in the pill bottles, you can wait until the material has settled, and then have the students pour off the water from above the settled material. They can then shake the bottles once or twice again and pour another sample into each pill bottle. If the activity is started at the beginning of a class period, you may be able to do this before the end of the period. Otherwise, have the students save the water in which the rocks were shaken until the next period, just in case.

12.7 The solubility of calcium carbonate

The solubility of calcium carbonate in water is an equilibrium reaction. It is expressed as

$$CaCO_3 \rightleftarrows Ca^{++} + CO_3^{=}$$

Because $CaCO_3$ is nearly insoluble, the concentrations of the two ions in the water, Ca^{++} and $CO_3^{=}$, are very low. Removal of either ion will change the equilibrium and cause more $CaCO_3$ to dissolve (Le Chatelier's Principle).

Adding carbonic acid (CO_2 dissolved in water) to the solution makes it more acidic and the hydrogen ions produced will combine with the $CO_3^{=}$:

$$H^+ + CO_3^{=} \rightleftarrows HCO_3^-$$

In effect this reaction decreases the original $CO_3^{=}$ ion in concentration. More $CaCO_3$ will then dissolve to equalize the $CO_3^{=}$ concentration.

MATERIALS

12.4 Mechanical weathering can happen under water.

CLASS OF 30

limestone, crushed 13 lb
bottles, plastic, about 250 ml (8 oz) 30
corks to stopper bottles 30
grease pencils 30
tape 1 roll
pill bottles, transparent, about 50 ml ($1\frac{1}{2}$ oz) 30
record player (optional) 1
record, fast music (optional) 1

12.7 Chemicals act as weathering agents.

TEACHER

hydrochloric acid, dilute 50 ml
evaporating dish 1
limestone 1 chip
Bunsen burner 1
ring stand 1

PREPARATION

Dilute hydrochloric acid (HCl) can be prepared by pouring concentrated HCl slowly into four times as much water, stirring slowly with a glass rod. **Never add water to acid.**

12.8 What does soil contain?

CLASS OF 30

pill bottles, as in Section 12.4
soil

PREPARATION

Dig soil from several inches below the surface in order to get a heterogeneous mixture. Try the activity beforehand to make sure the soil settles in several layers. Sand or sawdust may be added if necessary.

12.10 There are many types of soil.

CLASS OF 30

pill bottles, as in Section 12.4
soil, supplied by students

REFERENCES

Leet, L. Don, and Sheldon Judson. *Physical Geology*. Prentice-Hall, Inc., Englewood Cliffs, New Jersey, 1971.

Strahler, Arthur N. *The Earth Sciences*. Harper & Row, New York, 1971.

Chapter 13 Erosion

OBJECTIVES

After completing this chapter, the student should be able to

(1) explain and demonstrate those factors that limit the water available for eroding.
(2) explain and describe conditions that increase or decrease a stream's ability to erode.
(3) list the features that distinguish old streams from young streams.

BACKGROUND

Investigating a stream

This can be an individual study, or you can organize a field trip around it. The probable findings are as follows:

(1) Stream velocity is greatest at the outside of a stream bend and least at the inside of the bend. Refer students to the meander study. Erosion is greatest where stream velocity is greatest.

(2) Velocity is greatest at the top of the stream because this is where friction is least. The stream carries the greatest load where its velocity is greatest.

(3) The bottom sample contains the most sediment, because velocity is least at the bottom of the stream. Sediment from the top settles to the bottom, where the velocity is not great enough to carry material along.

(4) Temperatures vary because some areas are shaded and others sunlit. Unless there is a large amount of mixing, bottom water would be cooler, because solar radiation warms the surface, and cold water is denser than warm water and thus tends to sink.

(5) Pebbles are smallest near the shore and largest in the middle of the stream. This is because friction is greatest near the shore and stream velocity therefore least. This means that only small pebbles are carried near the shore. The stream bottom in the middle has large pebbles because small ones have been carried away. Sediment at the inside of a stream bend is fine-grained because this is where velocity is least. Sediment at the outside of the stream bend is coarser because velocity is greater.

(6) Plant life may vary because of depth (affecting the amount of available sunlight) and the degree of pollution.

(7) What happens to a stream as time goes by will vary from stream to stream. Have students save their water samples until you have covered Chapter 19, Man in His Environment. (The color of the samples may change as materials settle out. Also, strong odors may develop.) Over the years you will build up a very interesting record.

MATERIALS

13.2 Glaciers push, drag, and carry rocks and soil.

TEACHER

snow
dirt or sand
stream table

13.5 More water falls on the land than runs off.

CLASS OF 30

graph paper, 5-plus squares to the inch 30 sheets
colored pencils, two different colors 60

13.7 You can measure pore space and particle size.

CLASS OF 30

screen sieves, two different sizes 30
sand, beach 25 lb
test tubes, large 30
dishpans or other large containers 15
grease pencils 15

13.8 Flow rate and water held back can be measured.

CLASS OF 30

stone, crushed, small 15 lb
stone, crushed, large 15 lb
tubes, plastic, 2 × 24 in, fitted with drain and clamp 10
stop watches 5
tin cans, #10 or 1 qt size 10
graduated cylinders, 1000 ml 10

PREPARATION

You can start saving tin cans at the beginning of the school year. Crushed stone is sold for aquariums and by lumber yards. Use a common supply and have each student fill his own container.

13.10 Small streams are usually parts of larger streams.

CLASS OF 30

stream tables 10
sand 180 lb
bottles, plastic, gallon size 10
food coloring, squeeze bottles 10
blocks for props 20
nails (to make the hole in the bottle and use as a stopper) 10

PREPARATION

A wooden box about 50″ × 15″ × 4″ lined with plastic will do for a stream table. Make a drain high on one end. See Figure 13/13. Use a gallon-size plastic bleach bottle or milk jug. With a nail make a hole the diameter of a pencil lead in the bottom.

13.12 Distributaries develop at the ends of some streams.

CLASS OF 30
stream tables, as in Section 13.10 10
sand
bottles, plastic, gallon size 10
food coloring, squeeze bottles 10

13.13 The faster the stream, the more it erodes.

CLASS OF 30
stream tables, as in Section 13.10 10
sand
bottles, plastic, gallon size 10
food coloring, squeeze bottles 10

13.14 Stream speed depends on the angle of the stream bed.

CLASS OF 30
stream tables, as in Section 13.10 10
sand
bottles, plastic, gallon size 10
food coloring, squeeze bottles 10
markers, sticks or pencils 20
protractors 10
stop watch 1
nails 10

13.16 Changing stream volume changes stream velocity.

CLASS OF 30
stream tables, as in Section 13.10 10
sand
bottles, plastic, gallon size 10
markers, sticks or pencils 20
protractors 10
stop watch 1
nails 10

13.17 Stream erosion decreases with age.

CLASS OF 30
stream tables, as in Section 13.10 10
sand
bottles, plastic, gallon size 10
food coloring, squeeze bottles 10
protractors 10

REFERENCES

Leet, L. Don, and Sheldon Judson. *Physical Geology*. Prentice-Hall, Inc., Englewood Cliffs, New Jersey, 1971.

Strahler, Arthur N. *The Earth Sciences*. Harper & Row, New York, 1971.

Chapter 14 Deposition

OBJECTIVES

After completing this chapter, the student should be able to

(1) explain and demonstrate how sediment can be sorted into different particle sizes and mineral types.
(2) explain how turbidity currents carry sediment beyond river deltas.
(3) demonstrate that an irregular sea floor does not prevent sediment from being deposited in horizontal layers.
(4) explain the theory of geosynclines and give such evidence for their existence as coral reefs, sediment depth, and deep-sea trenches.

BACKGROUND

14.5 Comparing a slurry with a saturated salt solution

You can carry the investigation of differing water densities one step further with an activity that gives a striking result.

Clamp a column at an angle. Pour in a saturated salt solution, until the column is half full. Fill it the rest of the way with fresh water. Be careful not to let the two kinds of water mix; pour the fresh water down the inner surface of the column.

Dump in a slurry made from soil (carefully, so as not to disturb the water layers). The slurry will remain above the salt water, with a sharp line of demarcation between the two layers.

MATERIALS

14.2 Different particle sizes settle at different rates.

CLASS OF 30

screen sieves at least 2 sets of 4
stop watches 5
ring stands and clamps 10
columns, transparent, 3 ft × 2 in, fitted with drain and clamp 10
paper cups 60
sand 10 lb
spoons, plastic 10
graph paper, 5-plus squares to inch 30 sheets

PREPARATION

Get 100% quartz sand if possible. White sandbox sand works well. Transparent columns are available from Hubbard.

14.3 Some particles are heavier for their size than others.

CLASS OF 30

same materials as 14.2
lead shot 3 lb

PREPARATION

Lead shot is available from sporting goods stores and Cenco.

14.5 Currents carry sediment far out to sea.

CLASS OF 30—suspension activity

dust from sieves, left over from Section 14.2
spoons, plastic 10
test tubes, large 10

CLASS OF 30—turbidity current activity

column, transparent, 3 ft × 2 in, **fitted with drain and clamp or earth box or aquarium** 10
ring stands, if column is used 10
salt water
food coloring, squeeze bottles 10
test tubes, large 10

14.6 Turbidity currents can be influenced by conditions in the sea.

CLASS OF 30

aquariums or earth boxes 10

funnels 10
clamps and ring stands 10
food coloring, squeeze bottles 10
beakers, 500 ml 10
hot water 500 ml

14.7 A strong current may spread sediment over the sea floor.

CLASS OF 30
stream tables 6
plaster of Paris 6 lb
Portland cement 6 lb
soil, fine, dark 3 lb
sand, coarse, unsorted 5 lb
troughs, 3 ft long 6
hose, 2 or 3 ft long 1
pail 1
corks (optional) 6

PREPARATION
Gutters from a house work well as troughs. The transparent column (without its drain and clamp) can be used, too.

14.9 Sediment deposition can change the shape of the sea floor.

CLASS OF 30
jars, gallon size, wide mouth 10
crushed stone, pieces $\frac{1}{2}$–1 in long 10 lb
paper cups or 250-ml beakers 20
soil, dark 10 lb
sand, light colored 10 lb
spoons, plastic 10

PREPARATION
Gallon pickle or mayonnaise jars can be obtained from the school cafeteria. Driveway gravel works well as crushed stone.

REFERENCES

Leet, L. Don, and Sheldon Judson. *Physical Geology*. Prentice-Hall, Inc., Englewood Cliffs, New Jersey, 1971.

Strahler, Arthur N. *The Earth Sciences*. Harper & Row, New York, 1971.

The magazine *Sea Frontiers* is recommended.

Chapter 15 Mountain Building

OBJECTIVES

After completing this chapter, the student should be able to

(1) explain how convection currents may account for the existence of geosynclines.
(2) cite evidence in support of the continental drift theory and how it supports the convection theory.
(3) explain recent changes in the convection and continental drift theories.
(4) explain, with the aid of diagrams, how earthquake data provides information about the earth's interior in support of the convection theory.

BACKGROUND

15.4 Convection currents

In the activity on convection currents, test tubes may be substituted for bottles. Anything with one-ounce capacity will work. Larger bottles may be used for a demonstration. For the best results, cool the turpentine in a refrigerator. Do not use spray paint. It contains materials that will not dissolve in turpentine.

A complex of cells can be set up by laying a bottle on a hot, wet washcloth. An orderly network of convection cells will result.

A model made with gasoline or carbon tetrachloride will produce fast currents when held in the hand. However, these liquids are dangerous.

MATERIALS

15.1 The continents are in balance with the ocean floor.

CLASS OF 30

aquarium or earth box (optional) 1 or more

blocks, wooden, 2″ × 4″ × width of aquarium or earth box enough to make 3 layers in each aquarium

grease pencils 2 different colors for each aquarium or earth box

15.2 Mountain roots rise as the sea floor sinks.

CLASS OF 30

copies of Figure 15/5, mimeographed 30

15.4 A theory of drifting continents was proposed many years ago.

CLASS OF 30

turpentine 1 qt

gold or silver paint, model type 1 oz

pill bottles 30

15.7 New evidence from the sea floor supports the theory.

CLASS OF 30—question 2, page 442

paper cups 30

corks 90

15.12 The earth has an inner core and an outer core.

CLASS OF 30

boards, about ½″ × 8″ × 36″ 15

cloth, enough to cover half of each board lengthwise 15

thumbtacks 90

round pencils or dowels, 6 in long 15

REFERENCES

The following series of articles in *Scientific American* will provide an excellent background on the continental drift theory. The first article outlines the historical development of the theory and the state of knowledge of continental drift at the time the article was written. If you read the other articles in the order in which they appeared, you can see how more data have been added and more approaches explored.

Hurley, Patrick. *The Confirmation of Continental Drift*. April, 1968.

Dietz, Robert and John Holden. *The Breakup of Pangaea*. October, 1970.

Anderson, Don. *The San Andreas Fault*. November, 1971.

Dietz, Robert. *Geosynclines, Mountains and Continent-Building*. March, 1972.

Dewey, John. *Plate Tectonics*. May, 1972.

Newell, Norman. *The Evolution of Reefs*. June, 1972.

Unit 5 The Earth's History

Chapter 16 Geologic Time

OBJECTIVES

After completing this chapter, the student should be able to

(1) describe four methods of dating objects older than 2500 years.
(2) solve simple problems related to tree-ring dating, carbon-14 dating, varve dating, and uranium dating.
(3) locate events in the earth's past on a time line representing the 4.5 billion years of the earth's existence.
(4) demonstrate an acquaintance with the geologic time chart and its use.
(5) explain the truth of analogies like "one million years ago in the life of the earth is like yesterday in the life of a man."

• *PREPARE NOW*

Preparation for Section 19.5 activities

The activities in Section 19.5 illustrating the effects of pollution on nylon and cloth are planned to take a month. Especially if Chapter 19 falls at the end of your school year, plan to set up these two activities about now.

MATERIALS

Opener

CLASS OF 30
ditto master 1
typewriter 1
scrap paper, blank on back 150–200 sheets
masking tape 1 roll
felt-tip marker 1

16.5 Carbon atoms place events in time.

CLASS OF 30
cardboard boxes, shoe-box size, with lids 6
pennies or washers painted on one side 600
BB shot 3000

16.7 Important events help organize the earth's history.

CLASS OF 30
adding machine tape $2\frac{1}{2}$ rolls
colored pencils 30
meter sticks 15

REFERENCES

Dott, Robert H. Jr., and Roger L. Batten. *Evolution of the Earth*. McGraw-Hill, New York, 1971.

Kummel, Bernhard. *History of the Earth: An Introduction to Historical Geology*. (2nd ed.) W. H. Freeman, San Francisco, 1970.

Chapter 17 Stories in Stone

OBJECTIVES

After completing this chapter, the student should be able to

(1) explain how to determine whether or not a rock layer has been overturned.
(2) explain and demonstrate how fossils can be used to spot where rock layers are missing from a series of layers.
(3) explain and demonstrate with the aid of sketches how several outcrops can be used to construct a geologic history of an area.
(4) explain and demonstrate how to interpret geologic maps.
(5) interpret the geologic history of the North American continent, given a time line illustrated with maps of North America during the various time periods.

BACKGROUND

"Go ahead, Holmes. I'm listening." Dr. Watson put the puzzle back on his thumbs. Holmes had made it look too simple to resist trying it again.

"The clue is in the soil, where all the digging took place." Holmes spoke louder toward the end of the sentence to get his friend's attention. "The sand was on top before any digging. But, Watson, what kind of soil would be at the bottom of the burrow if the dirt that filled it had been brought from elsewhere?"

"Why, the sand, of course," answered the doctor. "He had to have dug that up first and it would have been in the first load dumped in."

"Good, Watson! And what kind of soil would have been at the top of the filled burrow?"

"The clay, naturally." The question seemed almost childish to Watson, now that he was paying attention.

"Right. But the police said that the pile they saw next to the hole had clay on top. When the thief removed the soil from the hole, what would he have removed first, Watson?"

"Hmm. The clay. And the sand last. The sand should have been on the top of the pile. Why, that's amazing, Holmes." Watson looked like a quiz-show contestant who had just won.

"The real thief is the passerby. The pile of dirt the police saw was never in the burrow. It was exactly the way it was brought from behind the bush; sand in the first jacketful, covered by the clay brought later. The passerby nearly put the blame on some make-believe gang and got off scot-free."

"Better take an earth science book to the inspector, to rub it in a little," Dr. Watson laughed. Sherlock Holmes treated himself to another pipeful of tobacco.

17.1 Interpretation of Figure 17/3

The history can be outlined in a series of stages, starting with the oldest indicated events and ending with the most recent events.

Stage I Cambrian near shore deposition of gravel and sand (layers 4 and 5) followed by a retreat of the shoreline, with Ordovician to Devonian marine deposition away from the shore (layers 1 to 3).

Layers 1 to 5 are overturned. Layers 4 and 5 were originally below 1 to 3 so they must be older. Since 3 is Ordovician, 4 and 5 must be Cambrian (or older).

Stage II Folding of rocks to the point of overturning, and accompanying uplift. The uplift is indicated by the unconformity at the top of layer 5.

The fault presents two possibilities here.

(A) If fault ends at the top of layer 5, the faulting must have taken place after the overturning but before the unconformity was formed.
(B) If fault goes up through layer 14, the crustal stress that formed it must have been among the latest events, or Quaternary in age.

Answer to Sherlock Holmes mystery

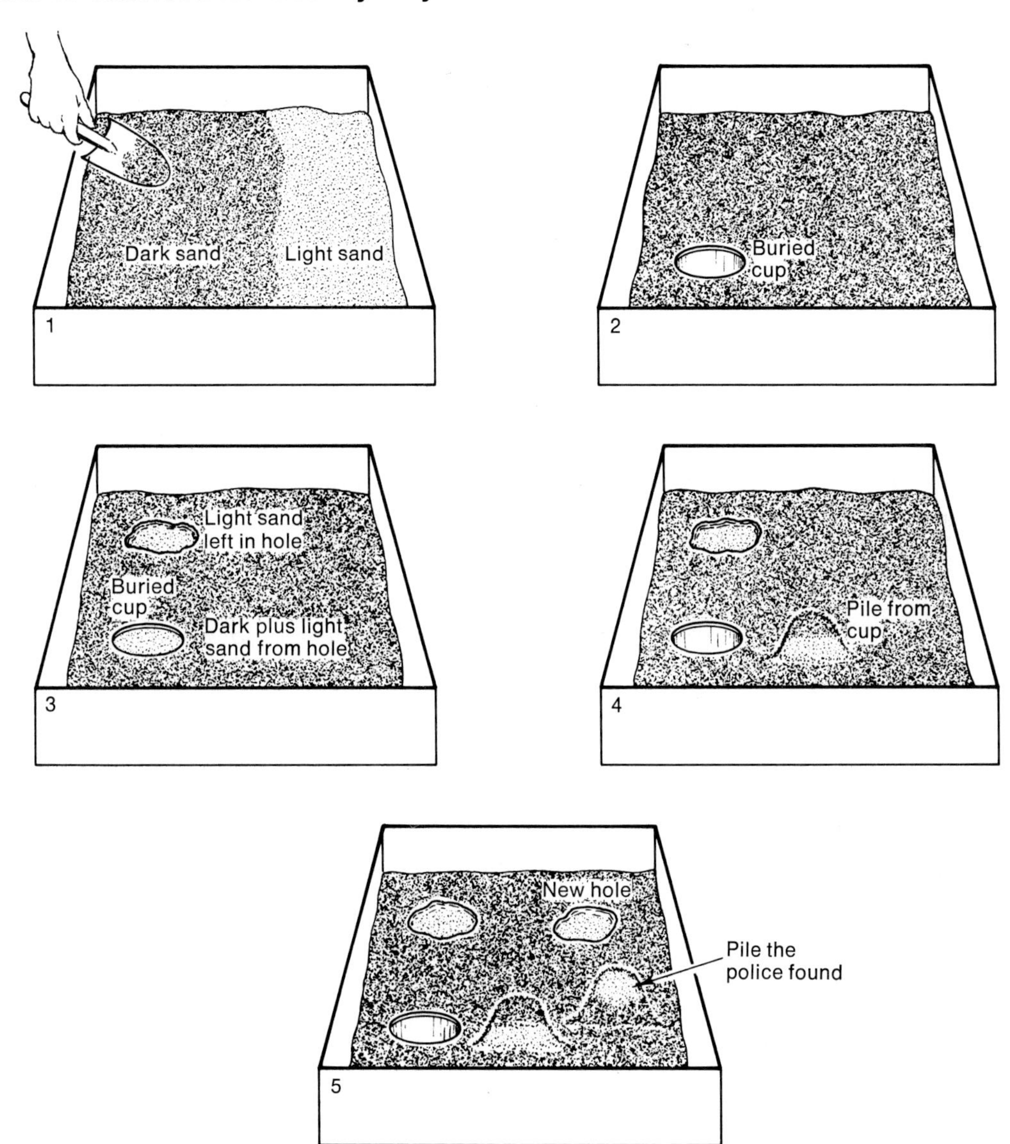

According to possibility (A):

Stage III Intrusion of magma, followed by faulting. The intrusion, faulting, and overturning could be parts of a large scale mountain-making event, during the Mississippian.

Devonian rock is overturned, and the next youngest unchanged layer whose age is known is Pennsylvanian in the simplified version. The folding and uplift must have occurred during the Mississippian. Actually, the time could have been late Devonian or early Pennsylvanian.

Stage IV Erosion of the mountains to form the unconformity between layers 5 and 6.

Stage V Deposition of layer 6, probably in the ocean away from the shore.

If the shale layer is as extensive as the limestone

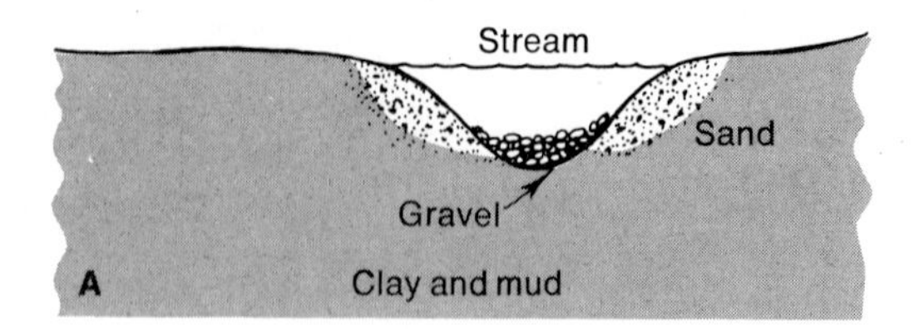

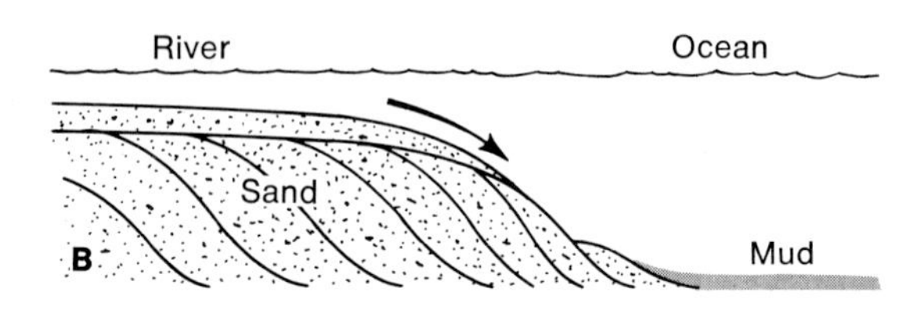

layers, it is probably too extensive to have been a lake bottom. The fact that there are no fossils is a reminder that sometimes rock formations are found that offer little obvious information. Their understanding may require much detailed study and careful comparing with other areas.

Stage VI Uplift and erosion to form unconformity between layers 6 and 7.

Stage VII Marine deposition during the Pennsylvanian, Permian, and Triassic Periods forming layers 7, 8, and 9. The gradation upward from limestone through shale to sandstone indicates a shoreline slowly approaching the area.

Stage VIII Uplift and erosion. In the simplified version, this would have occurred during the Jurassic, since rocks of that age are missing.

Stage IX Cretaceous marine deposition, resulting in layers 10 and 11.

The gradation upward from shale to limestone might indicate that the nearest land was being reduced by erosion, and weakening streams had become unable to send mud out into the ocean in large enough amounts to form shale layers. Another possibility is that an obstruction on the ocean bottom or a current change sent the mud somewhere else and permitted the limestone to form.

Stage X Possibly Tertiary uplift and erosion, forming the unconformity between layers 11 and 12.

Stage XI Quaternary terrestrial deposition (layers 12 to 14). The sediment is no doubt from nearby rising mountains.

According to possibility (B):

Stages I–XI The history would be the same except for the removal of the faulting process from Stage III and the addition of the faulting to Stage XI. The faulting could be associated with the formation of the new mountains.

Note: Crossbedding has been omitted, along with clues regarding old stream beds. However, you may want to discuss them with your classes. Some states require covering them in earth science courses.

(1) Wandering stream beds: As a stream meanders across its flood plain, it leaves gravel and sand deposits along its stream bed. As the flood plain is built up, a "stream-bed trail" like the one shown in Figure A will result.

(2) Crossbedding: Sediment dumped over the far end of a developing delta may produce slanted layers like those shown in figure B. If there are sedimentary rocks in your area, students may have noticed crossbedding in sandstone layers.

17.7 Question 1, page 501: Interpretation of Figure 17/8

Column A The first five layers, counting from the bottom, indicate an overturned fold similar to the one diagrammed in Figure 17/5. Note that Silurian rocks are missing. The history of this area probably occurred in the following stages:

I. Cambrian and Ordovician ocean deposition.
II. Uplift and erosion in the Silurian.
III. Devonian ocean deposition.
IV. Folding, uplift, erosion.
V. Jurassic, Cretaceous, and Tertiary ocean deposition.
VI. Uplift, bringing the marine rocks above sea level.

Column B Devonian to Permian rocks are found on top of Mesozoic rocks. This indicates a low-angle fault (called a thrust fault), with older rocks pushed on top of younger rocks as the diagram shows. A famous example of this sort of movement is at Chief Mountain in Glacier National Park. The history of this area probably occurred in the following stages:

I. Devonian to Cretaceous marine deposition.
II. Tertiary faulting and uplift.
III. Quaternary terrestrial deposition.

Column C This section includes three unconformities, represented by missing layers. The history of the area probably occurred in the following stages:

I. Devonian ocean deposition.
II. Uplift and erosion, Mississippian.
III. Pennsylvanian ocean deposition.
IV. Uplift and erosion, Permian.
V. Triassic ocean deposition.
VI. Uplift and erosion, Jurassic.
VII. Cretaceous ocean deposition.
VIII. Uplift, promoting Tertiary terrestrial deposition.

Column D The lower portion of this section consists of an overturned series of layers, at the top of which is an unconformity. The history of this area probably occurred in the following stages:

I. Cambrian to Devonian marine deposition.
II. Folding and overturning, uplift and erosion. When this began is not sure, since you don't know what is below the Devonian layer in the given section.
III. Triassic and Jurassic ocean deposition.
IV. Uplift and erosion, Cretaceous.
V. Tertiary ocean deposition.
VI. Uplift, with Quaternary terrestrial deposition.

17.7 Question 2, page 502: Interpretation of Figure 17/9

Section A As the question implies, the histories of the sections in Figure 17/9 should be more detailed than the histories of Figure 17/8. In section A, the first five layers (starting at the bottom) indicate marine deposition. The first three suggest a shoreline moving back and forth, possibly due to erosion and then minor uplift in the nearest land mass. The fourth and fifth layers suggest a retreating shoreline, or lowering of the land, in the change from sandstone to shale to limestone.

A period of uplift and erosion is shown by the unconformity at the top of the limestone.

A later uplift of the land must have caused the deposition of gravel on the lower portions of the land surface. The gravel ultimately solidified into conglomerate.

Erosion must have continued, to form another surface, that cuts across both the limestone and the conglomerate. This indicates a long period of erosion, with the land well smoothed off (old age).

The basalt in the upper part of the section may have come from either on land or under water. A land volcano suggests that the sandstone layers above and below it are terrestrial. The volcano could be associated with nearby faulting and mountain formation. The erosion of the mountains would explain the sand and mud that became the sandstone and the top layer of shale.

If the lava flow was extruded from an underwater volcano, the shale and sandstone might be marine. Fossils could prove the origin of the layers.

This section is similar to what is actually found in the western Great Plains area, where Cenozoic terrestrial volcanic and sedimentary rocks overlie Paleozoic and Mesozoic marine sedimentary rocks. The Cenozoic rocks were formed as the Rocky Mountains rose and were eroded.

Section B The first five layers probably represent ocean bottom material. The marine sandstone and shale were then raised, faulted, and eroded. The uplifted block on the right formed a hill.

After the period of erosion, the area was lowered (or the sea level rose) and limestone was deposited.

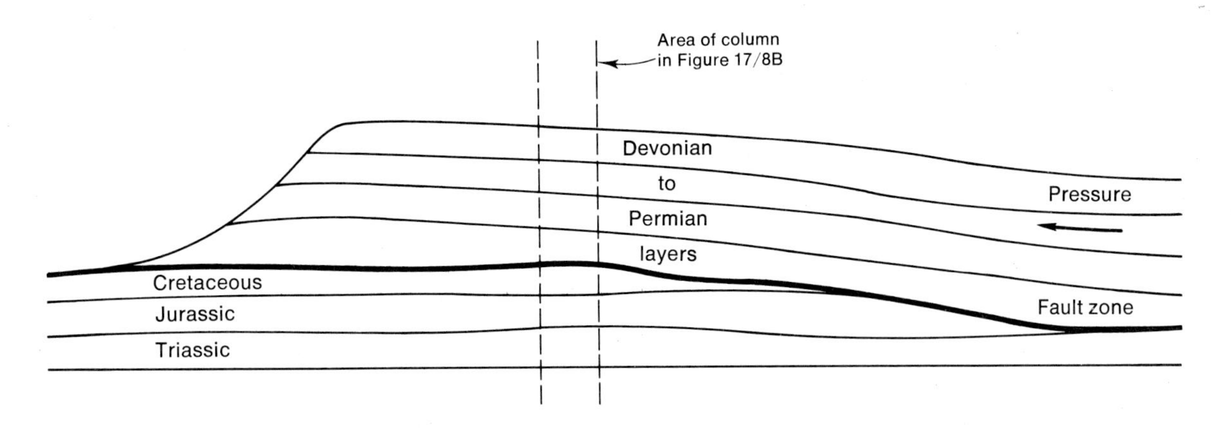

The conglomerate indicates land was rising nearby, and rivers dumped gravel along the shore. The rising process continued until the conglomerate was strongly eroded.

The sandstone and shale above the conglomerate are probably terrestrial, and indications of further rising and mountain making. The last event would be the volcanic activity. The granite represents the last magma that rose before the volcano became extinct.

The last three layers could be marine, and recently raised. However, if these layers are terrestrial only, this section represents what one might find in New Mexico or Arizona.

17.8 Interpretation of Figures 17/10 and 17/11

Layer	Interpretation
0	Granite basement
1	Ocean near shore—deeper at D
2	Retreat of shoreline—ocean bottom away from shore all across area
3	Advance of shoreline—beach gravel throughout area
4	Retreat of shoreline—beach or shallow water sand at A, grading to mud away from shore at B to D
5	Ocean bottom away from shore
6	Ocean bottom away from shore but sand is washing into area at A
7	Ocean bottom away from shore
8	Advance of shoreline—beach gravel at A and B, grading to beach or shallow water sand at C and D
9	Clean ocean bottom, away from shore
10	Beach gravel
11	Beach or shallow water sand

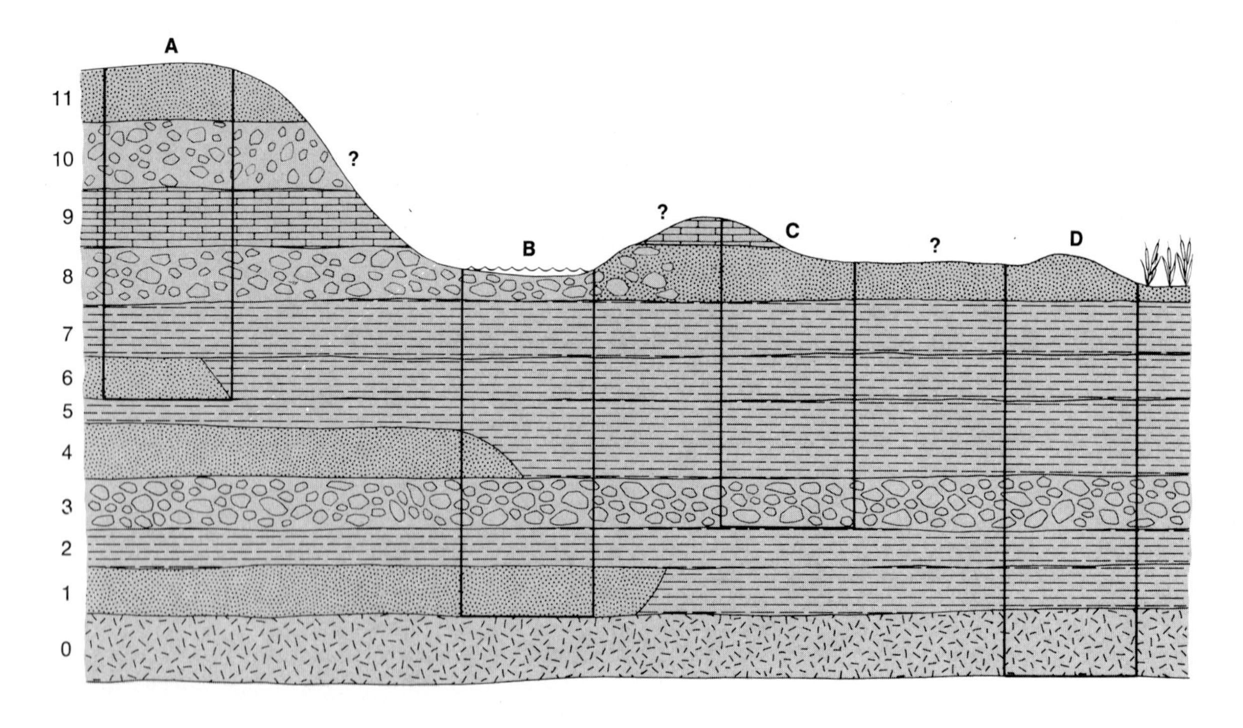

The history should include a last stage of uplift and erosion to form the present landscape.

17.9 Making a modeling clay map of New York State (with Figure 17/14)

To make a layered clay model of New York, push the Adirondack area upward to form a dome, and slice it horizontally. Except for the eastern edge and the southeastern corner, the rest of the state consists of layers gently tilted downward toward the south. The rocks along the eastern margin have been folded and faulted upward, bringing older rocks to the surface.

The structure of Ohio also lends itself to modeling in clay, with layers dipping gently away from each side of the Cincinnati arch. This can be seen to connect with the basin structure of lower Michigan. The rock structure underneath Illinois forms a basin although the upper surface is relatively flat.

MATERIALS

17.8 Hidden rocks can be used as clues.

CLASS OF 30
tracing paper (cut in half) 15 sheets
scissors several pairs
transparent tape 1 roll

17.9 Rock patterns can be used as clues.

CLASS OF 30
clay, 2 different colors any amount
knives, plastic 6

Pennsylvania border through Lake Ontario

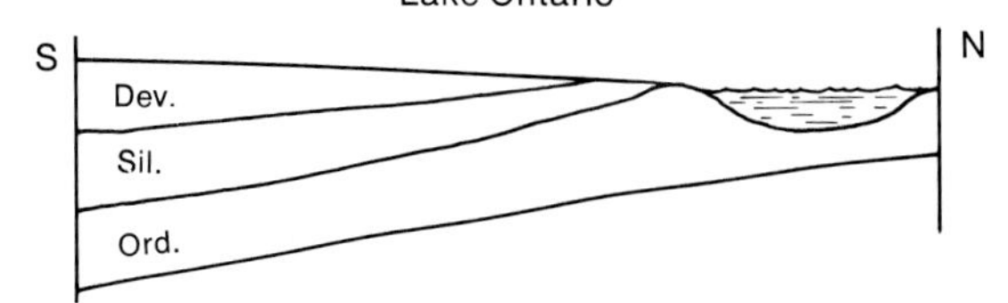

Adirondack area

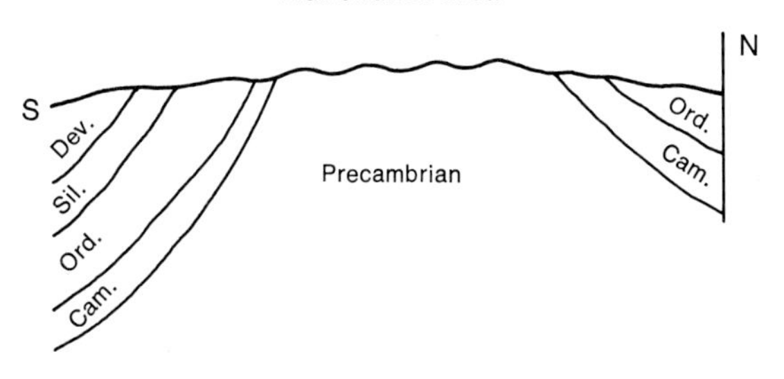

Eastern New York to Connecticut border

W
E
Dev. & Sil.
Cam. & Ord.
Ord.

REFERENCES

Dott, Robert H. Jr., and Roger L. Batten. *Evolution of the Earth*. McGraw-Hill, New York, 1971.

Kummel, Bernhard. *History of the Earth: An Introduction to Historical Geology*. (2nd ed.) W. H. Freeman, San Francisco, 1970.

Chapter 18 Development of Life

OBJECTIVES

After completing this chapter, the student should be able to

(1) describe how records in the rocks are clues to the development of life.
(2) describe the richness of the biosphere, and the influence of the environment on the geographical distribution of organisms.
(3) explain how mutations may have helped produce the wide variety of life forms now on the earth.
(4) explain how environment and mutations may have determined which creatures have survived.
(5) show familiarity with the concept of extinction.

BACKGROUND

18.3 Animal geography

This activity should produce a good discussion and promote thought. It also lends itself to library research. Obviously, some animals are more restricted in habitat than others, so opinions may vary when the class decides which animals belong in which sections of the map.

The basic reasons for the geographic distribution of a species are climate, food supply, and evolutionary history. A species that evolved in one area may be prevented by an ocean or a mountain range from occupying similar territory in another area. Alligators in China and the United States illustrate the type of history in which the ancestors spread in two directions and then became extinct in the middle. Their descendants remain widely separated. Or, an animal type can become extinct in a part of its original range as exemplified by the modern case of the musk-ox.

What follows is a list of significant requirements of the various animals that should serve as a basis for discussion.

Bison—(A1)—temperate climate, extensive grasslands. This refers to the American plains bison, as opposed to the woodland bison or the European bison.

Capybara—(C1)—high altitudes, grass and mountain vegetation.

Dromedary—(D2)—desert, coarse desert vegetation.

Giraffe—(D1)—scattered trees, hard ground.

Gorilla—(D1)—forest.

Kangaroo—(F)—flat, smooth plains, grasslands.

Mongoose—(E, D1, D2)—warm wet climate, thick vegetation near rivers.

Musk-ox—(A2)—Arctic, small plants. Found in American arctic region, but killed off by man in Eurasian arctic.

Penguin—(G)—seashore, swimmer, eating sea animals.

Sea snake—(E, F)—warm water, eats fish. This poisonous snake cannot migrate to the Atlantic because the water is too cold in the Cape of Good Hope and Cape Horn regions. (It will pose a problem if a sea level canal is dug through Central America.)

Spider monkey—(C1)—warm climate, trees.

Tiger—(E)—jungle.

Tuatara—(F)—small islands (no predators), eats insects, earthworms, snails.

Vicuña—(C2)—high altitude, semi-arid grasslands.

Walrus—(A2)—edge of polar ice, shallow water, eats mollusks.

Wild (Przewalski's) horse—(B1)—temperate grasslands.

White-tailed deer—(A1)—woods.

18.5 The Birdland puzzle

In the puzzle, the **woodpecker** (F) was the only bird in the colony at location A that could get at the insects inside the trees. The **flamingo** (I) survived at location B because it had the only beak suited for digging in the mud. At location C, the beak of the **merganser** (E) was just right for gripping and holding slippery fish. The hard seeds at location D required the nutcracker-like beak of the **parrot** (D). The **hawk** (A) was probably the easiest to place at location E. Its beak is adapted to tearing meat.

The **crossbill** (B) aptly named, might give the class some trouble. It is the only bird that could survive the location F diet. It spreads parts of pine cones apart by twisting. This exposes the seeds. If the students have ever tried to scoop a bug from the surface of a pond, they probably placed the **black skimmer** (G) at location G.

The **hummingbird** (C) is better than the others at getting nectar from flowers. So it would survive at location H. The **toucan's** (J) beak is built like a fruit slicer. So it belongs at location I. The **whippoorwill** (H) with its bristle net can catch the flying insects at location J. Ask the students what would happen if all birds ate the same kind of food.

MATERIALS

18.5 Only fittest organisms survive.

CLASS OF 30
tracing paper (cut in half) 15 sheets
scissors

Skullduggery

CLASS OF 30
copies of dot puzzles 30

REFERENCES

Dobzhansky, Theodosius. *Genetics and the Origin of Species*. (3rd ed.) Columbia University Press, New York, 1951.

Romer, Alfred S. *The Vertebrate Story*. University of Chicago Press, Chicago, 1959.

Simpson, George G. *Tempo and Mode in Evolution*. Hafner Publishing Co., Inc., New York, 1966.

Chapter 19 Man in His Environment

OBJECTIVES

After completing this chapter, the student should be able to

(1) explain what is meant by and give examples of dynamic equilibrium.
(2) explain how dynamic equilibrium operates in the environment to give the living conditions on which man depends.
(3) explain how a system in dynamic equilibrium can be upset to create new equilibriums.
(4) characterize the world's population growth.
(5) give examples of what is meant by water pollution, air pollution, noise pollution, visual pollution, thermal pollution, and people pollution.
(6) demonstrate how the amount of pollution in the immediate environment can be measured.
(7) explain some of the things that can be and are being done to help restore old equilibriums.

BACKGROUND

19.2 Equilibrium demonstrations

You can reproduce the plastic foam ball fight in a box. For a permanent demonstrator, make a box like the one illustrated on page T62. The partition in the middle is fitted with one-way doors over every other opening. The doors should be easily removable. Use whatever kind of small spheres are available. (This can also be done in an earth box or a small aquarium.)

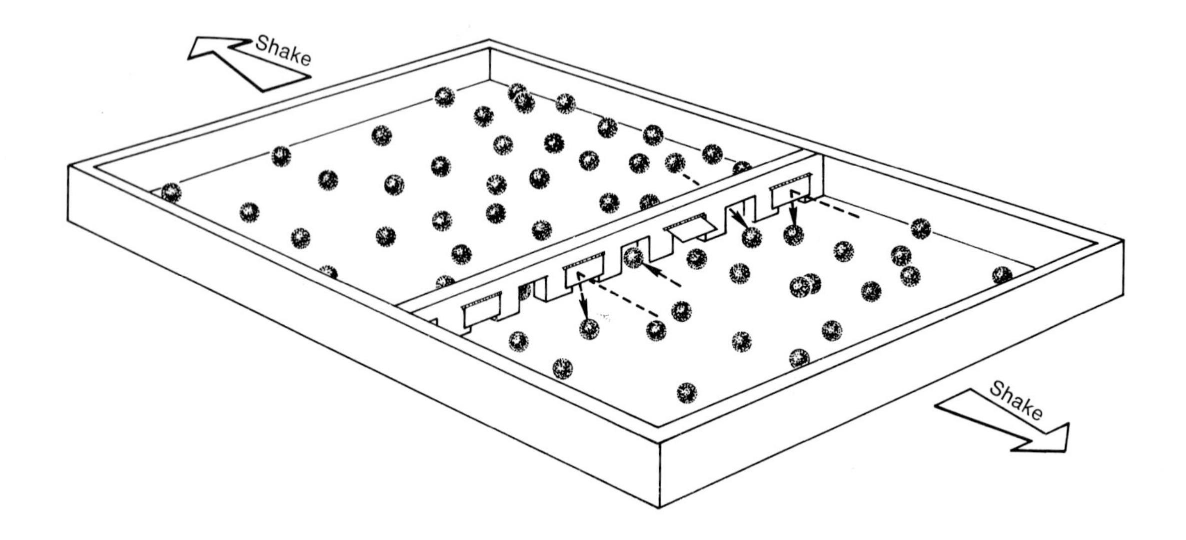

If the doors are used, you can model the fight as it is described in the text. If the box is oriented as it is in the drawing, the left side represents the large team and the right side represents the small team. Put 100 or more balls in the left side, and tilt slightly or shake the box toward the right. Then reverse the procedure. (The large team has thrown the balls over the line, and the small team has returned its share.) Count the balls on the right side. Repeat nine more times. Graph the results. An equilibrium will be reached with more than half the balls on the right (small team's) side.

To change one of the conditions, remove the doors from the partition. This would be the equivalent of having both teams the same size. Start with all of the balls on the left side, as before. After the box has been shaken or tilted ten times, graph the results. An equilibrium should be reached with half or less than half the balls on the right side.

The activity can be successfully performed in a shoe box. Make the partition from the top of the box, and the doors of paper, attached with masking tape. Dried peas make excellent, appropriate-size spheres. Use 100 for each box. If you lift one end of the box five to ten cm off the table, the peas will roll toward the opposite end and come to rest. If the box is not moved, the peas can be counted easily, since they are irregular enough to stay in place.

The most sophisticated demonstration involves water. A pan is separated into two parts by a watertight partition. One half is labeled A and the other B. The partition contains three holes, one to insert a small hose and two to act as water returns. Side A is fitted with a water pump operated by two $1\frac{1}{2}$ volt batteries. The pump is wired to a switch. Pushing the switch down allows $1\frac{1}{2}$ volts to power the pump. Pulling the switch up allows 3 volts to power the pump. The middle position on the switch is "off." With this setup, the following can be demonstrated:

(1) Static equilibrium: Fill the pan up to a predetermined water line. This is static equilibrium. The water levels on each side of the pan are equal and unchanging. A few drops of food coloring in either side will show that the water in side A stays in side A and the water in side B stays in side B.

(2) Dynamic equilibrium: Fill the pan with water up to the water line. Switch on one battery. Observe the shift of water levels on the two sides. When the water levels have stopped changing, mark the new water levels. The water levels are balanced; they are in equilibrium.

Watch as food coloring is added to one side. Here, the water molecules involved are continually changing sides. Water is moving from side A to side B as fast as water is moving from side B to side A. This accounts for the unchanging but different water levels. When the switch was first turned on, one of the rates was faster than the other. (A to B was faster than B to A. Slowly, rate A to B decreased while rate B to A increased.)

When the system is in dynamic equilibrium, a disturbance that increases either reaction will cause a shift in the equilibrium. Flip the switch up and watch as the rate from A to B is increased. A new equilibrium is established. Mark these two water levels and repeat the food coloring experiment with a different color.

Give the students a quiz after emptying the pan. For instance, what would happen to the water levels if a third battery were wired into the circuit?

19.5 Temperature inversion activity

The accompanying drawing suggests how an effective smoke generator can be made, to use with a cigarette. If you blow gently into the rubber tubing, smoke will flow through the glass tubing into the box.

You can also use punk for the smoke source, and lead it into the box via the funnel and tube arrangement illustrated in Chapter 7. If you set the punk inside

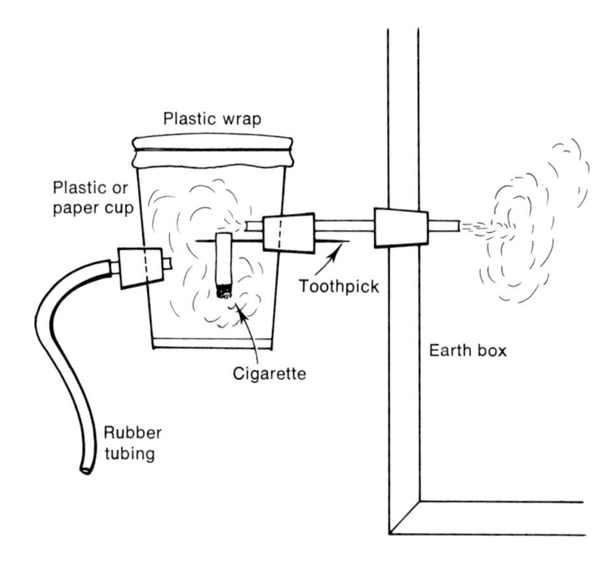

the box, the column of smoke tends to rise so strongly that it will not stay in the cold layer of air.

If you are a pipesmoker and this activity is done as a demonstration, you can make a fine layer of smog by rolling up a piece of paper and blowing smoke through the tube thus formed. Have the end of the tube close to the bottom of the box. This method and the smoke generator method are the best for use with a modified aquarium or an earth box without holes in the ends.

19.8 Comparison of heat radiated through smoky air and clear air

The theory that air pollution might tend to cool the earth can be impressively illustrated by an activity. Install a reasonably airtight transverse partition in the middle of an earth box or modified aquarium. The partition should fit tightly against the top of the box. Place a thermometer on the bottom of each chamber. The thermometers should be insulated from the bottom, and placed so that their bulbs face each other, close to the partition. Record the temperatures and put the top on the box. Fill one chamber with smoke from a smoke generator or put in a piece of punk. When a thick cloud of smoke has accumulated, direct the light from a 300-watt bulb at the middle of the box from a height of about 35 cm. The temperatures in the two chambers will rise considerably during the first minute or two, but the one in the clear side will go up several degrees more than the other.

If the bulb is too close to the top, or if it is used for more than four minutes, the temperature in the smoky side may catch up with that of the clear side.

SKULLDUGGERY

Answers to the pollution awareness test

1. A dump.
2. Rats or mice.
3. Individuals change, but quantity does not.
4. Changing the quantities or available space. (A temperature change will often change both.)
5. Examples: CO_2 and soot in the air, people and food supply, water and sewage.
6. Examples: cars, trash burning, power plants, industry, jet planes.
7. Examples: develop cleaner transportation, burn trash in better incinerators, use nuclear reactors instead of coal-burning power plants, install filters to trap pollutants in smokestacks, pass new laws on the use of devices that pollute the air.
8. Examples: strip mining, leaching around dumps, sewers, industry, insecticides, oil spills.
9. Examples: replanting stripped land, waste-disposal plants, better sewage treatment, stricter regulations on industry, safer methods of controlling insects, stricter regulations on oil-carrying tankers and stricter inspection and enforcement of laws regarding offshore oil wells.
10. Examples: noise, visual, thermal.
11. "People pollution" (overpopulation).
12. Cooling or heating of the earth. (See Section 19.8)

MATERIALS

19.1 A classroom battle can illustrate what happened.

CLASS OF 30

balls, plastic foam **or wadded up sheets of paper,** 1 to 2 in diameter 800
tape, colored (to show up on classroom floor) long enough to go across classroom
stop watch (optional) 1
graph paper, 5-plus squares to inch 30 sheets

19.4 Population increase affects the environment.

CLASS OF 30

hot plate 1
frying pan, small, with cover 1
popcorn, unpopped 1 cup

19.5 We pollute the air.

CLASS OF 30—temperature inversion

aquariums or earth boxes 6
bags of hot water 18
bags of cold sand 18
smoke generators or punk, funnel, and tube sets 6

CLASS OF 30—first air pollution activity

slide frames, cardboard, 35 mm 30
nylon stockings 2
stapler 1 or more
slide or overhead projector 1

CLASS OF 30—second air pollution activity

microscope slides 30
double-faced tape 1 roll
microscope, low power (optional) 1

CLASS OF 30—third air pollution activity

can or plastic jug, gallon size 1 or more
plastic bag, to fit can 1

CLASS OF 30—fourth air pollution activity

cloth, colored cotton 1 sq yd

CLASS OF 30—fifth air pollution activity

snow
paper cups 60

19.6 We pollute the water.

CLASS OF 30

eyedroppers 30
glass plates, 4″ × 4″ 30
water, pure and impure

REFERENCES

Cox, George W., ed. *Readings in Conservation Ecology*. Appleton-Century-Crofts, New York, 1969.

Harte, John, and Robert H. Socolow, editors. *Patient Earth*. Holt, Rinehart and Winston, Inc., New York, 1971.

Shepard, Paul, and Daniel McKinley, editors. *The Subversive Science: Essays Toward an Ecology of Man*. Houghton Mifflin Co., Boston, 1970.

Smith, Robert Leo. *The Ecology of Man: An Ecosystem Approach*. Harper & Row, New York, 1971.

Wagner, Richard H. *Environment and Man*. W. W. Norton & Co., New York, 1971.

If you and the students become involved in water pollution activities, the following two books will prove very useful:

Needham, J. G., and P. R. Needham. *A Guide to the Study of Fresh-Water Biology*. Holden-Day, Inc., San Francisco, 1969. This is an excellent reference work for the subject.

Standard Methods for the Examination of Water and Wastewater. (13th ed.) American Public Health Association, New York, 1971. This gives test procedures, collection methods, and illustrations of appropriate organisms.

Spaceship Earth

Earth Science

Teacher's Edition

The following pages of the Teacher's Edition are an annotated copy of the student text.

Spaceship Earth

Joseph H. Jackson

Edward D. Evans

HOUGHTON MIFFLIN COMPANY | BOSTON

ATLANTA
DALLAS
GENEVA, ILL.
HOPEWELL, N.J.
PALO ALTO

Earth Science

AUTHORS: **Dr. Joseph H. Jackson** is a former Professor of Philosophy at the University of Connecticut. He was Project Editor, *Investigating the Earth,* Earth Science Curriculum Project, Houghton Mifflin Company. An enthusiastic amateur astronomer, he is also author of *Pictorial Guide to the Planets* (Thomas Y. Crowell).

Edward D. Evans is a high school science teacher at Hilton Central School, Hilton, New York. The author of numerous investigations for the NSTA Earth and Space science Education Committee "Syllabus in Earth Science," he served also on the revision committee for the New York State Syllabus in earth science.

CONSULTANTS: PHYSICAL GEOLOGY
Dr. Robert E. Boyer, University of Texas at Austin

HISTORICAL GEOLOGY, PALEONTOLOGY, EVOLUTION
Dr. Bernhard Kummel, Harvard University, Cambridge, Mass.

ENVIRONMENT
Dr. Daniel McKinley, State University of New York at Albany

PHYSICAL AND HISTORICAL GEOLOGY
Dr. Richard S. Naylor, Massachusetts Institute of Technology, Cambridge, Mass.

ASTRONOMY
Dr. William H. Pinson, Massachusetts Institute of Technology, Cambridge, Mass.

METEOROLOGY
Dr. Frederick W. Ward, Air Force Cambridge Research Laboratory, Hanscom Field, Bedford, Mass.

The illustration on page 422 was prepared by Paul Hartung, 8th Grade student at Weston Junior High School, Weston, Mass.

We would like to thank Alice Neilan, Superintendent of the New London Public Schools, New London, Connecticut, and Walter Eccard, Principal of New London Junior High School, for allowing us to photograph student investigations.

We also wish to thank the Geology Departments of Harvard University and Wellesley College for permitting us to photograph rock samples from their collections.

Library of Congress Catalog Card Number: 72-2752

Student's Edition ISBN: 0-395-20685-5
Teacher's Edition ISBN: 0-395-20686-3

Preface to the Student

Before you begin, you have to know where to start your study of the earth. Should you start from way out or way in? The next page will help you to decide. Before you turn the page, be ready to cover it up with a sheet of paper. It is important to see only one picture at a time. Paper ready? Then turn the page and cover up everything but the first few lines of instructions. Slide the paper down the page as you read.

What do you see? What is this a picture of? Record a few guesses in your notebook. Tell as much about the picture and the probable surroundings as possible.

Uncover the next picture. It is of the same thing as the first photograph but a little bit farther away. Do you have a better idea what the camera is taking a picture of?

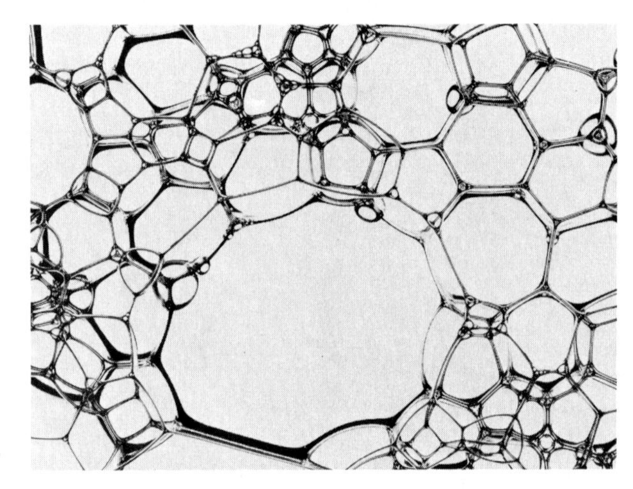

Try looking at the last picture. Does this make your guesses any easier? Now, what do you think this is a picture of?

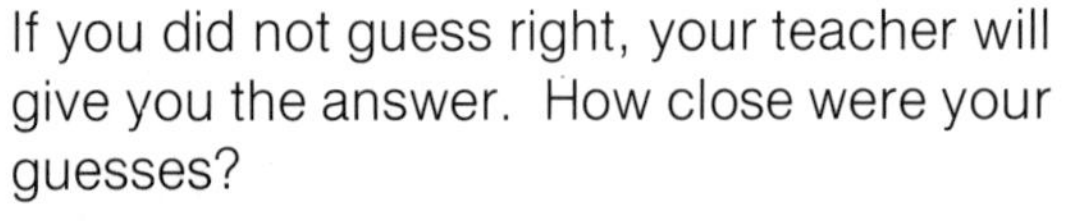

If you did not guess right, your teacher will give you the answer. How close were your guesses?

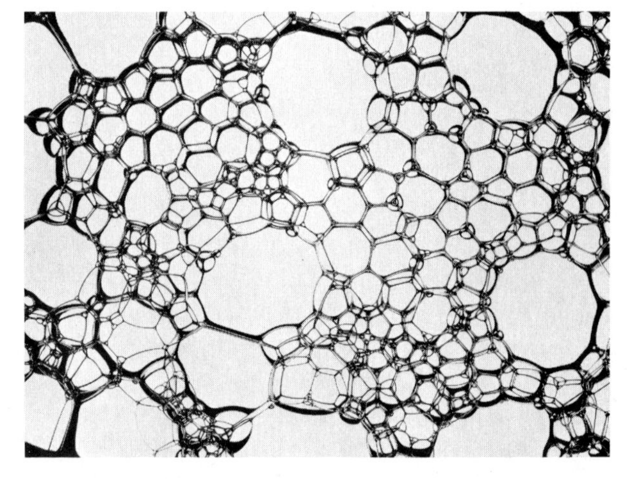

The pictures you have been looking at start close up and move away. Would it be easier to start with a more distant picture and move closer? How would you begin a study of something you didn't know much about? Would you start from afar and move in or start close up and move out? Perhaps you should begin the study of the earth from afar. We'll do it that way!

Contents

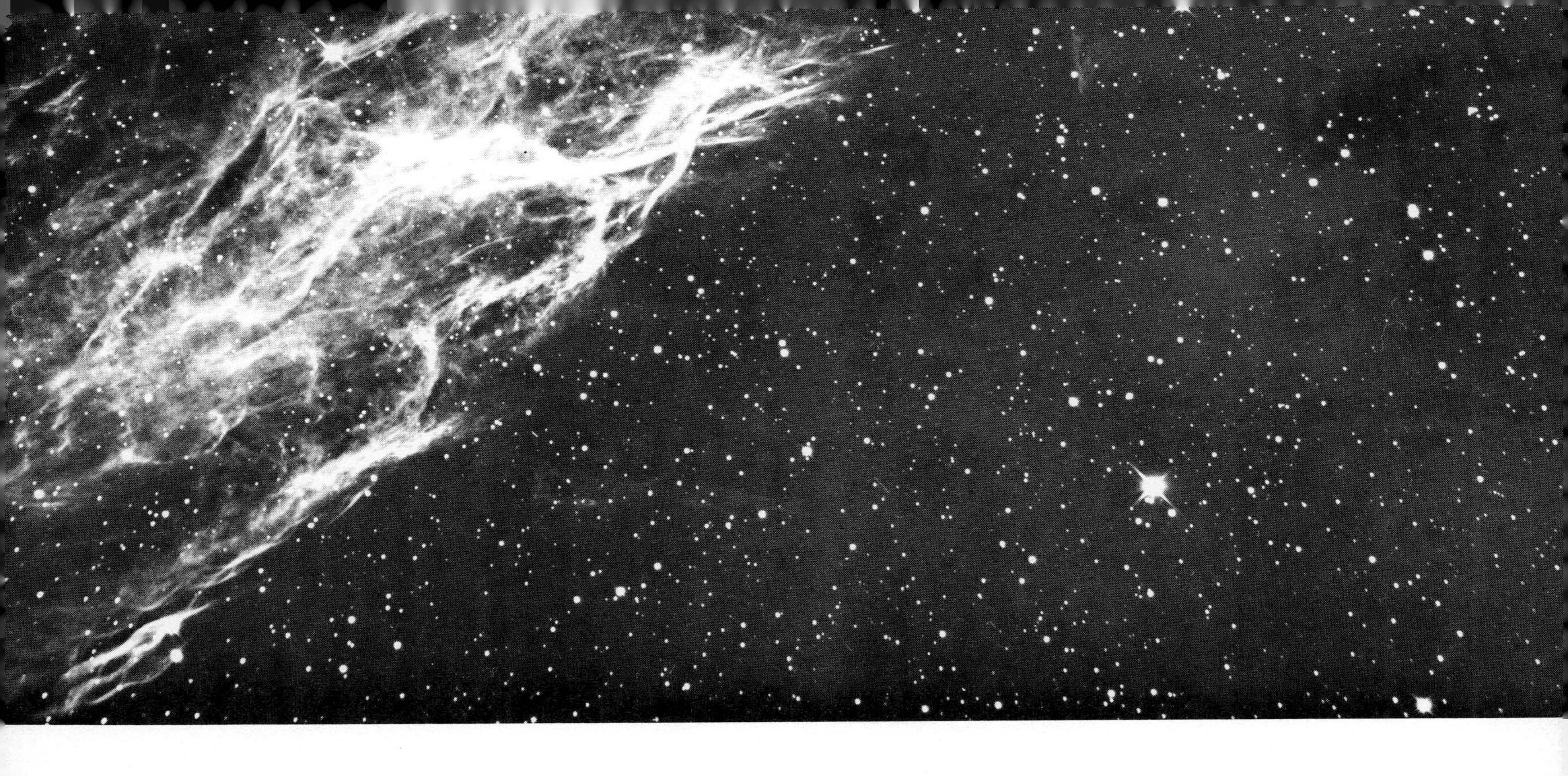

Unit One
Earth in the Universe

Chapter 1 Stars

Opening can be used to start discussion giving students a chance to tell what they know about what's in the sky and to speculate freely about deep space and the universe. The greater the variety of beliefs, guesses, and speculations offered and discussed, the better.

"Star light, star bright, first star I see tonight. I wish I may, I wish I might, have the wish I wish tonight!" Rachel chanted. She whirled round and round beside the waves breaking on the shore.

The group sprawled around the campfire watched her leaps and turns. Sparks showered as Larry stirred the fire with a stick. Leaning back against a driftwood log, Bill strummed on his guitar. The flames soared higher, casting long, flickering shadows into the gathering dusk.

Suddenly Rachel stopped and pointed upward to the blue-black eastern sky. "First star I see tonight," she cried. "I wish . . ."

"Wish not granted," Larry said from beside the campfire. "That's no star. You wished on a planet. This time of year, it's probably Mars . . . or maybe . . ."

"Listen to the big expert!" Wendi scoffed. "I'm not so great in science, but even I know that's Jupiter. The biggest planet. It's got to be—it's so bright."

"Who cares what it is?" Rachel said. "I just like to look at it, it's so beautiful. It makes you wonder, though . . ." She waved her hand across the sky. "I mean . . . what it's like up there in space . . . how the stars got there, and everything."

"It's like trillions and trillions of miles out there where we see the stars!" Wendi chimed in.

"Big mistake, Wendi," Larry said. "We don't really *see* the stars. We just see their light after it's traveled through space maybe a hundred or a thousand years. That's just old light we're seeing, not stars."

"But that old light tells us how the stars used to be," Kathy said. "It's sort of like having a time machine."

"Wow, look at all the stars that have come out now!" Bill said. "And I've read that there are trillions of stars we can't even see."

"Do you suppose there's any end to them?" Rachel asked, brushing her hair away from her eyes.

"They haven't found any yet, not that I've read," Bill said.

"And there may be lots of other universes beyond the one we happen to live in," Larry said.

"Say, have you heard about those fantastic stars that explode?" Bill asked.

"Hey . . . what's going on around here?" Al murmured sleepily from the piece of driftwood he was using for a pillow. "Did I hear something blow up?"

"Relax, Al. Kathy just popped her bubble gum," Larry said. "But you just missed out on a talk about stars and stuff."

"Yeah, it's time for you to wake up and join the universe, Al," said Bill.

What Do Stars Look Like?

1.1 How big are stars?

Relations between metric and English scales are given in Appendix A. Great Britain is now turning to metric system, as United States probably will eventually.

When you look up at the sky on a starry night, what do you see? Just lots of little twinkling points of light, if the moon and planets aren't out. These are the stars. But our sun is a star too. Why is the sun so large in the sky, and why is it so much brighter and hotter than other stars? You've probably guessed the reason. The sun is the closest star to us, just 93 million miles (150 million kilometers) away. A star can be much more than just a tiny speck in the sky, if you're close to it. (See Appendix A for a discussion of the English and metric systems of measurement.)

Largest known star, Epsilon Aurigae, is an eclipsing double star. Its main star is a supergiant with the given diameter and 35 times the mass of the sun. Its other star, which moves around the main star in 27 years, may be a black hole, or *collapsar,* whose gravity has become so strong that it is collapsing into itself and gives off no radiations.

How big are stars? The sun seems huge, but it's not especially big as stars go. Its diameter of almost 900,000 miles (1.4 million kilometers) is just about average. The largest star known is about 2300 times larger than our sun with a diameter of about 2 billion miles (3.2 billion kilometers). It's hard to imagine anything even as large as the sun. But if you think of the sun as 1 foot across, this monster star would be over 2000 feet across.

In photographs, stars look like round dots or rings because radiations are spread by lenses or mirrors of telescopes. Actually, stars look like points, and not disks. Only sun, moon, and planets look like disks, and only nebulas and galaxies show shapes.

If stars are so big, then why do all of them but the sun look so small to us? Because they are trillions and even trillions of trillions of miles out in space. So far away that even in a **telescope,** an instrument that makes faraway things appear closer, they still look like little points of light.

Figure 1 / 1 *The sun's radiations light and warm the earth, move the air and ocean currents, and provide energy for life on spaceship earth.*

Yet **astronomers,** the scientists who investigate stars, have found out a lot of facts about these strange, distant objects. How? By studying the different kinds of energy they give off.

1.2 Stars give off light and heat.

Have you ever lighted a sparkler on the Fourth of July? If you have, you know that it glows red as you hold a match to it. When it catches fire, it becomes white hot, giving off dazzling light and heat.

Think of stars as huge, bright hot sparklers far out in space. When you look at the blazing sun, our star, in Figure 1/1, you have some idea of just how hot and bright a star can be. The sun and all the stars give off vast amounts of energy.

Radiations are of two kinds: (1) energy or "wave" radiations and (2) particle radiations. Light and heat are energy radiations discussed in this chapter; particle radiations discussed here are protons and electrons in the solar wind. Radiations of whole electromagnetic spectrum are mentioned in Section 3.7. Atomic particles are discussed in Appendix G.

Energy has the ability to change or move things. The energy in a toaster changes a slice of bread from white to brown. The energy from gasoline makes the wheels of a car turn. Work is being done as energy is given off and used in such ways. So energy is called the ability to do work.

Particles produce the solar corona or "crown" in a total solar eclipse. Solar wind beyond this "blows" out past the earth and moon.

Radiation is all the different kinds of energy that hot objects give off. It includes the light, heat, and moving particles from sparklers, from the distant stars, and from our sun. Most of the time the tiny atomic particles the sun shoots out can't be seen. The sun is so bright and the particles are so small. But sometimes the moon moves in front of the sun's face. Then you can see the brightness from the particles the sun sends out in all directions.

Encourage students to look at stars during clear evenings, if they can get away from brightly lighted areas. Work with them on sky maps, Appendix C, identifying some constellations and stars. Make a trip to planetarium, if feasible. (See Background Material, Teacher's Manual.)
Process: observing

1.3 Stars make patterns in the sky.

Astronomers use their telescopes to measure these radiations from stars. They use cameras, as in Figure 1/2, to record what they see. But if you go outdoors on clear evenings you can learn a few things about stars just by using your eyes. A pair of binoculars will help. Comet

Figure 1/2
An astronomer about to photograph the stars from the observing cage atop the tube of the 200 inch Hale telescope.

Figure 1/3 *Comet Mrkos blazes in the sky as it nears the sun. The photograph is a time exposure held on the comet. The dashes are stars that appeared to move with the turning of the earth.*

hunters like to use binoculars to sweep the sky for the bright heads and streaming tails of comets, like that in Figure 1/3.

While you're stargazing, you may notice that stars seem to make shapes or patterns in the sky. Some of them may look like squares, or triangles, or little circles to you. Others may look like cars or owls or kittens. Since ancient times, people have imagined that groups of stars formed pictures of animals or people or gods and goddesses. These picture groups of stars are called **constellations** (kon-stell-AY-shuns).

Constellations are not always easy to locate. But Ursa (ER-sah) Major, which means the Great Bear in Latin, is one that people can sometimes find easily. Not because it looks much like a bear, though. In Figure 1/4 you see that a part of it looks like a dipper with a long handle, so most people call this group of stars the Big Dipper.

See Teacher's Manual for suggested illustration of this point.

You might think that stars in a constellation like Ursa Major are close together in space. Not so. They are actually very far from each other and at many different distances from the earth.

Following these old patterns imagined by ancient people, astronomers have divided the whole sky into 88 different constellations, each with its own name. This helps them to tell clearly where any star is located. To say that a star is in Ursa Major is like locating the city of New Orleans by saying, "Oh, yes, that's in Louisiana." Many of the brightest stars in a constellation have names, too. Some star names are given in Appendix B.

Notice that the names of the stars in the Big Dipper are given in Figure 1/4. If you look very closely at Mizar (MY-zahr) you will see that it is really a double star. The other star is called Alcor. Mizar and Alcor move around each other. If you can see it as a double star in the sky, your eyesight is very good. North American Indians called Mizar "the horse" and the smaller star beside it "the rider."

Constellation Cassiopeia in Figure 1/5 also appears in Figure 1/7 for student estimation of magnitudes of stars.

Another part of the sky with the constellation Cassiopeia (cass-ee-oh-PEE-ah) is shown in the photograph in Figure 1/5. Cassiopeia can be seen in the northern sky throughout the year. The way people pictured this constellation is shown in Figure 1/6. Find Ursa Major and Cassiopeia on the sky maps in Appendix C. See if you can locate these and other constellations when you're stargazing.

Stars are useful for finding directions. Two of the stars that you saw in Figure 1/4, Dubhe and Merak, are used to locate the North Star. This is important, because whenever you point to the North Star you know you are pointing north. And from north, you can tell all the other directions.

1.4 How many stars can you see?

★ Don't worry—you're not going to have to count all the stars you can see in the sky! You can find the answer the way polltakers do in a survey. Polltakers don't question everyone in the nation. They take a typical sample. And you can sample the number of stars you see the same way.

Figure 1 / 4 *The handle of the Big Dipper is the Great Bear's tail. Most of these stars were named by Arabian astronomers.*

Figures 1 / 5 and 1 / 6 *Can you see how the brightest stars in Cassiopeia trace an "M", or a "W", or a chair (called the Queen's chair) in the sky? The drawing shows her sitting in her chair as imagined in* A.D. *1574. Note the position of a rapidly brightening new star called a nova.*

Materials

paper tube 11 inches x 1.5 inches diameter, paper and pencil

Activity Time: 15 to 30 minutes, in evening. This activity is a good one to start off a star party or for students to do at home. Tell students to stop their tubes at random, not at bright stars. Eyesight, lights nearby, smog, and haze are variables affecting results. Students may estimate from 100 to 2500 stars visible above horizon. Use for earth-science notebook, if you wish.
Process: observing

Get a tube like the one in a roll of paper towels. Take it outdoors with you on a clear night. Get as far away from lights as you can. Let your eyes get used to the dark. Look through the tube at the bit of sky you see at the other end. Count the number of stars in that bit of sky. (Count it as 0 when you can't see any stars.) Jot down the number of stars in Sample Number 1.

Now follow several paths across the sky. Stop at several places along each path to count more samples. Just stop when you happen to, not when you see a bright star in the end of the tube. Count and record the stars in the tube at each stopping place. Do this about 25 times to get a fair sample of the sky.

Back indoors, add all the stars you counted in the samples and divide the total by the number of samples you took. Include those samples where you saw no stars. Now you have the average number of stars for each sample bit of sky.

About 700 of your tube-end samples would cover the whole sky from the ground up. So multiply your average number per sample by 700. This gives you about the number of stars you can see in your sky.

Bright lights, haze and dust in the air, or mist and fog, all cut down the number of stars you can see. Notice sometime that you can see only a few very bright stars near the full moon. So you're lucky if you can see as many as a thousand stars. ★

Stars in our Milky Way galaxy are estimated from 100 to 200 billion, and 20×10^{22} stars in the universe are currently observable with optical and radio telescopes.

Telescopes gather the light radiations even from very dim stars, so they detect many more than you can. A rough guess gives 100 billion stars alone in our neighborhood in space. And there may be 200 billion times 100 billion stars in the universe that astronomers have seen so far!

Did You Get the Point?

Stars are bright hot bodies that radiate energy in the form of light and heat and give off moving particles.

You can learn many things about stars just by looking at them in the sky. Astronomers learn even more by observing stars with telescopes.

Stars appear in picture groups, or constellations, in the sky. Constellations help to locate and name different stars.

Although only a few thousand stars can be seen with the naked eye, telescopes tell us of an almost countless number of stars.

Answers:
1. Light and heat, plus particles.
2. Fire—yes—with light, heat, and smoke particles. Flashlight—no—light, a little heat, but no particles.
3. Vega (VEE-gah), Arcturus (Ark-TOOR-us), Capella (Cap-PELL-ah), and Deneb (DEN-ebb).
4. In constellations of Lyra (LY-ruh)—Vega; Bootes (Boh-OHT-eez)—Arcturus; Auriga (Or-REE-gah)—Capella; and Cygnus (SIG-nus)—Deneb.

Check Yourself

1. What kinds of radiations do stars give off?

2. What are stars like, besides sparklers?

3. Find the names of the four brightest stars toward the north horizon in the fall sky map (Appendix C).

4. Name the constellations of these brightest stars. If you have a chance, find these constellations and stars in the night sky. Wait for a clear night, and get as far away from lights as you can.

Stars Are Bright and Hot

Moon and planets do not radiate light but reflect it from the sun. They do radiate heat, which astronomers measure. Largest planet, Jupiter, one-tenth diameter of the sun, may have been close to conditions for forming a star, radiating energy like the sun.

1.5 Stars look dim or bright.

Looking at the stars in the sky, or even in photographs, how do they differ from each other? Right, in brightness. Some are so dim you can barely make them out. Others almost seem on fire.

The moon and planets are the brightest lights in the night sky, of course. But they do not radiate light. They are not hot enough. Instead, they reflect some of the light that comes to them from the sun. But the stars radiate light. Many of the brighter stars in the sky are named on the sky maps in Appendix C.

Materials

paper and pencil, sky maps

Activity Time: one or two hours on one or more clear evenings. Another activity for a star party or homework. The more stargazing, the more the night sky (and this chapter) will mean to students.

★ If you can do some stargazing, pick out some of the bright stars at night along with their constellations. Use the sky maps as a guide. List the names of the stars you find. Bring the list to school, and see how your list compares with those of your classmates. You might have a contest for a week. Each named star counts for 1 on your score, each named planet for 5. ★

Figure 1/7
The magnitudes of some stars in this photograph are numbered from 1 (bright) to 6 (dim). Judge the magnitudes of stars lettered A through L by comparing them with the numbered stars.

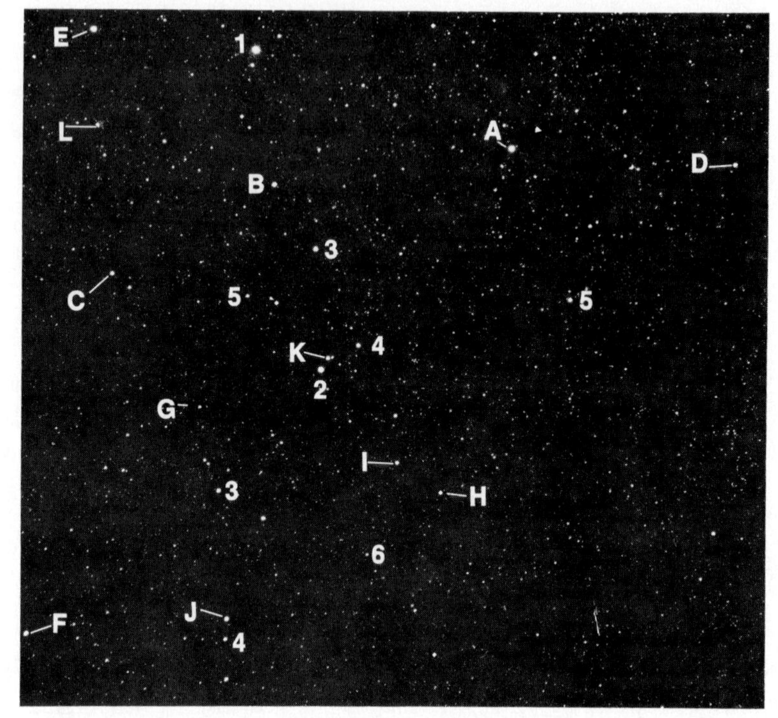

Magnitudes are the most basic astronomical data. Astronomers use comparison stars of standard magnitude just as students do here. Full magnitude scale of the various sky objects appears in Teacher's Manual. Magnitude of some stars is in Appendix B. Magnitude of the sun is −26.75, of full moon −12.2.

Astronomers give a number to the brightness of stars. This number is called the star's **magnitude.** Look at the photograph of a bit of sky in Figure 1/7. The numbers given to some of the stars are their magnitudes. The dimmest star numbered in the photograph has a magnitude of 6. This is the magnitude given to stars that people can just barely see with the naked eye on a very dark, clear night. Notice that the brighter the star, the lower its magnitude number. These six different star magnitudes are shown in Appendix C with different-sized dots.

Materials

pencil and paper

Activity Time: 20 minutes for students to estimate and record magnitudes of stars pictured and to check their estimates against the list you give them. An opportunity to work on tables. If students need help, refer them to Appendix D. Another possible use for the earth-science notebook. Since this is a telescopic photograph, stars with magnitudes less than 6 can be seen.
Process: measuring

★ Other stars in the photograph in Figure 1/7 are lettered from A through L. Make up a table with these letters in one column. Put your estimates, or ideas, about the magnitudes of the lettered stars beside their letters, in another column. (See Appendix D.) Make your estimates by comparing the brightnesses of the stars lettered A through L with the brightnesses of the stars whose magnitudes are given. Astronomers make estimates of the magnitudes of stars in much the same way.

After you have made your magnitude estimates, your teacher will give you the actual magnitudes. How many did you get right? ★

1.6 Stars have different real brightnesses.

Materials
pencil and paper

Activity Time: 15 minutes, followed by discussion of results and difference between visual magnitude and luminosity. Figure 1/8, left to right: A = 30 watts, B = 200, C = 150, D = 300, E = 100, and F = 60. In figure 1/9, G and H have been moved toward viewer and J has been moved back. Magnitudes: A = 6, B = 2, C = 3, D = 1, E = 4, F = 5, G = 5, H = 1, I = 3, J = 3, K = 4, and L = 5. After students have estimated the magnitudes of model stars in Figures 1/8 and 1/9, tell them the correct magnitudes and the wattage of each bulb.
Process: measuring

★ You have judged the brightnesses or magnitudes of stars in the photograph in Figure 1/7. And some of you may have practiced with stars in the sky, too. But there's a funny thing about the brightnesses of stars. Look at the model stars in Figures 1/8 and 1/9. The pictures were made from light bulbs in a back yard at night. Six light bulbs stand for stars.

Call the model stars in Figure 1/8 A, B, C, D, E, and F, from left to right. Give the brightest star a magnitude of 1 and so on, to the dimmest, which you give a magnitude of 6. This is the way you judged the magnitudes of stars in Section 1.5. Compare your answers with those of your classmates.

The *same* light bulbs, or model stars, are shown in Figure 1/9. But there have been some changes made. Call these model stars G, H, I, J, K, and L, from left to right. Judge their brightness again. Compare your answers with those of your classmates. Which star here is the same as star F in Figure 1/8? Try to match up all the stars in the two photographs. Remember they are the same six light bulbs. Your teacher will give you the correct answers when you are through.

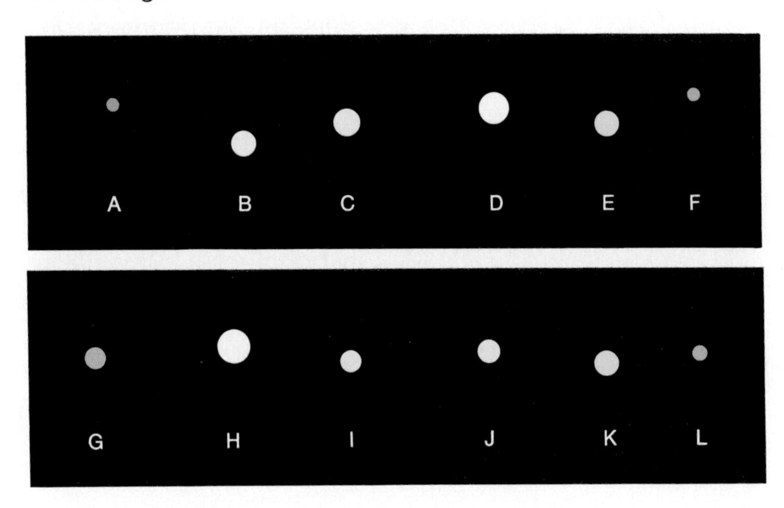

Figure 1/8
Judge the magnitude of these six model stars.

Figure 1/9
Judge the magnitude of these model stars, the same stars as those in Figure 1/8. What has changed?

It looks as though the model stars changed their left-to-right position when the picture in Figure 1/9 was made. But something else changed. Can you figure out what happened? ★

You know that when a streetlight is very far away from you, it is dim. When you are close to it, it is bright. The light's brightness as you see it is its **visual magnitude.** You judged visual magnitude in Section 1.5. But both light bulbs and stars have their own *real* brightness. This is called **luminosity** (loom-in-OSS-it-ee). A 50-watt bulb is a 50-watt bulb no matter how far away it is from you. That is, its luminosity does not change with distance. The bulbs in Figure 1/8 are all at the same distance from the viewer. You can tell how bright they *really* are. But the same bulbs in Figure 1/9 are at different distances from the viewer, so you can only tell how bright they look *to you.* The luminosity of a star is the total amount of energy the star radiates, no matter how bright it may look to you.

Magnitude of stars seen in the sky with naked eye is visual magnitude, given for a number of stars in Appendix B. Table also shows many star names, distances in light-years, surface temperatures, and luminosities. See Teacher's Manual for background on magnitude and luminosity.

To tell how much energy stars really radiate, astronomers must learn their distances. Stars are too far away to express their distances in miles. Instead, the distances are stated in light-years. Light energy from stars travels through space at 186,300 miles (300,000 kilometers) a second. If a plane could go that fast, it could go around the earth almost eight times in a second! In one year, light from a star travels almost 6 trillion miles. This distance that light travels in one year is called a **light-year.**

In Section 1.8 you will use the luminosity and color of stars as astronomers do to learn something about what stars are like and how they change. First, let's look into the color of stars.

1.7 Colors tell star temperatures.

The stars that sparkle in the sky differ in color as well as in brightness. Betelgeuse (BET-ell-gerss), in the constellation Orion (oh-RYE-yun), is a bright red point of light. Our sun, we know, is mainly yellow. Sirius (SEER-ee-us), the brightest star in the northern sky, looks bluish, almost purple, as it twinkles in the sky.

Some people see star colors better than others. Colors seen with binoculars and telescopes may be deceiving because the lenses may be uncorrected for colors, so colors will be different. Most people see Sirius as bluish-purple, but some see it only as a bright light.

Red, yellow, blue—sounds like the colors in a rainbow, doesn't it? This series of colors, called a **spectrum,** is found

Figure 1 / 10 *This spectrum was photographed by pointing a camera toward the sun. The camera was fitted with a grating of fine lines that breaks up light like a prism. The large object nearby is a radio telescope.*

A hexagonal transparent plastic pen will give a good spectrum of sunlight. Hold pen in sunlight close to a white sheet of paper and turn. A bottle with square sides filled with water will give spectrum under same conditions. You can also demonstrate a spectrum with a prism or small spectroscope, whichever is available. The full electromagnetic spectrum is given in Section 2.6.

in rainbows. Ordinary sunlight has a spectrum, although you don't see it. But if you let sunlight shine through a three-sided piece of glass or plastic, called a **prism,** you can see the spectrum. The prism breaks up the light into its various colors. Raindrops act like prisms and break up sunlight to make a rainbow.

Astronomers use spectroscopes to break up the colored light of stars into their spectrums. A **spectroscope** is a special instrument for breaking up light. The spectrum of light from the sun, a yellow star, is shown in Figure 1/10.

Why do astronomers study the spectrums of stars? For one thing, they can tell a star's surface temperature from its spectrum. Though you might not expect it, stars with blue light are hotter than those with red light. Sirius is hotter than Betelgeuse. The temperatures of yellow stars, like the sun, are in between. In the same way, the blue part of a flame from a torch or Bunsen burner is hotter than the yellow or red parts of the flame.

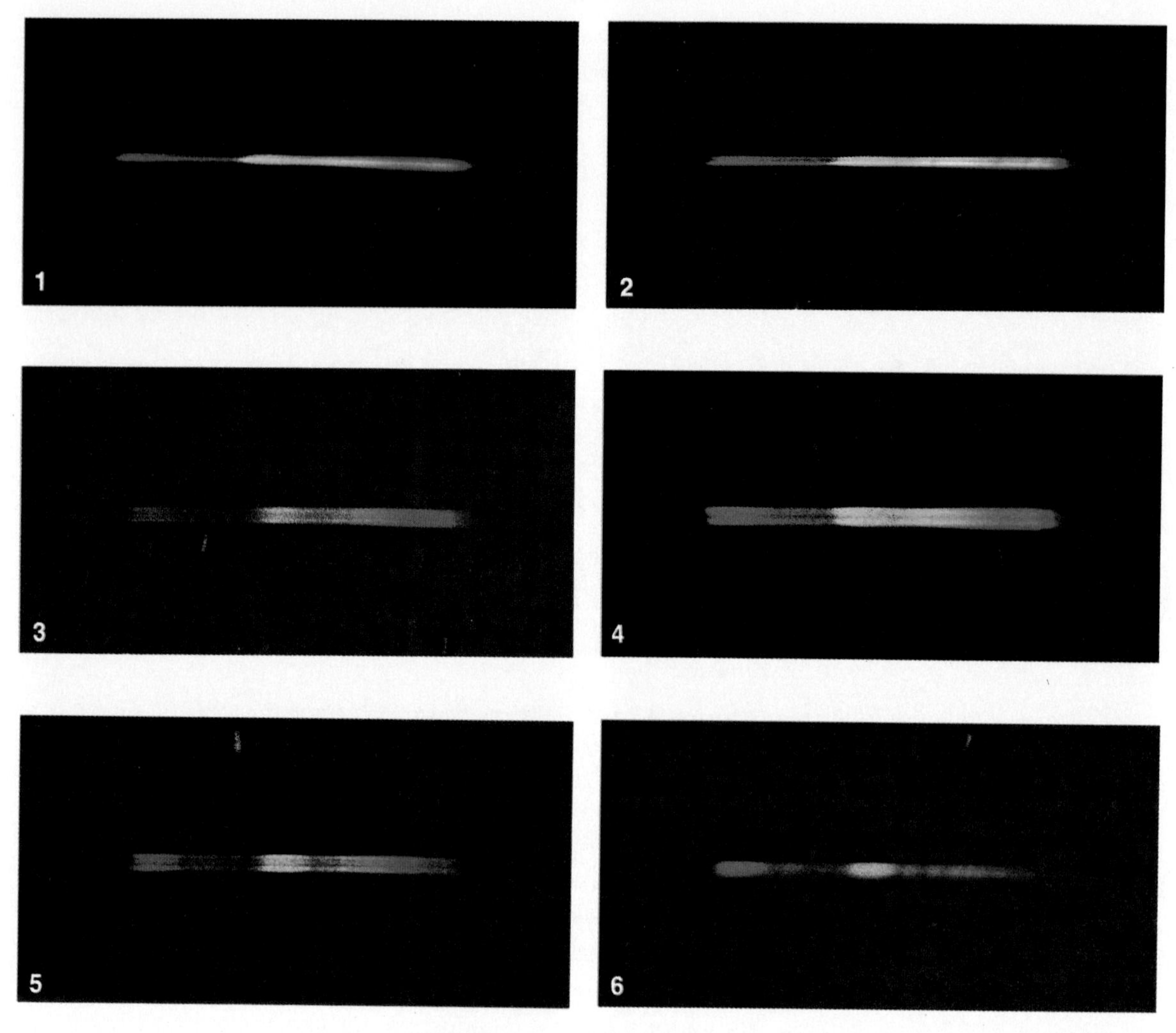

Figure 1/11 *Spectrums of stars can be used to tell how hot they are. Star 1 is a group B star, 2 is group A, 3 is group F, 4 is group G like our sun, 5 is group K, and 6 is group M.*

The letters of groups or classes of stellar spectrums are mixed up, with gaps, because groups were added, and some of original groups, such as C and D, were found inadequate and stars in them were placed with other groups. Many subgroups or subclasses of these groups have been worked out, such as subgroups of A stars from A_0 to A_6. **Process: classifying**

Figure 1/11 shows spectrums of the hottest stars. Look at all the blue in these spectrums. As you go from spectrum 1 to spectrum 6, the stars become cooler and cooler. So you see more and more red in their spectrums. Notice, too, that each of these groups of stars is given a letter. O and A stars are very hot. M and N stars are very cool. F and G stars are in between. Our sun is a group G star. The basic temperature groups of stars, with their colors, are shown in Figure 1/12. The temperatures are in the Kelvin scale astron-

Figure 1 / 12 *Temperature groups of stars.*

Star group		Star surface temperature (degrees Kelvin)	Typical star
O		45,000	Alpha Carinae
B		25,000	Spica
A		11,000	Fomalhaut
F		7,500	Canopus
G		6,000	Sun
K		5,000	Aldebaran
M, R, N, S,		up to 3,000	Alpha Centauri C

omers use. Kelvin degrees are compared with Fahrenheit and Celsius degrees in Appendix E.

Translation between Kelvin, Fahrenheit, and Celsius scales is given in Appendix E. Astronomers use Kelvin scale, which adds 273° to Celsius temperatures.

What difference does it make how hot a star is? Any one of them would burn you to a crisp in an instant from a few million miles away. Why is the surface temperature of a star important to astronomers? You'll find out in the next section.

Did You Get the Point?

The brightness of a star is called its magnitude.

A star's visual magnitude is a measure of its brightness made with the eyes alone, or as you see it in the sky.

The real brightness, or luminosity, of a star is the actual amount of energy that the star radiates, no matter what its distance.

The colors of stars tell their surface temperatures. Spectrums of stars, made by breaking up their light with spectroscopes, show their surface temperatures more precisely.

Stars are so far from the earth that astronomers express their distances in light-years.

Answers:

1. Easier to see a star of visual magnitude 1 than one of visual magnitude 6. Increasing positive visual magnitude means decreasing brightness.
2. Star with high luminosity, giving off greater radiations, would warm you more.
3. Answers given in Figure 1/16.
4. Different observed brightness, or visual magnitude, explained by varying distances or luminosities of light sources.

Check Yourself

1. Would it be easier to see a star of visual magnitude 1 than a star of visual magnitude 6?

2. Would a star with a high luminosity warm you more than a star at the same distance with a low luminosity?

3. Find the groups to which the stars in Figure 1/13 belong by using the groups of stars given in Figure 1/12.

4. What reasons can you think of for the different brightnesses of some of the lights in Figure 1/14?

Figure 1/13
Surface temperatures of stars.

Star	Surface temperature (degrees Kelvin)
Beta Centauri	21,000
Deneb	9,900
Sirius	10,400
Van Maanen's Star	7,500
Barnard's Star	2,800
Antares	3,400

Figure 1/14
The lights of Boston seen from an airplane form an unusual man made constellation. Pick out some of the brightest and dimmest "stars."

How Do Stars Change?

Materials
25 paper clips, graph paper

Activity Time: 30 minutes. Have students place graph paper on pads or newspapers. Scattering of clips is random, without any repeating pattern, in contrast to plotting data on graph on basis of surface temperature and luminosity. Points for making good graphs are in Appendix F.

The relation of these two variables on the scatter diagram of the graph shows students one of most basic of scientific processes, used in many activities in this book. The graph arrangement of temperature, increasing from right to left, has been traditional in astronomy.

1.8 Scatter stars on paper.

★ You use about 50 stars in this activity, scattering them on graph paper. First, though, you might want to see what happens when you scatter stars any old way. Let about 25 paper clips stand for the stars. Hold them in your hand about a foot above a sheet of graph paper. Let them all drop at once. Do the paper clips make any pattern on the graph paper? Try it a few more times. Does any pattern keep appearing? ★

Astronomers have scattered stars on graph paper, but not by chance, as you've just done. But if you'd done the next activity about 60 years ago, you would have made history in astronomy. Two astronomers beat you to it, though. In 1912, they put temperature and luminosity of stars together in a graph. The results were startling.

Materials
graph paper, pencil

★ You are going to make a graph, just the way the astronomers did. What they did was to mark stars on the paper with the luminosity shown on the vertical line and the temperature shown horizontally. The setup for your graph is shown in Figure 1/15. (Appendix F tells you something about making graphs.)

You will find the data for your graph in Appendix B, which gives the temperatures and luminosities of a number of stars. First, where does our sun belong on the graph? Its temperature is 5800 degrees Kelvin and its luminosity is 1. Move your pencil along the horizontal line at the bottom of the graph until you come to just under 6000 degrees on the temperature scale. Then move your pencil up to 1 on the luminosity scale and mark a circle for the sun. Make an X on the graph for each of the other stars, according to their temperatures and luminosities.

When you have finished, look over your graph. Do you see a fairly clear diagonal line of stars from one corner to another? What does this show about the temperatures and luminosities of stars? As temperature increases from right to left, what happens to luminosity from bottom to top? Does luminosity increase or decrease?

Line or band of main sequence shows that stellar luminosity increases with stellar temperature, the relation of these two variables.
Process: interpreting data

Figure 1 / 15
Set up the grid for your star graph to look like this one.

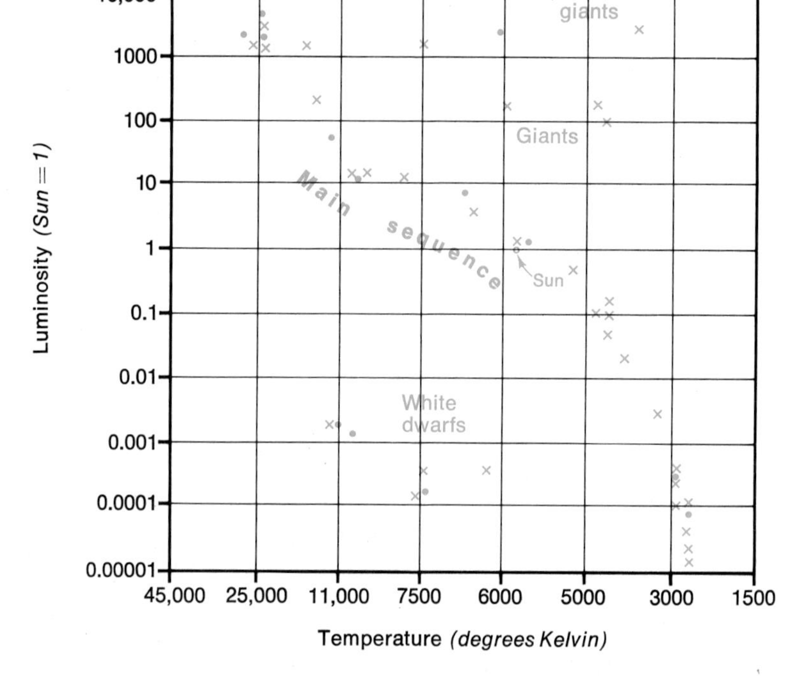

Points plotted with x's are for those stars with asterisks in the tables. Points plotted with dots are optional.

Check the temperatures given at the bottom of your graph with the temperature groups of stars in Figure 1/12. Put the letters for star temperatures under the correct temperatures on your scale. Now you can see that each temperature group of stars has a different place along the band of stars you have made. This band is called the **main sequence** (SEE-quence) of stars. But some stars do not fit on the main sequence, as you see. ★

0 group of stars fits 45,000 degrees on temperature scale along bottom of graph, as shown in Figure 1/16. Students note how grouping stars by temperature has led to more information about them.

You have just made a famous graph. The names of the astronomers who made it first were Hertzsprung and Russell. So it is called the **H-R diagram** in their honor. An H-R diagram with many more stars than yours is shown in Figure 1/16.

Whether you have students do activity in Section 1.8 or not, Section 1.9, with Figure 1/16, gives them a chance to interpret a scatter diagram or graph. The last half-century of astronomy has stemmed from this graph.

1.9 What does your star diagram mean?

Now you come to the surprises in the H-R diagram. Before it, about all astronomers knew about stars was what you

Figure 1 / 16
The stars plotted on an H-R diagram fall into several groups. Our sun is on the main sequence band.

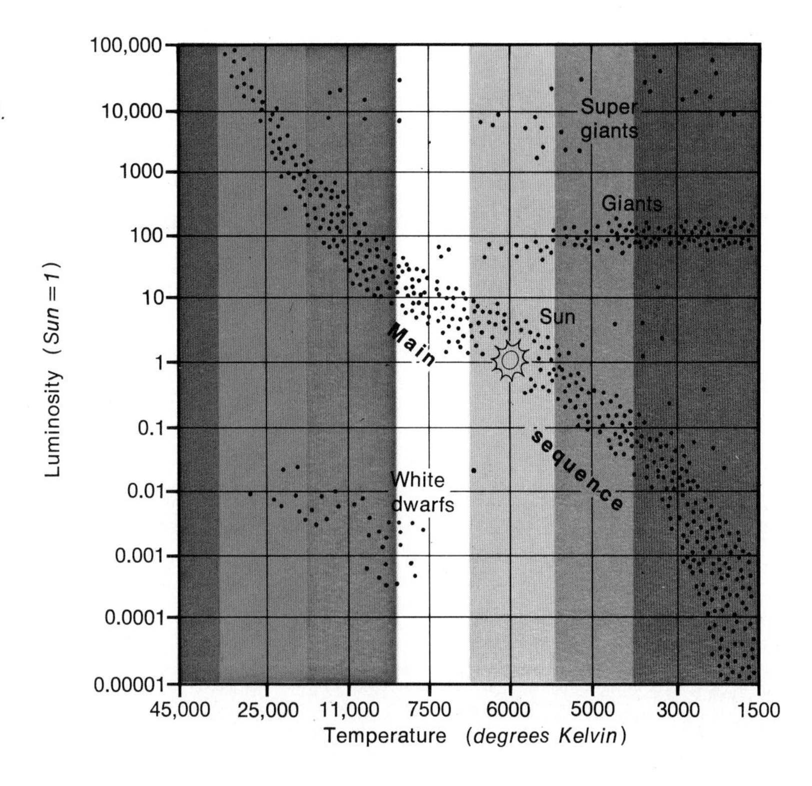

have learned so far! When astronomers compared the surface temperatures and luminosities of stars, all at once they saw that the stars could be grouped into different types.

Look at your H-R diagram and the more complete one in Figure 1/16. Most of the stars fit along the band of the main sequence. These stars run all the way from very bright, very hot blue stars (upper left corner) to very cool, dim red stars (lower right corner).

But look at the other clumps of stars in other places on the graph. In the upper right corner, for instance, is a group of very luminous stars. But they have very low surface temperatures! So how can they possibly give off so much light energy? Answer: Because they are big, so big that astronomers decided to call them *supergiants.* Below the supergiants you see another clump of very luminous stars called *giants,* not quite so big. Giants and supergiants can be from 50 to 500 or more times larger than the sun. That is, they may be from 40 million to 400 million miles (65 to 650 million kilometers) across.

The group of stars at the lower left of the diagram is about 1000 to 10,000 times less luminous than the sun. They are very, very dim stars, then. But look how hot they are. Twice as hot as the sun! These stars are called white *dwarfs* because they are so hot and yet so tiny. They may be only 1 or 2 hundredths the size of the sun, 10,000 to 20,000 miles (16,000 to 32,000 kilometers) in diameter. Compare this with the earth's diameter. And the stars at the lower right of the diagram are small, but cooler, so they are called red dwarfs. Our sun is a yellow dwarf star between the giants and the red and white dwarfs.

Diameter of earth is 7926 mi. (12,956 km).

1.10 Stars, like people, have life histories.

Stars are probably born in great swirling clouds of gas and dust billowing in space like the one you see in Figure 1/17.

Figure 1/17 *Gravitational attraction may cause great clouds of gas and dust to collapse into stars.*

Figure 1/18
A star may begin as contracting gas and dust, and expand to become a red giant. Then it may explode to become a nova and fade to a small white dwarf.

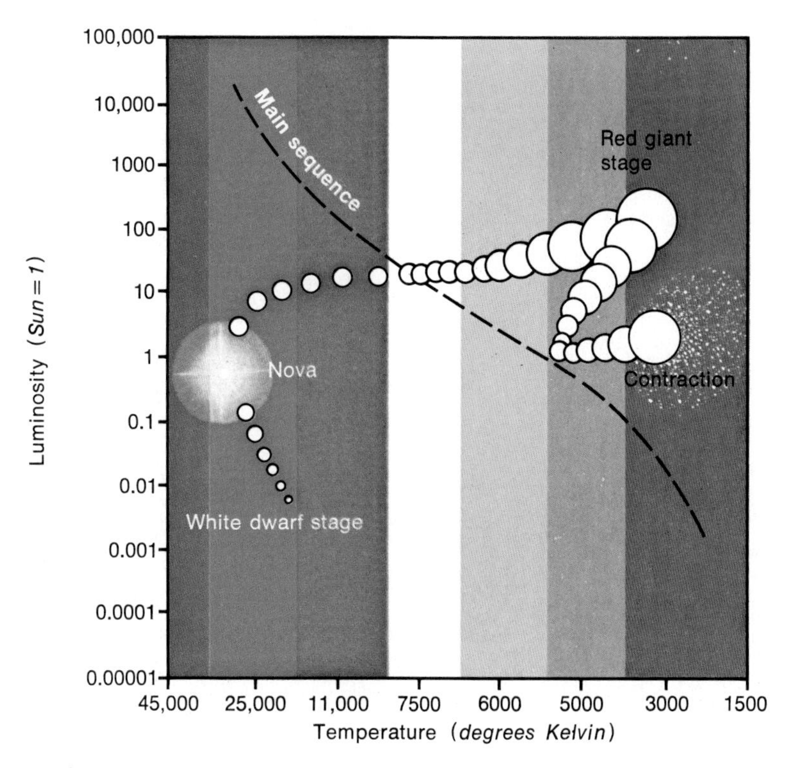

Process: inferring

Introduction of concept of gravitation on qualitative basis. Concept further developed in Chapter 4. Just as things near the earth fall toward its center, so masses of gas and dust in space clouds fall toward their center, gradually forming a larger, heavier globe.

The light from, and magnitude of, many stars varies. Some variable stars brighten and dim, or pulse, by themselves. Others are systems of two or more stars that move about each other. When they come in front of each other, as seen from the earth, they are called eclipsing variables, with their light brightening and dimming. You might use transparencies on binary (two-star) or multiple (three or more star) systems. More than half of stars in our neighborhood are probably binary or multiple stars.

These clouds are ghostly masses that blot out what's behind them like a heavy fog. Gravitation pulls together masses of gas and dust into these clouds at many places in space.

Gravitation is something you experience every day. Because of it, your pencil will fall to the floor if you let it go. Gravitational attraction brings a pole vaulter back to earth and makes it hard to learn to ride a bike. **Gravitation** is the attraction that pulls things toward one another.

The huge masses of gas and dust, as they form stars, collapse to become very compact and very hot. When the developing star is hot enough, it begins to give off light, heat, and other energy radiations. Millions of stars may be formed in this way from one great space cloud.

After a developing star has settled down, it becomes a main sequence star. You would find it along the band shown in Figure 1/16. When the forming star has much more material than usual in it, it may grow even bigger later in its life and go off the main sequence. It may become a red giant for a while, as shown in Figure 1/18. Later the red giant may become a **variable star,** a star whose light brightens and then fades over and over again. Betelgeuse in Orion is a red giant variable star. The circles around white dots in the sky maps mean variable stars. (See Appendix C.)

After a long time being just a normal star on the main sequence, some stars don't become red giants. They explode. Just why, nobody knows. These exploding stars are called **novas,** or "new" stars, because they appear suddenly in the sky. Perhaps the Star of Bethlehem was a nova.

A nova like that shown in Figure 1/19 blasts quantities of its matter into space. Novas brighten very fast as they explode. Sometimes a nova explosion turns a star into a tiny, hot white dwarf. You would then put it in a different place on the H-R diagram.

Supernovas are much bigger explosions. In this explosion and afterward, a star may become millions of times brighter than it was before. Then it fades to a small white dwarf or even tinier star. Only three explosions thought to be supernovas have ever been recorded in our neighborhood in space. One of these was observed in A.D. 1572 in Cassiopeia and recorded as you saw in Figure 1/6.

Figure 1/19
Nova Persei blew up in 1901. Gas and dust were still expanding around the star in 1949.

What Happened to That Star?

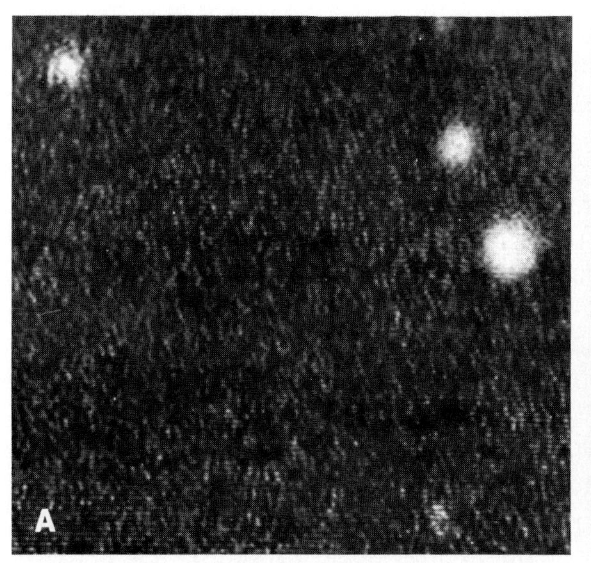

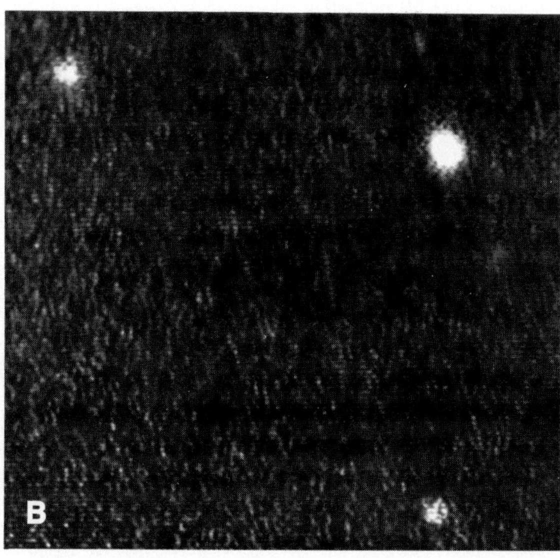

Star is pulsar in the Crab Nebula, NP 0532. See Teacher's Manual for background on pulsars.

Look at these two photographs, A and B, of the same part of the sky in the constellation Taurus. What is the difference between the two photographs? Can you guess what happened to that bright star?

The bright star that appears nearest the center of A but has disappeared in B is a pulsar. **Pulsars** are stars whose radiations pulse, or turn off and on like a blinking Christmas tree light. In one second, the pulsar in A turns on 30 times and off 30 times! This star is thought to be only about 10 miles (16 kilometers) in diameter, and made mostly of particles called neutrons. It's all that's left after a gigantic supernova explosion that was seen about 900 years ago.

The shutter of the camera that took these pictures was fixed either to open when the bright flashes of the pulsar came or to open between the bright flashes. When the shutter opened at the time of the bright flashes, the camera made the picture in A. When the shutter opened between flashes of the pulsar, the camera made the picture in B, in which no light is visible from the star at all.

This pulsar is about 6500 light-years distant. That's pretty close as space distances go. Most of the other pulsars that have been discovered are nearby in space. Could these blinking lights be signals from other intelligent beings in the universe? At first astronomers thought they might be. But the pulses were so strong that they could not imagine how any living being could make them.

Astronomers now think that the strong pulsing radiations may somehow be produced as these little stars spin very fast. How fast? All the way around as often as 30 times a second!

1.11 What gives stars their energy?

Where do stars get the energy to shine so brightly and the much greater energy to tear themselves to shreds? To explain this you need to know what materials stars are made of and what happens to these materials inside stars.

Appendix G sketches atomic theory, lists the elements, and describes chemical reactions, fusion, and fission. You might also use transparencies on atomic particles and processes if students are unfamiliar with atomic theory. **Concept: particulate nature of matter.**

Like all things in the universe, stars are made up of a few basic materials called elements. **Elements** are materials that are made up of only one kind of atom. **Atoms** are the smallest wholes of material that can exist alone or with other similar small wholes. These wholes are made up of many parts, but the wholes are different from the parts. Appendix G tells more about elements and atoms and the parts of which atoms are made.

The materials in stars are mostly the lightest element, hydrogen, some helium, and a little carbon. Stars also have tiny traces of most of the other elements. The kinds and amounts of elements a star contains depends on the group of stars it belongs to.

When gravitation attracts great masses of hydrogen gas into a star, the temperature deep inside the star goes up millions of degrees. The atoms of hydrogen combine in a process called **fusion**. During fusion, two hydrogen atoms combine to make one atom of helium, another kind of gas. The atoms lose a very small amount of their material in fusions, and a tremendous amount of energy is given off. Once started, fusion keeps on going, giving us continual radiations from stars. Fusion is going on all the time in our sun. A hydrogen bomb explosion is another example of fusion. Figure 1/20 shows how much energy fusion frees.

Stars changing hydrogen into helium by fusion may shine brightly for billions of years. But when much of the hydrogen in the center of a star has been used up, the star may collapse and become smaller. Then it may begin to fuse the helium or carbon it has made into heavier elements. This makes the star's radiations even brighter and hotter than before. The star leaves the main sequence and may become a giant or a great red supergiant.

Particles in solar wind are mostly protons and electrons. "Cosmic ray" particles range into the heavier, natural-element nucleuses.

The energy lost by fusion in stars stirs them up so much that stars constantly shoot out some of their materials into space. This makes the brilliant halo of hot atomic particles seen near the sun. These tiny particles go on out into space

Figure 1 / 20
A hydrogen bomb explosion is like the fusion taking place in the sun.

and even reach the earth. They are moving so fast that all of them together are called the **solar wind.**

Did You Get the Point?

Stars can be plotted by temperature and luminosity on a graph called an H-R diagram.

The H-R diagram shows a broad band of stars, like our sun, on the main sequence. It also shows other groups of stars, like giants, supergiants, and white dwarfs.

Gravitation brings together clouds of gas and dust made up of atoms of different elements to form stars.

Hydrogen atoms fuse at the high temperatures inside stars to form helium, giving off great amounts of energy.

Stars also shoot out particles into space in a solar wind.

Check Yourself

1. How are stars located on the H-R diagram? What do the main sequence and other places on the diagram mean?

2. Tell the story of how a star is born, how it gets its energy, what radiations and particles it gives off, and how some stars blow up as they age.

Answers:

1. Stars are located on H-R diagram (horizontally) by surface temperature and vertically by luminosity. Main sequence and other groupings mean these properties of stars are related, increasing or decreasing together. Giants and white dwarfs are located on H-R diagram in Figure 1/16.
2. Stars are born of gravitation-gathered gas and dust in space, get energy from atomic fusion, radiate light, heat, other radiations, and atomic particles into space. Stars are gradually destroyed in novas and blow up in supernovas.

What's Next?

You know about stars now, how bright and hot and distant and numerous they are. You know, too, how stars form, what materials they contain, where they get their energy, and how a few of them explode. But amazing as stars are, they are the most *ordinary* things in the sky. In the next chapter, you look into some of the *extraordinary* things to be found there, like galaxies and quasars.

Skullduggery

Code of message: (1) A = F; (2) B = H; (3) C = R; (4) D = C; (5) E = M; (6) F = W; (7) G = G; (8) H = V; (9) I = O; (10) J = B; (11) K = K; (12) L = U; (13) M = Q; (14) N = I; (15) O = D; (16) P = Z; (17) Q = N; (18) R = Y; (19) S = J; (20) T = T; (21) U = X; (22) V = L; (23) W = S; (24) X = A; (25) Y = P; (26) Z = E.

Message decoded: "Live on third planet out from small dwarf star. Have two eyes and backbone. Bouncing radio signal off only moon. Live in city named Los Angeles. April Fool."

Decoded answers to questions: (1) constellations; (2) Ursa Major; (3) magnitude; (4) Mizar; (5) spectrum; (6) K; (7) Betelgeuse; (8) luminosity; (9) H-R diagram; (10) main sequence; (11) dwarf; (12) atom; (13) gravitation; (14) supernova; (15) pulsars neutrons; (16) fusion; (17) galaxy.

On the night of April 1, 1971, at the National Radio Observatory in Greenbank, West Virginia, astronomers were astounded to pick up radio pulses from some place in the sky near the moon. The pulses were long and short as if in the dots and dashes of a code. After working for weeks making pulses into letters, the astronomers finally broke the code. Can you?

Message Received at National Radio Observatory
UOLM DI TVOYC ZUFIMT DXT WYDQ JQFUU CSFYW JTFY. VFLM TSD MPMJ FIC HFRKHDIM. HDXIROIG YFCOD JOGIFU DWW DIUP QDDI. UOLM OI ROTP IFQMC UDJ FIGMUMJ. FZYOU WDDU!

Directions for Breaking Code of Message
The answers to the questions and the missing words in the statements below are given in the code. When you have the first few answers, the ones you *know* are correct, you will have enough of the code to help you figure out the answers to the other questions. The code is simple. Most of the letters of the alphabet have been rearranged. A *Z* is written as an E, but a *K* is a K. Since answers to all questions give you the whole alphabet, you can use the code for your own messages. Answer the questions and away you go.

1. What are the picture groups of stars in the sky? RDIJTMUUFTODIJ

2. XYJF QFBDY is a pattern of stars, part of which is called the Big Dipper.

3. What is the brightness of stars called? QFGIOTXCM

4. The name of one of the stars in the Big Dipper is QOEFY.

5. Starlight gives a JZMRTYXQ when it is broken up.

6. A group of stars that are orange in color is called K stars.

7. One of the red giants which is also variable is named HMTMUGMXJM.

8. If the temperature of a star is much higher than that of another star, its UXQOIDJOTP is probably higher too.

9. What is a diagram that astronomers often use? V-Y COFGYFQ

10. Where does our sun fall on this diagram? QFOI JMNXMIRM

11. As a G-group star, our sun is called a CSFYW star.

12. What is the smallest part of an element that can exist alone? FTDQ

13. The materials of stars are pulled together by GYFLOTFTODI.

14. In a JXZMYIDLF, a star blows itself to bits.

15. ZXUJFYJ are stars probably made up largely of IMXTYDIJ.

16. Atoms are forced to combine in the process of WXJODI.

17. A GFUFAP is one of the extraordinary things in the sky taken up in Chapter 2.

For Further Reading

Asimov, Isaac. *To the Ends of the Universe.* New York, Walker & Co., 1967. What's in the universe beyond our solar system, who has found out about it, and what they have thought.

Boeke, Kees. *Cosmic View: The Universe in 40 Jumps.* New York, The John Day Co., 1957. Each "jump" is a view from ten times as far as the previous one. You will see our earth become a tiny point, then not visible at all in relation to the sun, and then even the sun not visible.

Bok, Bart J., and Priscilla F. Bok. *The Milky Way.* Cambridge, Massachusetts, Harvard University Press, 1957. A tour of the Milky Way by two astronomers, with an introduction to radio astronomy.

Rey, Hans. *The Stars.* Boston, Houghton Mifflin, 1967. Charts of constellations, plus just what you need to know about when to see them best.

Chapter 2 Galaxies

The light of dim stars near the moon is actually overwhelmed by moonlight scattered in the sky, just as even bright stars cannot be seen in sunlight during daytime.

Students can see the wavy band of the Milky Way in the Sky Maps, Appendix C. Maps show the dark "islands" and how the Milky Way changes as we view different parts of it through the seasons, as earth moves around the sun. Although moonlight or city lights diminish its brightness greatly, in most places students can find Milky Way and trace its path through night sky for themselves.

The sky is full of puzzling things. But the largest puzzle in the sky runs right across it like a wavy bright band. Since ancient people didn't understand what it was, they made up all sorts of stories about it. In most of the stories it is called a river of heaven or a river of light. Look at a part of it in the photograph, and you'll see why. Compare it with the other photograph of an ordinary bit of sky. It *is* a puzzle.

This river in the sky does not look like an even ribbon or streak of light. It has dark spots and patches, like islands, in it. It looks enough like a stream wandering across the sky, so that the ancient Chinese named it the Celestial River. They imagined that fish in this river, the stars, were frightened by the crescent moon. The fish thought it was a hook. This "fish story" explained, the Chinese said, why small, dim stars disappear near the bright moon.

Some North American Indian tribes called the band in the sky the White River. Other tribes thought of it as a path or road that the souls of the dead must travel from this world to the spirit world. Some Indians called it ashes or smoke from a fire.

Siberian Indians were sure that the sky had once broken in two. They thought that the bright band was the seam where the sky was sewed together! And even in the first century A.D., astronomers thought of the band as marking the place where the two halves of heaven were joined.

If you had lived 2000 years ago, you might have believed one of these stories about this band in the night sky. Or you might have made up an even better story of your own. But the truth turned out to be stranger than any of these stories.

Investigating Bright Blurs

Explanation of the Milky Way is given in Section 2.4, after students have studied nebulas and galaxies. This sequence was also followed historically.

2.1 What are the fuzzy spots in the sky?

Some of you may already have guessed that this "river of light" in the sky is the Milky Way. But you may not know what the Milky Way is made of. Let's look at some photographs that will help to explain what the Milky Way is.

Besides the Milky Way there are several other mysterious blurs in the sky that you can see. With your eyes alone, or with binoculars, you can see the fuzzy spots in the sky, the ones photographed in Figures 2/1 and 2/2. Pick out

Figure 2 / 1
Pick out the blurred, fuzzy spot in this photograph of the constellations Monoceros and Orion. This spot is a nebula.

Figure 2/2 *Find the blurred, fuzzy spot in this photograph of the constellations Pegasus and Andromeda. This spot is a galaxy. Compare it with the nebula in Figure 2/1.*

See note on galaxies and nebulas in Teacher's Manual,
Process: observing

the fuzzy spots in these photographs. Now, look at the star maps in Figures 2/3 and 2/4. They show where you will find these fuzzy spots, if you look for them in the sky.

When telescopes were invented about 350 years ago, astronomers soon found more of these strange bright blurs. And as telescopes became more powerful, astronomers saw thousands and thousands of them. Finally, early in our own century, they found what these spots or blurs are.

There are two kinds of fuzzy spots. The first kind is like the one shown in Figure 2/1. This kind is a nebula, which is the Latin word for cloud or mist. A **nebula** (NEBB-you-lah) is a vast cloud of gas and dust as fine as smoke. Nebulas

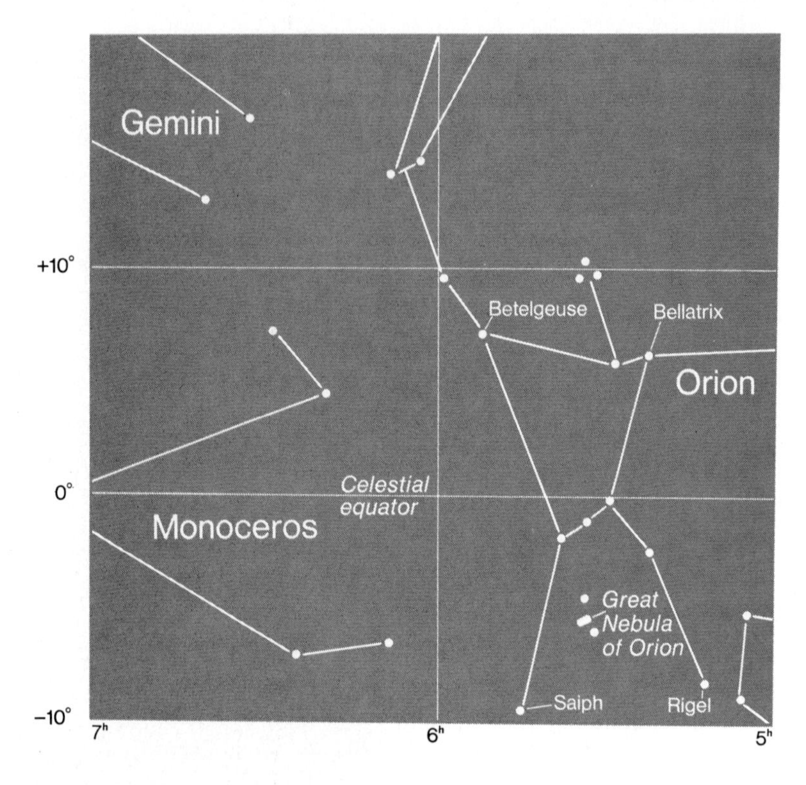

Figure 2/3 *Locate the fuzzy spot in the sky in the constellation Orion (Figure 2/1) with this chart and the star maps in Appendix C.*

do not radiate light but are lighted up by one or more stars within them or nearby. The Great Nebula of Orion, as photographed by a large telescope in Figure 2/5, is about 1500 light-years away. This means that its light takes 1500 years to reach us. But for fuzzy spots this is close. Nebulas are nearer to us than the other kind of fuzzy spot.

Where do nebulas come from? Often from a star that exploded, a nova or a supernova. You might say they are like the clouds of smoke drifting around after a Fourth of July rocket has gone off. Sometimes, though, a nebula is formed by gravitation when it pulls together the material in a thin cloud of gas and dust. This packs the material into a smaller, thicker cloud, the nebula.

Concept: levels of organization

In Figure 2/2, you saw the other kind of fuzzy spot as it looks to you in the sky. It doesn't look much different from the nebula in Figure 2/1, but it is *very* different. It is a **galaxy** (GAL-ax-ee). A galaxy is a collection of billions

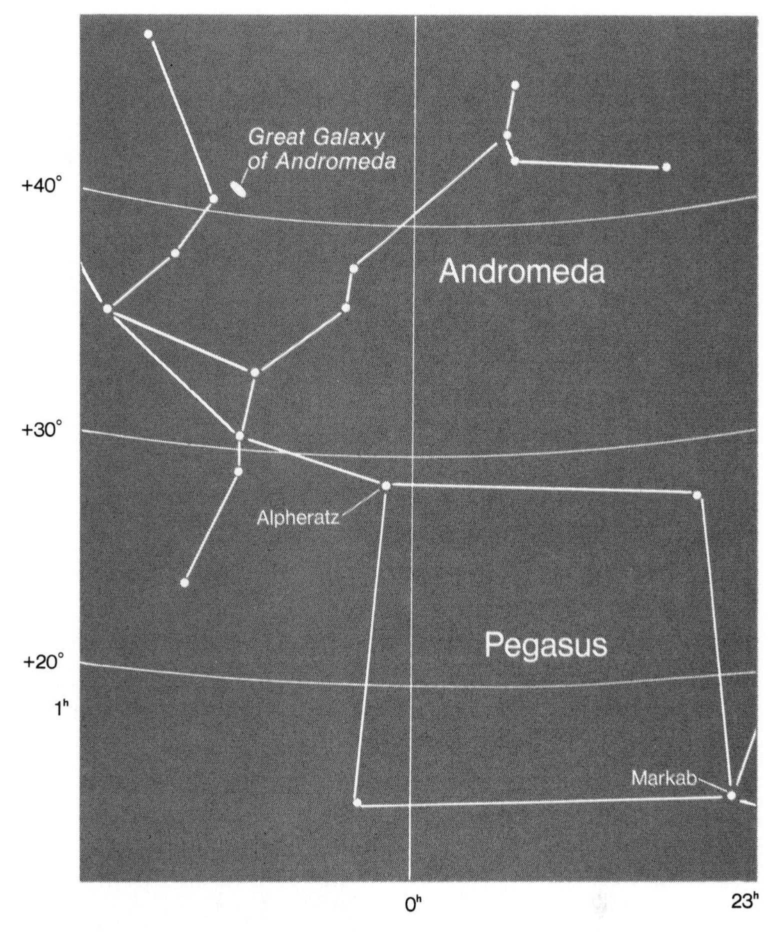

Figure 2/4 *Find the fuzzy spot in the sky in the constellation Andromeda (Figure 2/2) with this chart and with the sky maps in Appendix C.*

and billions of stars. Some galaxies have gas and dust clouds mixed in between their stars, but others are nearly free of such clouds.

Look at the picture of a galaxy in Figure 2/6, made with the world's largest telescope. This is the Great Galaxy of Andromeda (an-DROM-uh-dah). Andromeda is the constellation where this galaxy is located. The galaxy is over 2 million light-years away! Remember that the Great Nebula in Orion is only 1500 light-years from us.

Draw on chalkboard, or have students draw, or show on overhead projector some of the many shapes of a coin held in different positions. Have students find similar shapes in Text Figure 2/7. Not all galaxies are flattened like a coin in disklike shape.

The fuzzy spots in Figure 2/7 are galaxies, too. Notice the many shapes they come in. The round spots in the photograph are nearby stars.

Figure 2/5
The Great Nebula in Orion photographed with the 100 inch Hooker telescope on Mount Wilson in California.

Figure 2/6
The Great Galaxy of Andromeda with its two nearby, or satellite, galaxies.

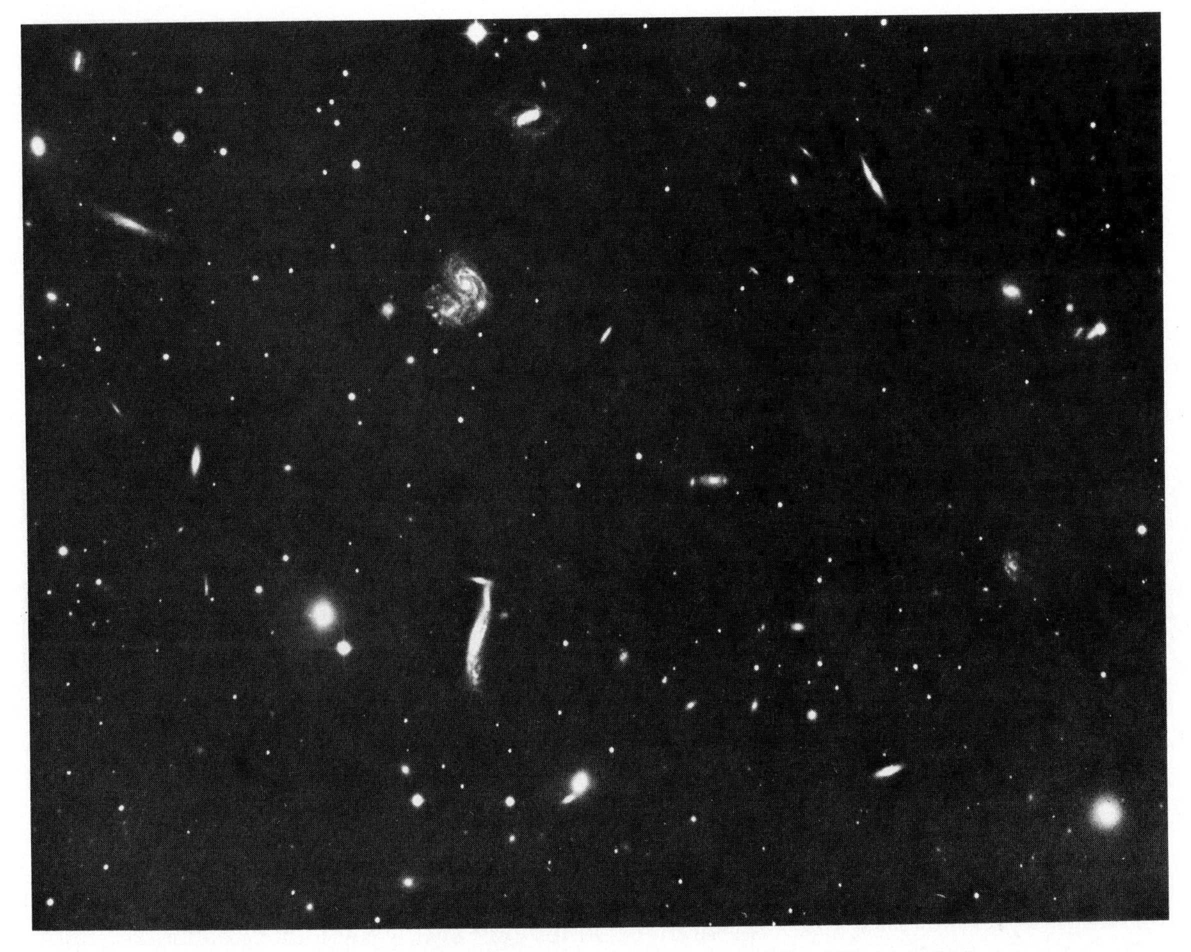

Figure 2/7 *The camera of the 200 inch Hale telescope on Mount Palomar recorded this large cluster of many different types of galaxies in the constellation Hercules.*

Ancient Chinese called new stars appearing in sky "guest stars," and astronomers divined the future from them. The astronomer who reported the guest star of Crab supernova predicted a successful reign for the new Chinese emperor, because the starlight was yellow, the imperial color. Crab Nebula is now invisible to naked eye with magnitude of +8.6, although its magnitude was probably −6 at its height. Answer to question: 3600 years − 1054 years (A.D.) = 2546 years (B.C.)

2.2 Model nebulas and galaxies.

To many people, the big cloud of gas and dust in Figure 2/8 looks like a crab. That's why this cloud is called the Crab Nebula. It is located in the constellation Taurus (TORE-us), the Bull. The supernova explosion that made this nebula was first seen in August, A.D. 1054. The event was recorded by Chinese, Japanese, and Italian observers. The explosion was so great the star could be seen in the daytime for 23 days! The Crab is about 3600 light-years from us. When did the supernova actually happen then?

Figure 2/8
This may look like a smashed custard pie but it is really the Crab Nebula. Near the center of this blob is the pulsar shown in What Happened to That Star?

Materials

10–12 sheets paper, newspaper or paper towel, small hammer, drop-bottles of food coloring, salt (shaker or tablespoonful), fixative

Activity Time: 15 to 20 minutes. Individual students. To save time, some students could model nebulas, others galaxies. Force of hammer blows may squeeze coloring through paper undersheet, so protect table with newspaper, paper towel, wood, or cardboard. Hammer blows should be made quickly before drop soaks into paper.

Remarkably realistic models of nebulas can be made. Extra salt along one side of coloring drop yields nebula similar to Great Nebula in Orion. Salt in drop itself and around its edge gives a Crab Nebula with moderate size drop and blow of hammer. Collect a series of nebula photos for students to model. Transparent spray or fixative will coat models for preservation.

The pulsar pictured in *What Happened to That Star?* is the remnant of the supernova that made the Crab Nebula. It is now, perhaps, the most intensely studied object in the sky.
Concept: energy in motion

★ You can make models of nebulas like the Crab yourself. A hammer blow as in Figure 2/9 will represent the supernova explosion. Squeeze out a drop of food coloring onto a sheet of paper placed on a paper towel or newspaper. The drop represents a star ready to explode. Make the drop between a quarter-inch and a half-inch in diameter. Sprinkle a little salt into and around the drop to make your star more solid. Place another sheet of paper over the drop. Quickly strike the drop squarely in the middle with the hammer. Take off the covering paper. There's your nebula!

Make several nebulas. Try different-sized drops of coloring. Sprinkle salt in different places. Change the hammer blows. Can you make a nebula that looks like the Great Nebula in Orion in Figure 2/5? Let the nebulas dry, then spray them with a fixative. You can keep the best ones in your notebook. ★

After the explosions of their stars, real nebulas shoot out in all directions and keep on expanding. After 900 years, the Crab Nebula is still expanding at about 715 miles a second. Some explosion!

Figure 2/9
Make your nebula models this way.

Materials

dish or pie plate (round), turntable, cup or test tube, food coloring, water

Activity Time: 15 to 20 minutes. Individual students, sharing equipment in groups of three. Some students can make model galaxies while others make nebulas, sharing results. Slow turning is better than fast, which sends coloring to side of dish.

★ To model nebulas, you let the food coloring represent masses of gas and dust from an exploding star. To model galaxies, you let the tiny particles of food coloring stand for billions of stars. They are in a pan of water representing space.

Fill a round pan or dish about half full of water and put the dish on a turntable, as shown in Figure 2/10. Have the center of the plate over the center of the turntable so the pan will spin evenly, without wobbling. Let the plate stand for a minute or two until the water is quiet.

Figure 2/10 *Make your galaxy models like this.*

When spinning stops, currents move dye toward center of dish again, so turning can be resumed, but clear, spiral shapes may be lost. Diluted coloring gives better galaxies than undiluted coloring. Several drops or lines yield varied shapes, like barred spirals.

Dilute a few drops of food coloring in a cup or test tube of water. Pour some coloring into the middle of the dish and turn it slowly counterclockwise. Watch what happens to the coloring. You turn the dish to start the coloring moving. Many galaxies turn, or rotate, in just this way.

What happens when you stop the spinning? Add more coloring near the center and turn the dish again. Dump the water when it gets too dark, refill, and start again. Try several drops close to each other, or a line of drops. Some of your models will look very similar to galaxies pictured in Figure 2/11. ★

2.3 Galaxies come in different shapes.

Materials

paper and pencil

★ Examine the galaxies shown in Figure 2/11. Pick out the ones that look alike in many ways. Group the letters of galaxies that look alike in your notebook. Some galaxies

Activity Time: 15 to 20 minutes. Individual students. Practice in grouping or classifying. When students disagree on groups, ask them to describe feature of group that distinguishes it from others. Students may try to group by size or brightness, but galaxies, as pictured, are at many distances, so these characteristics do not help in grouping them. **Process: observing, classifying**

may be hard to group. They may be in between other groups. So you may have to leave some galaxies out.

When you have finished, compare your groups with those of others. One way to do this is to put all the groups on the chalkboard. See if there are a few groups on which almost all of you agree.

Astronomers group galaxies by their shapes, too. One of their groupings, called the Hubble classification, appears in Appendix H. Are your groups the same as those given in this classification? Did you put galaxies in the right groups? ★

Basic Hubble classification: elliptical, normal spiral, barred spiral, irregular, and lenticular galaxies. Edwin Hubble first developed this classification at Mount Wilson Observatory, California, in 1935. Spiral pattern appears in about half of all galaxies and could be (1) a permanent characteristic of some rotating galaxies or (2) a temporary stage through which most galaxies pass. Answers: Unlikely students arrived at the Hubble classification; some galaxies are hard to identify and place in their proper groups.

Why do galaxies have so many shapes? No one knows yet. You saw some model galaxy shapes change in Section 2.2. And remember how stars change during their lifetimes? (See Section 1.10.) Astronomers think that galaxies take on many different shapes as they age. This happens to people, too! But in galaxies, great explosions in their centers may also cause changes in shape.

2.4 We live inside the Milky Way.

When astronomers explored the Milky Way with larger and larger telescopes, they saw something like Figure 2/12. They found that the Milky Way, too, contained many billions of stars. Sounds like a galaxy, doesn't it? Well, it is a galaxy. The wavy band of faint light with its stars, and our sun and all the stars we can see, form part of our galaxy. We see it *from the inside,* so we see just a few of its stars. Many more can be observed only with telescopes.

Apt analogy is band you see with your head within a hoop or bicycle tire. See Teacher's Manual for a demonstration that can be used here. Why do we see the band of the Milky Way, which we know contains billions of stars, and also see some bright stars outside of it? Use bottle filled with rice or other small objects, marked with an "X" for our position in Milky Way, as described in Teacher's Manual.

Astronomers have figured out that the Milky Way has a spiral shape, very much like the shape of the Andromeda galaxy you saw in Figure 2/6. If we could see it from the outside, the Milky Way would look like the drawings in Figure 2/13.

Our sun is located at the inner edge of one of the spiral arms of the galaxy, as pictured in Figure 2/13. The sun and all the other stars move around the center of the Milky Way. The whole galaxy is spinning like a top or a merry-go-round. But our galaxy is so huge that it takes about 200 million years for the sun to carry us all the way around it. The earth completes one turn around the sun in just one year!

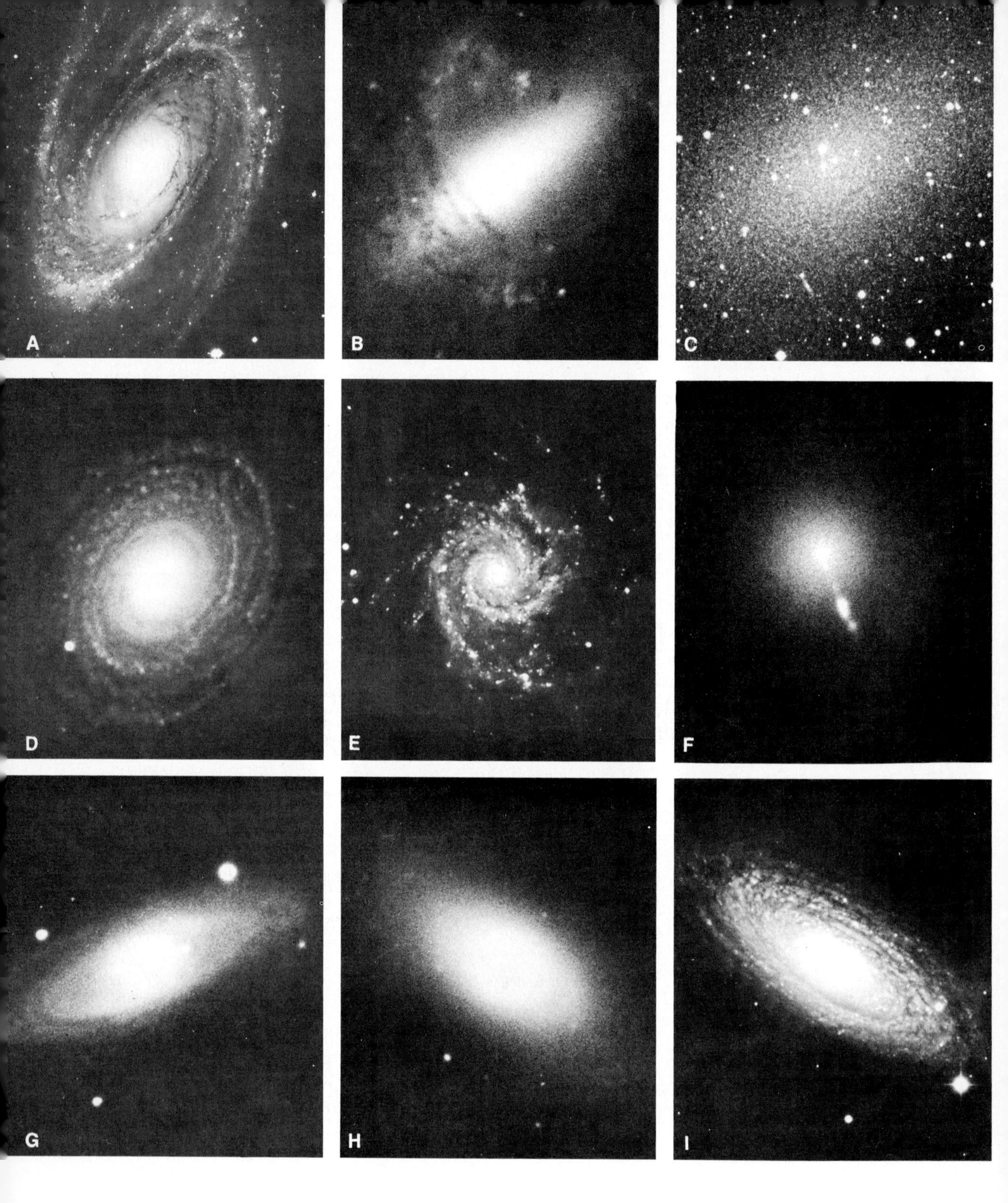
A
B
C
D
E
F
G
H
I

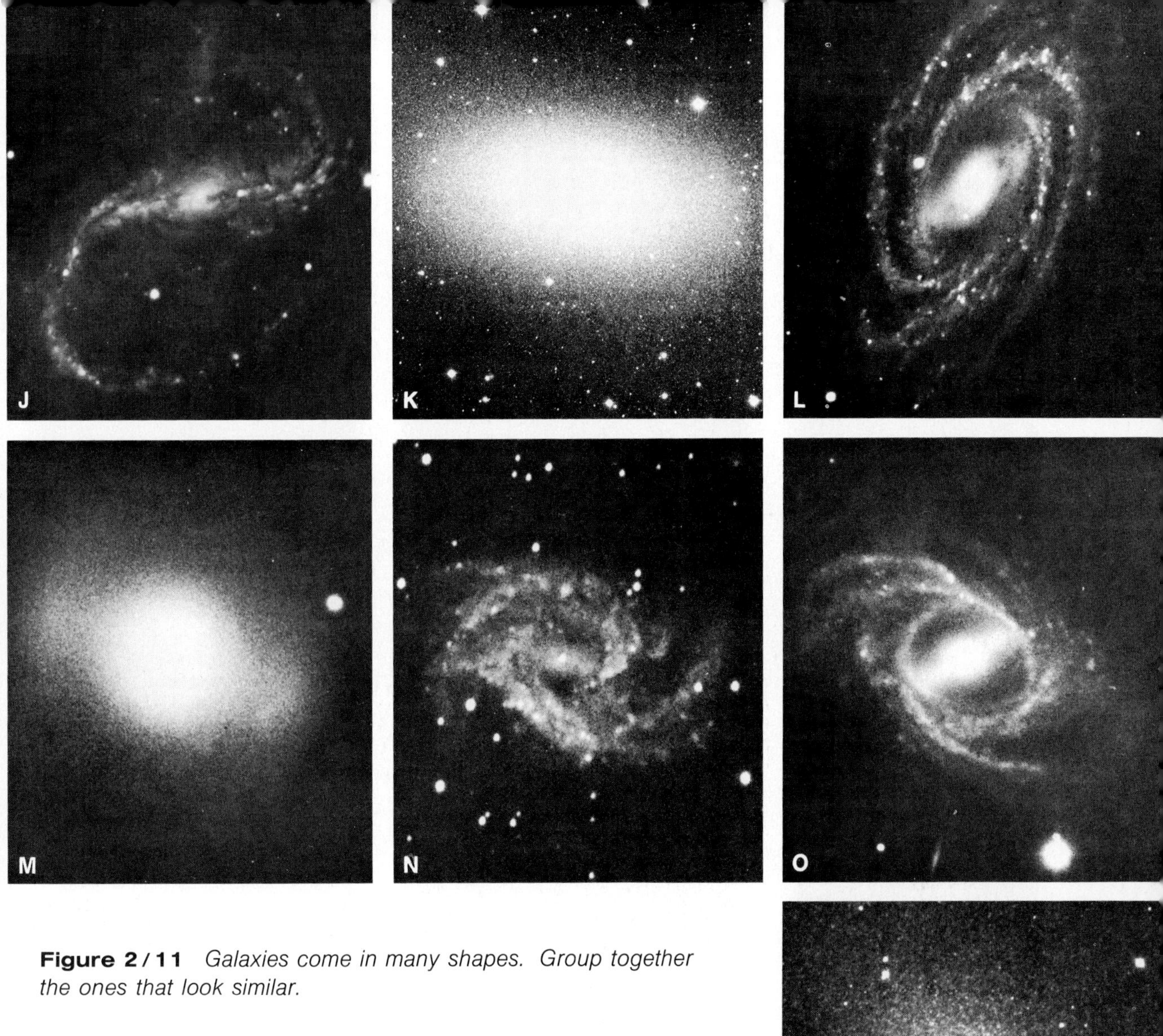

Figure 2/11 *Galaxies come in many shapes. Group together the ones that look similar.*

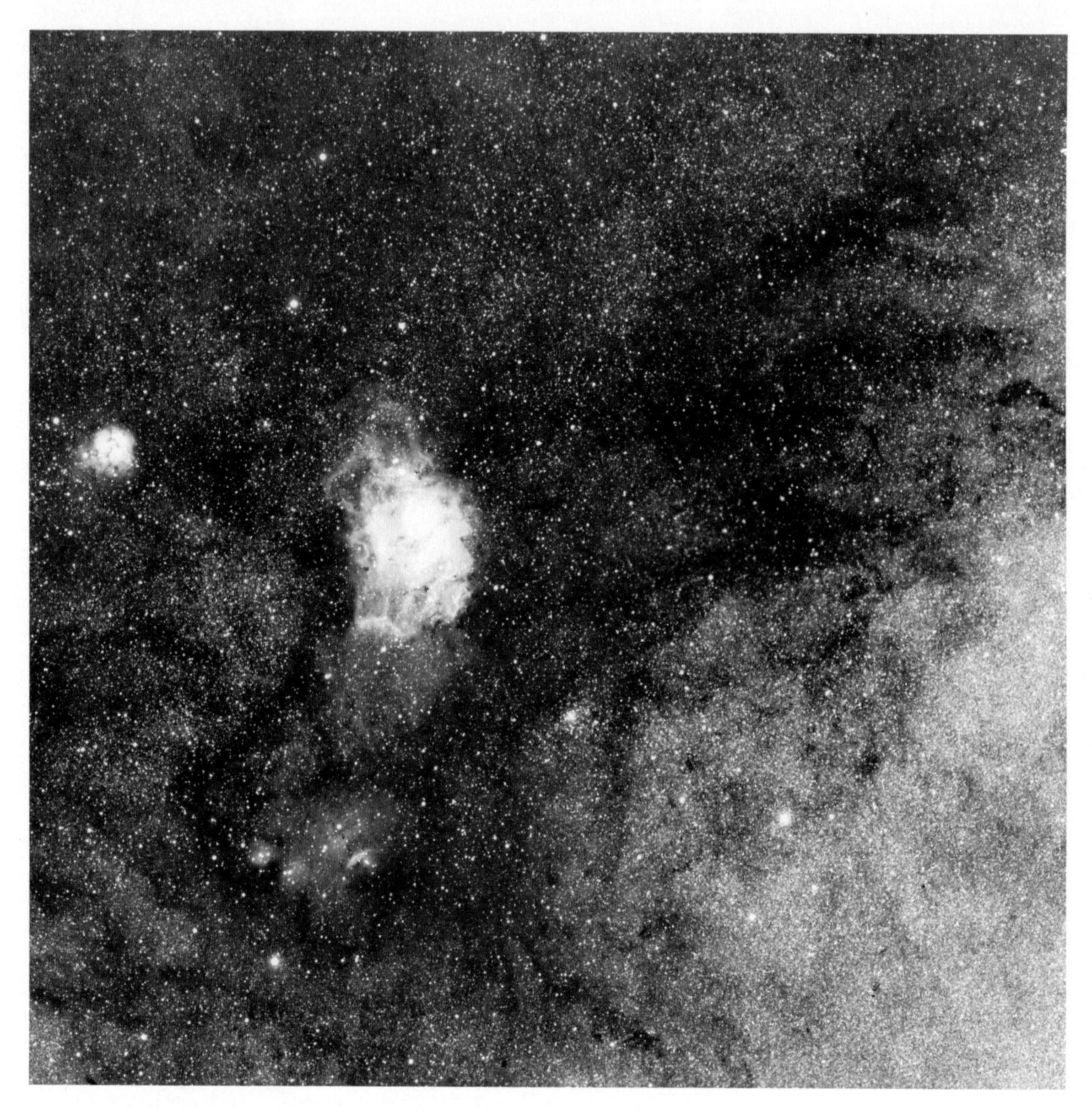

Figure 2/12 *Clouds of stars in part of the Milky Way, the region of constellation Sagittarius. Notice also the two large gas nebulas and several star clusters.*

The spiral arms and the center of our galaxy are filled with billions of stars like our sun: more than 100 billion stars, perhaps 200 billion. The space between the arms has fewer

Figure 2/13
The Milky Way spiral galaxy as seen from the side in A and as seen from above in B. Most stars we see in the sky are in the box around the sun in A.

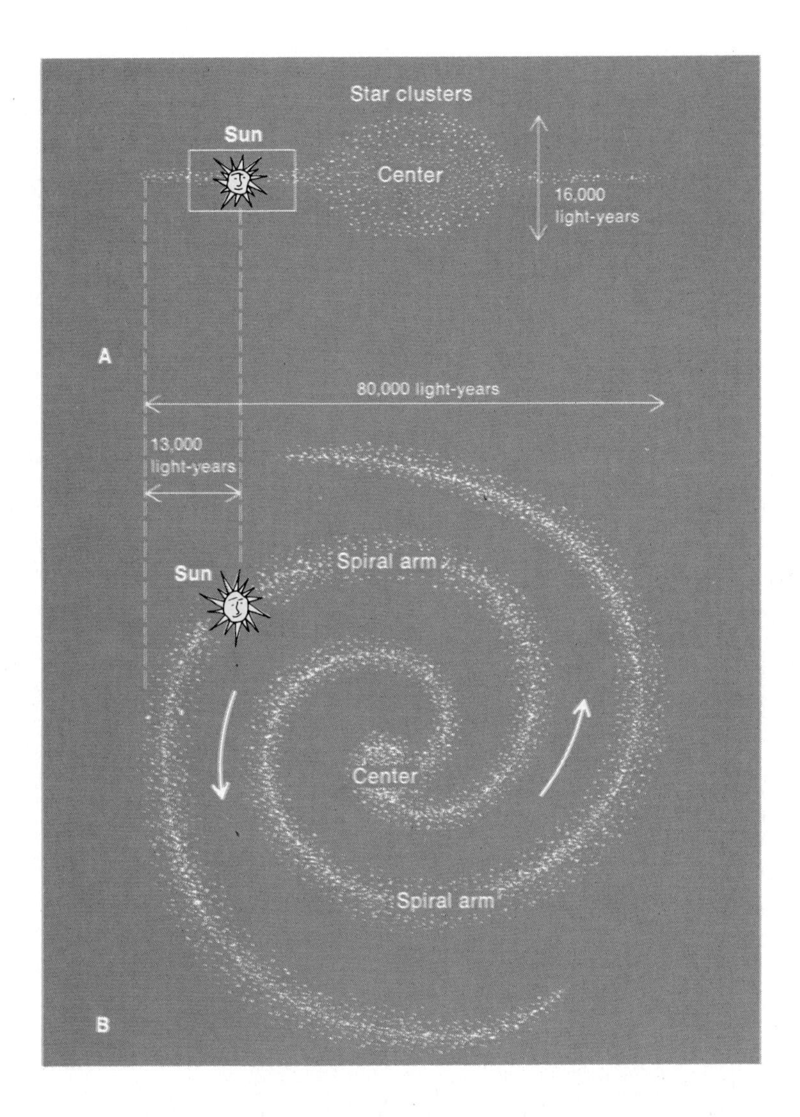

Short waves, (such as X-rays and light) vibrate vertically to their direction much more rapidly (or have greater frequency) than long waves (infrared, microwave, and radio); so short waves are more likely to run into dust particles and be stopped or scattered by them.

Knowledge of the center and distant parts of our galaxy is still sketchy, although the structure of a spur projecting above the center is fairly well established.

stars and more gas and dust. In some places the dust is scattered so thickly that light cannot get through it. Notice the dark spots and streaks in Figure 2/14. In these dark places there are no light radiations to show astronomers what lies beyond.

Radio waves, another kind of radiation that stars give off, can pass through these thick clouds of dust even when light can't. Instruments called **radio telescopes** can gather and record radio waves. A radio telescope is shown in Figure 2/15. By studying these records, astronomers have learned what the hidden part of our galaxy is like.

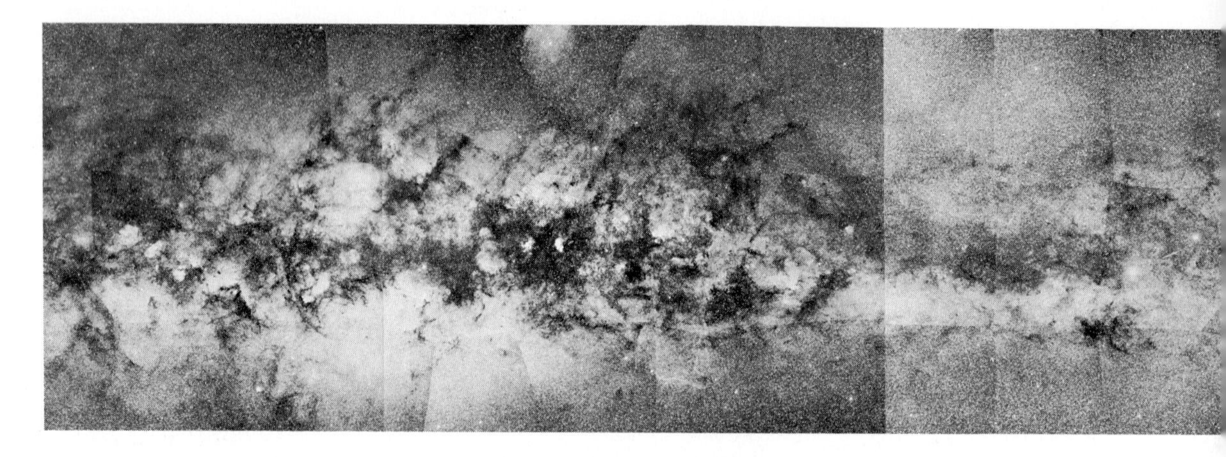

Figure 2 / 14 *This long photograph of the Milky Way was made by piecing together many smaller pictures. The dark spots in the middle of the Milky Way are dense clouds of dust.*

Figure 2 / 15 *The radio telescope at the National Radio Astronomy Observatory, Green Bank, West Virginia. The dish, 140 feet in diameter, can gather radio waves from the hidden parts of our galaxy.*

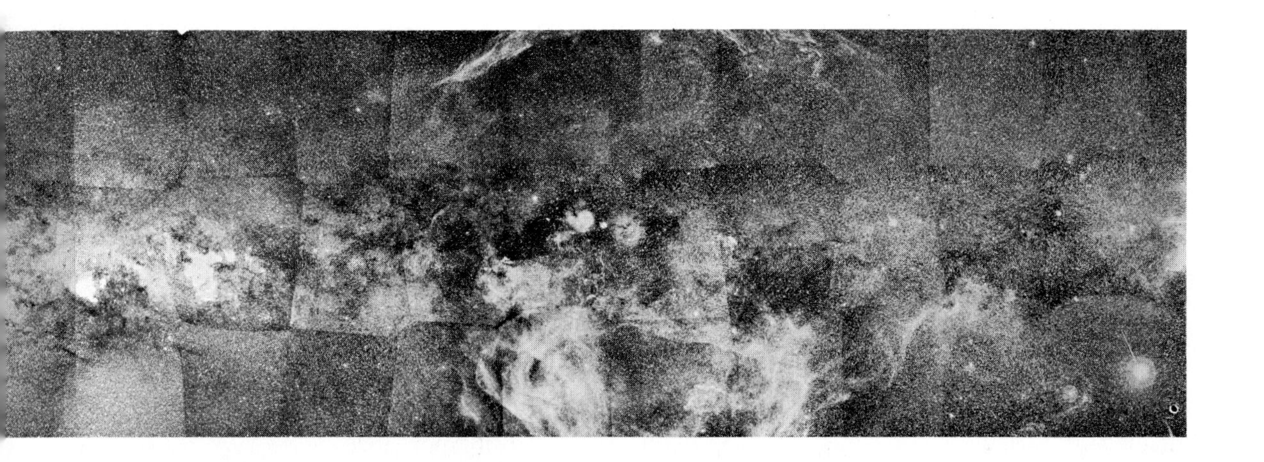

Did You Get the Point?

Better and stronger telescopes have shown astronomers that the fuzzy spots in the sky are nebulas and galaxies.

Nebulas are nearby clouds of dust and gases pulled together by gravitation or made by star explosions.

Galaxies are distant collections of billions of stars, with a little gas and dust between the stars.

Galaxies have been placed in many groups according to their shapes.

Our Milky Way is a spiral galaxy. Our sun revolves around the center of the spinning Milky Way, along with all its other stars.

Answers:
1. Stars are large masses of gases undergoing fusion; constellations are groups of stars pictured in relation to each other as viewed from earth; nebulas are relatively small masses of gas in our galaxy; and galaxies vast collections of stars of which the Milky Way is one collection.
2. From within Milky Way, we look out at many stars along its disk, or we may look up and down from the disk, where fewer stars are visible.
3. Within an elliptical galaxy, shaped roughly like a football, we'd see heaviest band of stars toward ends from center and stars diminishing in density from ends to middle, with fewest stars in a band around the middle.
4. From crowd's edge, we'd see many people directly in front and on both sides, making one large oblong spot of people; As you moved into crowd, you'd see few people behind you, increasing on the sides to more in front; at center you would see many people around you in all directions.

Check Yourself

1. What are the differences between stars, constellations, nebulas, and galaxies? Name examples of each.

2. Explain why the Milky Way looks the way it does in the sky.

3. If you lived on a planet of a sun near the center of an elliptical galaxy, how would the stars in your sky look?

4. Suppose you are on the edge of a large crowd listening to a rock concert. You can't hear the groups very well, so you move in toward the center of the crowd. Describe how you see the people in the crowd from its edge and as you move into the center. Compare the people in the crowd with the stars in a galaxy.

Mysterious Broadcasts from Deep Space

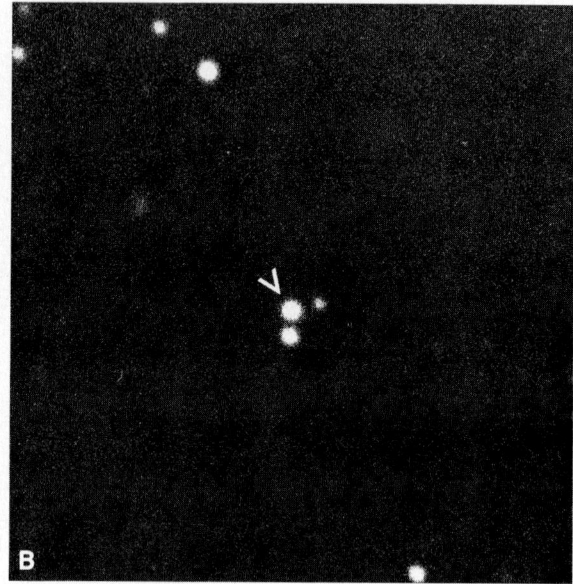

About 8000 places in the sky are known to broadcast radio waves. Just a few of these radio sources, like the Crab pulsar, are in our own galaxy. Others have been discovered in galaxies nearby. The galaxy shown in A is nearby and its radio waves are very strong. Do you think something big happened in this galaxy?

During the early 1960's, some bright but slightly fuzzy objects thought to be ordinary Milky Way stars were found to be very strong radio sources. The object in B is such a source. Study of their waves showed that they were not stars and were not in the Milky Way. Astronomers do not know just where those radio sources are. They may be nearly at the limits of the universe we know or much closer to us. Further, they were more powerful radio sources than any ever even imagined. Are they galaxies, or what? Since astronomers don't know, they call them "quasi-stellar radio sources," or "quasars" for short. "Quasi-stellar" means "like stars."

Some nearby radio-source galaxies give off a million times more energy than our galaxy or Andromeda. But the energy of quasars may be a billion times greater than that of the usual galaxy! So they're a billion times larger than any other galaxy, you might think. Not true. They're so tiny that astronomers haven't yet been able to measure their size. To add to the mystery, the radiations from some of them may change in strength over just a few months. Galaxies just don't behave like this.

What are these mysterious things called quasars, then? Are they very far away or are they close to us? Astronomers are still trying to answer these questions.

Investigating Radiations

More strictly speaking, electromagnetic radiations are simultaneous periodic variations of electric and magnetic field intensities, the two kinds of oscillations being perpendicular to each other.
Concept: energy in motion

Activity Time: 30 minutes. Students can do this activity individually or in groups of 2 or 3. This activity can be done as a demonstration by attaching spring clip to overhead projector table with tape and having students wiggle chain and measure projected waves. Activity in Section 2.8 can be done with this one if necessary or convenient. Chain waves model transverse wave forms, like the graph of the fluctuating field of radiations from space in the electromagnetic spectrum. Longitudinal waves are illustrated by seismic waves in Chapter 11. Particles in a transverse wave fluctuate transversely (or perpendicularly) to the line of their motion, while particles in a longitudinal wave (sound or seismic waves) fluctuate longitudinally to (or along) the line of their motion.

2.5 Model energy radiations.

Remember that hot, bright things like light bulbs and stars pour out floods of radiations. You know how a guitar string moves back and forth, or vibrates, when you pluck it. The vibrations in the string make vibrations in the air, which you hear as sounds. The atoms in stars and galaxies vibrate, too, only much, much faster, because of all the energy released by fusion. So heat radiations, light, radio waves, and other radiations are given off, instead of sound vibrations. All these atomic energy radiations are both electric, like current in a wire, and also magnetic, acting like magnets. So these radiations are called **electromagnetic radiations.** All the kinds of energy radiated to us by stars and galaxies are electromagnetic.

In the experiment you are about to do, your moving hand gives the energy to make radiations in the form of waves in a chain. You can observe how changes in the amount of energy (the distance your hand moves) and the speed of vibration (how fast your hand moves) make changes in the waves in the chain. These waves model the energy radiations from stars and galaxies.

Materials

paper pad, spring clip, paper clip, ball chain and eyelet, ruler, watch or clock, pencil

Table or bench should be horizontal, not slanting, so that regular wave form will be made. Having made and traced a wave, student can tear off his sheet from pad to measure, leaving clean sheet for next student.

Answers: Waves are clearly seen as chain moves on paper. Vibrating motion of hand is source of waves. Waves clearly move away from this source to other end of chain. Waves become smaller when rapidity of motion increases. Other end of chain moves up and down just as fingers do, and so does any one ball on chain. You can define transverse waves in terms of this up-and-down motion of balls, vertical to line of motion.

★ Put together the equipment for the experiment on a long paper pad, as shown in Figure 2/16. Place the pad flat on a table or desk. Have the spring clip on one side and the free end of the chain on the other. You make waves across the paper with the ball chain. Be sure the eyelet end of the chain is free to move back and forth on the paperclip.

Stand or sit facing the pad that is flat on the table as shown in Figure 2/16. With your hand on the table, hold the ball chain between your thumb and forefinger, slightly above the edge of the paper. The chain should be a little loose across the paper. Move your hand rapidly up and down, keeping your hand on the table as shown in Figure 2/16.

Can you see waves in the chain on the paper? What is the source of the waves? In what direction are the waves moving? When you vibrate the end of the chain faster, what

Figure 2/16
Set up your pad and ball chain like this to make model waves.

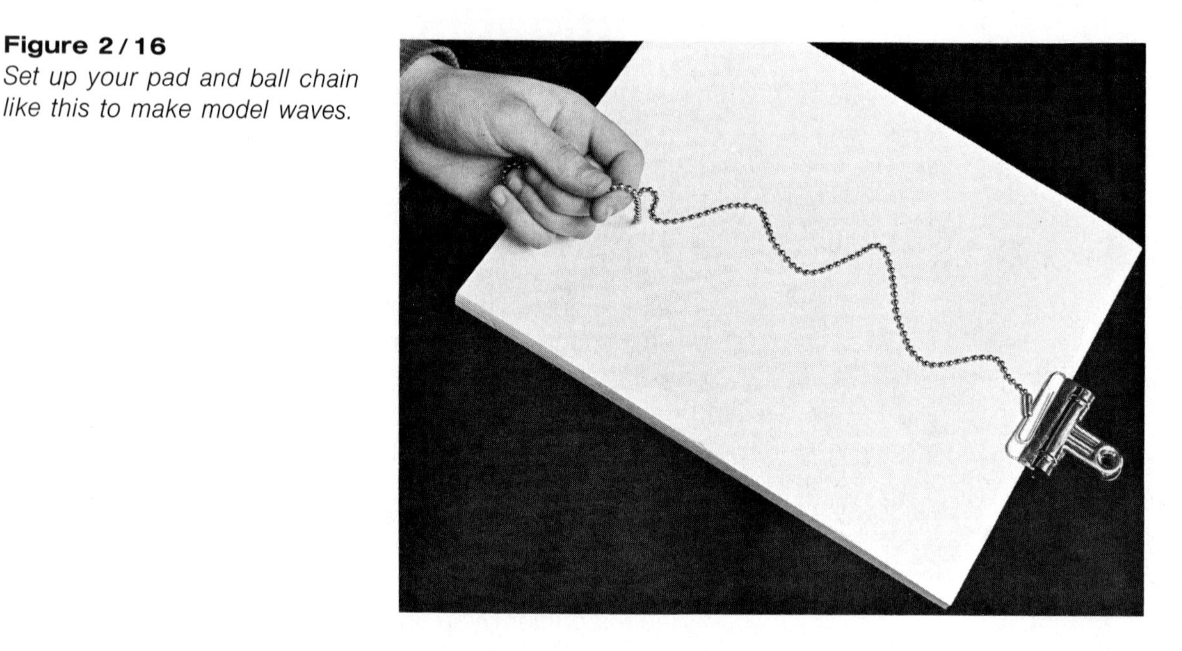

happens to the waves? How does the other end of the chain move? How does any one ball on the chain move?

Students may need a little practice to obtain good waves, adjusting such variables as looseness of chain, height of fingers holding chain above paper, speed of hand movement, and stopping chain to yield smooth wave forms. Since the chain rolls after the shaking stops, waves may have to be "restored" when they are traced.
Process: measuring

Stop moving the chain. Start and stop again, until the chain stops with fairly even wave shapes on the paper when you stop moving it. Draw along the chain with a pencil. Then mark and measure the waves as shown in Figure 2/17. The top of the wave is called the **crest,** and the bottom the **trough** (TRAWF). Mark two troughs and a crest on your wave line, or two crests and a trough. Measure the distance between two troughs or two crests next to each other. This is called the **wavelength.** Measure the distance of the up-and-down motion and divide it in half. This is called the **wave amplitude.**

Make smaller and larger waves, mark them, and measure their wavelengths and amplitudes in the same way.

One vibration, or one full wave crest to crest, is movement of hand up and down; up and down counts for just one vibration, not two. Students can easily count movements of hand against 5 seconds on watch, but counting actual passing crests or troughs of chain wave is nearly impossible.

Now vibrate the end of the chain up and down rapidly, keeping the rate the same. Your hand makes one vibration, or one full wave crest to crest, in moving up and down. Count how many vibrations you can make with your hand in 5 seconds, watching a second hand on the watch. The number of waves you make per second is called the **frequency** of the waves. Counting the full vibrations and dividing by 5 gives you the frequency of the waves produced.

Figure 2 / 17
Mark and label your model waves.

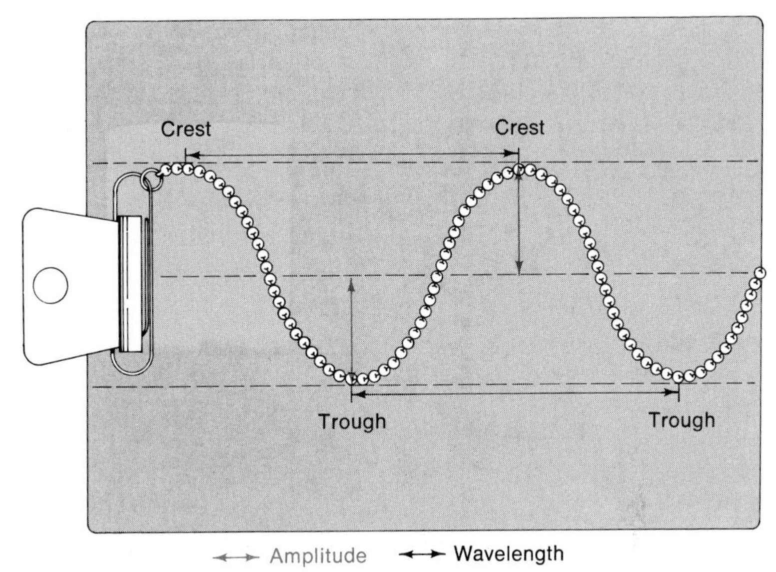

Students here derive wave speed, or velocity, on basis of the equation: Velocity(V) = Frequency(F) multiplied by Wavelength(W). Wave velocities (V) range from 10 to 30 inches per second.

After 5 seconds counting the frequency, stop moving the chain. Measure the wavelength of the waves this frequency produced. Multiply the frequency by the wavelength. This gives you the speed of motion of your waves, in inches per second. How fast were your waves moving? ★

All these things you have measured in the waves you made are the things astronomers measure in electromagnetic radiations from stars.

2.6 Many kinds of radiations come in from space.

Demonstration of distance of nearest star, Alpha Centauri, actually a multiple star system, on chalkboard or overhead projector exemplifies how distant things in space are: 186,000 miles per second, times 60 = miles per minute, times 60 = miles per hour, times 24 = miles per day, times 365 = miles per year or 1 light-year, times 4.3 = distance to Alpha Centauri. Light from sun reaches earth in about 8.3 minutes; 93,000,000/186,300 = 499 seconds or 8.3 minutes.
Concept: energy in motion

Think of energy radiations from the stars and galaxies as waves traveling through space. They are a lot like the waves you made along the chain. These electromagnetic waves really zoom through space. Remember that the speed of light is 186,300 miles (300,000 kilometers) per second. This is also the speed of the other electromagnetic radiations in space, because light is just one form they take. Light comes the quarter of a million miles from the moon in about a second and a quarter. The sun is about 93 million miles away. How long does it take its light to reach us?

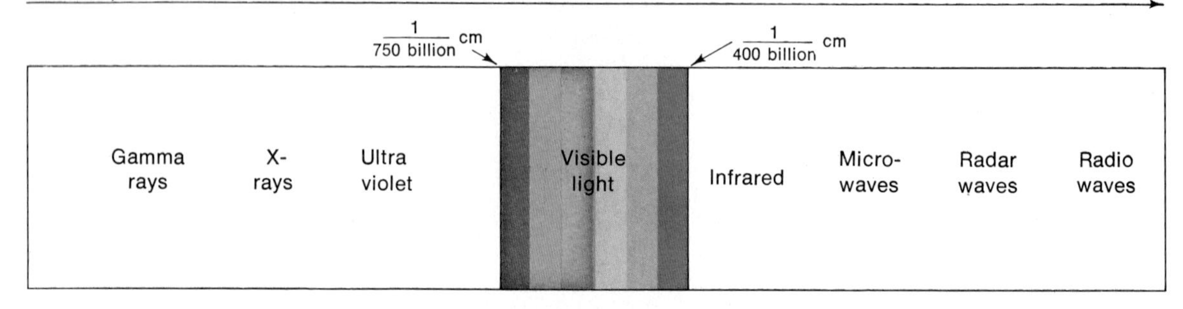

Figure 2/18 *The major kinds of waves, arranged by wavelength and frequency in the electromagnetic spectrum. All of these waves come in from space and give us information about the objects that made them.*

Various instruments astronomers now use for different portions of the electromagnetic spectrum make excellent subjects for student reports, since popular articles appear on all of these instruments.

Light waves, heat waves, and radio waves are only a few of the many kinds of radiations that come from stars and galaxies. These waves have different frequencies and wavelengths. Each has its own spectrum, like the light spectrum from the stars that you read about in Section 1.7. All these spectrums, when put together in a row, are called the **electromagnetic spectrum**, shown in Figure 2/18. Note that the light we can see comes in different wavelengths. This makes the different colors. You can see that red light has longer wavelengths than blue light.

Did You Get the Point?

The vibrating atoms in stars and galaxies radiate electromagnetic energy, a combination of electrical and magnetic energy.

Electromagnetic radiations may take such forms as light, heat, radio waves, and X-rays.

Wavelength of radiations is the distance from trough to next trough, or crest to next crest.

Frequency of radiations is the number of troughs or crests per second.

Electromagnetic radiations are arranged in the electromagnetic spectrum by wavelength and frequency.

Answers:
1. Distance between cars (wavelength) is short when frequency of cars passing house is high; as distance between cars (crests) increases, frequency of cars decreases.
2. An answer to your question would take a minimum of 20 years, 10 years for your question to reach the planet and 10 years for answer to return. Blue and red light, radio waves, and X-rays would carry messages at same rate, since all electromagnetic radiations travel at the speed of light. The longer radiations, red light and radio waves, would have the best chance of passing through any atmosphere the planet might have.

Check Yourself

1. Let's suppose that cars whiz past your house at about the same speed. When the frequency of cars passing your house is high (lots of cars per minute), what must be true of the distance between the cars (wavelength)? As the distance between the cars increases, what happens to the number of cars passing your house (frequency)?

2. Suppose you suspected that a star 10 light-years distant has a planet inhabited by intelligent beings and you wanted to send a message to them. How long would it take to ask them a question and to get their answer? Would blue flashing light carry your message faster than red flashing light? Would a radio signal carry it faster than an X-ray signal? What kind of wave would be likely to carry your message best?

Exploring with Radiations

Nearest of the galaxies in local group are the Large and Small Magellanic Clouds, at distances of 160,000 and 180,000 light-years, but visible in the southern hemisphere as blurs to the naked eye. Until the Maffei possible galaxies were discovered, the Triangulum galaxy in Scorpius was the most distant member of the local group at about 2.3 million light-years. Answer: Light travels at about 6 trillion miles per year (6×10^{12}) times 3 million years (3×10^{6}) equals 18 followed by 18 zeroes (18×10^{18}).

Concept: levels of organization

Students at college and graduate school levels often make contributions to science, as with the discovery of Maffei I and II. Amateurs can do considerable work in astronomy, such as observing meteors, timing periods of variable stars, and hunting for new comets and asteroids (minor planets).

2.7 Galaxies get together in groups.

Stars are grouped into galaxies. But how are galaxies related to each other? Astronomers have found that galaxies also form groups or clusters, as in Figure 2/7. At least 17 nearby galaxies have been found in what is called our **local group,** or **cluster,** of galaxies. The Milky Way, the Magellanic (madge-ell-ANN-ik) Clouds, and Andromeda belong to this club. The Magellanic Clouds are twin galaxies visible in the southern hemisphere. These and other local group galaxies are all around us in a blob of space about 3 million light-years in diameter. How many miles is this? The galaxies in our local group may be spinning, too, just like the Milky Way.

An Italian astronomy student recently found two faint hazy patches on photographs he was making of a part of the Milky Way. His photographs were made in infrared light,

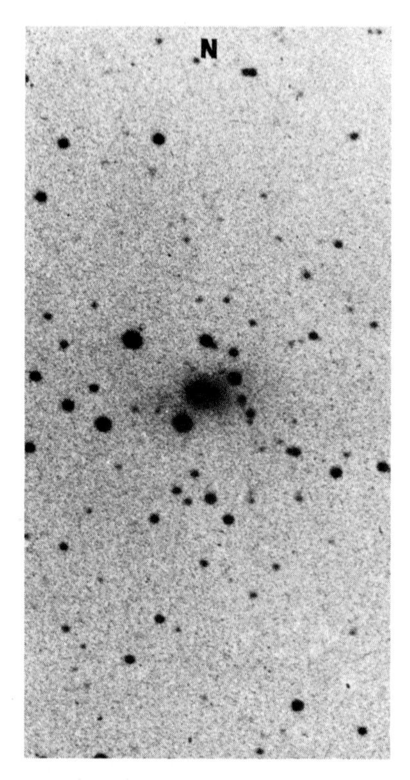

Figure 2 / 19
This great elliptical galaxy, Maffei 1, may be a member of our local group of galaxies. The film was exposed to light for three hours and printed as a negative; what is black in the negative is really white.

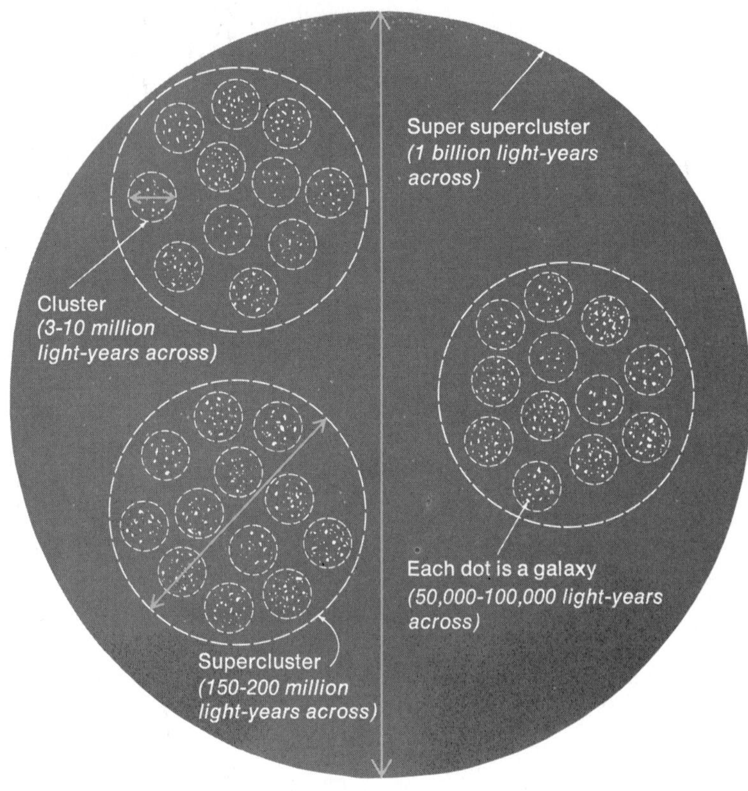

Figure 2 / 20 *If galaxy clusters form superclusters and superclusters form super superclusters, as some astronomers believe, this may be a polka-dot universe.*

which has the longest wavelength of any light. American astronomers took more pictures of these patches and decided that at least one of them is a galaxy in our local group.

The largest patch, named Maffei (mah-FAY-ee) after the Italian student, is probably a giant elliptical galaxy, shaped something like a squashed ball. It is shown in Figure 2/19. Only its long infrared rays can get through the dust of our Milky Way, which is in front of it. It's thought to be about 3 million light-years away, near the edge of our local group. It may be nearly as large as the Milky Way or Andromeda, and about 200 billion times more massive than the sun.

Astronomers are seriously tackling the questions whether our local group is rotating, how it may be moving in relation to other clusters, and whether superclusters and larger clusters actually exist.

Do local groups of galaxies get together in larger groups called **superclusters**? Some astronomers think so. A sketch of this idea is shown in Figure 2/20. You can see why this has been called a "polka-dot" model.

The center of our own supercluster may be in the direction of the constellation Virgo as we see it in the sky. (See sky map in Appendix C.) A supercluster is really gigantic—it may be 150 to 200 million light-years across. Some astronomers are even thinking about super-superclusters with diameters up to 1 billion light-years!

Everything that exists has been called the **universe.** It will be a long time before astronomers know whether the universe follows a polka-dot model, some other pattern, or no general pattern at all.

Activity Time: 15 minutes. Individually or in groups of 2 or 3 students. Activity can be done, if convenient, with that in Section 2.5 or as a demonstration of redshift using the table of an overhead projector.
Concept: energy in motion

2.8 What is the redshift?

All the radiations from stars zoom away from them through space at the speed of light. The stars, as you know, move in their galaxies. Our sun, for example, travels around the center of the Milky Way. What happens to the radiations when stars or galaxies are moving? Does motion of the radiation source change the radiation in any way?

Materials

paper pad, spring clip, paper clip, ball chain and eyelet, ruler, watch or clock, pencil

★ You can answer this question yourself with the chain model of radiation. Set it up as you did in Section 2.5. Start moving the chain with your hand, the source of the radiation, at about the same rate as you did before. Let the chain stop on the pad and measure the wavelength.

The next time, again move the chain at the same rate as before. Then as you keep moving it, slowly move your hand two or three inches away from the edge of the pad along the table, and stop. Measure the wavelength of the waves. What has happened to the wavelength?

The wavelength lengthens, when hand (source of radiation) is moved away from pad. Wavelength shortens as hand moves in over pad. The cause of this Doppler effect, the shift in wavelength, is that the source of vibrations is in motion in relation to the observer's frame of reference. The source (hand) is in motion in this activity, but the effect is produced by the pulling out of the chain or letting it move in, which is not analogous to the cause of the Doppler effect.
Process: measuring

Again, as you move your hand up and down at the same rate, bring your hand *in* two inches from the edge of the pad onto the pad, and stop. What happens to the wavelength this time?

Would you say these next two statements are true? When the radiation source is going away from the direction of radiation, the wavelength is lengthened. When the source is moving in the same direction as the radiation, the wavelength is shortened. ★

Both statements true. Students obtain effects like the Doppler effect, giving the redshift of light and other electromagnetic radiations. The redshift implies the expansion of the universe discussed in Section 2.10. You can use familiar example of sound waves rising to higher pitch (shorter wavelength) as a sound source like a car horn approaches and dropping to a lower pitch (longer wavelength) as the source passes and moves away.

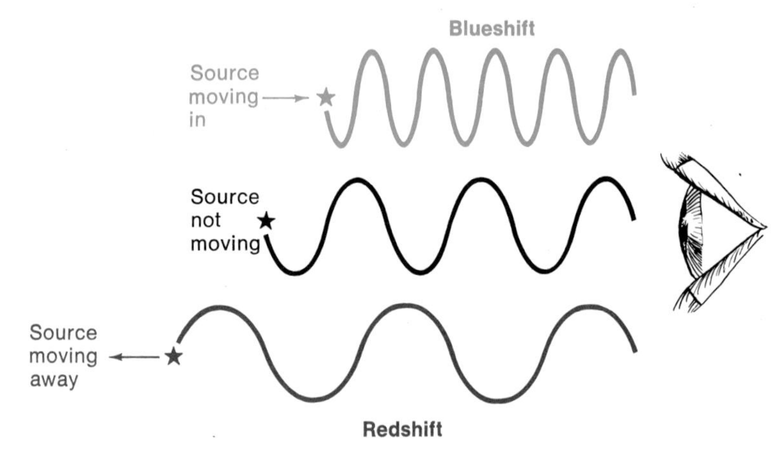

Figure 2/21 *The motion of light sources toward or away from the observer shifts the wavelengths of the radiations to the blue or red end of the spectrum.*

The redshift is the only way at present to estimate distances of faraway galaxies. Greater shift toward red end of spectrum implies greater receding velocity, increasing with distance; so large redshift implies great distance. Background material on the use of parallax for measuring relatively short distances to nearby galaxies is contained in the Teacher's Manual.

When a source of radiation is moving, the motion stretches out (lengthens) or pushes together (shortens) the waves of radiation, changing their wavelengths. These changes are shown in Figure 2/21. In this way, the wavelengths from light radiation sources moving toward us become shorter (bluer) and longer (redder) from sources moving away from us. The lengthening of wavelengths from sources moving away from us is called the **redshift.** This redshift appears in the spectrums from the radiations of moving stars or galaxies.

Suppose a source is radiating sound waves. When the source moves toward us, the sound waves are shortened and they sound higher in pitch. When the source moves away from us, the waves are lengthened and the sound is lower in pitch. You can hear the sound of a train's horn become higher as the train approaches and lower as the train moves away.

In the experiment with the ball chain, you did not move your hand away from or toward the pad at different speeds. But if you had, you would have found that the shift in wavelength was greater when your hand was moving faster. So the greater the redshift in a radiation spectrum from a star or galaxy, the faster it is moving away from us.

2.9 What kind of a universe is this?

What do the redshifts in spectrums from galaxies tell us? They say that galaxies are moving away from us and that the galaxies farthest away probably are moving fastest. So the redshift is a way to measure great distances in the universe. Some scientists question the value of this measure, because they think motion may not be the only cause of such redshifts.

Materials

balloon, felt-tip pen

★ Why do these faraway galaxies move faster and faster away from us the farther away they are? One explanation is that the whole universe is expanding, like a balloon that is being blown up. Take a balloon and mark spots at equal distances on it. Blow up the balloon. As the balloon grows bigger and bigger, do you see that the dots move faster and faster away from each other? This is what astronomers think the galaxies are doing. ★

If the astronomers are right, what made the universe start to expand like this? Perhaps, billions of years ago, the whole universe was one super-heavy ball of matter. Probably it was made up mostly of protons. (See Appendix G.) Did an enormous explosion blow this matter apart, making a huge, expanding fireball as shown in Figure 2/22? It doesn't sound very scientific, but astronomers call this the **big bang theory.** You were demonstrating the big bang as you blew up your balloon.

The big bang idea isn't as wild as it seems. Recently a telescope was sent high up into the sky in a balloon, to measure the microwave radiation from space. **Microwaves** are very close to heat (infrared) radiation in the electromagnetic spectrum, as you can see in Figure 2/18. The telescope recorded microwaves coming in *from all directions* in space. Scientists believe that microwaves may be everywhere in space and that they may be the little bit of heat left over from the big bang.

Spotted balloons expanding and contracting give a rough idea of the oscillating theory.

An addition to the big bang theory is the theory that the universe may expand to a certain point. Then it contracts and falls back to being one ball of matter again. After this, it explodes and expands again. This is called the **oscillating** (oss-ill-late-ing) **theory.** You can see this happening if you blow up your balloon full, let the air out, and then blow it up again.

Figure 2 / 22
Three theories of the way the universe works. With the evidence we have now, no one of these ideas is more than a good guess.

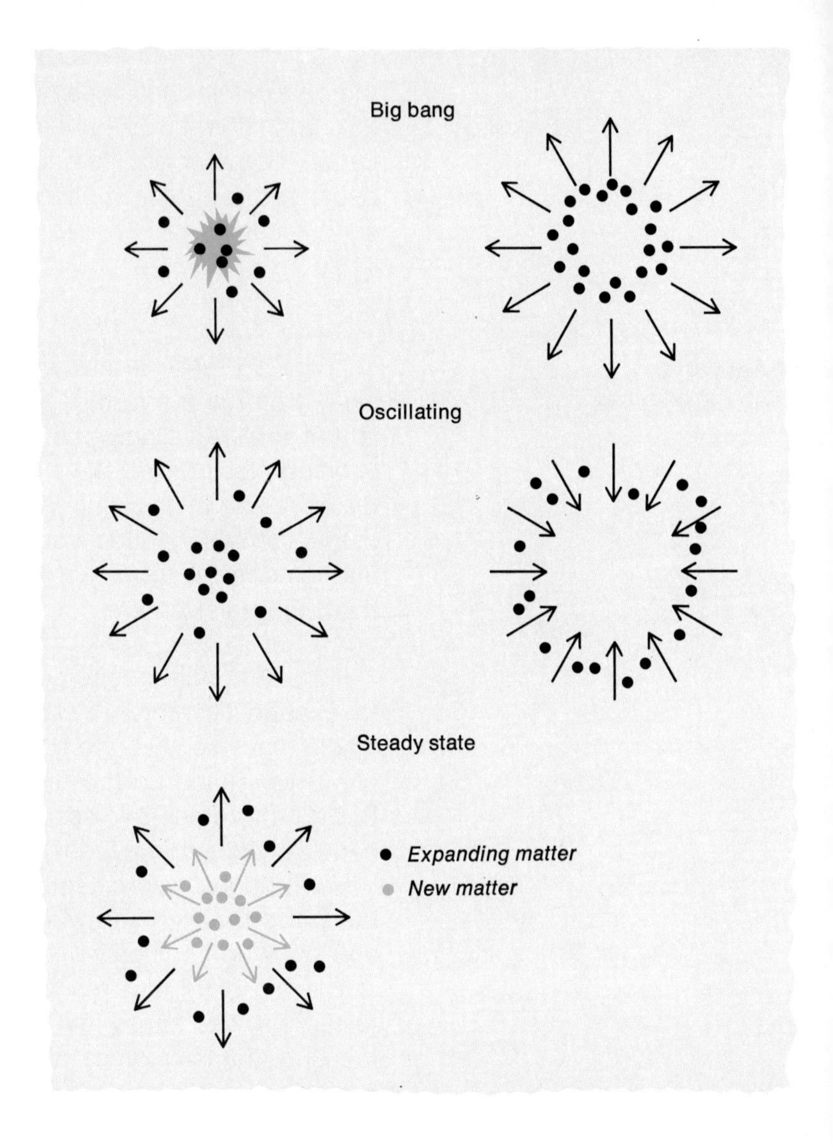

There's another entirely different explanation for the expanding universe. This theory is that the universe has always been the way it is now, so it's called the **steady state theory.**

If everything is the same as it always was, how can the steady state theory explain the distant galaxies speeding faster and faster away from us? And how can a steady state universe be expanding?

Steady-staters believe that these faraway galaxies may eventually go so fast that their speed is equal to the speed of light. Then all their radiations would be moving away from us as fast as they are moving toward us. The faraway galaxies would just disappear! Meanwhile, say the steady-staters, protons are popping into existence throughout the universe. Gradually these protons form new galaxies to replace the ones that vanished. This keeps the universe in a steady state.

The "polka-dot" universe of super and super-superclusters of galaxies (Section 2.7) could fit under any of these three theories of the universe. The big bang theory is most widely accepted at present, but new discoveries like quasars have opened up astronomy, and their effects on theories of the universe can hardly be predicted.

No one has proved that any one of these ideas about the origin of the universe is the right one. But each year brings discoveries of things never dreamed of. So perhaps some brand new theory, not yet imagined, will turn out to be better than any thought up so far.

Did You Get the Point?

Just as stars form galaxies, galaxies form clusters and perhaps superclusters in larger and larger groups.

The motion of a source of radiation changes the wavelength of the radiation. When the source is moving away, the wavelength is lengthened toward the red end of the spectrum.

The increasing redshift of things more and more distant from us suggests that the universe may be expanding.

The big bang, oscillating, and steady state theories try to explain the origin and development of the whole universe, but this remains a great mystery.

Check Yourself

1. Make a drawing or diagram that shows where you think the earth is located in the universe.

2. Tell what happens to the sound waves of a car horn when the car is coming toward you. Compare this with the light waves from a star moving toward us and another star moving away from us.

Answers:

1. Drawing can show sun and planets, Milky Way Galaxy, local group, and a supercluster or rest of the universe.
2. Sound is high in pitch (short wavelengths) as car approaches, then becomes low in pitch (long wavelengths) as car passes and moves away. Light waves from star moving toward us are shifted toward short wavelengths (blue), and waves from star moving away from us are shifted toward long wavelengths (red).

What's Next?

Stars and galaxies give off energy in electromagnetic radiations around them. These radiations form the whole electromagnetic spectrum. From observations of these radiations, astronomers have found our place in the universe.

We know our general location in the Milky Way. But where exactly is our earth in space, and where is the earth among the planets? We turn now to our exact location in space and time.

Skullduggery

Directions: The answer to each of the questions is given directly above the question, but you have to unscramble it. Read the question *before* trying to unscramble the answer. Make a list of the unscrambled words in your notebook.

Answers to scrambled words: (1) electromagnetic; (2) spectrum; (3) trough, crest; (4) frequency; (5) wavelength; (6) amplitude; (7) redshift; (8) light-years; (9) nebula; (10) galaxy; (11) cluster; (12) supercluster; (13) big bang; (14) quasar.

1. atteeegnmorilcc
What kind of energy makes the radiations in space?

2. cmuptesr
What includes all the different kinds of radiations in space?

3. gutohr trcse
Waves have a ______ and a ______.

4. qyfeenruc
______ is the number of waves passing a point per second.

5. thewevngal
This becomes longer when sources of radiation are moving away.

6. dipetmaul
Half the distance from trough to crest of a wave is its ______.

7. erdfisht
The greater their ______, the more distant the galaxies.

8. gilth-ersay
Light travels just five of these in five years.

9. ubnael
One of these is found in the sky in the constellation of Orion.

10. yaxlag
Astronomers finally discovered that a ______ contains many stars.

11. rtcslue
The Milky Way galaxy is a member of a ______.

12. retsulcrepus
Our local cluster may be a part of a ______.

13. gbi gbna
The ______ ______ theory starts the universe off like the Fourth of July.

14. rqaaus
A ______ is a highly energetic source of radio waves.

For Further Reading

Bova, Ben. *In Quest of Quasars.* New York, Crowell-Collier Co., 1970. Here you can read about these superenergetic mystery stars and also about stars in general.

Branley, Franklyn M. and Helmut Wimmer. *The Milky Way: Galaxy Number One.* New York, Thomas Y. Crowell Co., 1969. What's ahead for our galaxy? How do astronomers measure distances? What have we found out through radio astronomy?

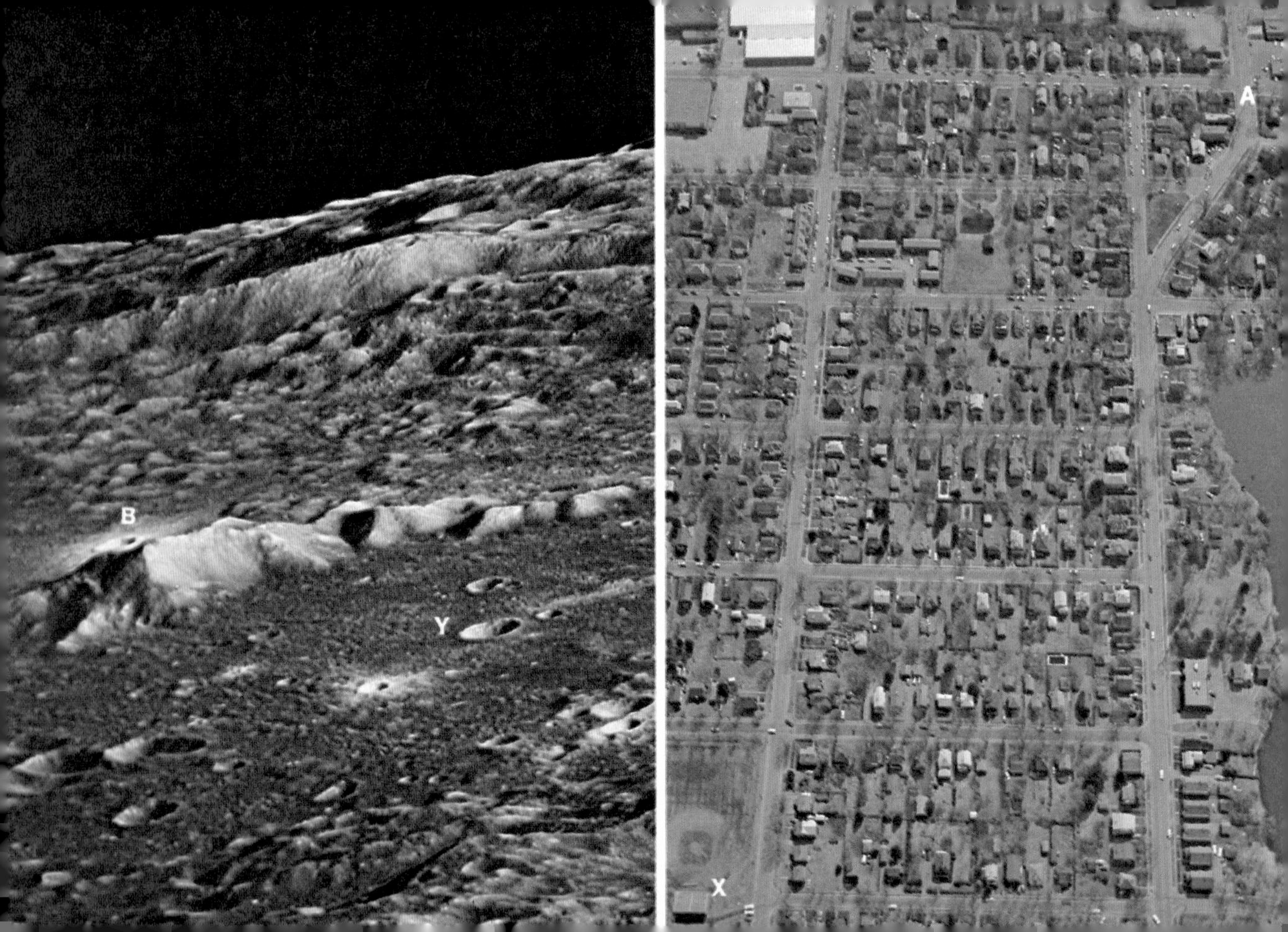
B
Y
A
X

Chapter 3 Place and Time

This chapter deals with locating places or things on the earth and in the sky and with placing events in time. Activities stress measuring with reference points and lines.

Suppose you are walking along the street when a car pulls up beside you. The driver rolls down her window and asks, "Can you tell me where Ms. Ersa Mynor lives?"

Let's say Ms. Mynor is on your paper route, so you know where her house is. But how do you tell the lady?

"Well, she lives up top of the hill in a white house." You wave up the hill.

"But I don't know her address," the driver protests.

Reference point is place at which directions are given; then directions follow reference lines of streets.

"Oh. You go up this street the way you're headed. Then you take the second street to the right, drive up it . . . let's see . . . three streets, and go left on the next. She lives in the fifth house, I guess it is, from the corner on the right. Big white house to the north."

"Thanks so much. Um . . . turn right, then left, then . . . white house." Up shoots the window. The car straightens out and streaks up the hill right past the second street right and into the next one—which is a dead end. Oh, well, you tried to tell her how to get there, anyway.

Students give directions similar to those of story, of how to go from reference place or point A to point X along streets.

Giving directions isn't always easy. Especially if you have no map. How would you tell someone at **A** in the photograph how to get to **X**? What would be the best directions you could give him?

Reference point for astronauts on moon is the sun; so tell them to go directly toward the sun. Then, they turn 90° right, then go away from the sun. Shadows on the moon made by the sun also give astronauts reference lines to follow.

Here's one that's a little tougher. How would you tell astronauts at **B** on the moon, where they have landed in their spacecraft, how to get to the crater at **Y**? The crater is there, but they can't see it. What directions would you give them to get from **B** to **Y**?

Often we have maps of places to show us where things are and how to get from one place to another. But how do we make the maps?

Locating Places on Earth

3.1 How are places located?

The lady in the car got directions to Ms. Mynor's house from the place where her car was. You tell someone how to get to X from a point A by going along certain streets. You tell the astronauts how they can get to Y from B by going away from the position of the sun, or along the line made by their shadows from the sun. Notice how places, or points, or lines are used in giving directions to other places.

Activity Time: 10 minutes. Asking students to locate their desks (or other places) in the classroom can lead to definition, use, and discussion of reference points and lines.

Process: measuring

How do you locate the place of any object? Try describing where your desk is. You may say it's behind Dick's desk. Or in the second row from the back of the room. Notice that you have to use something else to locate your desk. The something else you can use to locate a thing like your desk is called a **reference point.** A **reference line** is a line used in locating other places. They're called *reference* points and lines because you *refer* to them in locating other things.

Materials

ruler, pencil

Activity Time: 10 minutes. Measurements from reference lines of edge and top or bottom of page give exact location of letter A or of other letters on page, a two-dimensional surface. Exact measurements require a 90° angle of ruler from edges. One measurement is not enough; two are required. You might ask students if they can think of other ways to locate A. One way is by using a drawing compass from two corners of page, making arcs intersecting at A; another way is by measuring from side of page and two corners to A with a protractor and ruler. Drawing compass, protractor, and ruler are needed.

★ To locate places or things exactly, you need to make measurements to these reference points or lines. Using a ruler, can you locate the position of this letter **A** exactly on this page? Try it and see. Write down your answer. Could you tell anyone where the letter is, using only one measurement? Can you locate any place on the page with just two reference lines and measurements? Try it and see. ★

Horizontal measurement and location are made from vertical reference line, and vice versa. You can refer back to this operation, if students are puzzled in Section 3.4 by the measurement and location of latitude from horizontal parallels on earth and measurement and location of longitude from vertical meridians.

How could you locate the letter **A** exactly on the page? You had to measure how far the letter is from one of the side edges of the page (horizontal measurement) and from the bottom or top edge of the page (vertical measurement). The edges of the page are the reference lines to which you refer or relate the letter **A**. Notice that from the vertical edges you make the horizontal measurement. And from the horizontal edges of the top or bottom you make the vertical measurement. With such reference lines and measurements, you can locate places exactly.

3.2 Locate places on a ball like the earth.

More references are needed on a three-dimensional curved surface like earth than on a two-dimensional flat surface.

It's easy to locate places on a flat surface like this page with a couple of reference lines. But how could you locate a place on a curved surface like a ball—or like the earth? To answer this question, use a ball to stand for the earth.

Materials
ball, marker (grease pencil or felt-tip pen)

Activity Time: 20 to 25 minutes. Spinning the ball gives its poles of rotation, the basic reference points. A circle through poles gives vertical reference line (like Prime Meridian) and a circle halfway between poles gives horizontal reference line (like equator). Then positions of X's and Y's can be located roughly (or with measurements) in relation to these reference lines.

★ Mark an X on the ball to stand for the place where you are. Mark a Y around on the other side of it for the place you need to locate, as in Figure 3/1. How can you give directions how to get from X to Y? There aren't any streets (reference lines) here as there were on the photograph at the beginning of this chapter. Try any way you can think of and see if it works.

You might try spinning the ball on a flat surface. Or try turning the ball between your thumb and forefinger. Suppose you mark the points where you hold the ball. Can you use these as reference points to make some reference lines?

Without letting him look at the ball, try telling one of your classmates how to place the X and Y on the same locations on his ball. And have him try to do the same thing with you. ★

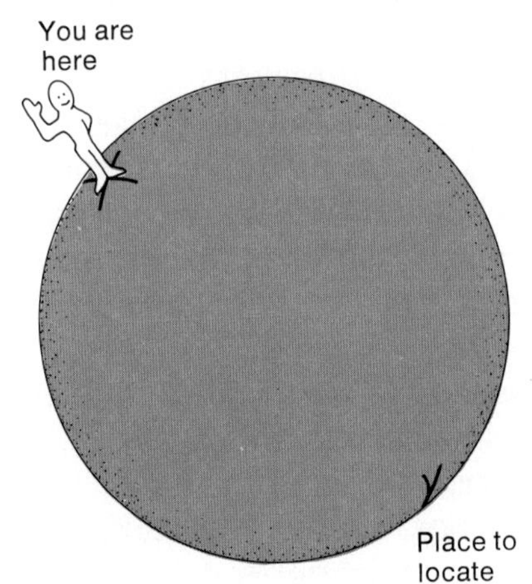

Figure 3/1
Imagine a friend is standing at X on the ball. How can you give the friend directions how to get to Y?

3.3 The earth is another spinning ball.

Look at the photographs of the earth in Figure 3/2. The time each was taken is given below the photographs. What do you notice about the earth?

Note the shapes and positions of some clouds in the first photograph. You can see that the clouds in the first photograph had moved by the time the second photograph was taken. They had moved more in the third photograph or perhaps had disappeared to the right. Photographs like these show that the earth is turning, if we know that the camera was not moving. It's true that the camera *was* moving slightly, since it was mounted on a Surveyor spacecraft that had landed on the moon. But the moon did not change position enough in a few hours to account for the changes you see.

Turning around a center, or turning on an axis, is called **rotation.** A phonograph record is rotating around its center on a turntable. The front or back wheels of a cart rotate

Figure 3/2 *The earth photographed from the moon at the given times. The N is above the North Pole.*

These Surveyor 7 photographs of the earth from the moon, as well as orbiting satellite photographs of the earth, give direct proof of its rotation with the motions of features from west to east. The moon moves only about 13° in 24 hours, as seen from earth. You can discuss or demonstrate here older proofs of the earth's rotation, such as Foucault pendulum, Coriolis effect (Teacher's Manual, Chapter 7), or earth's flattening, with its equatorial bulge (Section 5.3).

Process: observing

You can demonstrate why stars make tracks on photos with compass and ball. Point of compass in ball-earth stands for pole; compass-pencil for star. Hold compass (and pencil-star) steady and turn ball. Compass-pencil traces curved track on ball (as stars do on photo). Although pencil-star has not moved, ball-earth has turned.

Appendix I is a supplementary activity for school or homework on finding the north celestial pole, measuring the angle of earth's turning per hour, and calculating time for a complete rotation. Activity time: 20–25 minutes. Units of angular measurement are given in Appendix J. Activity familarizes students with, or gives them review of, angles and their measurement.

around a rod, or axis. A pencil or knitting needle through the center of a ball is an axis on which the ball can turn. The earth turns, too, just like the ball you spun between your fingers. The earth's axis is an imaginary line through its center called its **axis of rotation.** To locate places on the earth and make maps of it, you treat it like your ball.

How could you locate the earth's axis of rotation? What reference point could you use? How about the stars?

Look at the curved, bright star tracks in Figure 3/3. To take this photograph, the camera was aimed at Polaris (poh-LAR-iss), also called the Pole Star, or the North Star. The camera shutter was left open for several hours. As the earth turned on its axis, the stars seemed to move and made the star tracks on the photograph. The North Star made the shortest bright track near the center of the circles.

Note that the stars all seem to be moving in circles around a center. This center is the earth's imaginary axis of rotation extended up into the sky. It's so close to the North Star that for rough estimates of location, the North Star is used as the center.

This center in the sky is called the **north celestial** (sell-ESS-chill) **pole.** "Celestial" means heaven or sky. The places on the earth directly under the north and south celestial poles, where the axis goes through the surface of the earth, are called the **geographic poles.** Figure 3/4 shows how these poles are located by referring to the celestial poles in the centers of the star circles in the sky. As you see in Figure 3/4, looking down on the earth from above the North Pole, the earth turns counterclockwise (opposite to the hands of clocks) from west to east on its axis.

Figure 3/3
If the star tracks were complete circles, the earth's axis of rotation would pass through the point at the center. To take this picture, the camera shutter was left open for 12 hours.

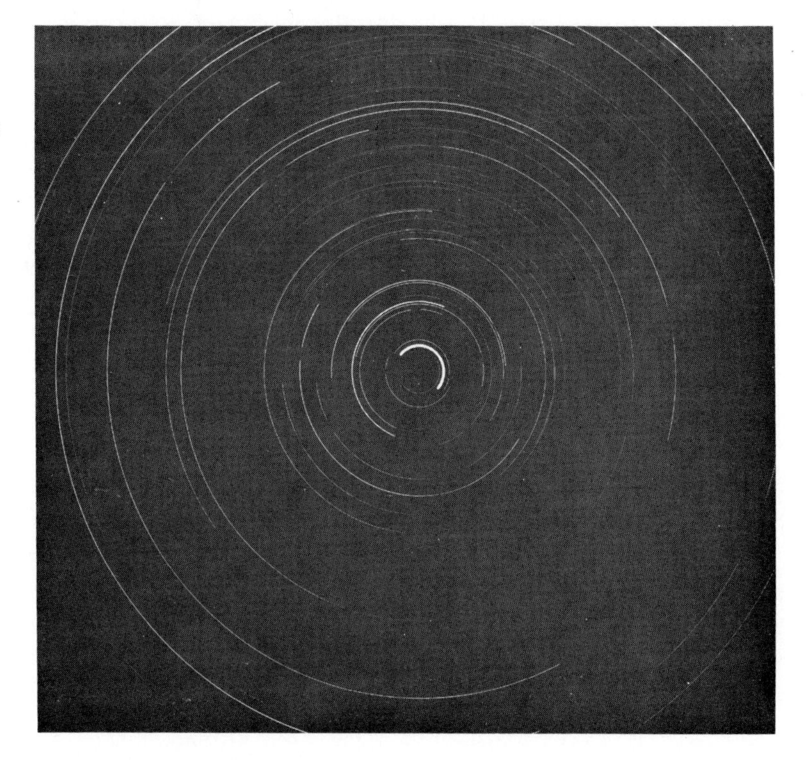

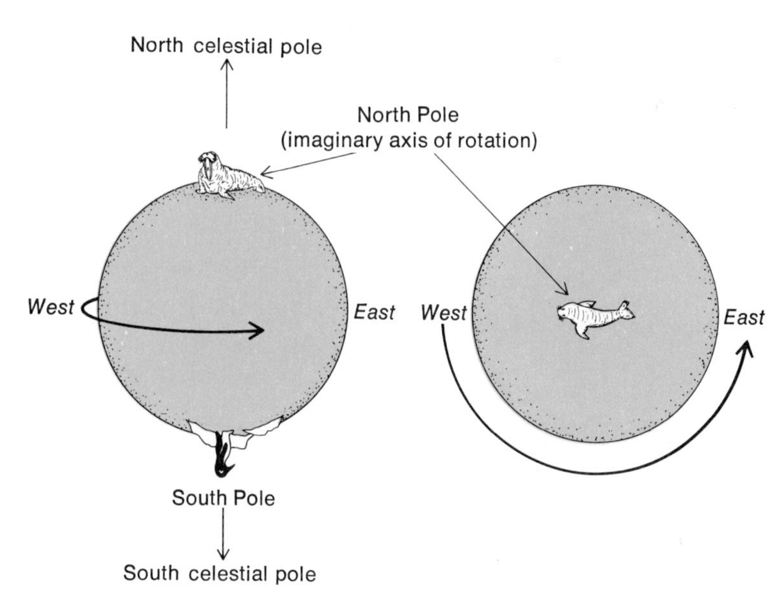

Figure 3/4 *The earth's North and South Poles are located from the celestial poles. Looking down at the North Pole you can see that the earth rotates counterclockwise around its axis.*

(Appendix I gives an activity in which you use Figure 3/3 to find the north celestial pole and calculate the time for one complete rotation of the earth.)

You start from the earth's North and South Poles to locate places and to measure distances between places on the earth. The geographic North and South Poles are the reference points used in locating *all* other places on earth. Without these reference points, like the places where you spun the ball, reference lines could not be made on the earth.

Sizable globe is handy equipment for this section. Overhead transparencies are available for reinforcement of geographic reference system, although this will be a refresher for some students.

Process: measuring

3.4 Latitude and longitude tell where you are. Remember when you located the letter A in Section 3.1 you needed two reference lines? One to measure horizontal distance and one to measure vertical distance. You used the edges of the page. Reference lines like the edges of the page have to be imagined on the earth, though you see these lines on maps and globes.

On the earth, the reference line like the top or bottom edge of the page, from which you made vertical measurements, is called the equator. The **equator** is the line around the earth halfway between the North and South Poles, as shown in Figure 3/5. Notice that the equator is at a 90-degree angle from each pole. (You can review angles and degrees in Appendix J.)

A **meridian** (mer-RID-ee-an) is a line running from pole to pole. The reference line like the up-and-down edges of the page, from which you made horizontal measurements, is called the Prime Meridian. Look at Figure 3/5. The **Prime Meridian** is the line running through both the poles and through Greenwich (GREN-itch), England. Greenwich was picked by scientists as the place from which these horizontal measurements would be made probably because there was an observatory at Greenwich. But the measurements could be made from any other place just as well.

A good place to discuss great circles, whose planes pass through the center of earth; all meridians are great circles. The only parallel that is a great circle is the equator. Orbits of satellites must always follow great circles, since they are determined by the force of gravity acting as if from earth's center.

The reference lines of the equator and the Prime Meridian tell us where north and south and east and west are on the earth. Up and down from the equator is **north** and **south. East** from any place is toward the Prime Meridian in the direction the earth spins. **West** is away from the Prime Meridian in the opposite direction.

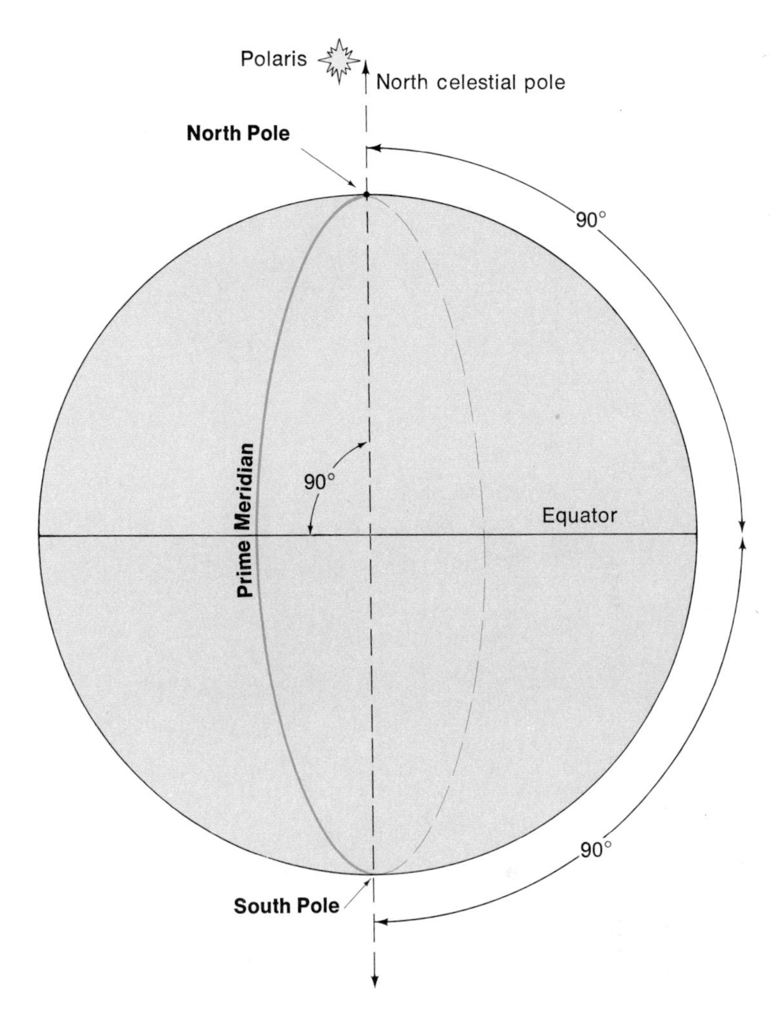

Figure 3/5 *The equator and Prime Meridian are the reference lines for locating places on the earth. Both lines are based on the reference points of the poles.*

Remember that you could locate the letter A or any other place on the page from the reference lines of two of the edges of the page. In the same way, you can locate any place on earth by using the reference lines of the equator and the Prime Meridian.

Maps and globes show circles going around the earth both north and south of the equator. These circles are called **parallels of latitude** as in Figure 3/6. Parallel lines are lines that always stay at the same distance from each

Figure 3/6
Parallels give the position of a place in degrees of latitude north or south of the equator.

Figure 3/7
Meridians give the position of a place in degrees of longitude east or west of the Prime Meridian.

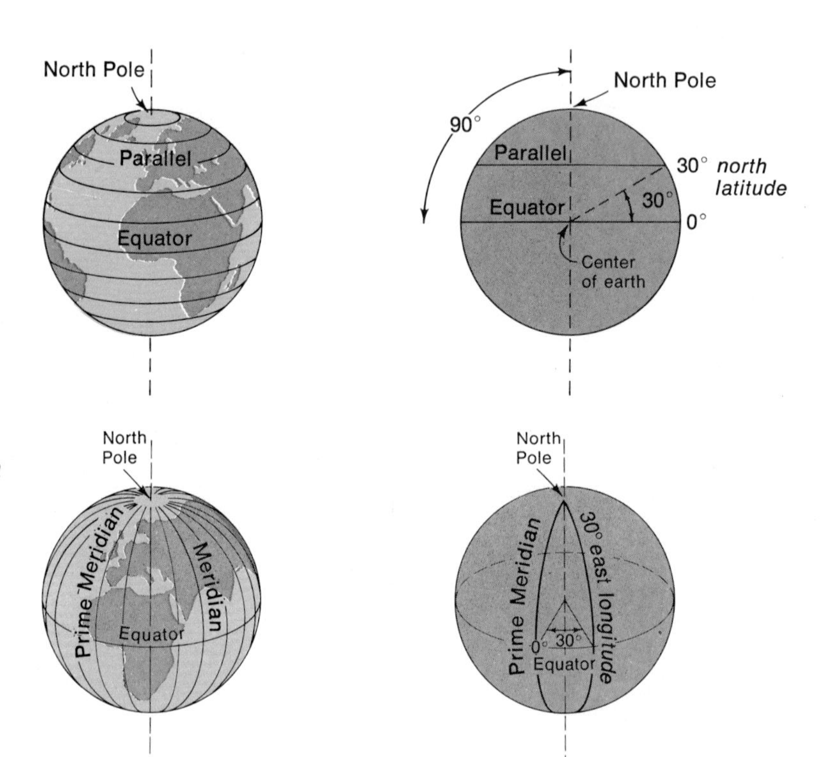

other. The parallels measure **latitude,** the distance in degrees north and south of the equator. With the equator at 0° latitude, the poles are at 90° north and south latitude.

The meridian lines enable you to measure **longitude,** the distance in degrees east and west of the Prime Meridian, shown in Figure 3/7. The meridians are at right angles (90°) to the equator.

The meridians east of the Prime Meridian (toward Asia) run to 180° east longitude. There in the Pacific Ocean they meet the meridians running west of the Prime Meridian (toward America). This meridian on the side of the earth opposite the Prime Meridian is called 180° east *and* west.

Places on earth are located by measuring the degrees of latitude north or south of the equator and their degrees of longitude east or west of the Prime Meridian. A degree of latitude measures about 69 miles (111 kilometers) on the surface of the earth. Why do degrees between meridians, the parallels measuring longitude, give shorter distances toward the poles?

Distances between degrees along parallels become shorter toward the poles because the lengths, or circumferences, of the circles of the parallels become shorter. If a globe or globes are handy, have students measure the distance between major meridians (15°) (A) at the equator, (B) at 30°, and (C) at 60° north or south latitude to demonstrate this situation and arrive at or support their explanations.

Materials

globe or map

★ How could you address a letter to a friend in Anchorage, Alaska, without giving the name of the city? Locate Anchorage on a globe or a map. Find its latitude (by degrees north of the equator) and its longitude (by degrees west of the Prime Meridian). Now you can write down your friend's address in degrees north latitude and degrees west longitude. All places on land or sea are located this way. Try locating other places on a globe or map by their latitudes and longitudes. ★

Anchorage, Alaska, is at 61° 13′ north latitude and 149° 54′ west longitude. If you ask for greater precision than degrees, students can work out minutes.

3.5 Make a space flight around the earth.

Activity Time: 25 to 30 minutes. Groups of 4 to 6 can use one globe together, taking turns reading latitudes and longitudes from globe, each making a table and plotting the flight path on his own map. Or you can work with the whole class, having students read locations and write table on chalkboard. Each student then makes his own map. **Process: observing**

Imagine that you are about to make a space flight around the earth in an Apollo spacecraft. You will blast off from the Space Center at Cape Kennedy, Florida. First you should plot your path on a map of the earth, so you will know what places you are going to pass over. Here's one way to make your own map.

Materials

globe of earth, rubber band, pencil, graph paper

★ Snap a large rubber band around a globe of the earth as shown in Figure 3/8. Adjust the band to pass through southern Florida, where you take off. Make the band slant down across the equator just north of South America. Then the band passes over the southern tip of South Africa, and so on, as shown in Figure 3/8. The rubber band crosses about 34 degrees south latitude in the Indian Ocean and about 28 degrees north latitude at Cape Kennedy, Florida.

Follow the rubber band around the earth from Cape Kennedy and record the positions of your flight in a table like that in Figure 3/9. The blastoff position and the first two flight positions are given in the table. Observe and record the latitude of the flight for each following 15 degrees of longitude around the earth.

The 15° longitude intervals for positions of the spacecraft are those used to mark the time meridians and time zones around the earth, as discussed in Section 3.8. Completed table is in the Teacher's Manual.

Then on a piece of graph paper draw latitude and longitude lines like those shown in Figure 3/10. This is your map. Use the latitude and longitude observations from your table to mark dots on the map. Join the dots with a smooth line. This line is your flight path around the earth on a flat map. If you wish, sketch in on this map the continents you will pass over or near in your flight. ★

An outline map of the continents can be used for this activity.

Figure 3/8 *Make the path of your space flight with a rubber band around a globe.*

Figure 3/9
Use a table like this to record your latitude at every 15 degrees of longitude on your space flight.

Positions of Spacecraft in Orbit

Position	Longitude	Latitude
1	81°W	28°N
2	75°W	26°N
3	60°W	23°N
4		

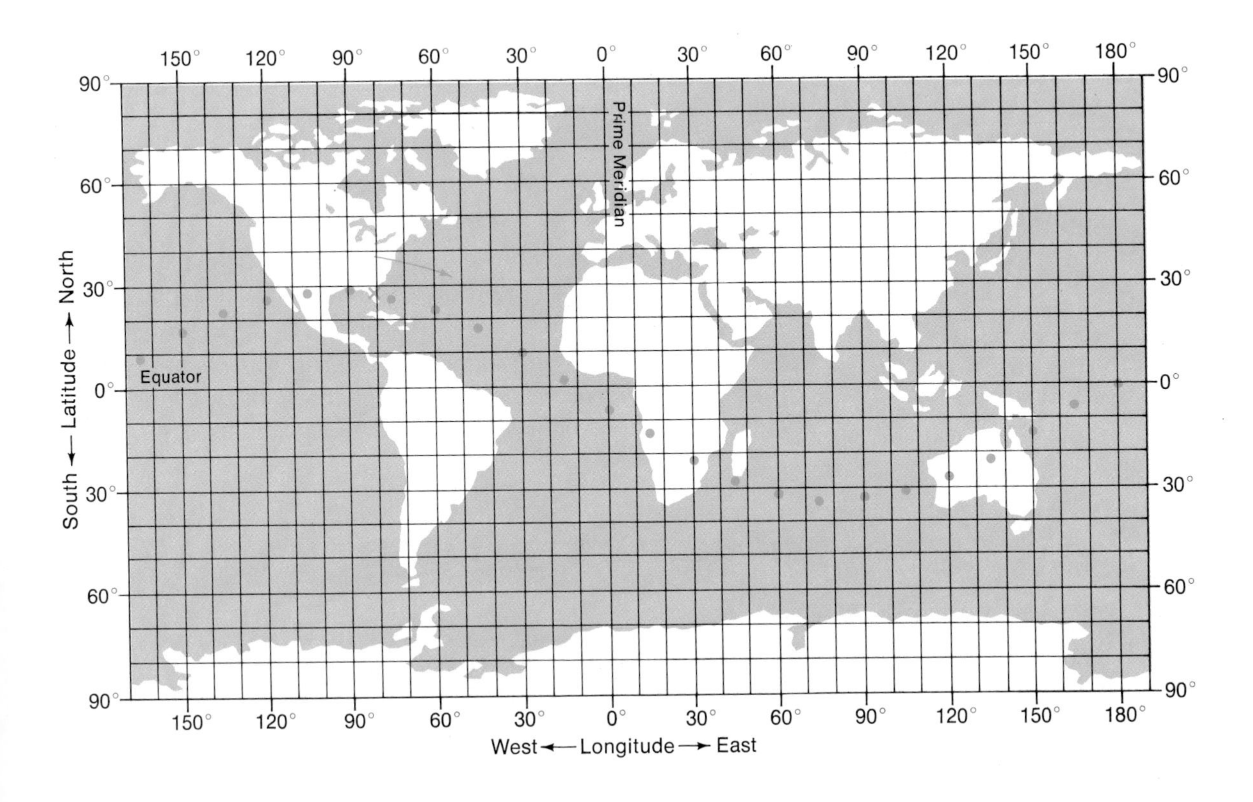

Figure 3 / 10 *Label a piece of graph paper like this to map your space flight around the earth.*

Did You Get the Point?

Movements of earth features in a series of photographs made from about the same position in space show that the earth is rotating.

The center of star tracks around the North Star is the position of the north celestial pole. The earth's geographic North Pole is directly under the north celestial pole.

The reference points of the North and South Poles and the reference lines of the equator and the Prime Meridian are used to locate places on earth.

The position of any place on earth is given by its latitude and longitude, measured from these reference lines.

Answers:

1. On flat surface, locate positions with 2 reference lines, one sideways (horizontal) and one up and down (vertical to other). On ball like earth, you need same reference system as earth's, points of axis of rotation and lines (sideways–parallels, up and down–meridians).
2. Something inside box, or head in room, needs 3 reference lines to locate it, since it is not rotating: (A) along one side of box or room, (B) along the other side next to it, and (C) vertical in box or room, or up a corner between sides.
3. Answer obtained from rubber band around globe. You might also demonstrate equatorial synchronous and polar orbits.
4. Students use globes or atlas to answer.

Check Yourself

1. What reference points or lines do you need to locate the position of an object on a flat surface? Its position on the surface of a ball?

2. What reference points or lines do you need to locate the position of something inside a box? Of your head in the room?

3. What other parts of the earth will you see if you make a round-trip space flight starting from Cape Kennedy, Florida, and passing directly over the state in which you live?

4. Find the latitude and longitude of the place where you live. Then find some other places around the earth with the same latitude and some other places with the same longitude.

Measuring Time

Activity Time: 25 minutes. Groups of 4 to 6 students. Each group needs a watch with a second hand. The classroom clock will do if it has a second hand in a position in which all timekeepers can see it. If a clock in the classroom "clicks" every minute, play a radio to prevent the clicks from being heard.

Materials

watch or clock with second hand, pencil, paper

Activity starts students thinking about time, shows need for timepieces to measure time since time estimates are so rough, and gives practice in setting up tables and recording data. If students can work with percentages, have them compare estimates of several different time intervals by figuring: Percentage error = error (seconds)/period estimated (seconds) × 100.

Process: measuring

3.6 How well can you judge time?

You have seen how places are located by using the rotation of the earth, its poles, and latitude and longitude. Rotation and longitude are also used to tell what time it is around the earth. **Time** is when things happen and how long they last.

★ First, let's get better acquainted with time. You can see and touch objects to find their length and width, but you can't see or touch time. Without using a clock, how can you tell how much time has passed? How well can you judge, or estimate, time?

Form groups to test this out. Each group appoints one person to be the timekeeper. No one but the timekeeper should look at a watch or clock. The timekeeper will record everyone's time estimates, as shown in Figure 3/11.

When the timekeeper says "Start," each person waits for what he thinks is three minutes and then raises his hand. Everyone should cover his eyes, so that he won't be influenced by other time estimates. Each time a hand goes up, the timekeeper puts down the real time in the table next to that person's name.

After everyone has judged, the timekeeper tells each person how much time actually passed during his estimated three minutes. Each person then describes the way he or

Estimates of Time

Name	Trial Number	Time Period (seconds)	Time Estimate (seconds)	Error of Estimate (seconds)
Katie	1	180	200	+20
Juan	1	180	175	−5
Peter	1	180	150	−30
Jennifer	1	180	75	−105

Figure 3/11 *The timekeeper should record each person's time estimate in a table like this one.*

People use a variety of ways to judge time. Students may just guess time, count pulse beats, count seconds, judge each 5 or 10 seconds, and so on. Percentage errors are needed to compare estimates of different periods of time, although students can make rough comparisons, for example, by doubling the errors for estimating 3 minutes to compare with errors for estimating 6 minutes.

Group results can be put on chalkboard to show estimates for whole class, for discussion and answering questions. Usually time estimates improve with practice. Some people are much better than others at estimating time. Methods counting brief periods in some way, such as "one thousand and one" for a second, are normally the best, but some students may do better just "guessing." Counting brief periods of time is acting like a clock, but real clocks are needed because estimates are so variable, as students' results show.

she judged the time. Do you find that people in your group estimated the time in more than one way?

Now do the same thing again, with each person using the method he thinks best. Then estimate three minutes a third time, or try estimating one or two minutes, or even four or five.

Look at the whole table, now (Figure 3/11). How many improved their estimates? Got worse? Stayed about the same? Was counting seconds more precise than just letting time pass by?

Do time estimates improve with practice? Are some people better than others at estimating time? What methods are best? Do you see why clocks and watches are needed to measure time? ★

As everyone knows, length is measured by yardsticks or meter sticks. Time is measured by clocks or watches. Yardsticks and meter sticks measure feet and inches or meters and centimeters. Clocks and watches measure hours, minutes, and seconds. Some antique gadgets that people used to measure time are shown in Figure 3/12.

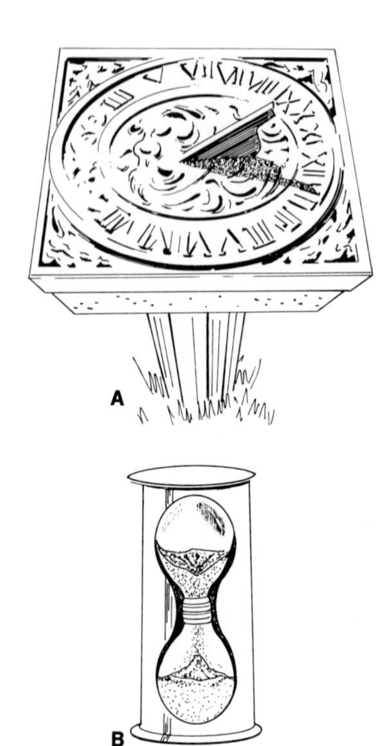

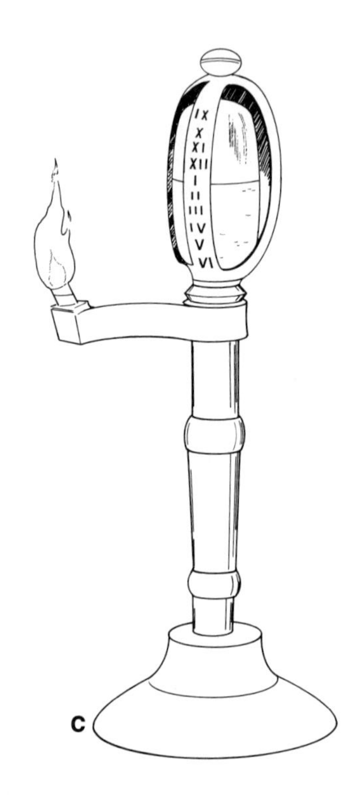

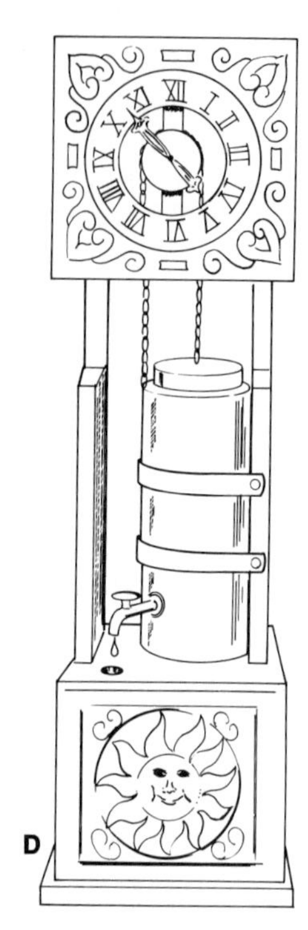

Figure 3/12 *Old ways of measuring time: (A) sun's shadow on a sundial, (B) sand falling in an hourglass, (C) falling oil level in an oil lamp, and (D) dropping water level in the tank of a geared clepsydra.*

Large globes or overhead transparencies can be used to discuss time zones with students. Globes and flashlights for students to do activity in groups demonstrate time zones best with rotation of earth. You can use Figure 3/7 to show meridians, Figure 3/10 as made up by students, and Figure 3/15 with time zones resulting.

3.7 Time is measured and set.

As the earth rotates counterclockwise from west to east in 24 hours, half of it is in sunlight and half in darkness. (See Figure 3/13.) So when it is daytime on the half of the earth facing the sun, it is nighttime on the other half.

Materials

globe of earth, flashlight

★ If you stand a globe on a table, darken the room a little, and shine a flashlight on the globe, you can see the daytime and nighttime halves of the earth clearly. And if you turn the globe, west to east, you can see how day and night move around the earth. (See Figure 3/14.)

Sunrise comes earlier in place east of place where you live or any other place. Sunset comes later in place west of another place.

See how the night goes and day comes, with sunrise, as the line between night and day moves around the turning globe. Does the sunrise come earlier or later in a place

Figure 3/13
Daytime and nighttime on the earth photographed from an Apollo command spacecraft orbiting the moon. In the foreground is a LEM about to go down to the moon. In which direction is the sun?

Figure 3/14
You can see how day and night move around the earth as it turns.

to the east of where you live? And as the globe turns, see how the sunset comes and night falls on the side of the globe opposite the sunrise. Does the sunset come earlier or later in a place to the west of the place where you live? Let's see how the clock times around the earth are different because of the earth's rotation on its axis.

The Prime Meridian or 0° (Figure 3/7) is the reference line for time as well as for longitude. The time at this meridian, which passes through Greenwich, England is called **Greenwich Mean Time,** or GMT for short.

When the sun in the sky is directly over the Prime Meridian, it is **12:00 noon,** GMT. Make it noon, GMT, with your globe-and-flashlight model, too. The earth rotates through a full circle of 360 degrees in 24 hours. If you divide 360 degrees by 24 hours, you will see that the earth rotates 15 degrees in one hour. This means that the sun *seems* to move 15 degrees through the sky in an hour. An hour after the sun has crossed the Prime Meridian, the noon position of the sun in the sky will have moved along to the meridian 15 degrees west of the Prime Meridian. It will then be noon at all the places north and south along that meridian. The meridians every 15 degrees east and west of the Prime Meridian are called **time meridians.** Figure 3/15 shows the time meridians.

With your globe at noon, GMT, notice where on the earth the sun is rising and where it is setting. Then turn the globe along 15 degrees from west to east, just as the earth turns in an hour. What time is it now at Greenwich, and all along the Prime Meridian? See how the sun has risen at all the places along a new meridian and set at all the places along the meridian opposite it.

The time is now 1:00 P.M., GMT.

On your globe what time is it now at the time meridian 15 degrees west of the Prime Meridian? The sun is directly above it. As the earth turns, the clock time must be changed to keep up with the position of the sun. So the time is now noon at this meridian. ★

The time is 12:00 noon in the time zone of this meridian.

3.8 Time zones adjust time to the sun.

A **time zone** is a band that includes about 7.5 degrees on either side of a time meridian. When it is noon at each time meridian, it is also called noon everywhere else in the time zone made by that meridian. Take the meridian at 30 degrees west of Greenwich, for example. It makes a time zone that includes from 22.5 degrees through 37.5 degrees west longitude, roughly. Time is the same all through this zone. There are 24 of these time zones around the earth, as shown in Figure 3/15. What time zone are you in?

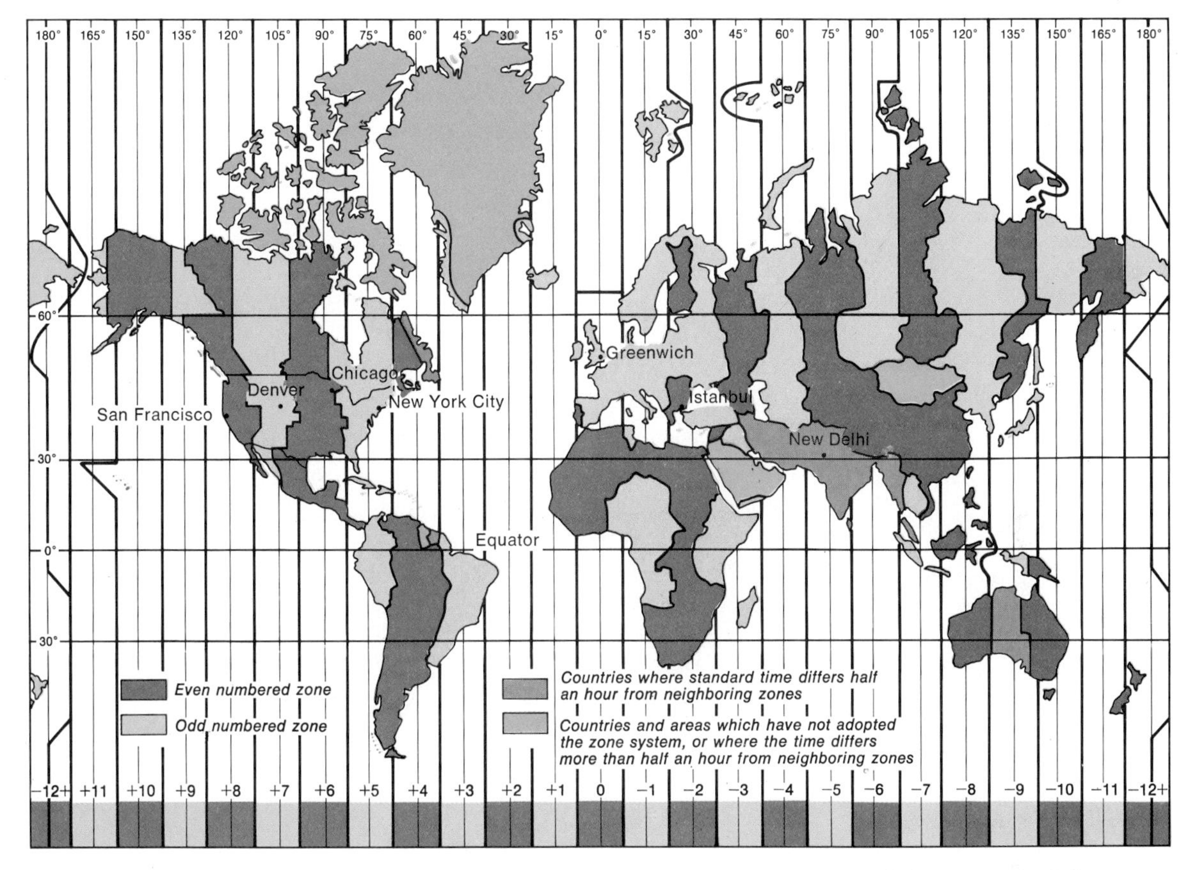

Figure 3/15 *The time meridians and time zones of the earth.*

International Date Line, more or less along the meridian at which east and west meet at 180° east and west longitude, may be discussed here. See Teacher's Manual for Chapter 3.

As the earth turns, the sun takes one hour to move from the crossing of one time meridian to the crossing of the next one. Suppose you could transport yourself instantly to the next time zone to the west. You'd find that the clocks there were one hour earlier than in the zone you came from. If you went one zone to the east, the clocks would read an hour later.

Time zones across the United States give Eastern, Central, Mountain, and Pacific times, based on the time meridians at 75, 90, 105, and 120 degrees west longitude. Hawaii and most of Alaska are based on the 150 degrees west time meridian, two hours later than Pacific time. How much later than Greenwich Time is each of these United States time zones?

United States' time zones are 5, 6, 7, 8, and 10 hours later than Greenwich Mean Time (GMT) respectively.

See Appendix K for an enrichment activity for school or home on local sun time and your longitude; activity is discussed in Teacher's Manual for Chapter 3.

If you can catch the sun out at noon, you can see how much the sun time at the place where you live differs from your time-zone time. You can also check your longitude with this activity. (See Appendix K.)

Did You Get the Point?

Time is when things happen and how long they last.

Time is measured by clocks in units of hours, minutes, and seconds.

Time on earth is measured from the position of the sun directly above the Prime Meridian, passing Greenwich, England, when it is noon, Greenwich Mean Time (GMT).

Time around the earth is set by time zones, each 15 degrees of longitude farther east and west of the Prime Meridian. Since the earth makes a complete rotation in 24 hours, the sun seems to move about 15 degrees in an hour in the sky. So the time from one zone to the next is changed by one hour.

Answers:

1. Daytime and nighttime are each 5 hours on Jupiter. Sun would appear to move 360°/10 hours, or 36° per hour, on Jupiter, as opposed to 15° per hour on earth.
2. Sun would come up earlier for A than for B. Time of sunset could vary up to nearly 60 minutes (1 hour), since this is the time taken for the sun to appear to travel the usual 15° across the time zone.
3. According to GMT from Greenwich, England (Prime Meridian), one must call Denver at 12:00 midnight to 1:00 A.M.; San Francisco at 11:00 P.M. the day before to 12:00 midnight; Chicago at 1:00 to 2:00 A.M.; New York City at 2:00 to 3:00 A.M.; Honolulu at 9:00 P.M. the day before to 10:00 P.M.; Istanbul at 4:00 to 5:00 A.M., and New Delhi at 1:00 to 2:00 A.M. (Actually Indian time is 5 hours, 30 minutes, ahead of GMT, but students can figure 5 hours from map.) Great Britain actually now goes by the time of the next zone to its east, or Continental Time, but question should be answered on the basis of GMT. Shift of Britain to Continental Time is based on desire to be on the same time as countries across the Channel.

Check Yourself

1. The great planet Jupiter makes one complete rotation on its axis in about 10 earth hours. How long is daytime on Jupiter? Nighttime? How many degrees would the sun seem to move in an hour in Jupiter's sky?

2. Suppose one person A lived near the east edge of a time zone and another person B lived near the west edge of the same time zone. Would the sun come up for person A earlier or later than for B? How many minutes could their time of sunset differ?

3. Imagine you are on a trip around the world and have reached Greenwich, England. According to GMT, when would you have to call friends in Denver, San Francisco, Chicago, New York City, and Honolulu in order to catch them between 7:00 and 8:00 A.M., before they had left for school? When could you call ahead to make reservations in Istanbul, Turkey, and New Delhi, India, to catch hotel clerks between 6:00 and 7:00 A.M., when they would not be busy?

Locating Sky Objects

3.9 Measure the altitude of Polaris.

When you look up at the sky, doesn't it seem huge compared with the small part of the earth you can see around you? How can the location of things be measured in such great stretches of sky? Are there reference points and lines for the sky like those for the earth? Again, the north celestial pole next to the North Star, Polaris, is the best reference point for locating things in the sky, just as it was for locating places on earth.

Materials
astrolabe, map of local area, or atlas

Activity Time: about 25 minutes indoors, longer if students do it at home or on a star party. In classroom or lab, indicate a line behind a table at one end of the room, or line on floor at which several students can work at a time. Students sight with astrolabes on an object standing for Polaris on opposite wall or ceiling at a height that will give an angle of Polaris equal to your latitude. Concepts in Sections 3.9 and 3.10 are essential for locating places or things in the sky, just as earth's poles and equator are essential for locating places on earth. But these concepts are not emphasized or crucial in following chapters; so these sections can be skipped or read but not learned.
Process: measuring

Methods for constructing (A) a simple or (B) a more complex astrolabe are given in Appendix L, or you can use a ready-made simple astrolabe (A), which only measures altitude above horizon. This is adequate for activity here, but a complex astrolabe (B) is needed for Appendix M activity with horizon system and any activities involving the equatorial system you may wish to add.

★ First, find Polaris in the sky, or among model stars put up in your classroom. Then measure Polaris's height in degrees with an astrolabe (AS-troh-layb), as the student is doing in Figure 3/16. An **astrolabe** is an instrument that measures distances in the sky by their angles. Two kinds of astrolabes that you can use (or make) are shown in Appendix L with the directions for using them.

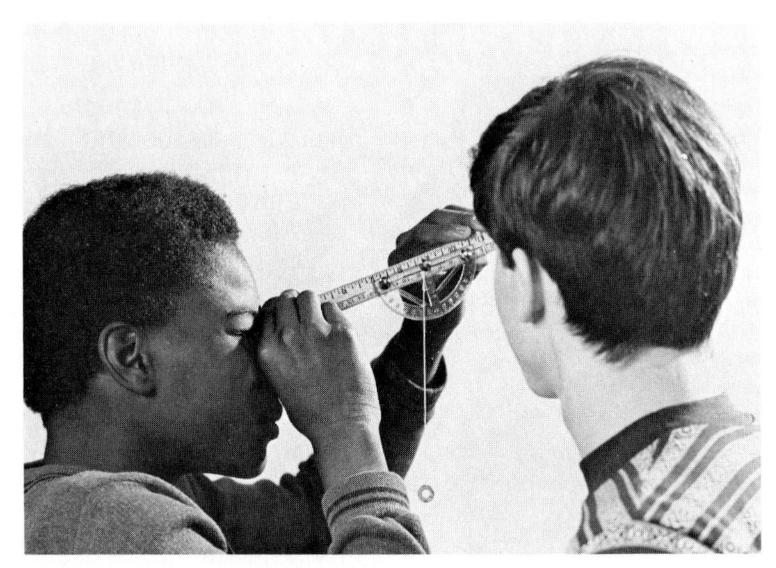

Figure 3/16 *Point your astrolabe at Polaris in the sky or a model star in the classroom to find its height.*

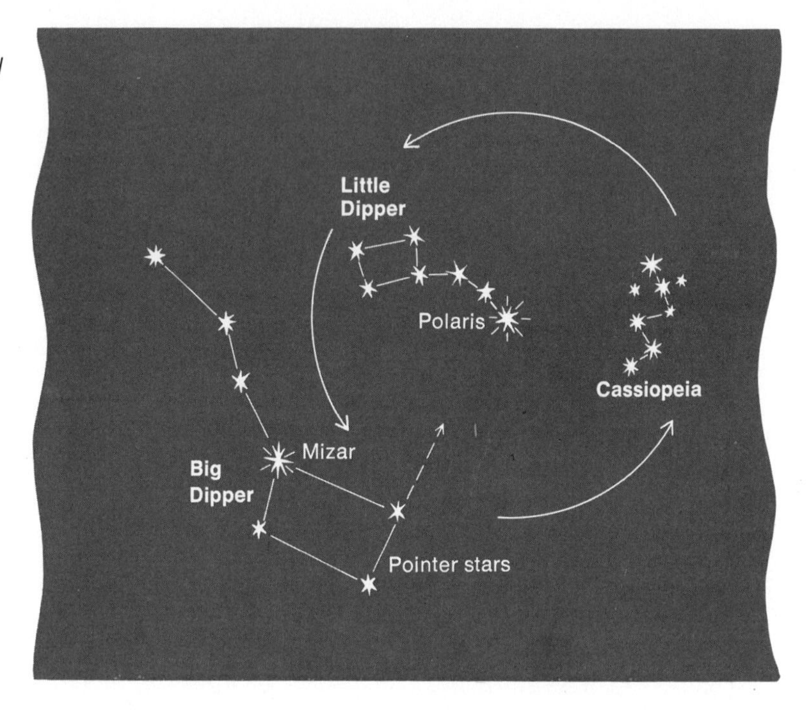

Figure 3/17
The two bright stars at one end of the Big Dipper point to the North Star, Polaris.

If you are outdoors in the evening, find Polaris in the sky by means of the pointer stars of the Big Dipper, as shown in Figure 3/17. Sky map A of Appendix C will help you. Indoors, stand at the line your teacher indicates. Sight carefully through the astrolabe tube on Polaris. Read and record the degree where the vertical string crosses the edge of the protractor. This gives Polaris's height in degrees above the horizon, the line where the sky meets the earth.

Look up the place where you live on a map of your state. Find your latitude as closely as you can. The height of Polaris at any place on earth is almost the same as the latitude of that place. (See Figure 3/18.) Remember that latitude is the number of degrees north or south of the equator. Does your observation of the height of Polaris come close to your latitude? If it is very different, what mistakes might you have made in your observations? ★

Sources of error: not sighting accurately on the star through the astrolabe; jiggling the astrolabe so that the string gave incorrect angle; not reading the angle where string crossed the protractor carefully enough; or holding astrolabe improperly so string rubbed on astrolabe. Actually, a small error of 1 degree is involved in sighting on Polaris rather than on the north celestial pole.

Navigators on ships and airplanes, and even astronauts in space, use observations of Polaris and other stars to locate their positions. If you wish, you can locate other

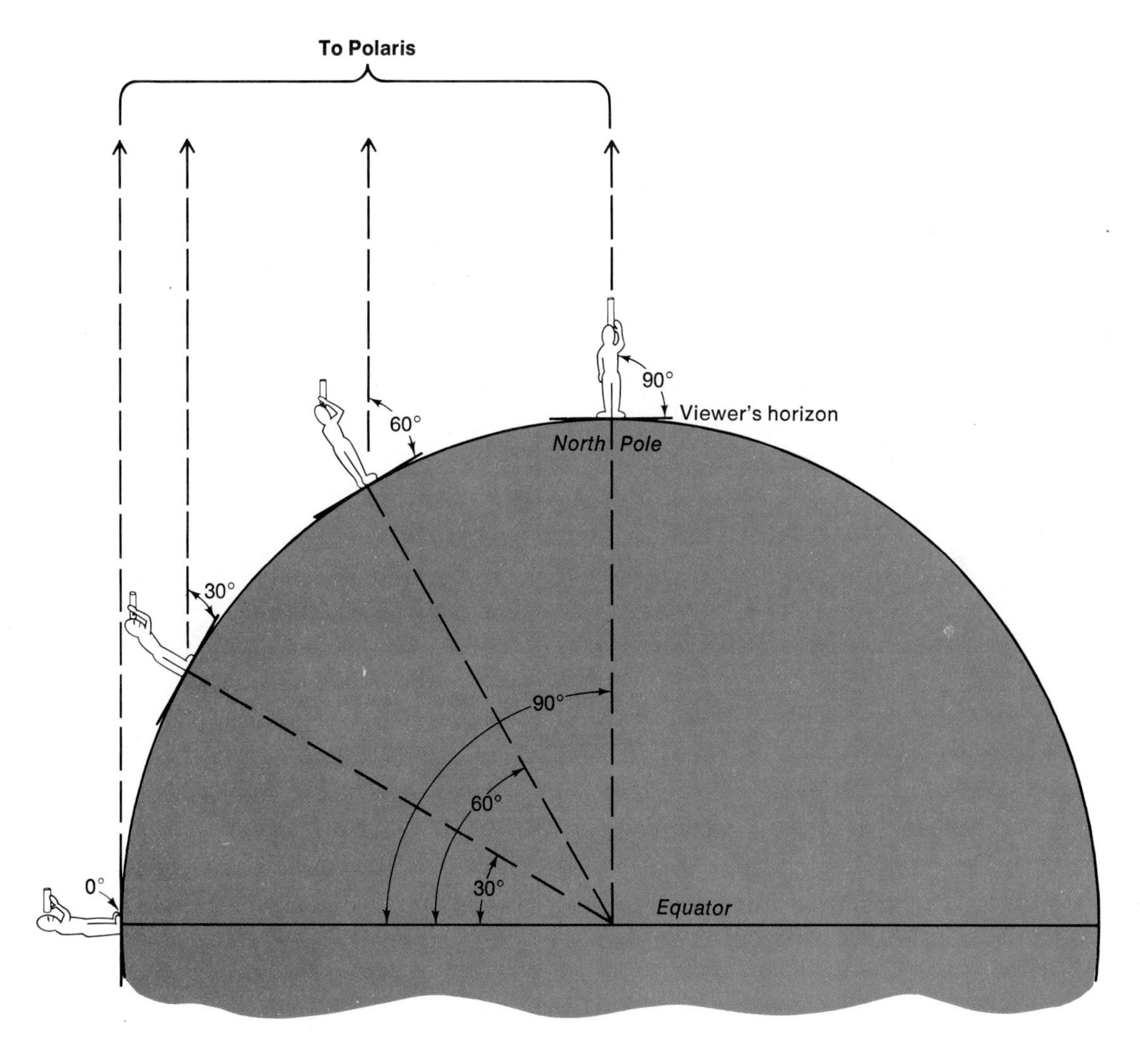

Figure 3/18 *The height of Polaris in degrees above the horizon is the same as the latitude in degrees from the equator in the Northern Hemisphere.*

See enrichment activity, Appendix M, on using astrolabe to locate stars in the local horizon system of coordinates in order to map constellations or stars in the sky.

stars with the astrolabe and map things in the sky where you live with the help of the activity in Appendix M.

As you saw in Section 3.3 and Figure 3/3, Polaris is very close to the north celestial pole, to which the earth's axis of rotation points. The north and south celestial poles are used to locate everything in the sky, just as they locate everything on earth. Here is how it's done.

Has the Earth Stopped Turning?

Here's an odd one! Look at the three photographs of the earth in A. Compare them. Aside from the fact that the earth is nearly round, partly covered with white clouds, and very beautiful, what do you notice about these three photographs?

Like the pictures of the earth in Figure 3/2, these photographs were taken at different times. The time is given under each photograph. Compare these views of the earth again. Can you tell why they all look almost exactly the same? Did the earth stop turning while these photographs were made?

No, the earth was still turning. It would take an awfully strong force over a very long time to stop the earth's turning. What was going on, then?

The photographs were made from a satellite orbiting the earth. Maybe the photographs were made by satellites in different places? Then as the earth turned, it would show the same face to the different satellites.

No, all three photographs were made from just *one* satellite, at the times shown. Impossible, you say?

Not at all. A satellite orbiting above the earth's equator, as shown in B, made the photographs. The satellite was moving at a speed that kept it up exactly with the earth's turning. So the satellite photographed the same part of the earth all the time. Look again at B. In reference to what was the satellite moving?

The diagram in C shows how just three of these "motionless" satellites could be used to broadcast TV and radio shows, or to send telephone calls, around the whole earth.

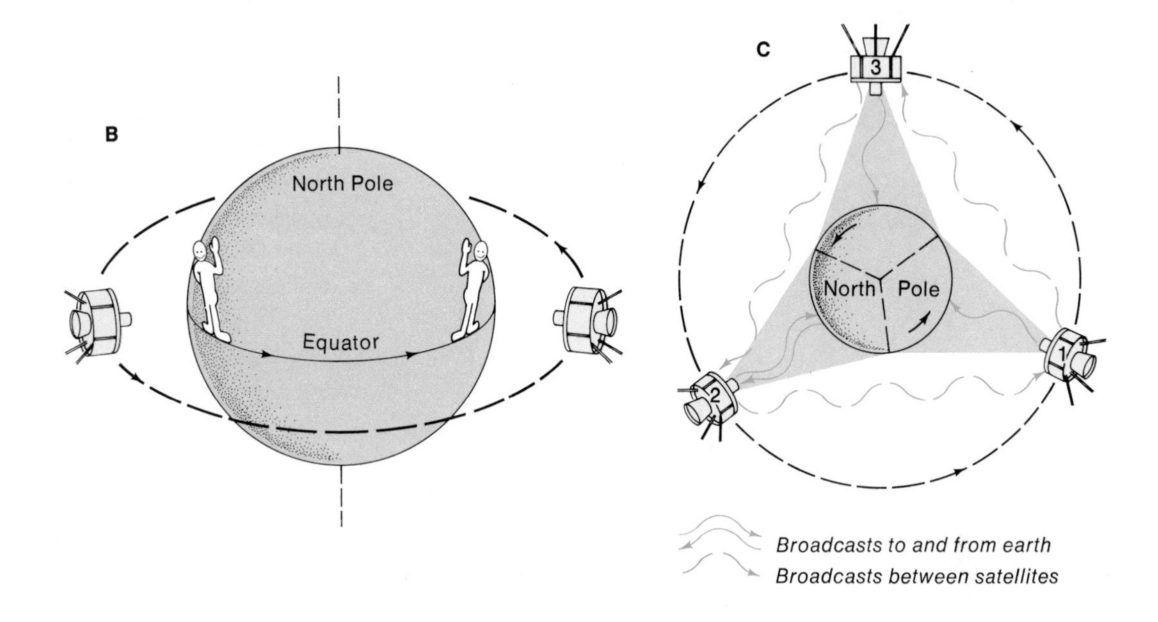

Use a celestial-sphere model with this section, if you have one, leading into study of earth and other planets in Chapter 4. Students can interpret and use most sky maps with equatorial system. Celestial equator, with some hours of Right Ascension marked on it, appears in sky maps in Appendix C.

See Teacher's Manual for suggestions for demonstration of celestial-sphere reference system.

3.10 Meet the celestial sphere.

Imagine that the sky above you, with all its stars, is a dome or a bowl turned upside down over you. Imagine that over the other side of the earth, the southern hemisphere, is another dome. Then you can picture the earth as being inside a sky ball, which is all around it. Astronomers call this sky ball, pictured in Figure 3/19, the **celestial sphere.** How do they find the places of stars on this sphere?

Imagine the earth's equator stretching out or expanding enough to go around the larger ball of the celestial sphere. And imagine the earth's axis of rotation lengthening enough to go through the dome of the celestial sphere at the top and bottom, as in Figure 3/19. Astronomers call the large equator the **celestial equator,** which is a reference line they use for locating stars. The long axis through the top and bottom of the sky ball gives them the reference points of the **celestial poles,** the second part of this reference system.

To see how this reference system works, look carefully at Figure 3/20. Just as longitude is measured from the Prime Meridian on earth, a reference point on the celestial equator is used to measure distances around the sky. This

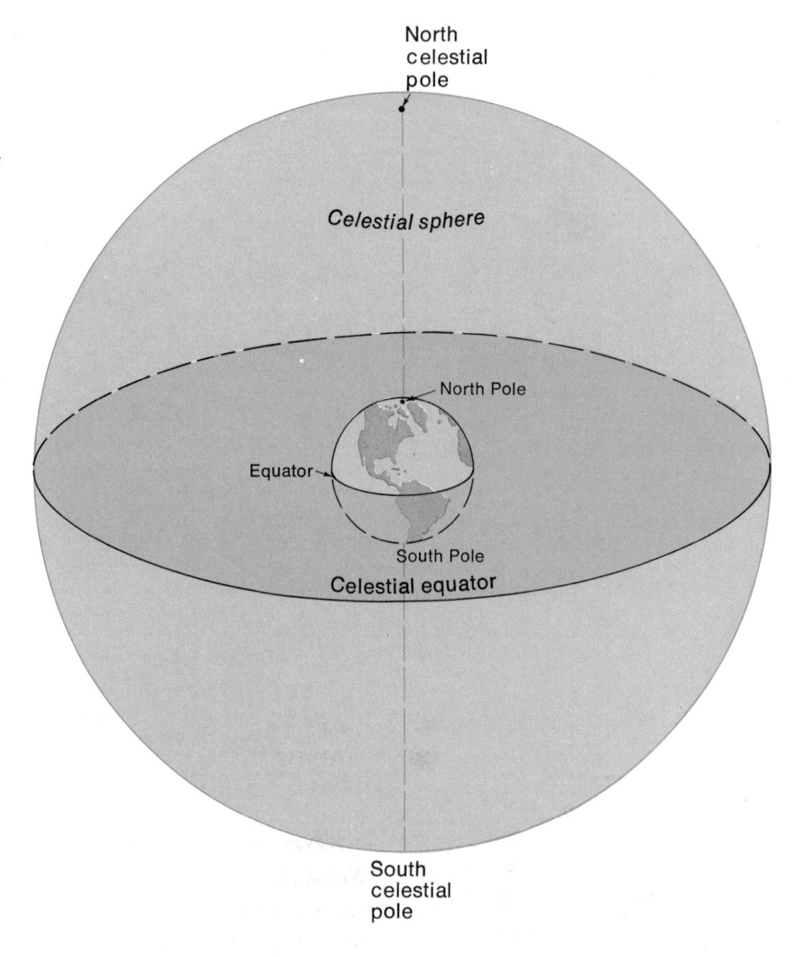

Figure 3/19
Extending the earth's poles and equator out into the sky makes the celestial poles and celestial equator. These reference points and reference line are used to locate places in the sky.

is called the **vernal equinox** (EE-quin-ox). This is the place in the sky where the sun is on the first day of spring. Distances east from the vernal equinox, called **right ascensions** (ah-SEN-shuns) are measured in hours and minutes. The full distance around the celestial equator is 24 hours, similar to the full distance of 360 degrees around the earth's equator. Distances north and south of the celestial equator, called **declinations** (deck-lin-AY-shuns) are measured in degrees, just as latitude is on earth.

Astronomers have used these big words since ancient times for places in the sky, just as longitude and latitude are used for places on earth. In the sky, right ascension means the same as longitude and declination the same as latitude.

Figure 3/20
The celestial sphere around the earth makes a globe for mapping the sky like the globes for mapping the earth.

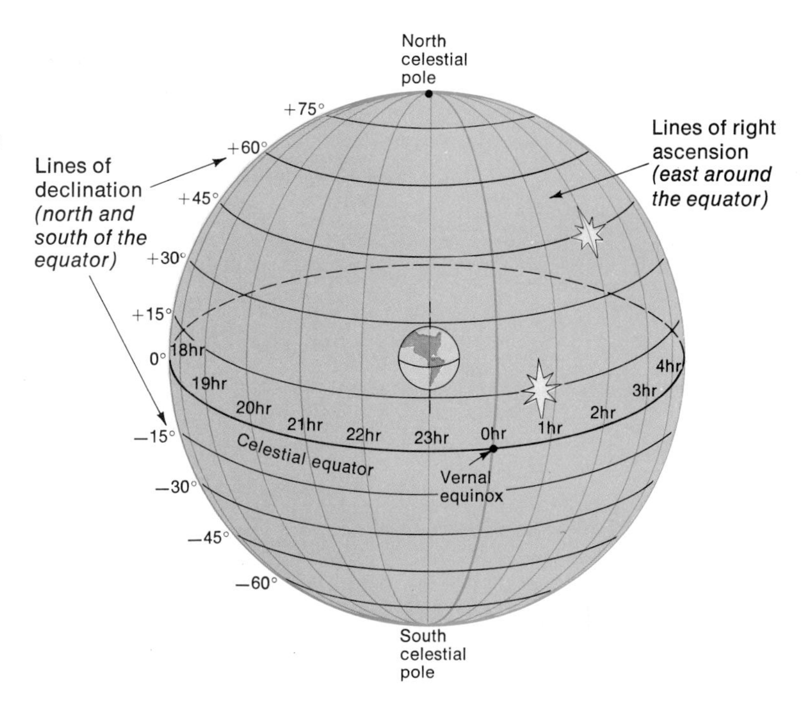

In addition to working with Figure 3/20, you might bring in a few sky maps from magazines, books, or other sources to familiarize students with equatorial system. Thus, you can show students on the celestial-sphere model how the vernal equinox, the 0-hour point on the celestial equator, is the place where the sun, moving north in spring, appears to cross the equator about March 21 to 22. The sun's other crossing of the celestial equator, moving south in the fall, occurs at the Autumnal Equinox about September 21 to 22.

Most star maps give declination and right ascension to locate stars, exactly as maps of the earth give the latitude and longitude of places. See if you can locate the stars in Figure 3/20. In the sky maps in Appendix C, "equator" means "celestial equator," and 0^h means 0 hours or the vernal equinox, just as 0° longitude means 0 degrees or the Prime Meridian.

Did You Get the Point?

Polaris, the North Star, is close to the celestial north pole, used with the celestial south pole to locate places in the sky.

The celestial sphere is the surface of an imaginary ball with the earth at its center.

The earth's poles and equator are extended out to the sphere to make the reference points and lines for the sky.

The position and motion of any object in the sky can be mapped on the celestial sphere.

Answers:

1. Altitude would be same as degrees latitude.
2. Capital letter formed is "L."
3. Making up a table of sky positions like this gives practice in the equatorial system. Students can swap tables with each other and draw the maps of the space trip.
4. Path around the north celestial pole at 70° will be a straight line on the map along the 70° declination line. On a flat map like this, the whole top of the map stands for 90° north, the north celestial pole.

Check Yourself

1. Suppose you saw an airplane pass in front of the North Star. What would be its altitude in degrees? How does this altitude as position in the sky differ from altitude as height above the surface?

2. What capital letter of the alphabet does a line on a map through the sky positions given in Figure 3/21 look like? Make a sky map like that in Figure 3/22.

3. Draw a wild space trip in the sky for your friends on a celestial-sphere map. Then make a table of positions located by declination and right ascension which they could use to guide themselves on the trip.

4. Imagine you have seen a UFO (Unidentified Flying Object) in the night sky recently. You observed it with an astrolabe and found it was circling the north celestial pole at a declination of 70 degrees. Draw a sky map of its path.

What's Next?

Places on the earth are located from the poles, the equator, and the Prime Meridian. Time is measured with units of regular motions, like the motion of the hands of a clock or the rotation of the earth. Objects in the sky are located on the celestial sphere with its own poles and equator.

Now that you know how to locate places in the sky, let's look more closely at some of the things in the sky that are nearby, like the sun, the moon, and the other bodies like the earth, the planets. When you locate them in the sky very exactly, you will find that their behavior is peculiar, to say the least!

Figure 3/21
Positions in the sky given by degrees of declination and hours of right ascension.

Sky position	Declination (degrees)	Right ascension (hours)
1	+40	5
2	+30	5
3	+20	5
4	+10	5
5	0	5
6	−10	5
7	−10	4
8	−10	3
9	−10	2

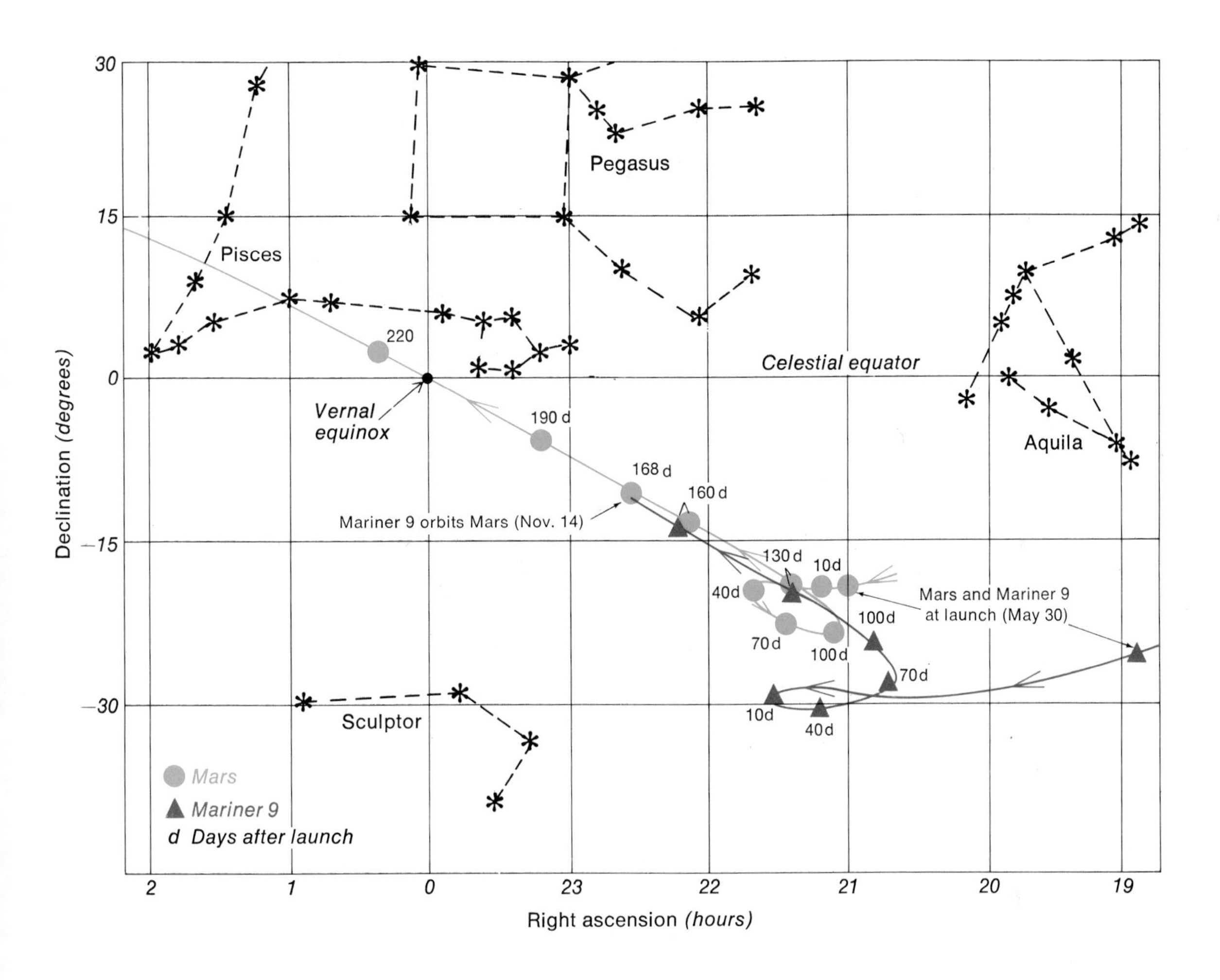

Figure 3/22 *The paths of the Mariner 9 spacecraft and the planet Mars as they were plotted from the earth. Mariner 9 journeyed to Mars and went into orbit around the planet.*

Skullduggery

After students have completed the message, you might ask them to trace the path of the treasure hunt on a globe or map of the earth.

Directions: You have found a tattered and dirty old piece of paper blowing about the street. It seems to contain a message about the location of a hidden treasure, but many words are missing. The answers to the following questions will help you read the treasure message. Copy the message in your notebook. When you get the answer to one of the questions, fill it into the gap with the same number in the message as the number of the question. See if the message tells you where and what the treasure is.

Treasure Message: Start at any one of the _______ (1) always the same number of _______ (2) from the _______ (3) and fly _______ (4) on this until you are opposite the _______ (5) _______ (5). Follow the _______ (6) by submarine and dog sled to _______ (7) _______ (8) _______ (8) _______ (8). Cross the _______ (9) _______ (9) _______ (9) and go down the _______ (10) _______ (10) meridian to _______ (11), _______ (11). At exactly 4:00 P.M. _______ (12) _______ (12) _______ (12), it will be 11:00 A.M. on the _______ (13) _______ (13) you now travel to by freighter. Let the earth _______ (14) until it is _______ (15). Treasure located in the _______ (16) _______ (16). Look directly _______ (17) and up from your _______ (18) _______ (19) _______ (19). At some _______ (20) you will find the treasure. What would life be worth without it?

Answers: (1) parallels; (2) degrees; (3) equator; (4) west; (5) Prime Meridian; (6) meridian; (7) latitude; (8) 90 degrees north; (9) geographic north pole; (10) 0 degree meridian; (11) Greenwich, England; (12) Greenwich Mean Time; (13) time meridian; (14) rotate; (15) noon; (16) celestial sphere; (17) south; (18) horizon; (19) reference lines; (20) declination. Treasure is the golden sun, of course.

Questions:

1. What do you call reference lines that are always the same number of degrees from the equator? _______

2. _______ of latitude are always the same length.

3. What do you call the reference line 90 degrees from both the North and South Poles? _______

4. _______ is the direction from which the earth rotates.

5. The line from which longitude is measured is called the _______ _______.

6. What is a line that runs around the earth to the poles? _______

7. The number of degrees north or south of the equator is called _______.

8. What is the latitude of the North Pole? _______ _______ _______

9. Directly under the north celestial pole is the earth's _______ _______ _______.

10. The _______ _______ _______ is called the Prime Meridian.

11. Name the place that it is agreed determines 0 degrees longitude. _______, _______

12. The time determined by the Prime Meridian is called _______ _______ _______.

13. A line on which a time zone is based is called a ________

14. If the earth did not ______, we would not need time zones.

15. When the sun is directly over a time meridian it is ______.

16. The ______ ______ has an equator and poles like the earth's.

17. ______ latitude is found from the equator to the South Pole.

18. Your ______ is the line where the earth seems to meet the sky.

19. Both the Prime Meridian and the equator are ______ ______.

20. ______ on the celestial sphere is similar to latitude on the earth.

For Further Reading

Asimov, Isaac. *The Clock We Live On.* New York, Macmillan, 1962. How man changed the moon calendar to one based on the sun.

Zim, Herbert S., and Robert H. Baker. *Stars—A Guide to the Constellations, Sun, Moon, Planets, and Other Features of the Heavens.* New York, Golden Press, 1963. More about how to locate places on earth and in the celestial sphere.

Chapter 4 Solar System

No photographs of whole solar system exist, because from the earth only a few planets can be seen at one time in the night sky along the ecliptic. The other planets are invisible in daylight on the opposite side of the earth. Reflected light of planets of other stars is too dim to register on telescopes.

The bright spot is *R* Monocerotis, one of a number of Tau Tauri variable objects, which astronomers believe to be preplanetary systems. These are studied particularly with new infrared telescopes. This opening ties in with the discussion in Sections 4.9 and 4.10 of other possible planetary systems being discovered. Current best estimate of age of solar system is 4.5 to 4.7 billion years, a recent study yielding 4.65.

Have you ever seen a photograph of our whole solar system, the sun and all the planets? Why not?

You've only seen artists' drawings or charts of the solar system, of course. And there are no photographs of other stars with the planets they may have around them. Telescopes aren't strong enough to show the very dim planets at such great distances from us. Telescopes can't even show the stars except as points of light.

There *is* a photograph of a forming solar system, though, a baby solar system that was probably just born. Look at the bright spot to which the arrow in the photograph points. The spot is beside a nebula in the constellation Monoceros (mon-oh-SAIR-ohs), the Unicorn. What is the bright spot? Mainly it is a mass of dust and gas. But there may be a star hidden inside the mass of dust. And a set of planets may be just forming around that star. Our solar system may have been like that 4.5 billion years ago.

This probable baby solar system is about 2250 light-years away. The dust and gases from much larger blobs have been pulled together into a thick ball by gravitational attraction. The pressure and temperature of the bright spot have shot high enough to start nuclear fusion in the star hidden at its center. Most of the dust may have formed into a disk that looks like a phonograph record spinning around the star. And some clumps of dust in the disk may be swirling and gathering into planets.

As more dust falls into the star and more is taken up by its planets, the star will settle down. It may become a normal G-type dwarf star, like our sun, on the main sequence of the H-R diagram. And it will have a whole set of planets circling around it.

It's all very well to have a photograph of another possible solar system. But how about our own? Can you think of any way a photograph might be made sometime of our whole solar system? There is a way. It has been planned, and sometime the plan may be carried out.

A spacecraft within the plane of the system far out by the outer planets where the sun is dim might photograph the whole system with a 360° lens. Even then some of the planets might be hidden behind the sun and bright space around it. A spacecraft traveling out above or below the plane of the solar system could photograph the whole system. A NASA research project proposes sending such an unmanned spacecraft far up above the solar system to learn more about the poles of the sun, among other reasons.

The plan is simple. Launch a rocket from earth more powerful than any yet made. Give the rocket enough thrust to blast it right up above our solar system. Then, looking far down on the sun and the planets, the spacecraft could make pictures of the whole system.

★ Try to imagine how this photograph would look. Make a drawing of what you imagine. Compare your drawing with those of others. ★

Look at the Whole Family

Figures 4/15 and 4/16 may be helpful in this activity. These are scale drawings of size of sun in relation to planets and of their distances from sun. In scale, drawings will look very different from those usually made of the system, which greatly exaggerate size of the planets and change scale for distances of outer planets.

4.1 Model the solar system yourself.

You've imagined what the solar system would look like from a spacecraft. Perhaps you were helped by drawings of the system you've seen before. But most of these drawings are not correct. Possibly your own is incorrect, too. Here you can make a correct model of the solar system that will show you what may be wrong with your drawing. First, a little about the system.

With the earth, the solar system has 9 known planets. **Planets** are the largest bodies in the system that move around the sun. And the planets have a total of 32 **moons,** or satellites. These are smaller bodies that move around the planets. Sometimes these moons are called *natural* satellites. *Artificial* satellites are man-made and sent into orbit by rockets.

Process: measuring

The distances of the known planets from the sun and their diameters are given in the table in Appendix N. Notice the distance of the earth from the sun, very close to 93 million miles. Astronomers call this distance one **astronomical unit** (A.U.). They use this unit to measure the solar system. In the same way, an inch is a unit to measure short lengths and distances. And just as you use yards and miles to measure greater lengths and distances, astronomers use light-years for distances beyond the solar system.

Materials

small ball, yardstick, pieces of paper, adding-machine tape

★ The sizes for two models, A and B, are given in the table in Appendix N. Use a ball about seven-eighths of an inch in diameter for the sun in model A, given in Appendix N.

Figure 4/1
Make a model of the solar system using a ball to represent the sun, and dots on a strip of adding machine tape for the planets.

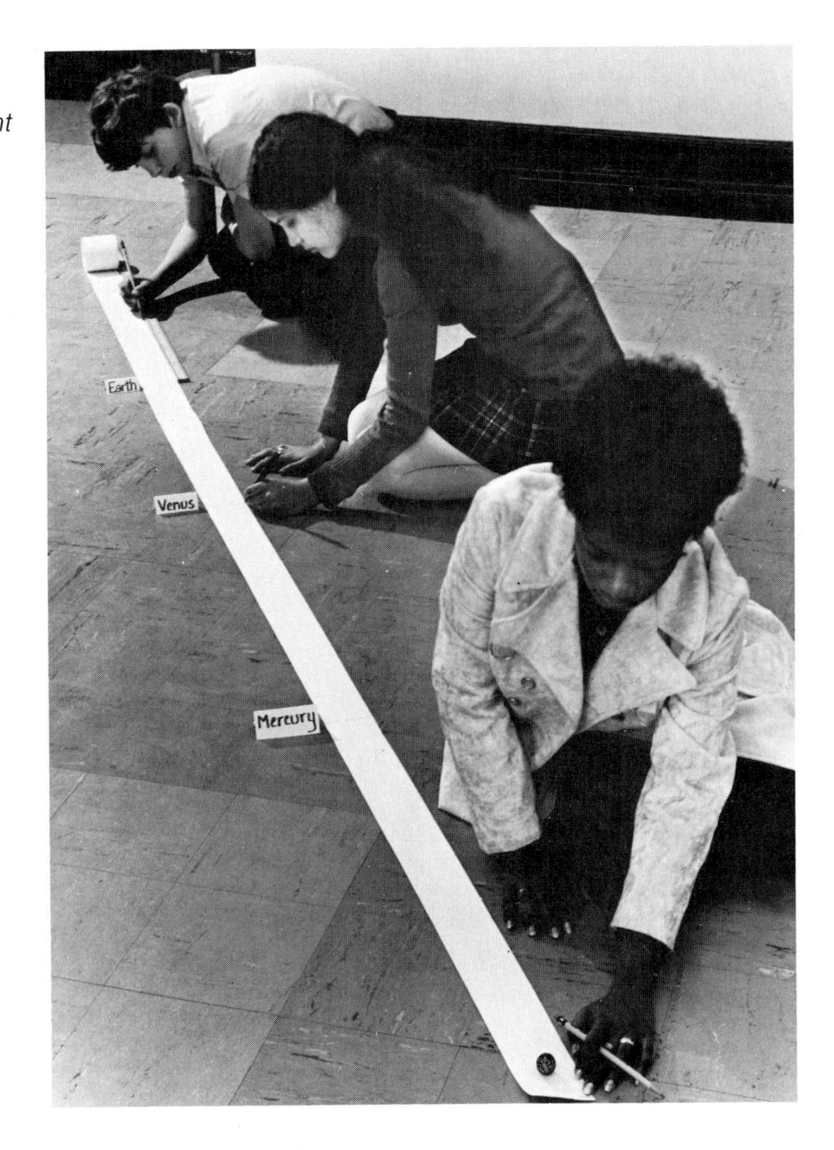

Activity Time: Indoors or outdoors, 20 to 25 minutes. Some groups of 2 or 3 can make model A, others can make model B; some may wish to design their own model. Those working on model A may find it too long for space available and turn to model B or own model.

Roll of adding-machine tape is 50–55 yards long. Two rolls needed for model A, and 1 roll for model B.

Use a piece of paper for each planet and mark a dot on it for the size of that planet. Most of the planets will only be tiny dots. Then measure how far away from the sun the planets should be, and place the planet papers there.

Another way to make the model A and B is to use a roll of adding-machine paper tape. For model A, use a seven-eighths-inch ball for the sun at the beginning of the tape. Then measure off and mark the places and sizes for each planet as you unroll the tape as in Figure 4/1.

Figure 4/2
The positions of the planets in their orbits around the sun and the constellations of the zodiac on the ecliptic in August, 1971.

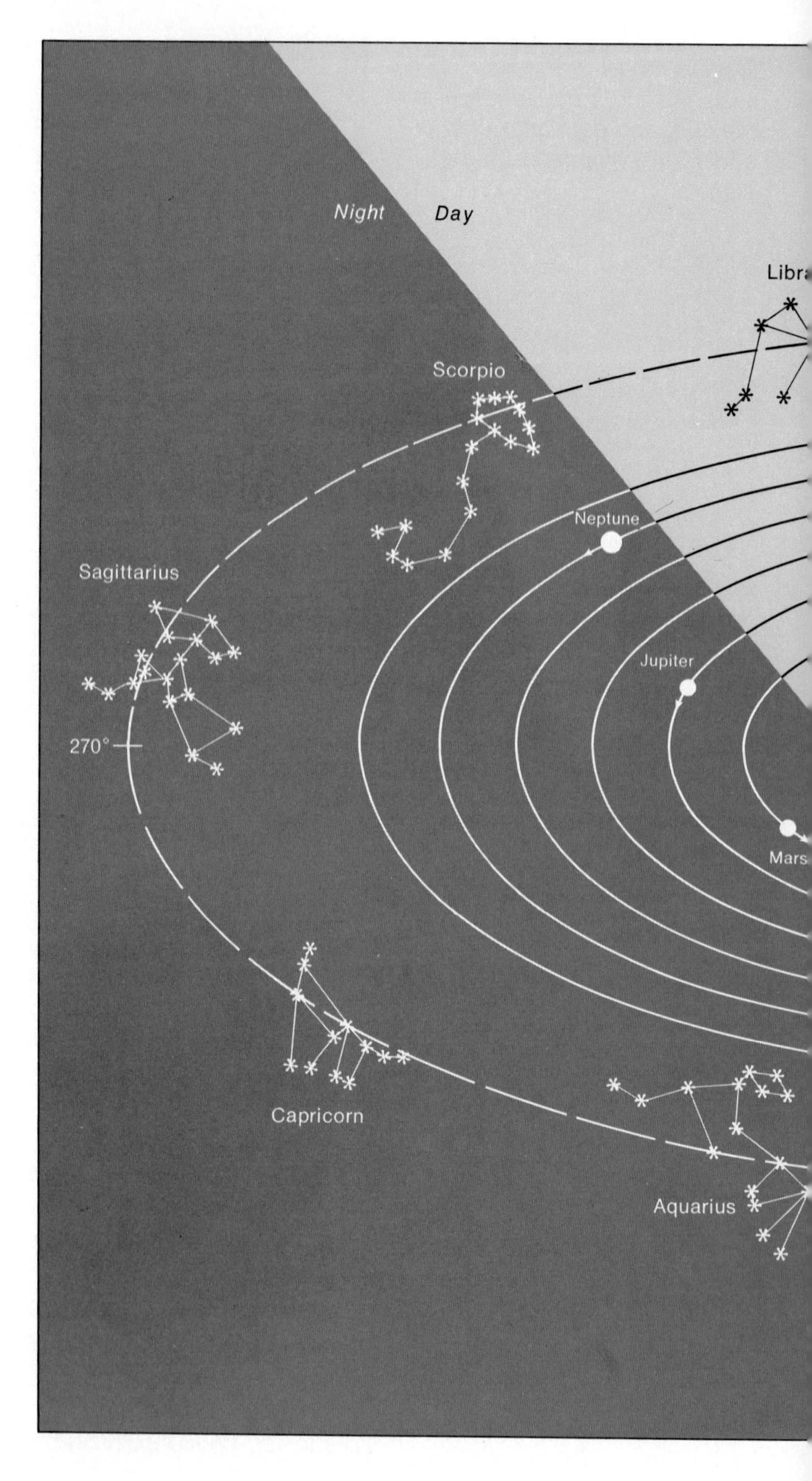

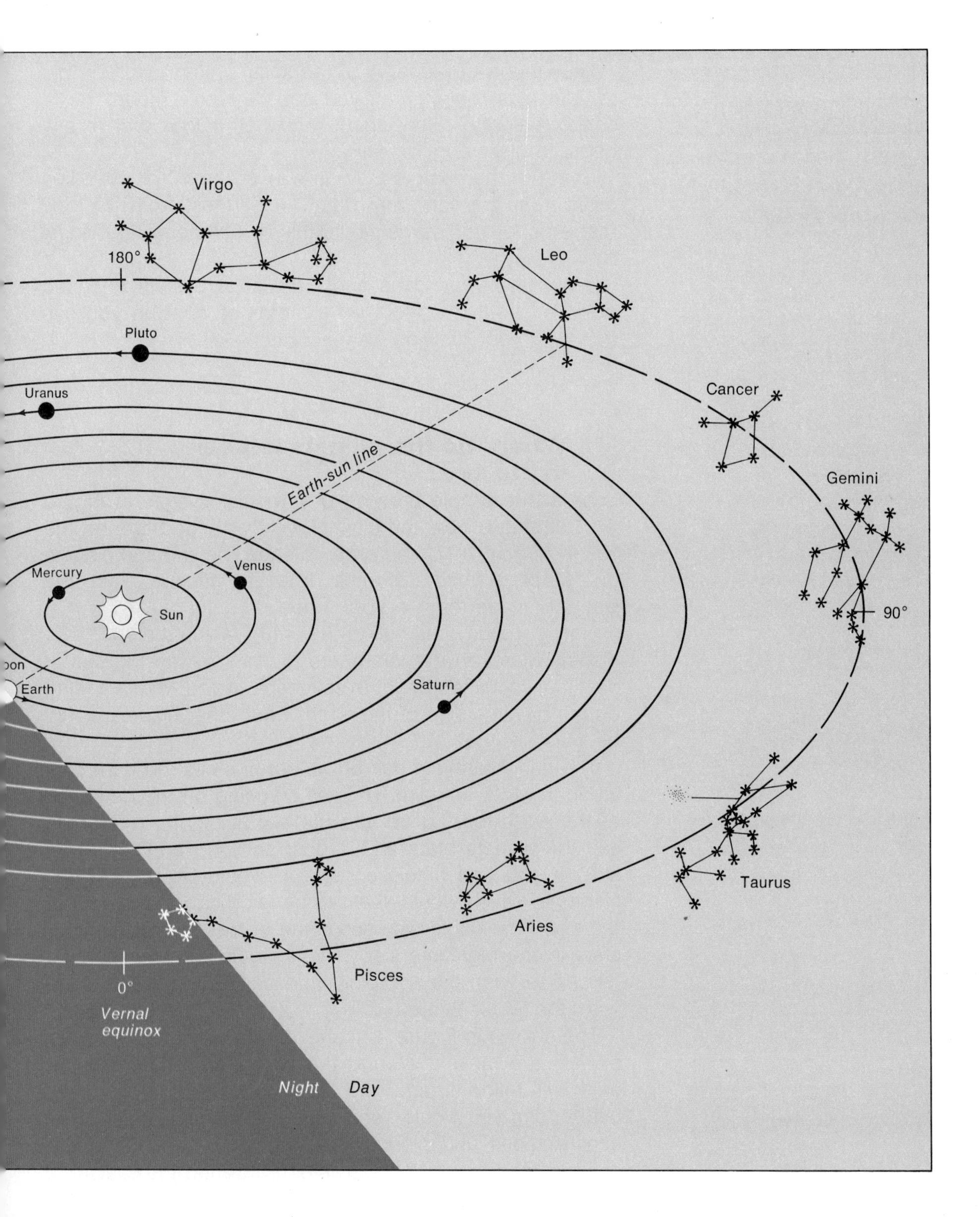

Virgo
180°
Leo
Pluto
Cancer
Uranus
Earth-sun line
Gemini
Mercury
Venus
Sun
90°
Earth
Saturn
Taurus
Aries
Pisces
0°
Vernal
equinox
Night
Day

Model A only out to Mars (12-foot radius) with sun in center may fit in classroom, possibly Jupiter (14-yard radius) can be added with sun at one end of classroom; full model A requires 100-yard corridor or football field. Model B may fit classroom with sun at one end (10 yards). See Teacher's Manual for another model.

Comets may be 10,800 inches (300 yards) to 21,600 inches (600 yards) in model A when farthest from sun; or 30 to 60 yards in model B.
Process: measuring

Good time to discuss scale of solar system, probably a new idea for students, and sizes of planets and their distances from sun.

Saturn, Uranus, and Pluto were also invisible in daytime sky in August, 1971.

When earth has moved along path to a place on a line between sun and Saturn, Saturn would be high in southern night sky and Mars, moving more slowly than the earth, would be far down in western sky.

How much of model A of the solar system can you get in your classroom? Can you get the whole model into a school hallway? How about model B? Does that fit any better in classroom or hallway?

Some of the comets may go way out to 10 or 20 billion miles from the sun when they are farthest from it. How many inches and yards would this be for model A? Model B?

How have your ideas about the solar system changed now that you've worked on a model of it? Can you see now why your drawing was incorrect? ★

4.2 Where do the planets appear in the sky?

Look at the simple drawing of the solar system in Figure 4/2. It shows how the planets were arranged around the sun in August, 1971. In order to show the sun and planets on one page, their sizes and distances are not exact, the way you made them in your model.

Notice the daytime half of the earth, facing the sun. The planets Mercury and Venus are in the sky with the sun in daytime, so it's too bright to see them. If you were standing on the daytime side of the earth, what other planets would be invisible, too?

The dark nighttime half of the earth is colored dark blue in Figure 4/2. Imagine yourself standing on the dark side of the earth looking up into the sky at night. You would see the planets Mars and Jupiter showing brightly in the sky. If you had a telescope, you would be able to see Neptune, which would seem to be fairly close to Jupiter in the sky. You could also see some of the constellations of stars in the nighttime sky. These would include Scorpius (SKORP-ee-yus), the Scorpion, and Aquarius (ah-QUARE-ee-yus), the Water-Bearer.

The arrows in Figure 4/2 show how the earth and other planets are moving around the sun. Trace the path the earth will follow around the sun in one year. Notice that before long you would begin to see Saturn at night. Other constellations and planets would begin to appear in the night sky, as month followed month through the year.

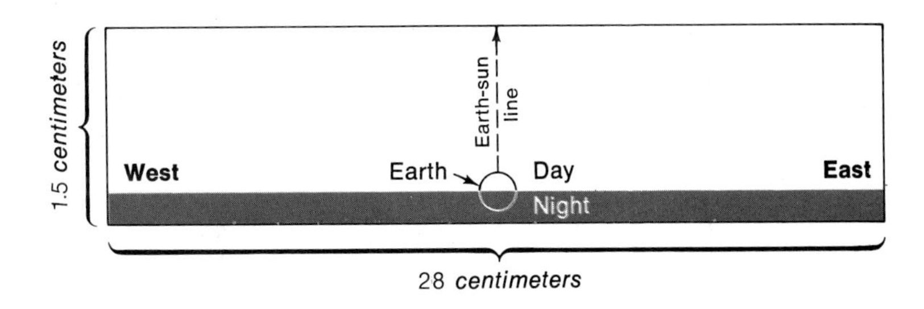

Figure 4/3 *By moving a piece of paper like this on Figure 4/2, you can see what objects appear in the night sky through the year.*

Materials

record, pencil, chalk, paper

Activity Time: 5 minutes. Students do activity individually. Demonstrates that stars and planets in sky change through the year because nighttime side of earth faces different ways as earth follows its orbit of the sun. "East" and "West" on paper strip refer to directions for observer on earth at night. After students have finished activity, you might ask all to identify which constellations of the zodiac will appear in the southern sky during fall, winter, spring, and summer by using sky maps in Appendix C, which students can relate to Figure 4/2.

★ You can see just which planets are visible from the earth at different times through a year. Cut and mark an oblong piece of paper to represent the daytime and nighttime sky, as shown in Figure 4/3. Then place the paper on Figure 4/2, with the earth on the paper over the earth in the figure, and the earth-sun line on the paper following the same line in the figure. The planets and constellations on the right side of the paper are the ones you can see in the sky at night.

Then rotate the paper as shown in Figure 4/3. At the same time move the earth on the paper farther along on the earth's path in Figure 4/2. Swing the earth-sun line to its new position. See how Saturn is now visible in the nighttime sky? Keep rotating the paper the same way through the whole yearly path of the earth in Figure 4/2. ★

How do you locate the planets in the sky? If you look up at the southern sky at night, you can usually see one or two of the brighter planets. They are found along an imaginary line across the sky called the **ecliptic.** The ecliptic is shown in the maps of the southern sky in Appendix C. All the planets move around the sun as if they were on a huge disk like a phonograph record. From the earth they appear to move along close to the same line in the sky, the ecliptic. Here's a way to see what this means.

Materials

paper strips from Figure 4/3

★ Make a model of the ecliptic yourself. Mark some white spots with chalk here and there along the rim of a phonograph record. Hold the record by putting a pencil in the center hole. Then put the pencil point through the edge

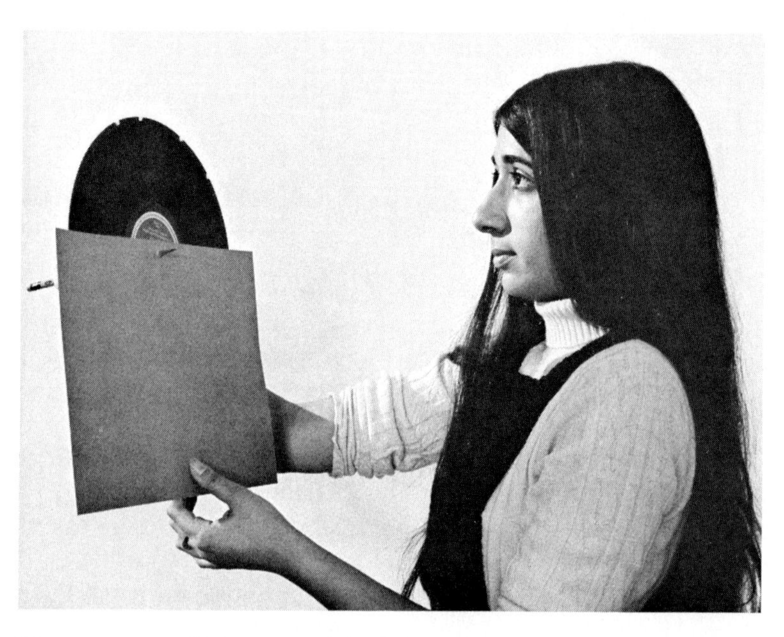

Figure 4/4 *Make a model of the planets along the ecliptic using a phonograph record and a piece of paper.*

Activity Time: 10 minutes. Class can take turns with a half dozen (old) records. (Chalk can be brushed off edge of record afterward.) Exact model of how ecliptic is traced by path of planets would have observer on earth *within* the record disk, rather than looking at it at an angle.

of a piece of paper, as shown in Figure 4/4. The edge of the paper stands for the horizon, the black record for the night sky, and the white spots for planets.

Turn the record slowly counterclockwise. Hold the paper motionless. Don't the spot-planets all follow a certain line in your model sky? This is like the line of the ecliptic marked on the southern sky maps in Appendix C. The planets all move around the sun in a thin disk like the phonograph record. Of course, the earth is in the same disk, too, just as if it were inside the phonograph record. ★

Demonstration: Tipping or inclination of the earth's axis can be demonstrated by student carrying globe of earth, with axis tipped always in the same direction, around another student standing for sun. Inclination of the axis is sometimes called "obliquity of the ecliptic." Text Figure 4/5 shows why.

The earth's axis of rotation is tipped about 23 degrees to the ecliptic, as shown in Figure 4/5. Since the celestial equator is an extension of earth's equator, it is also tipped to the line of the ecliptic.

The 12 constellations of stars through which the line of the ecliptic passes are called the **zodiac.** The word zodiac means "circle of animals." The zodiac makes a broad band across the sky along the line of the ecliptic. You can pick out the constellations of the zodiac by following the ecliptic through the year on the Appendix C sky maps.

Figure 4 / 5
The earth's equator and the celestial equator are at an angle to the ecliptic, the line along which the constellations of the zodiac appear to move throughout the year.

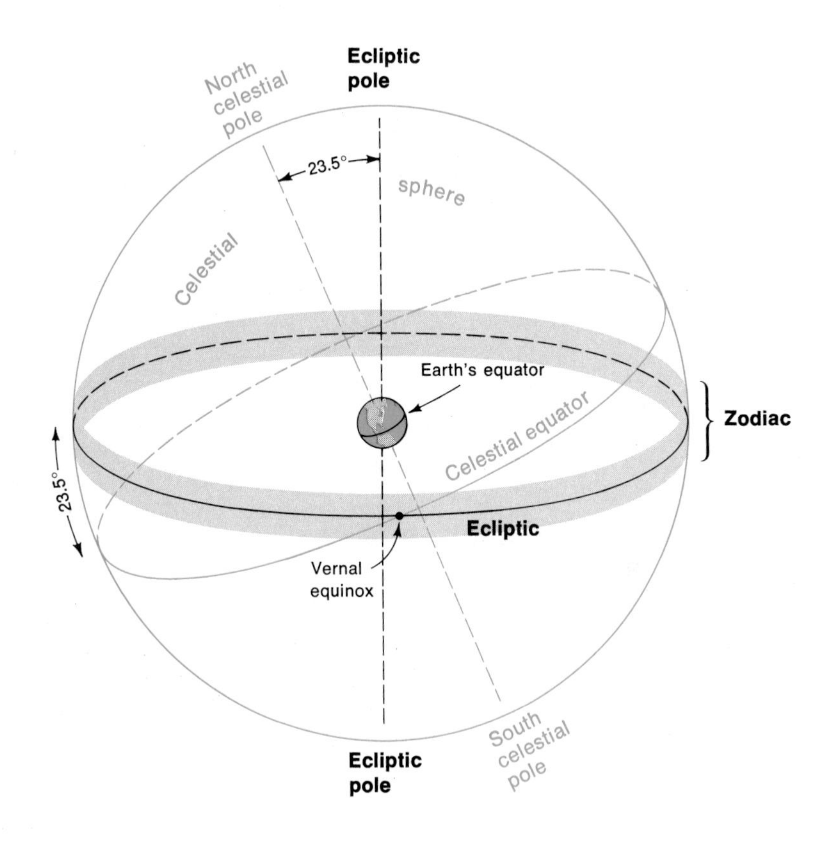

On a star party, or by using sky maps individually or in groups, students can find the ecliptic and the constellations of the zodiac, as well as any planets then visible in the night sky along the ecliptic. The only planets clearly visible to naked eye and only planets known before invention of telescopes, are Mercury, Venus, Jupiter and Saturn. (Uranus with magnitude 5.4 is lost among many stars.) Planets and moon are good objects to view with small telescopes.

4.3 Planets wander in the sky.

At their great distances, the stars appear to stay in the same positions in the sky and are called fixed. (Some of the stars can be seen to move just a little over many, many years.) But ancient people noticed how planets change their positions among the stars as they watched them for a few days or a week or two. Figure 4/6 shows such a change. So they called them planets, from a Greek word meaning ''wanderers.''

Ancient people noticed, too, that sometimes these wandering planets seemed to slow down or speed up. Sometimes they even seemed to stand still among the stars. This was a great mystery! What could the planets be doing?

Materials
graph paper

★ During the summer of 1971, the planet Mars became brighter and brighter, gleaming red in the early evening sky. And on nights in July and September Mars seemed almost

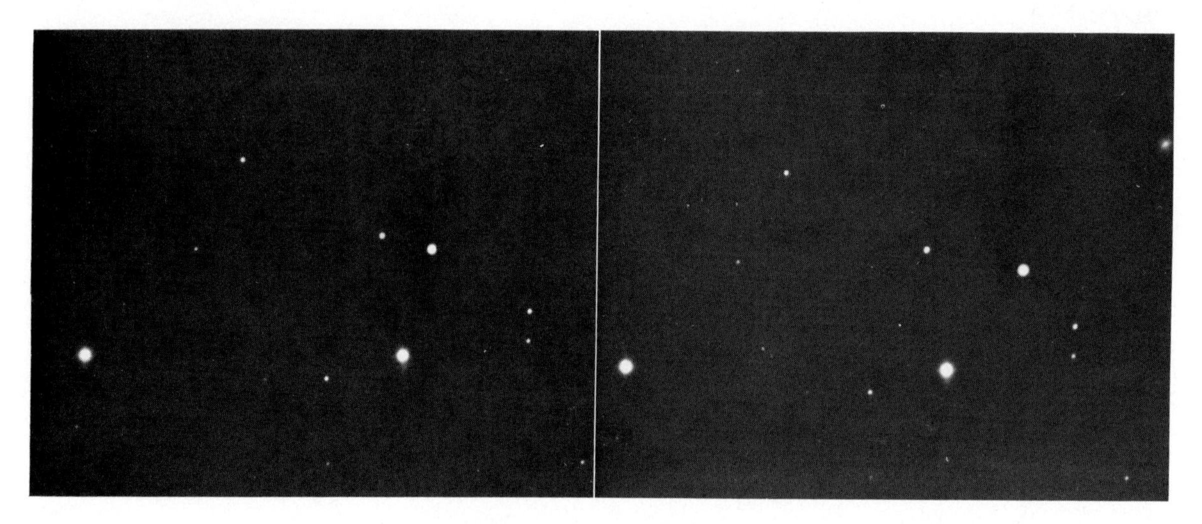

Figure 4/6 *The planet Pluto wandering among the stars. The photograph on the right was taken one night after the photograph on the left.*

Activity Time: 20 to 25 minutes, individual student activity. The more accurate student plotting of positions on map, the smoother Mars' path will be. *Retrograde* (as opposed to direct) *motion* is the technical term for Mars' backward motion. When earth passed Mars in 1971, a loop was traced by Mars. In 1969, as shown in Teacher's Manual, the retrograde path was pointed. The path retrograde motion follows depends on the relative positions and motions of the earth and other planet at the time earth passes it, crossing the straight line between the other planet and the sun. See diagram in Teacher's Manual.

Answers: Marks for Mars are farther apart in some places because Mars seemed to be moving faster in sky. Where the circles are close together earth is then moving relatively at same rate as Mars. Marks are nearly all 5 days apart, a few 6 days. Mars is not actually speeding up or slowing down enough to cause its apparent motion in sky. (Planets do speed up when close to sun in their orbits, *perihelion,* and slow down when farthest from sun, *aphelion.*) Mars' apparent backward motion is caused by earth catching up to and passing Mars.

to stand still in the sky. You can see what was happening by plotting Mars' path on a map.

Make a map of a part of the sky on graph paper, as shown in Figure 4/7. Mars moved through this sky region from March through October of 1971. Locate Mars' position with a cross or circle for each date shown in the table in Appendix O. The first position is marked in Figure 4/7.

When you are finished, look at the path Mars made in the sky during the eight months. Why are the marks on your map farther apart in some places than in others? Why do the circles get so close together in some places? Could Mars be speeding up or slowing down? What makes Mars seem to go backward? ★

The planets closer to the sun move faster than those at greater distances. So the earth moves faster than Mars. Sometimes the earth catches up with Mars and passes it in space. Then as seen from the earth, Mars appears to stop, move backward, stop again, and again move forward in the sky as it does on your map.

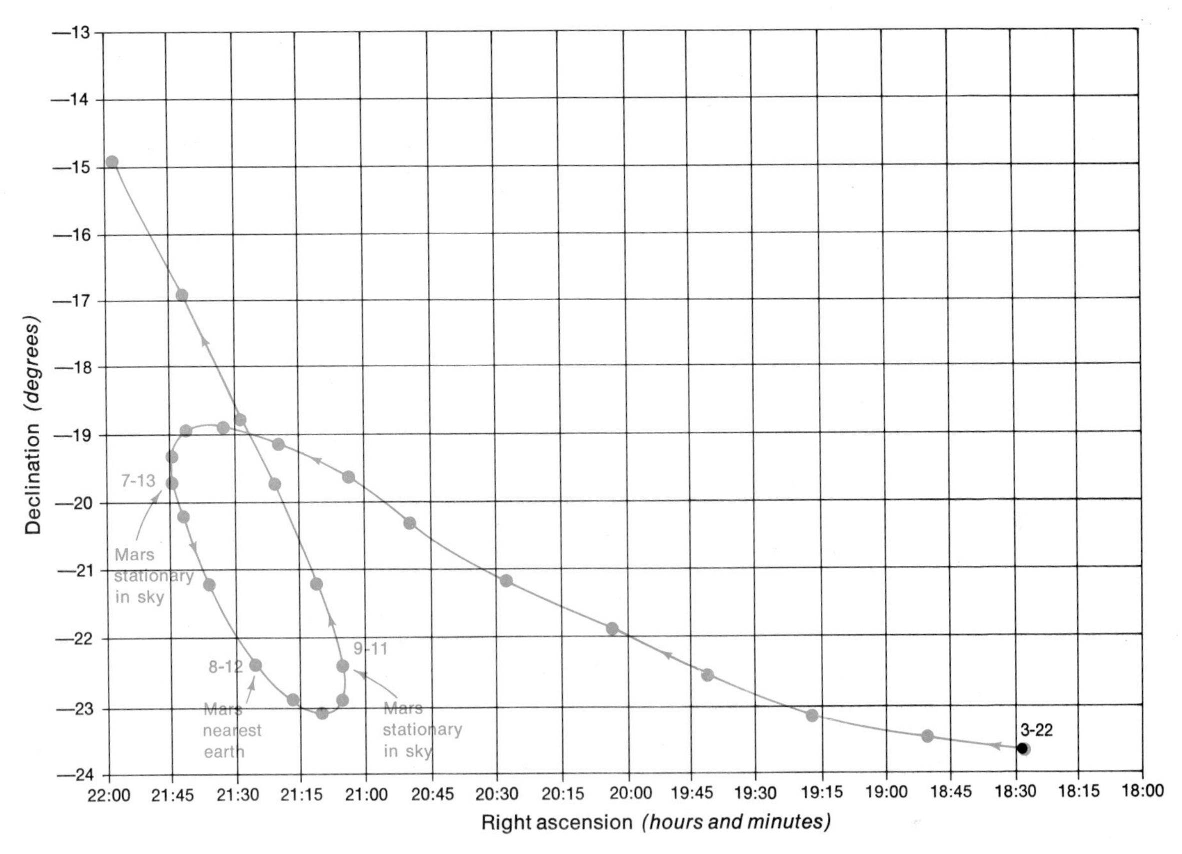

Figure 4 / 7 *Make a map like this on graph paper to follow Mars' motion in the sky.*

Materials

chair

When students watch both classmate's head and objects on wall, they can clearly see the retrograde motion.

Good place to contrast Ptolemaic (earth-centered) and Copernican (sun-centered) explanations of orbits of planets. Ptolemy had to use small, circular epicycles to explain backward motion. Loop made by Mars' retrograde motion in 1971 seems to fit epicycle explanation well; points of Mars' retrograde motion in 1969 are different. Ptolemaic system finds such variations in epicycles hard to explain—are there sub- and subsub-epicycles? Copernican sun-centered system offers simpler explanation for retrograde motions and their variations than Ptolemaic system.

★ Pretend you are the earth. Walk slowly around a chair standing for the sun in the center of the room. Have a classmate, standing for Mars, move around a much larger circle but more slowly than you. Watch your classmate's head as it moves past objects on the wall of the room behind him, when you go by him in orbit. Does his head appear to move backward? ★

4.4 How and why do the planets move?

For many hundreds of years everyone thought the planets went around the sky in circles. Then about 350 years ago a German astronomer, Johannes Kepler, worked with the

But Ptolemaic system was taught at Harvard University until 1750, although Copernicus announced his discovery in 1643.

Process: interpreting data

very exact observations of Tycho Brahe, (TEE-coh BRAH-hee), a Danish astronomer. Kepler found that Brahe's positions for Mars would not fit a circle at all. He wondered, what *is* the path of Mars, then? And it took him many years to discover the answer. Now you can make a path like that of Mars and the other planets in a few minutes.

Materials
thumbtacks, string, wood or cardboard base

★ Put two thumbtacks into a board or heavy cardboard, and arrange a string and pencil as shown in Figure 4/8. Pull the pencil out so the string is tight. Keep it tight while you trace with the pencil all the way around the thumbtacks.

Figure 4/8
You can make an ellipse like the orbits of the planets with a pencil, thumbtacks, and string.

Making an ellipse is a good activity for homework or class demonstration; it can also be done individually in class. Stick the thumbtacks into heavy cardboard or wooden board. Although they are sometimes nearly circular, all paths of natural and man-made satellites follow ellipses, as do those of planets.

What results do you get? Doesn't the path, or orbit, look like a flattened or squashed-down circle? The official name for this path is an **ellipse.** If you think of the path you have drawn as a planet's orbit, the sun is where one of the thumbtacks is. In the solar system, all the planets move around the sun in ellipses. There's a place in the orbit of each planet at which it's closest to the sun and a place at which it's farthest from the sun. Can you find these places on the path you made with your pencil? ★

Kepler found that all the planets moved around the sun in paths like the ellipse you made. Later the British scientist Isaac Newton worked out general rules or laws to explain why the planets move this way. His rules are called universal because they apply not only to the planets, but to the moon and stars. In fact, they apply to any moving things—a ball you throw in the air and your own walking or jumping.

Newton said that every body in the universe attracts every other body. If you let go of a ball, it is attracted toward the center of the earth and falls until the floor or the ground stops it. According to Newton's law of gravitation, the strength of the attraction changes with the masses of the bodies and their distance from each other. Very simple—mass and distance are all that count.

The **mass** of anything is the amount of material or matter that it contains. The more massive bodies are, the greater their attraction for each other. And the smaller the distance between bodies, the greater is their attraction.

What does a planet's orbit amount to, then? Imagine the earth in orbit around the sun, as shown in Figure 4/9. The massive sun tugs at the earth as the downward arrows indicate, so the earth is falling toward the sun. But the earth in orbit is also moving rapidly in a direction at right angles or sideways to the tug of the sun. These two motions—of falling toward and shooting sideways from the sun—keep the earth in an elliptical orbit. The earth moves sideways fast enough so that its gravitational fall toward the sun results in its elliptical orbit. The earth follows this orbit in space as it moves in the disk of the ecliptic with all the other planets.

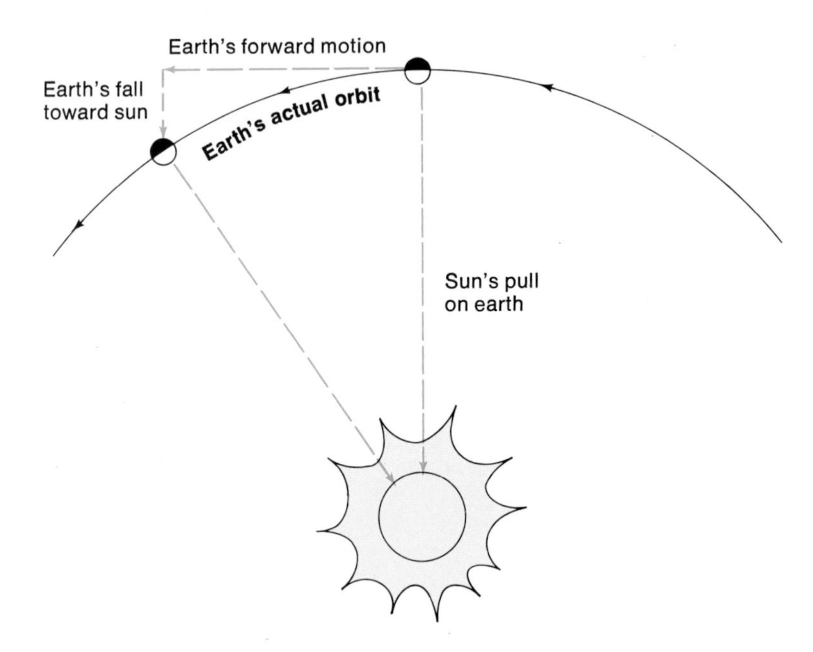

Figure 4/9 *The combined forward motion of the earth and the sun's gravitational pull on it makes the earth travel in an elliptical orbit around the sun.*

Did You Get the Point?

Models show that the planets of the solar system are very tiny compared with the sun, and they are at much greater distances from the sun than we usually picture.

The planets all move around the sun in a thin disk marked by the line of the ecliptic.

The constellations of the zodiac form a band along the ecliptic.

Because planets move at different speeds, sometimes the earth passes other planets and this makes them seem to move backward.

The attraction of gravitation increases with the masses and decreases with the distances of bodies attracting each other.

This attraction acting between the sun and the planets in the solar system keeps the planets in their elliptical paths around the sun.

Answers:

1. One car passing another is like earth passing Mars; other car seems to slow down, stop, go backward, and then move ahead against the objects in the distant background, corresponding to the stars.
2. From planet Mercury, moving faster than earth in orbit, you would see earth in backward motion as you passed it.
3. Decrease in mass of bodies decreases their gravitational attraction, so planets would be at much greater distances from the sun. Sun's radiation might be too weak at such distances to keep water from freezing and to support life as we know it on earth.

Check Yourself

1. Picture yourself in a car catching up to and slowly passing a car in another lane. Compare this with the motion of Mars in the activity in Section 4.3.

2. Suppose you are living in a new space colony just settled on the planet Mercury. Would you ever see the earth stop in the sky and move backward, then go forward again?

3. If the sun had been a much smaller star, with only half the mass it actually has, where would the planets be? Would there be any life on earth?

Look Inside the System

4.5 Measure the size of the sun.

Materials

meter stick or yardstick, cardboard (2 pieces 6″ × 8″ square), needle or pin, scissors

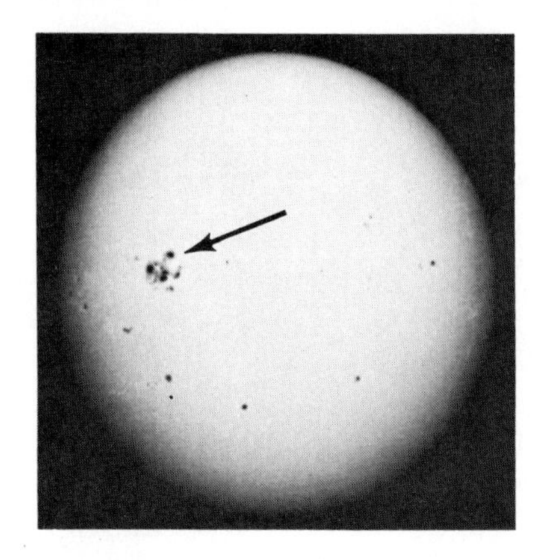

Figure 4 / 10
Sunspots on the blazing surface of the sun in a photograph made with a 12-inch telescope.

Activity Time: 25 to 30 minutes. Four to six students can make sunstick together, then each use it for making his own measurements and calculations. Pinhole should be as small as possible. Class can pool and discuss individual results, sources of error, vast dimensions of diameter and distance to sun. Point out that this distance is the astronomical unit (A.U.), the unit used to

★ Telescopes are needed to make observations of the sun like that shown in Figure 4/10. But you don't need even the smallest lens or mirror, only a pinhole, to discover things about the sun yourself. If you are given the distance to the sun, you can make observations to find its diameter yourself. Then you can tell how big are those black spots on the sun, called **sunspots.**

For your observations, pick a fairly clear day when the sun is out brightly at least now and then. You can do the experiment either indoors or outdoors. Cut slots in both cardboards to fit the meter stick or yardstick. Make a needlehole in one cardboard. Make two fine marks just 5 millimeters (a quarter-inch) apart on the other cardboard. Then assemble your sunstick as shown in Figure 4/11. Be sure the cardboard slots fit snugly enough on the meter stick so the cardboards will stay at right angles to the stick.

Stand sideways to the sun and point the stick in its direction. Have the cardboard with the hole in it toward the sun. CAUTION: Never look directly at the sun. This can badly injure your eyes or cause blindness. Wiggle the stick around until the bright spot made by the sun falls between the marks on the rear cardboard. Slide this cardboard back and forth until the spot of the sun just fills the space between the marks. Then figure the distance between the two cardboards on the stick as precisely as you can. Now you have your facts.

measure distances in the solar system. (See Appendix B.) Contrast A.U. with the light-year (l.y.), the unit used for distances to stars.
Process: measuring

Caution students *never* to look directly at sun, even when it is behind haze or clouds, from which the sun may emerge swiftly; particularly, *never* aim binoculars or telescopes directly at the sun, nor look at it through photographic negatives, whose density may vary greatly.

Students with little or no experience in solving equations may need help to perform and understand the calculations described in this paragraph. Students should reach a figure ± 5 to 10 percent of actual diameter of sun; if they do not, encourage discussion of their source(s) of error: arithmetic, inaccurate measurement of distance between cards, failure to have sun's image just fill lines on card, and so on. Using same formula with diameter given, if there is time, allows students to repeat work with simple proportion, even though they know what the answer should be.
Process: using numbers

Find the diameter of the sun (A in Figure 4/11), given its distance from the earth (B), which is about 150 million kilometers (93 million miles). Here's the way you figure it.

You know how the objects in an enlargement of a photograph are larger in all measurements than the same objects in the original photograph. This means the two photographs are *proportional* to each other. In the same way, the triangles from the pinhole cardboard to the sun and from the pinhole cardboard to the rear cardboard are proportional to each other.

The diameter of the sun (A) and its distance from the earth (B) are proportional to the diameter of the spot on the cardboard (C) and the distance between cardboards (D) in the same way as the enlargement of the photograph and the original photograph. In other words,

$$\frac{\text{A (diameter of sun)}}{\text{B (distance to sun)}} = \frac{\text{C (diameter of spot on cardboard)}}{\text{D (distance between cardboards)}}.$$

You know everything but A in this formula. So fill in the numbers for the letters B, C, and D. Multiply both sides of the formula by (B × D). The results will be:

$$A \times D = B \times C$$

Then multiply B by C. Divide your answer by D. That will give you the diameter of the sun, A.

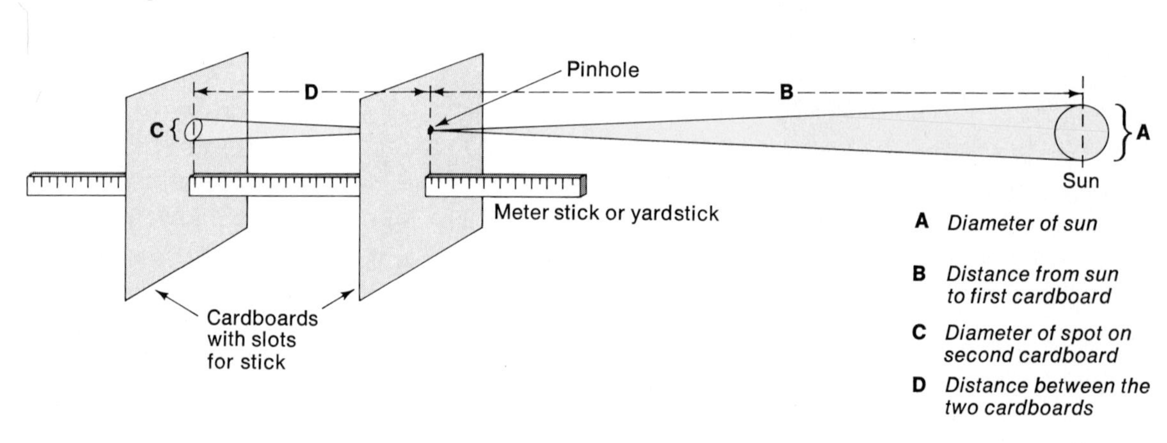

Figure 4/11 *You can find the diameter of the sun and the size of sunspots with a sunstick like this.*

Figure 4/12
If the earth were at the center of the sun, even the moon's orbit would look this small.

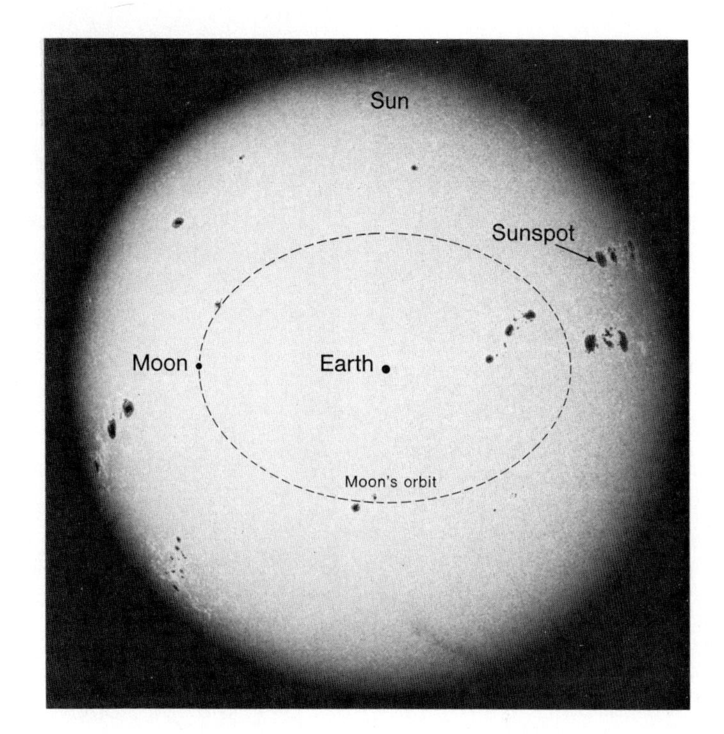

To figure diameter, students measure diameters of sun and sunspot indicated by arrow in Figure 4/10 and solve the proportion:

$$\frac{\text{A (Diameter of sunspot)}}{\text{B (Diameter of sun) (870,000 miles)}} = \frac{\text{C (Diameter of sunspot in photo)}}{\text{D (Diameter of sun in photo)}}$$

This gives the proportion:

$$\frac{\text{A (to be found)}}{\text{B (870,000 miles)}} = \frac{\text{C (4 mm)}}{\text{D (43 mm)}},$$

Then $A = \frac{B \times C}{D} = 80{,}930$ miles.

This optional activity also leads directly into any work students may do on the solar activity and sunspot cycle of ± 11 years.

Opportunity to present and discuss many features and structures of the sun and to ask students to bring in magazines or books containing pictures of the sun or features of the solar system.

It's a mighty big sun, isn't it? The size of the sun is compared with the earth and with the moon's orbit of the earth in Figure 4/12. Did you get anywhere near 870,000 miles (1,390,000 kilometers) for the diameter of the sun?

Now you've figured the diameter of the sun. Using the sun's diameter, see if you can figure out the diameter of the sunspot to which the arrow points in Figure 4/10. ★

4.6 The sun gives a continuous performance.

Something's always happening to the sun! One great sunspot after another, whole families of them sometimes, appear on the sun's disk. Vast solar flares shower space with intense radiations that can be dangerous for space travelers. Hundreds of areas of seething gases, up to 500 miles (800 kilometers) in diameter, make the sun's surface look like a pot of boiling breakfast cereal. Streamers of exploded gases rise hundreds of thousands of miles above the surface, as you can see in Figure 4/13.

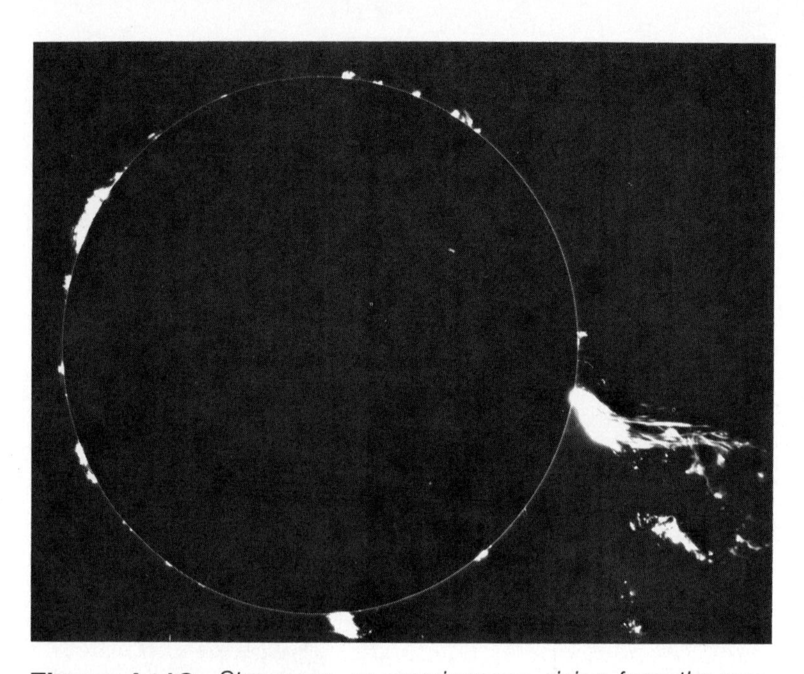

$$\frac{\text{A (Length of streamer)}}{\text{B} \quad 870{,}000 \text{ mi.}} = \frac{\text{C (25 mm)}}{\text{D (70 mm)}}$$

Then $A = \frac{B \times C}{D} = 310{,}700$ miles approx.

Figure 4 / 13 *Streamers, or prominences, rising from the sun, which was covered with a black screen for the photograph. Knowing the sun's diameter, how long are these streamers?*

Concept: particulate nature of matter

Particles radiated from the sun give it a bright halo, called the **corona,** always present but only visible during total eclipses. (See Figure 4/14.) From this corona, the **solar wind** is shot out like spray from a turning garden sprinkler or sparks from a fireworks pinwheel.

The sun puts on one dramatic act after another! And it's the only star close enough to study in detail. No wonder many scientists devote their lives to understanding what happens on the sun.

4.7 What are the planets like?

Imagine you are a space traveler from a planet of another star 15 light-years away. After 30 years of swift flight through space, you are only one-tenth of a light-year away from this solar system. You and others on your spaceship observe the planets and their moons to learn more about the solar system. Here are some of the things you find.

One-tenth light-year = about 600 billion miles from the sun, about 200 times the distance of the planet Pluto from it. A good distance for precise observations of the solar system.

You measure the sizes and masses of the planets and compare them with the sun. You arrange your data as

Figure 4/14 *The solar corona is created when the sun's light reflects from atomic particles and dust around the sun. The total light from the corona is about half the amount from the full moon.*

shown in Figure 4/15. You notice a place where the planets suddenly increase greatly in size and mass. Measuring the distances of the planets from the sun, you can tell from Figure 4/16 that the first four planets are very close to the sun, and the others way out.

Several planets have moons orbiting around them. The moons are mostly around the big outer planets. Jupiter's 12 moons look like a "mini" solar system. One of these moons, Io, was once believed to have an atmosphere of its own, but this has been disproved.

Mars' moons are so small and so close to it that you wonder if they could be artificial satellites. (See Figure 4/17.) Perhaps some space travelers came to Mars, left their ships to go down and explore the planet, and never got back! Then you use your most powerful telescope and find that Mars' moons are natural after all. Figure 4/18 shows one of them. (Continued on page 116.)

News stories, research reports, and articles will be discussing questions about Mars for several years, as returns from Mariner 9 and Soviet Mars craft are interpreted. Discuss the latest results with students, especially in relation to the possibility of life on Mars.

Sun

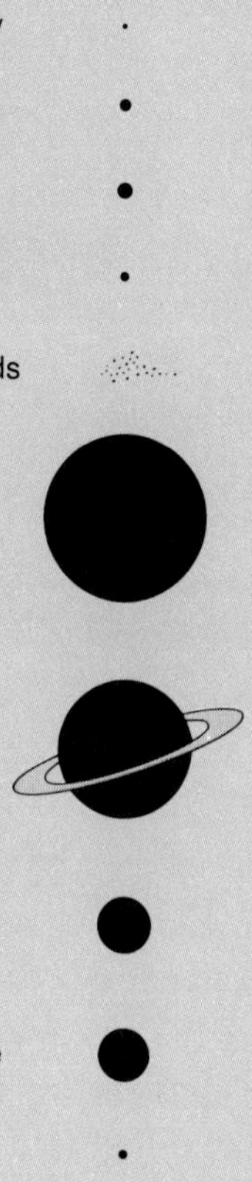

	Size	Radius	Mass	Number of moons
Mercury	·	0.38	0.05	0
Venus	•	0.96	0.85	0
Earth	•	1.00	1.00	1
Mars	•	0.53	0.11	2
Asteroids		—	—	—
Jupiter		11.19	319.00	12
Saturn		9.47	95.00	10 plus rings
Uranus		3.73	15.00	5
Neptune		3.49	17.00	2
Pluto	•	0.47	0.1	0

Figure 4/15 *Compare the size of the planets with the size of the sun. In this drawing, the radius and mass of the earth are each called 1.00 and the other planets are compared with them.*

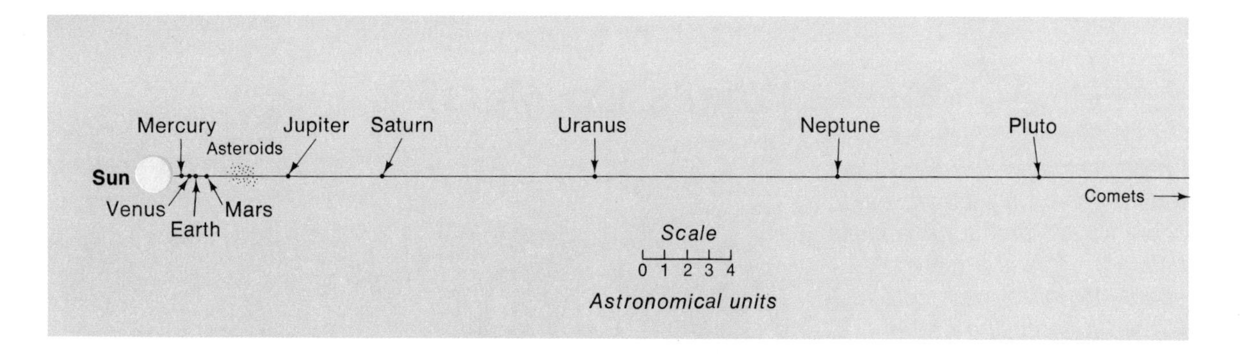

Figure 4/16 *Distances of the planets from the sun on a scale of astronomical units.*

Figure 4/17
Mars' satellites, Phobos (right) and Deimos (left). A separate photograph of Mars was added, because the original was overexposed to show the dim satellites.

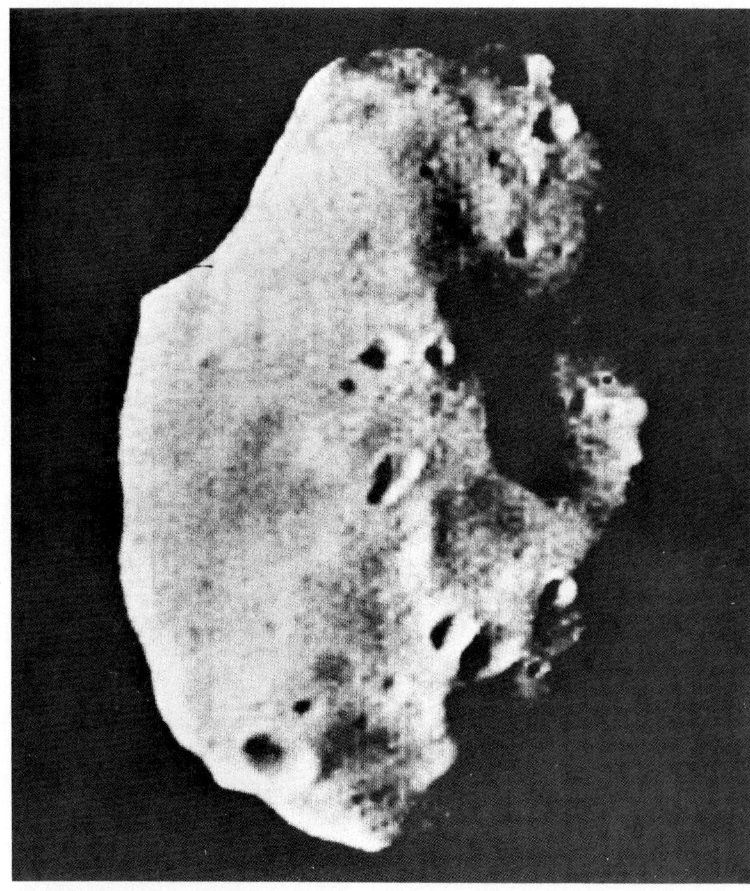

Figure 4/18 *Phobos, Mars' larger satellite, photographed by the Mariner 9 spacecraft orbiting Mars.*

What's On Mars?

There are craters on Mars like the craters on the moon, but in some ways Mars' surface looks different. This is a canyon about 45 miles wide, probably formed by cracking of the surface rocks, with a long strip of the surface dropping downward. Other cracks formed along the edge; they look like streams flowing into a river, but they probably are not. If you study the photograph closely, you will see that some of them do not actually open into the canyon, and some look like modified craters.

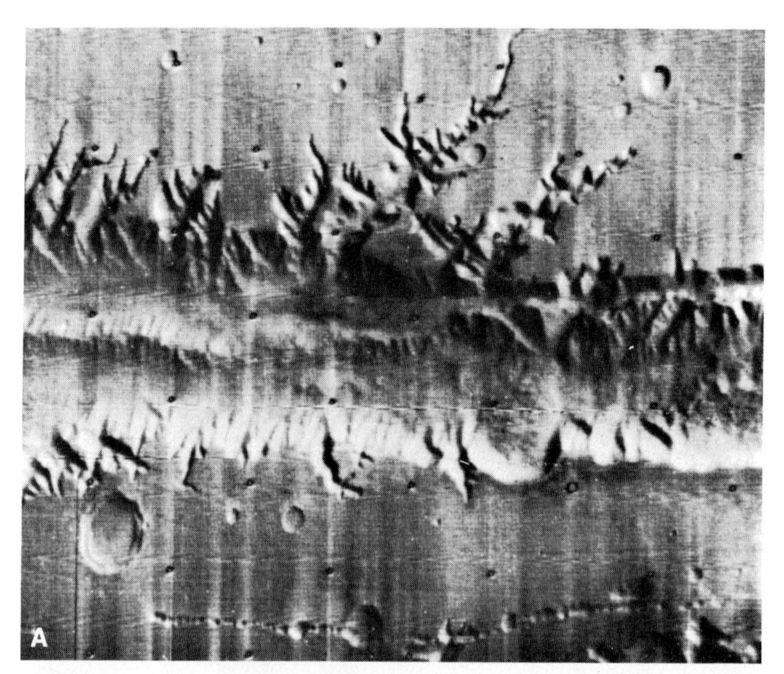

This area, 500 miles from the South Pole of Mars, is covered with pits and hollows. The large ones on the left are 10 miles across. The hollows might have formed from the melting of large blocks of ice, or they might be caused by wind blowing away loose dust and sand.

The surface of a plateau that rises $3\frac{1}{2}$ miles above most of the area around it. The surface probably is made up of lava rock from volcanoes. After the lava flows solidified, the surface cracked in a radiating pattern, leaving valleys $1\frac{1}{2}$ miles wide.

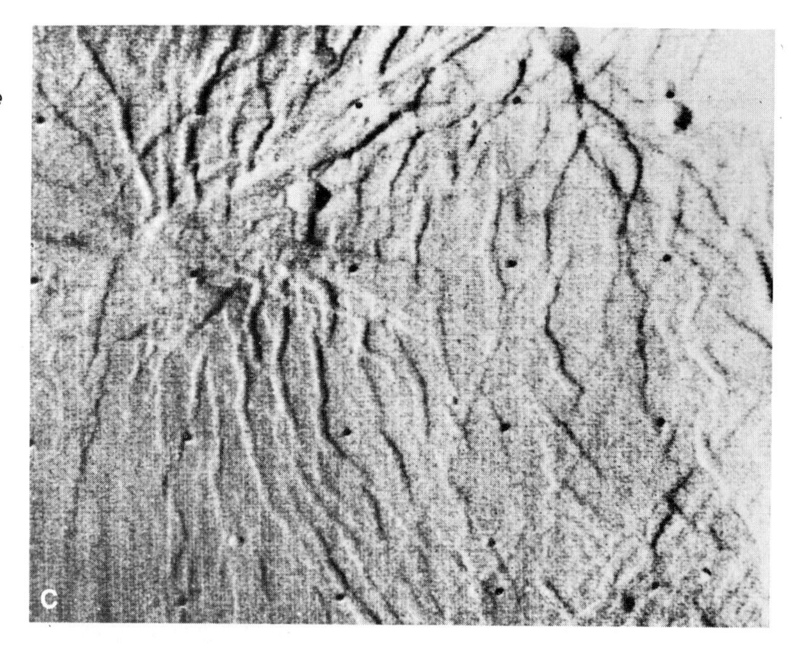

On November 13, 1971, an astronomer at the Goldstone Radio Observatory in California pressed a button. The 80-foot dish of the radio telescope aimed a radio signal out into space. After traveling 7 minutes at the speed of light, the signal was received aboard an unmanned spacecraft, Mariner 9, as it hurtled close past the planet Mars.

Mariner 9's rocket engine fired in response to the radio signal. A stream of hot, expanding gases, like the tail of a comet, slowed down the spacecraft. It went into an orbit around Mars, passing at its closest within 800 miles of the planet's surface.

A giant dust storm that first hid Mars finally cleared. TV cameras began to click. Infrared and ultraviolet telescopes began to measure radiations reflected from the planet. Mariner's radio began to send back a stream of pictures and other observations of Mars. Three photographs taken by Mariner 9 are shown in A, B, and C.

Mariner 9 was the first spacecraft ever to orbit another planet in the solar system. Its photographs from 800 miles were the closest ever taken of another planet. Later in November, two Soviet spacecraft also became artificial satellites of Mars.

Photographs and other data from all these spacecraft may help scientists answer such questions as the following:

Is there enough oxygen and water on Mars to support any life as we know it on earth?

Are the dark patches that are seen on Mars plants of some kind?

Do any of the spacecrafts' cameras, telescopes, and other instruments show any signs of life on Mars?

These spacecraft will give us better maps of Mars than we have ever had before. But probably we will have to wait some time before the question of life on Mars will be answered "yes" or "no."

Process: observing, inferring

What makes the planets close to the sun so different from those far out, you wonder. With further study, you find that the first four small planets from the sun are very massive for their size. They are made up mostly of rock material. So you decide to call them the **rocky planets.** But Jupiter and the planets beyond turn out to have much less mass for their huge size. So you decide that they must contain large amounts of gases, like the sun. These you call the **gas planets.**

You have a hard time trying to decide which planet to land on. With your strongest telescope you take the photographs in Figure 4/19. Jupiter's Great Red Spot and Saturn's rings are mysteries you would like to solve. With its rapidly changing clouds in its atmosphere and its great oceans, the earth is very unusual. And the earth's moon nearby might be worth studying, too.

Any of the planets would make good spaceships, you think, as you draw nearer to them. The inner, rocky planets zip around the sun as if they were racing each other. You find that the third planet from the sun, the earth, is speeding along at about 19 miles a second. But the outer gas planets move lazily along. Neptune creeps around the sun at only 3.4 miles per second. Its year is 165 earth years.

Figure 4/19
The planets (A)Jupiter, (B)Saturn, (C)Earth, and (D)Mars.

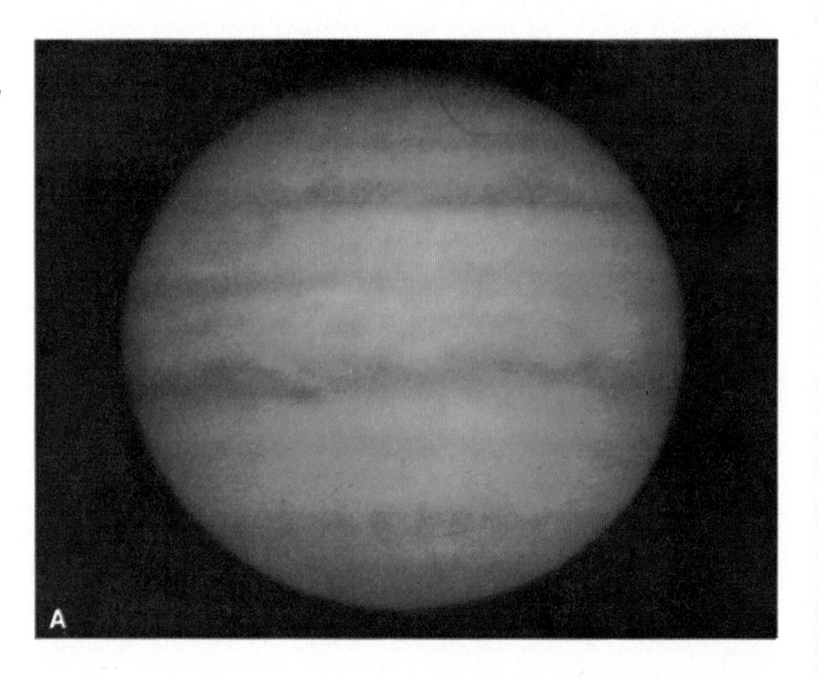

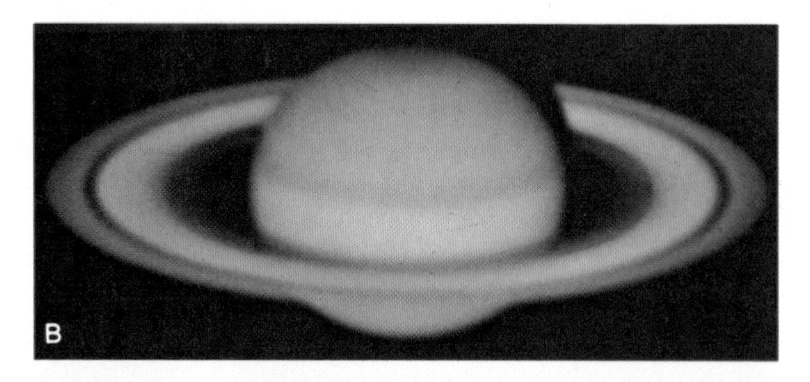

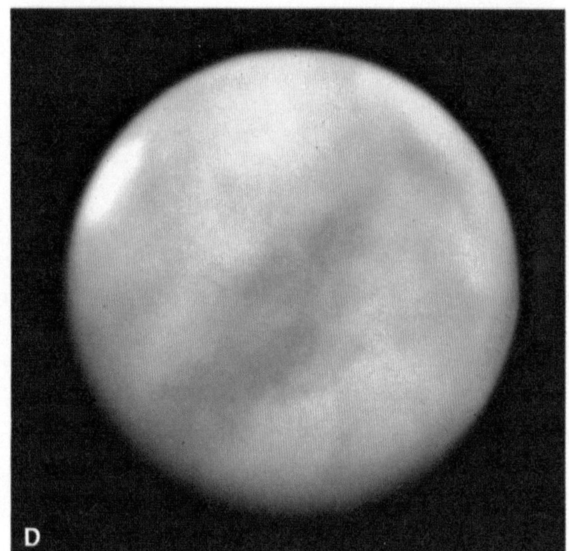

You compare the rotation periods of the planets with the time it takes them to go around the sun, which is called their **period of revolution.** Figure 4/20 compares the periods of rotation and revolution of the planets. Though the gas planets have longer periods of revolution, they spin much more rapidly than the rocky planets. You are curious about the strange rotation of Venus. One of the astronomers suggests that some giant object may have struck Venus very hard. The collision might have stopped its normal rotation and made Venus spin the other way.

You have a meeting on the spaceship to decide which planet to land on. Which planet would you choose? What reasons would you give for your choice?

Planet	Rotations per earth day	Period of rotation	Revolutions per earth year	Period of revolution
Mercury	6°	59 days	4+36°	88 days
Venus	1.5°	245 days	1+216°	225 days
Earth	1	24 hours	1	365.25 days
Mars	352°	24.6 hours	191°	687 days
Jupiter	2+144°	9 hours 50.5 minutes	30°	11.9 years
Saturn	2+108°	10 hours 14 minutes	12°	29.5 years
Uranus	2+72°	10 hours 49 minutes	4°	84 years
Neptune	1+216°	15 hours	2.2°	165 years
Pluto	60°	6.39 days	1.5°	248 years

Figure 4 / 20 *Compare the periods of rotation and revolution of the earth with those of the planets.*

4.8 The solar system is like a circus.

So far you've been looking at the sun, the planets, and their moons. But the solar system is full of many other things besides these.

Thousands of tiny **asteroids** (ASS-ter-oyds), very small planets, flock together mostly in the space between Mars and Jupiter. Once in a while, a stray asteroid passes close to earth. Probably asteroids are iron or rocks quite similar to those on earth.

Scientists have dreamed up several ways to account for the asteroids:

1. Asteroids are the pieces of several small planets that broke up and are no more. The pieces keep colliding in space, splitting into smaller and smaller bits.

All three theories of asteroid formation are "alive and kicking" today. The view that Jupiter captured 7 asteroids for its outer satellites is well established theoretically.

2. Asteroids are small chunks of matter that have been floating around since the early days of the solar system. Perhaps the attraction of gravitation would have pulled the chunks together to form a planet, but the mighty tugging of Jupiter nearby kept this from happening. The seven small outer moons of Jupiter are probably asteroids captured by the strong gravitational attraction of Jupiter.

The number of meteors seen increases greatly when the earth passes through meteor streams in its orbit. The resulting meteor showers are named for the constellations from which most of the meteors in the shower appear to come: October 10 (Draconids), October 21 (Orionids), November 4 (Taurids), November 10 (Andromedids), November 16 (Leonids), December 13 (Geminids), and December 22 (Ursids), to name only the fall showers.

3. A large planet broke up. Mars and possibly even the earth and moon were created from its larger pieces and the rest of the asteroids from its smaller ones.

As yet, none of these ideas has been proved to be correct, although the first theory is favored by some astronomers.

Large clouds of hydrogen gas have recently been found around the heads (or nuclei) of comets.

Many small pieces of matter, called meteors (MEE-tee-ors) flash in the sky when they come into the earth's atmosphere. Meteors may be as small as grains of sand or as large as buses. You can see bright streaks from meteors on a clear night. Figure 4/21 shows three unusually clear trails. When meteors are caught by the earth's gravitational attraction and fall to the ground they are known as **meteorites. Comets** have heads made mostly of dust and frozen gases, with gassy tails streaming far out into space, as you can see in Figure 4/22. They plunge in toward the sun, round it, and disappear into space. Most meteors are the remains of comets. Other meteors may be pieces of broken-off asteroids that have collided with each other. These are the stony and iron fragments that actually fall to the earth.

Figure 4/21
Meteor trails photographed from a balloon, 19 miles up in the sky. The meteors entered earth's atmosphere near the top of the picture and left the atmosphere near the middle.

Figure 4/22 *The comet Arend-Roland had an odd "spike" pointed toward the sun and a long, waving tail.*

Something new is constantly being discovered in our solar system. Who knows? Other planets may be sighted, and more satellites of planets may be discovered. Before long, spacecraft may be sent right through the heads of comets to measure and report what they are like, just as planes now fly through the eyes of hurricanes.

Sometime, astronomers tell us, rocket engines could push an asteroid into orbit of the earth, giving us another moon. Such asteroids might be a new source of iron and nickel, after earth mines run out. And sometime astronauts might hitch a ride on an asteroid passing close by, and go far out through the solar system on it.

Did You Get the Point?

If you know the distance to the sun and the proportions of the triangle it makes with a pinhole, you can find the diameter of the sun.

The first four small planets are rocky, and the giant planets are mostly gas. As you come to know more about the planets and their moons, you learn that each one has its own nature.

Other bodies in the solar system include small planets or asteroids, smaller bits called meteors, and huge gassy comets.

Check Yourself

1. How do the rocky and gassy planets differ from each other in size? mass? distance from the sun? rotation? revolution?

2. How many earth days are there in a Neptune year? How many Mars days in a Mars year?

3. Describe a space trip on which you explore Venus and Mercury, then go past the sun to Mars, to an asteroid, and still farther out. How far would you travel? What would you see and feel? What would you look for?

4. Make a list of what you would like to know about the sun, planets, satellites, and any other bodies in the solar system.

Answers:

1. Rocky planets are smaller, have less mass, are nearer the sun, rotate slower, and revolve faster than gas giants.
2. 165 earth years in 1 Neptune year, $165 \times 365 = 60{,}225$ earth days in a Neptune year. Mars' year = 657 earth days and Mars' day = 24.6 hours. Therefore, 657 days of 24 hours = x days of 24.6 hours; x = 640.6 days.
3. Details of trip and things seen on it can be gleaned from chapter to this point.
4. Question is designed to get students using material in chapter to think of what they don't know but would like to know about the solar system. You may be able to answer some of the questions, or you can encourage students to find the answers.

Jupiter's surface temperature has been found to be unusually high. Were Jupiter somewhat larger, with higher temperature and pressure conditions, it might have started fusion and become a small star. Then our solar system would have become a binary star system of the type that is so common, probably without planets.

Activity Time: About 25–30 minutes. Each student can pick, or be assigned, one of the three 10-year periods given in the data. Activity can be done as individual work in class or at home, or as class project. Units on graph stand for 1/1000 of an A.U. (93,000 miles) and sun itself has a radius of 435,000 miles.

Materials

graph paper

Look for Other Solar Systems

4.9 What is happening to the sun?

Our sun is a small star, classified as a yellow dwarf. But it is a huge body, compared with the earth and other planets that revolve around it in the solar system, as you see in Figure 4/23. The diameter of the largest planet, Jupiter, is only about one-tenth that of the sun. The planets are held in regular paths around the sun by its strong gravitational attraction. But the sun itself behaves in a very funny way.

★ You can investigate this strange behavior of the sun by using part of the data in Appendix P. Using these data, plot the predicted positions of the sun over a 10-year time on a graph. Set up your grid like that in Figure 4/24. See what you can conclude from what you find.

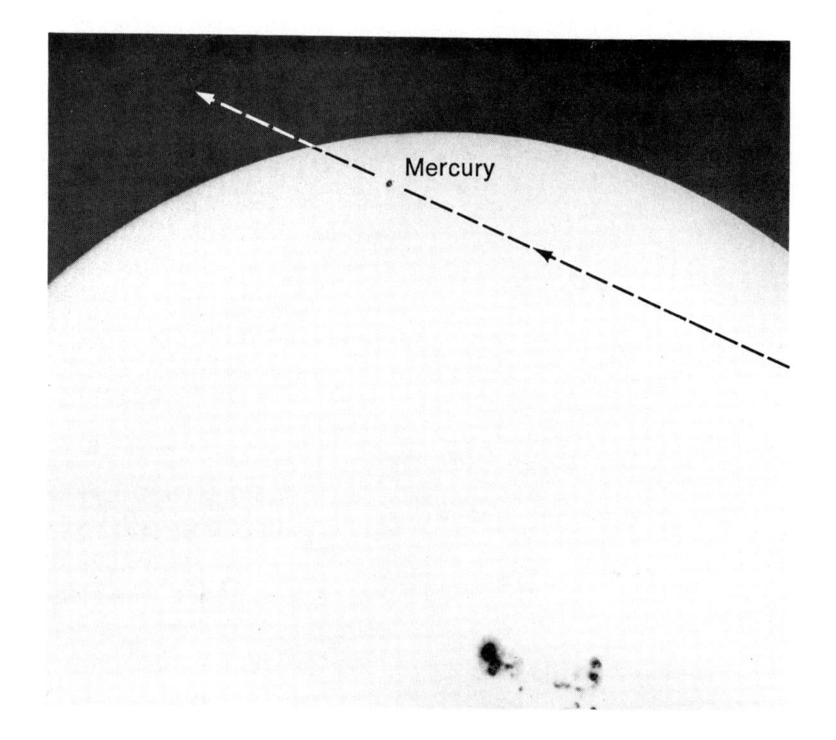

Figure 4/23
The planet Mercury appears to be a tiny speck crossing in front of the huge sun and its sunspots.

Figure 4/24
Make a grid like this to see the strange behavior of the sun. Imagine that you are looking down on the center of the solar system as you plot the sun's positions.

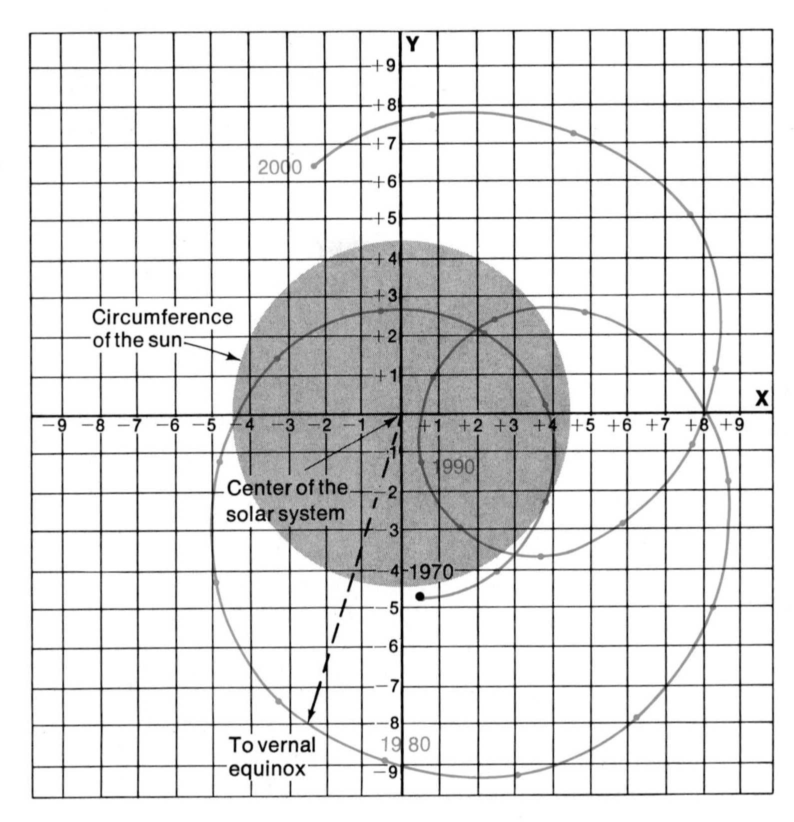

Sun's center is closest to center of solar system in early 1990, in this particular cycle of years.

Motion of sun's center is caused by varying gravitational attraction of the sun by the planets as their positions change in relation to each other and the sun. Sometimes the planets pull at the sun together, sometimes against each other.

Astronomers doubt if nuclear fusion, converting hydrogen into helium, could start up on stars of much less than 12% of the mass of the sun. Such stars would cool very rapidly and would not remain long on the main sequence, as does our sun, so they would hardly provide conditions for life on any planets.
Concept: interactions of matter

Use dots or small circles for the sun's positions. Connect the dots with a curved line to show how the sun moves along between positions. In what year will the center of the sun be coming closest to the center of the solar system, where the lines cross on the graph paper?

What can you conclude from this picture of the sun's path? What could make the sun behave this way? Is it just moving around by itself, or what? Think of any things that might cause the sun to follow such a strange path. ★

4.10 Can there be other planets like the earth?

Several other stars have been observed that show unusual motions in space, like those of our sun. (See Figure 4/25.) What makes them move like this? Astronomers think that one or more small dark bodies must be moving around these stars, tugging at them. They believe that these bodies

Figure 4/25
Barnard's star has the largest stellar motion known. The photograph at the bottom was taken 22 years after the one at the top. It also wobbles like our sun.

A

B

are not large or massive enough to start the nuclear fusion that makes stars radiate light. Could these dark bodies be planets?

The path of one of these stars, Barnard's star, has shown that it probably has at least one, if not two, planets circling it. You can see why the sun swerves in its path, because of the tugging of the planets in the solar system.

Opportunity for discussion of possibility for extra-terrestrial life on planets of other star systems, and, if such life should exist, possibilities for communications with it. Communications would be very slow because the distance in light-years equals the time for a one-way radio message.

Can you count to 300 in one minute? If you can count that fast, it will take you about 1000 years without stopping to count all the stars in the Milky Way galaxy right around us. It may have as many as 150 billion stars. And astronomers have figured out that about 6 billion of these may have planets around them. Could any of these planets have life on them?

To support life as we know it, a planet must have air around it. Only planets nearly the size of the earth or larger have gravitation strong enough to hold onto an atmosphere. About one billion stars may have planets this large. Think about this when you look up into the night sky. Of each 150 stars you see in the sky, perhaps one of them has an earthlike planet or planets moving around it.

One out of every 150 stars may have planets like the earth, on which intelligent beings may have developed. Would they be interested in communicating with us? Obviously, messages would have to be sent in signals that would have meaning for any intelligent being. When contact was made, each message back and forth might take thousands of years. But the information exchange might well be worthwhile.

Did You Get the Point?

The planets pull the sun about by their gravitational attraction, so the sun is pulled back and forth around the center of the solar system.

Other stars, like Barnard's star, move the same way as the sun, so they probably have planets, too.

One out of 150 stars may have planets that can support life. Perhaps sometime we will be able to communicate with intelligent beings on a planet of such a star.

Check Yourself

1. Pretend you just saw a TV science-fiction show. In it the earth has contacted intelligent beings from a far-away planet moving around a distant star. A friend of yours says that the program was stupid and that it couldn't possibly happen. What do you know about stars, their motions, and bodies around them that would help to convince your friend that it could happen?

What's Next?

Compared with the vastness and emptiness of galaxies, our solar system around one small star seems tiny and snug. This system with all its planets, moons, meteors, and comets makes up our surroundings, or environment, in space.

In this environment is our home, the earth, and our nearest neighbor, the moon, which men have just begun to visit. In the next chapter you take a closer look at these great spaceships, the earth and moon. What are they like, how do they differ, and what effects do they have on each other?

Skullduggery

Directions: Your teacher will supply you with a mumbo jumbo of letters like those in Figure 4/26. Within them there is a reminder of material covered in this chapter—a kind of secret message. You will recognize the message after you have crossed out some of the letters in the mumbo jumbo.

Figure 4/26
Copy this grid for the Skullduggery.

	13	14	15	16	17	18	19	20	21	22	23	24
1	P	E	G	A	E	V	R	R	K	A	L	A
2	E	P	I	S	V	E	I	T	L	A	T	L
3	P	I	A	E	W	C	D	O	O	T	N	M
4	I	A	O	A	N	L	W	N	M	D	A	A
5	T	S	T	K	O	E	R	O	P	S	A	Y
6	S	T	C	H	I	T	R	B	I	O	O	N
7	V	V	R	R	S	N	Y	E	A	A	W	U
8	A	T	S	R	I	O	W	E	S	U	W	I
9	T	O	L	B	T	F	I	O	H	M	A	S
10	R	N	S	E	A	S	S	E	M	S	A	N
11	S	O	D	N	W	I	A	M	D	N	L	G
12	I	M	O	S	T	N	A	N	C	E	E	S

Secret message is "Gravitational attraction varies with masses and distances." Words to be filled in to complete the statements are: (1) Kepler; (2) ellipse; (3) comet; (4) wind; (5) sky; (6) orbit; (7) Venus; (8) two; (9) Io; (10) Mars; (11) now; (12) one; (13) pass; (14) man; (15) solar; (16) Brahe; (17) two; (18) five; (19) day; (20) moon; (21) map; (22) sun; (23) law; (24) gas.

To find out which letters to cross out, fill in the blanks in statements 1–24 on a separate sheet of paper. Now find the row with the same number as statement 1. Cross out the first letters you come to in the row that spell "Kepler," the word that completes statement 1. Do the same for all other statements. Cross out letters from left to right or from top to bottom. Do all the *across* words first, then do the *down* words the same way.

Across

Example

1. (*Kepler*) was the astronomer who found that planets do not move in circles. (Cross out the letters KEPLER in order in row 1 across on your copy of the mumbo jumbo.)
2. The moon's path around the earth follows an ______.
3. A ______ is not an animal, but it has a head and a tail.
4. This kind of ______ is called "solar" because it comes out from the sun.
5. The ecliptic follows the zodiac through the ______.
6. An ______ is shaped like an ellipse, not a circle.
7. ______ is the name of the second planet out from the sun.
8. The planet Mars has just ______ satellites.
9. A small satellite of Jupiter, named ______, does not have an atmosphere of its own.
10. Most of the asteroids are found between Jupiter and ______.
11. What was once a large mass of gas and dust has ______ become the solar system of today.
12. The earth's distance from the sun is ______ astronomical unit.

Down

13. When bodies ______ stars in the sky, they must be in our solar system.
14. Natural satellites are not made by ______, but artificial satellites are.
15. The ______ corona flashes out in the sky during a total eclipse of the sun.
16. Kepler used the observations of an astronomer named ______ in studying the paths of the planets.
17. The planet Neptune has ______ satellites.
18. The planet Jupiter is about ______ astronomical units from the sun.
19. Mars' ______ is almost as long as the earth's because they have almost the same period of rotation.
20. The name of the earth's only natural satellite is the ______.
21. A ______ of the sky can show backward motion of a planet.

22. The single body around which the asteroids orbit is the ______.

23. Newton discovered the ______ of universal gravitation.

24. The giant outer planets contain large amounts of ______.

For Further Reading

Adler, Irving. *The Sun and Its Family.* New York, The John Day Co., 1958. The movements of the earth and other planets are clearly explained. There is also a discussion of ways to find our distance from the sun.

Jackson, Joseph H. *Pictorial Guide to the Planets.* New York, Thomas Y. Crowell Co., 1973. A guide-book through the solar system, with many photographs.

Sagan, Carl, and Jonathan Norton Leonard. *Planets.* New York, Time, Inc., 1966. A beautifully illustrated book on the planets, the people who have studied them, and their ideas.

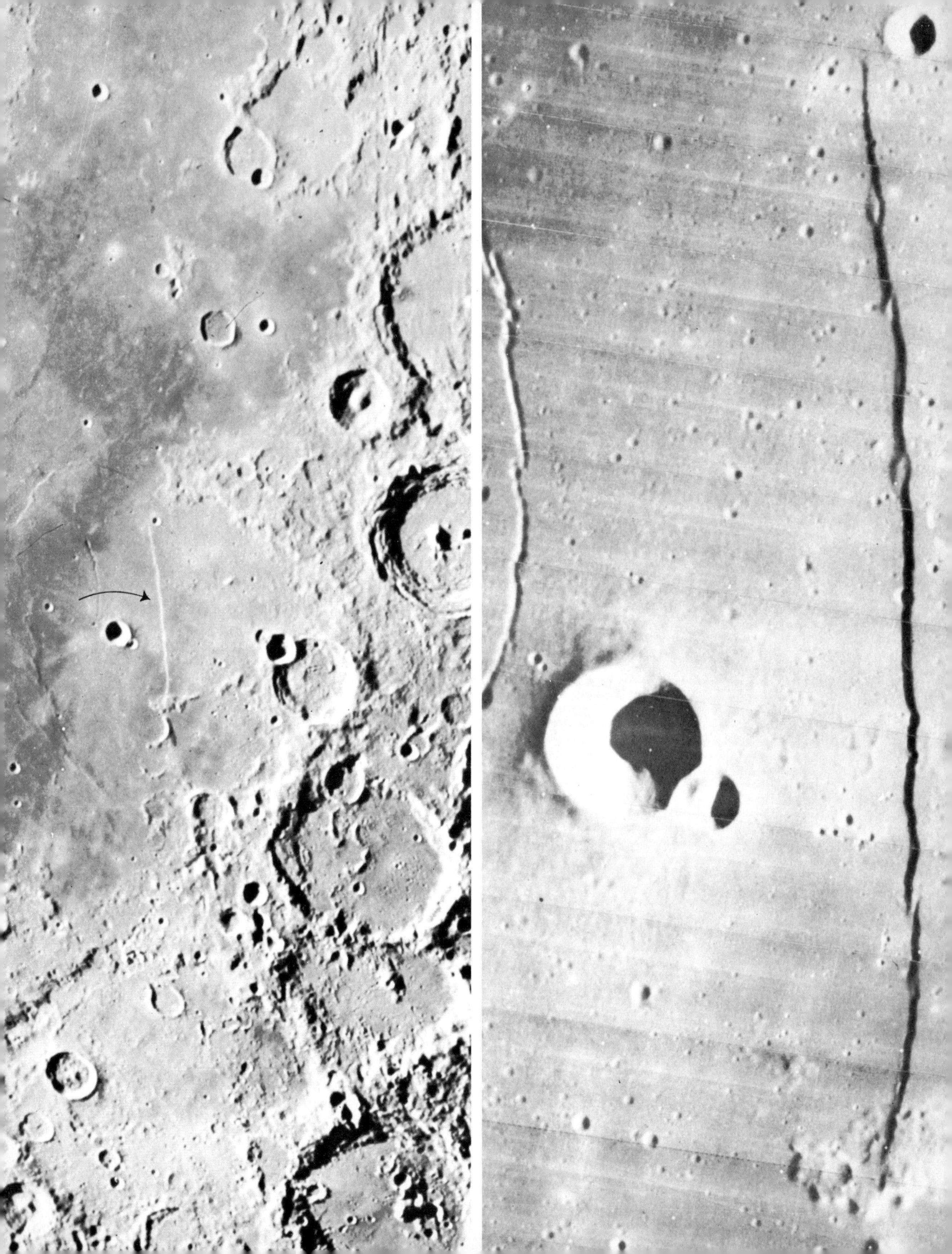

Chapter 5 Moon and Earth

People have always enjoyed looking at the moon and wondering about it. They have written songs and poems about the moon. Ancient people thought there must be human beings living on the moon. Or they imagined many strange moon people or fearsome beasts on the moon's bright continents and in its dark seas. And always the man in the moon, imagined from those bright and dark regions on the moon's face, looked down at us on earth.

Soon after telescopes were invented, astronomers began to map the moon. They discovered squares and triangles they thought were cities. They even saw a long, straight line that they named the Straight Wall. The arrow on the telescope photograph points to it.

For centuries many astronomers thought that the moon, all the planets, and even the sun under its surface, were the homes of strange people or other living things. These beliefs gradually weakened, as more was learned about actual unlivable conditions on the sun, moon, Venus, Mercury, and other planets. Mars seems to be the sole remaining planet on which microorganisms might have developed, even under its severe conditions.

That straight line is 75 miles (120 kilometers) long. It reminded people of the Great Wall of China. Intelligent beings must have built the Straight Wall on the moon, they believed. Nature herself contains no straight lines; they exist only in mathematics.

Scientists began to argue about the Straight Wall. The moon had no air, some said, so how could people or any other things live on it and build city squares or straight walls?

In modern times, of course, astronomers did not believe that people built the Straight Wall. But still they argued about what it actually was. Then, just recently, a satellite from earth was sent into orbit around the moon. This Lunar Orbiter took photographs from only 70 miles (112 kilometers) above the moon's surface. Look at its photograph of the Straight Wall. This close-up photograph finally settled all the arguments.

Lunar geologists regard the Straight Wall as a fault scarp on the moon, similar to those on earth, introduced in Section 11.2.
Process: observing

The Straight Wall is a place where the moon's surface was broken and one side was raised some 800 feet (250 meters) above the other side. Breaks like this on the earth are called faults. Something must have happened inside the moon to break its surface in this way. Could there have been a moonquake?

There's more in the photograph, if you look closely. Notice the uneven curve to the left of the big hole, or crater, in the photograph. This is a groove or trough. These troughs on the moon are called **rills.** And the astronauts of Apollo 15 studied the Hadley Rill, another big trough like this one.

Now astronauts are walking and riding across the moon's surface. They collect rocks, drill holes, and take pictures. They leave instruments to report moon events back to earth by radio.

In the telescopic photograph, the moon is in last quarter and the sun is on the left, lighting the vertical ridge of the Straight Wall. The sun is on the right in the Orbiter photograph, so the Straight Wall casts a dark shadow, the moon in first quarter.

Look at the two photographs of the Straight Wall again. There's a big difference in the Wall in the two photographs. A bright Wall in one photograph and a dark Wall in the other! The sun must be shining on the Wall from two different directions. Does the moon have day and night then? Does the sun move through its sky? And why does the moon show so many different shapes as we see it from the earth?

Answers: Moon has day and night, as sun appears to move through its sky, with the moon following its orbit of the earth. The moon's varying shapes, or phases, are studied in Section 5.1.

Earth and Moon Paths

Activity Time: 30–35 minutes. Groups of 4 to 6 can use each earth-moon-sun model and make their drawings of lunar day and phases. Bring the flashlight closer to the earth and moon, if the shadows on them seem too dim under your lighting conditions. If possible, use a slide projector for the sun.
Process: using space-time relationships

5.1 Make the moon's shapes.

You've seen in the photographs of the Straight Wall that the position of the sun above the moon changes. And you know how the shape of the moon as seen from the earth changes during a month. These changes in shape are shown in Figure 5/1 and are called the **phases** of the moon: new moon, crescent moon, half moon, full moon, and back to crescent and new moon again.

Figure 5/1 *The changes in shape, or phases, of the moon as seen from the earth during a month.*

Figure 5/2 *The earth waxing from the crescent phase to the half earth phase (left to right) as seen from the moon by the Surveyor 7 spacecraft.*

Figure 5/3
Set up a working model of the earth, moon, and sun. How do the model positions differ from the actual positions?

After completing the lunar phases with the model, students can also watch the earth go through phases as they watch it from behind the moon moving about the earth. "Half moons" are sometimes called "first quarter" and "last quarter," and phases opposite "crescent" moons are "gibbous" moons. Moon is said to be "waxing" from new moon to full moon and "waning" from full moon back to new moon.

The shape of the earth as seen from the moon changes, too. The earth has phases as seen from the moon. You can see a "half earth" in Figure 5/2. The phases of the moon, or of the earth, are caused by the positions of the sun, earth, and moon in relation to each other. The sun lights different parts of the moon as seen from the earth to cause the moon's phases. Let's see just how this works.

Materials

ring stand, clamp and rod, flashlight, ball (earth), ball (moon), thread

★ Set up the earth-moon model as shown in Figure 5/3. Mark an "X" on the earth ball to stand for the place where you imagine you are standing, looking up at the moon. Have the "X" on the side of the ball at right angles to the line from flashlight to ball. Swing the moon around to this same line. Now the moon is exactly between earth and sun in the phase called new moon. You may want to darken the room, if you can, or bring the flashlight closer to the earth and moon, so you'll be able to see the beam from the flashlight more clearly on the balls representing earth and moon.

For an observer at X (or behind X) it is just at or after sunset, in the early evening. The moon ball will be nearly lost to sight in the flashlight's glare, like the real moon in the sun's glare.

Stand behind the earth ball and look past the X and the moon to the sun. What time of day is it for you, the observer at X? What can you see of the moon? This is new moon. At new moon the glare of the sun is so bright that the moon is lost in it. Notice that the moon is directly in front of the sun.

Model moon's shadow moves across earth ball as in a total solar eclipse.

In a **solar eclipse** the moon covers the sun as seen from the earth. See if you can cause an eclipse by slowly moving the moon across the earth-sun line. Does the moon's shadow move across the earth?

In a **total solar eclipse** the moon's disk covers the whole sun as seen from the earth. In a **partial eclipse,** the moon

Figure 5/4
The moon was moving to the left at the end of this partial solar eclipse.

Appendix Q gives table of positions of sun and moon in sky for students to map with partial solar eclipse of August 20, 1971. Positions are given in declination and right ascension, as defined in Section 3.10. Completed map of this eclipse is in Teacher's Manual. To show partial eclipse clearly, students need to use full diameters of sun and moon, about 0.5 degree, or 30 minutes, as shown on completed map and given in Appendix table.

does not quite cover the sun, as you can see in Figure 5/4. In Appendix Q you can see how the moon crosses the sun in the sky in a partial eclipse.

Solar eclipses do not occur with every new moon. This is because the moon is usually higher or lower than the line between the earth and sun. In your model, the earth, moon, and sun are all on the same line.

Process: observing

Now you will see how the moon looks at eight different places in its orbit. Swing the moon around counterclockwise one-eighth of its orbit of the earth, as shown in Figure 5/5. Observe how the moon is lighted by the sun and how the moon looks to you as the observer at X. Make a drawing of what you see. Move the moon seven more times, making a drawing of what you see each time. From new moon back to new moon is a full lunar month of 29.5 days.

The terminator is the line dividing the day (lighted) from the night (dark) hemispheres of the moon. The moon has day and night, but its day lasts 14.5 earth days and its night the same, as the moon moves around the earth.

The line between the dark and lighted parts of the moon facing the earth is known as the **terminator** and marks the division between night and day.

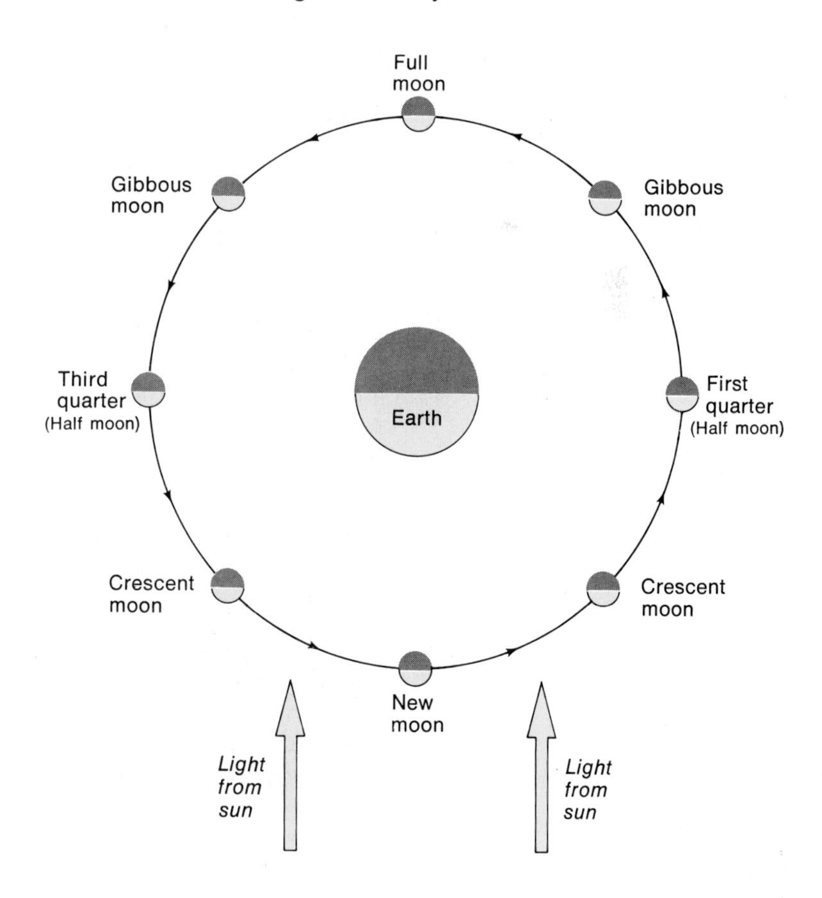

Figure 5/5
The positions of the moon during a month.

Answers: No part of the moon is always dark; one side of the moon always faces the earth and as the moon moves around earth all parts of it have their day and night in turn, as students can see from their drawings. So the farside of the moon has its daytime, too. The moon is often on the line with earth and sun at full moon. Then it passes through the shadow of the earth, giving a partial or total lunar eclipse, which the model shows.

Photograph at right in Figure 5/1 shows first crescent moon, and crescent at left is last before new moon. Moon goes through phases in the sky like those from right to left in photographs.

When students are finished you can use the setup to demonstrate the difference between the sidereal and synodic months, if you wish, using the drawing in Teacher's Manual. Sidereal month = 27.3 days; synodic month = 29.5 days, completing a cycle of phases.

Here are some questions to answer as you take the moon through its whole orbit around the earth:

1. Is any part of the moon always dark? Why or why not?
2. Does the farside of the moon, away from the earth, have its daytime, too?
3. What happens to the moon at full moon, when it is behind the earth on the line from sun to earth to moon?

Compare your drawings of the phases of the moon with the photographs of the moon in Figure 5/1. From your drawings, tell which of these photographs shows the first crescent moon, and which the last? Does the moon go through these phases from left to right on the photographs or right to left? ★

5.2 What paths do earth and moon follow?

The earth and moon models you used in Section 5.1 show the path of the moon around the earth. And you saw why the shape of the moon in the sky changed with the position of the sun in the model. The model is good enough for showing the moon's phases this way.

But the model does not show the exact shape of the moon's path around the earth. For the moon's path is really not circular around the earth, as it is in the model. The moon's path is really elliptical, as are the paths of all the bodies in the solar system. The moon's path is the kind of squashed circle you made in Section 4.4.

The relative closeness of earth and moon is shown by the fact that the moon's distance is only about 10 times around the circumference of the earth, 25,000 miles (40,000 km). Astronauts travel the equivalent of 10 round-the-world trips in going to the moon.

At the moon's closest approach to the earth, or **perigee,** it is only 221,000 miles (356,000 kilometers) from the earth. At its greatest distance, or **apogee,** the moon is out 253,000 miles (407,000 kilometers) from the earth. If the moon's path were circular, it would always be at the same distance from the earth. The moon's average distance in its elliptical orbit is 239,000 miles (384,000 kilometers).

Model in Section 5.1 gives lunar orbit in frame of reference of observer on earth; model here gives lunar orbit in frame of reference of observer above solar system.

In the model in Section 5.1, however, you didn't move the earth along while you moved the moon. Don't forget, though, that the earth moves in an elliptical orbit around the sun. At the same time, the moon moves in its elliptical orbit around the earth. Now just suppose that you are not on the earth but above the earth and moon looking down on them. Here's a way to model how the paths of the earth and moon would look from such a position.

Materials

board (wood or cardboard), yardstick or meterstick, drawing compass

Activity Time: 10–15 minutes. Individual students. Groove of earth's orbit would be slightly curved for exact elliptical path. Moon's path drawn by phases of moon is pictured in Teacher's Manual. Lunar orbit is actually always concave to the sun, as earth and moon together follow an ellipse. See diagram illustrating this point in Teacher's Manual.

★ Scratch a groove down the middle of a board. Do this by running the point of a compass along a meterstick placed on the board. Make the groove deep enough so the compass point won't jump out. Put the point of the compass in the groove at one end of the board. Adjust the compass so that the pencil doesn't quite reach the edge of the board.

The compass point stands for the earth, which moves along the groove. The groove represents the earth's path around the sun. The compass pencil represents the moon and traces the moon's path around the earth. Mark earth, moon, and direction to the sun at one end of the board, as shown in Figure 5/6. Set the compass point and pencil in the positions shown.

Ready to go? Hold the compass between thumb and finger and twirl the compass pencil *slowly* counterclockwise, as you move the point (earth) *slowly* along the groove. If the moon's path wobbles some, draw over it with a pen to smooth it out.

There's the path of earth and moon, from above, as they move together around the sun. Now mark the places on the board where both the earth and moon are at the phases of full moon, half moon, new moon, and half moon again on the board. ★

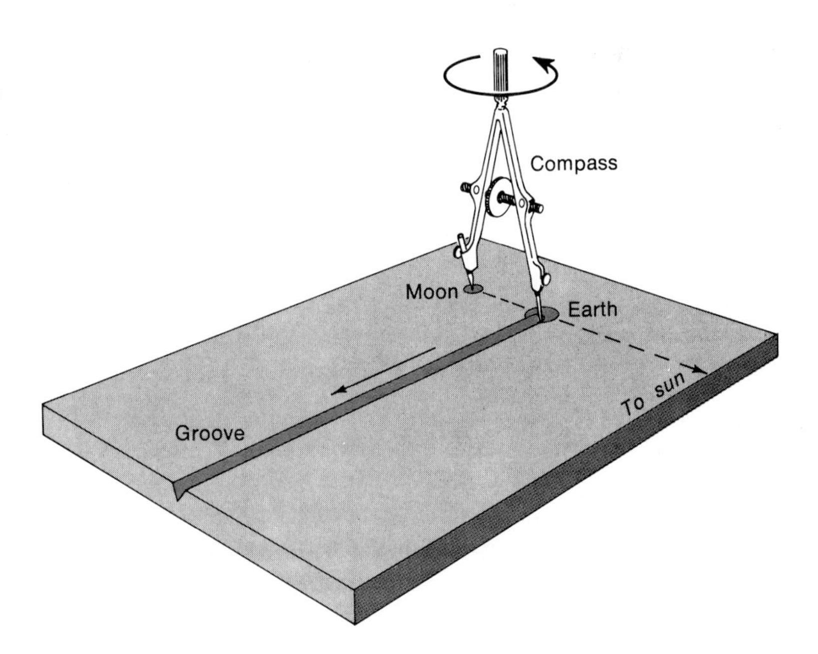

Figure 5/6
Use a compass to find the path of the moon around the earth as the earth moves around the sun.

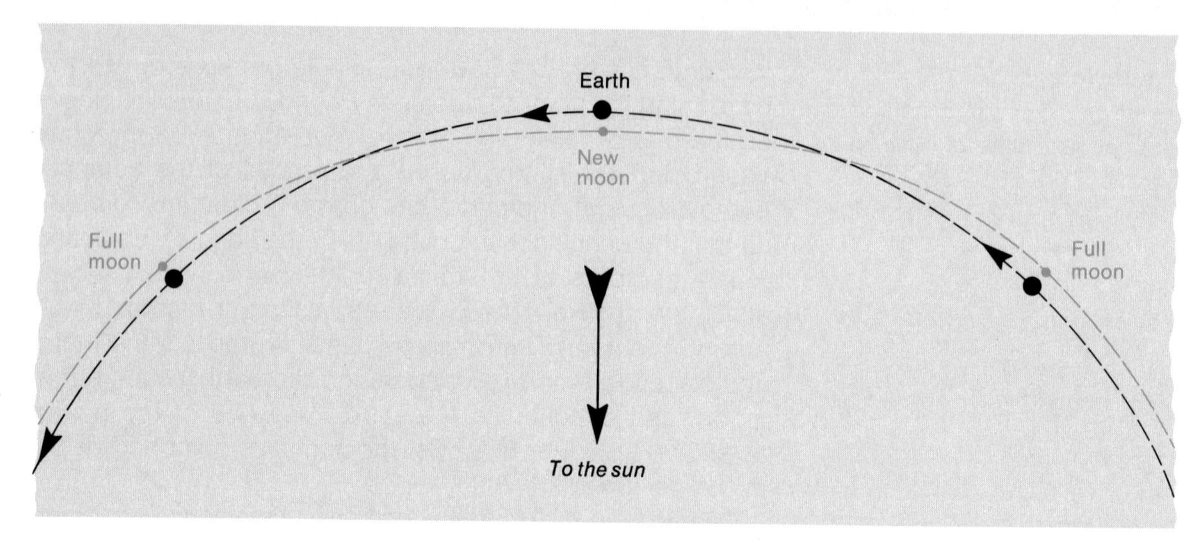

Figure 5/7 *The actual paths of the moon and earth look like this.*

The real earth and moon paths are stretched out a great deal longer than in your model, of course. So the wavy line of the moon following the earth's elliptical curve becomes much less wavy. The moon's path does not ever actually curve in toward the sun, as in your model, but is always curving out, away from the sun, as the moon follows close to the earth in the earth's curving path. (See Figure 5/7.)

5.3 Earth and moon are called a double planet.

The moon is only a little smaller than planet Mercury (diameter = 3030 miles —4880 km) and Mars (diameter = 4220 miles—6790 km). The surface features of both planets appear similar to the moon's, although Mariner 9 revealed that Mars has some features not found on the moon.

Our earth is about 7900 miles (12,700 kilometers) across and the moon about 2200 miles (3540 kilometers) across. The moon's diameter is a little more than a quarter that of the earth, then. No other satellite is nearly as big in relation to its planet as the moon is to the earth. The moon is almost a planet in its own right. For that reason, earth and moon are often called a double planet rather than planet and satellite.

Materials
quarter dollar, disk from paper punch, yardstick or meterstick

★ You can see how the sizes of the earth and moon compare with each other and how close they are for their size. Let a quarter of a dollar stand for the earth. Then a paper disk from a paper punch, about one-fourth the size of a

Activity Time: 5–10 minutes. Groups of two students. Relatively large size of earth and moon and short distance between them can be contrasted sharply with much smaller size of planets in relation to sun and much greater distances from sun, using solar system models from Section 4.1.

Process: measuring

quarter, can be used to stand for the moon. Put the earth and moon just 30 inches (76 centimeters) apart on a table or desk. The 30 inches is the right distance on this model for the 239,000 miles (384,000 kilometers) between earth and moon. On this same model, the sun would be 969 feet, nearly one-fifth of a mile, away from the earth and moon! ★

Since they are so close to each other for their size, as you see in this earth-moon model, the earth and moon affect each other a great deal. In Section 4.4 you saw that the closer two bodies are to each other, the greater their gravitational attraction. The earth is more massive than the moon, and its gravitational attraction is much stronger than that of the moon. How does the tugging of the earth on the moon affect the moon? And how does the moon's tugging back at the earth affect the earth?

The earth's gravitational attraction keeps the moon in the path around the earth that you saw in Section 5.2. But earth's gravitation has another effect. It keeps the same face of the moon, called the **nearside,** turned toward the earth at all times. We only see that one old familiar face of the moon. But at some points in its path the moon slowly swings one side or another around toward us a little. Then we see a bit of the other part of the moon, called the **farside.** In all, we see about 59 percent of the moon. We would see only 50 percent, if it kept exactly the same face toward us.

This effect, called libration, reveals more of the top and bottom of the moon as well as more of its sides, as the moon swings higher or lower in its orbit at an angle to the equator.

A map of the moon in Figure 5/8 shows you both its nearside and farside. Ever since man first looked up at the sky, he has known the nearside of the moon. Only in the last few years has the farside been photographed and mapped, first by unmanned spacecraft and then by astronauts orbiting it.

Lunar map indicates features named in chapter and identifies other prominent features, on both the nearside and the farside of the moon. Nearside has many more large "seas" or maria, but farside is more heavily spotted with craters. A large moon map on wall or bulletin board will help interested students develop more detailed knowledge of moon. See Teacher's Manual for place lunar map can be obtained.

How does the moon's gravitation affect the earth? Because of its spinning, the earth is not a perfectly round ball, but is flattened a little at its poles and bulges a little at its equator. The moon, as well as the sun, tugs most strongly at the earth's equatorial bulge. This pulling makes the earth wobble a bit as it spins on its axis. And this is just the beginning of the story of how the earth and moon interact with each other.

Wobble of earth's axis, called its *precession,* completes a circle in about 25,800 years. Wavy motion of this circle, due to lunar gravity, with a period of 18.6 years, is called *nutation.*

Figure 5/8A *The nearside (A) of the moon is the side we always see from earth. How does it differ from the farside?*

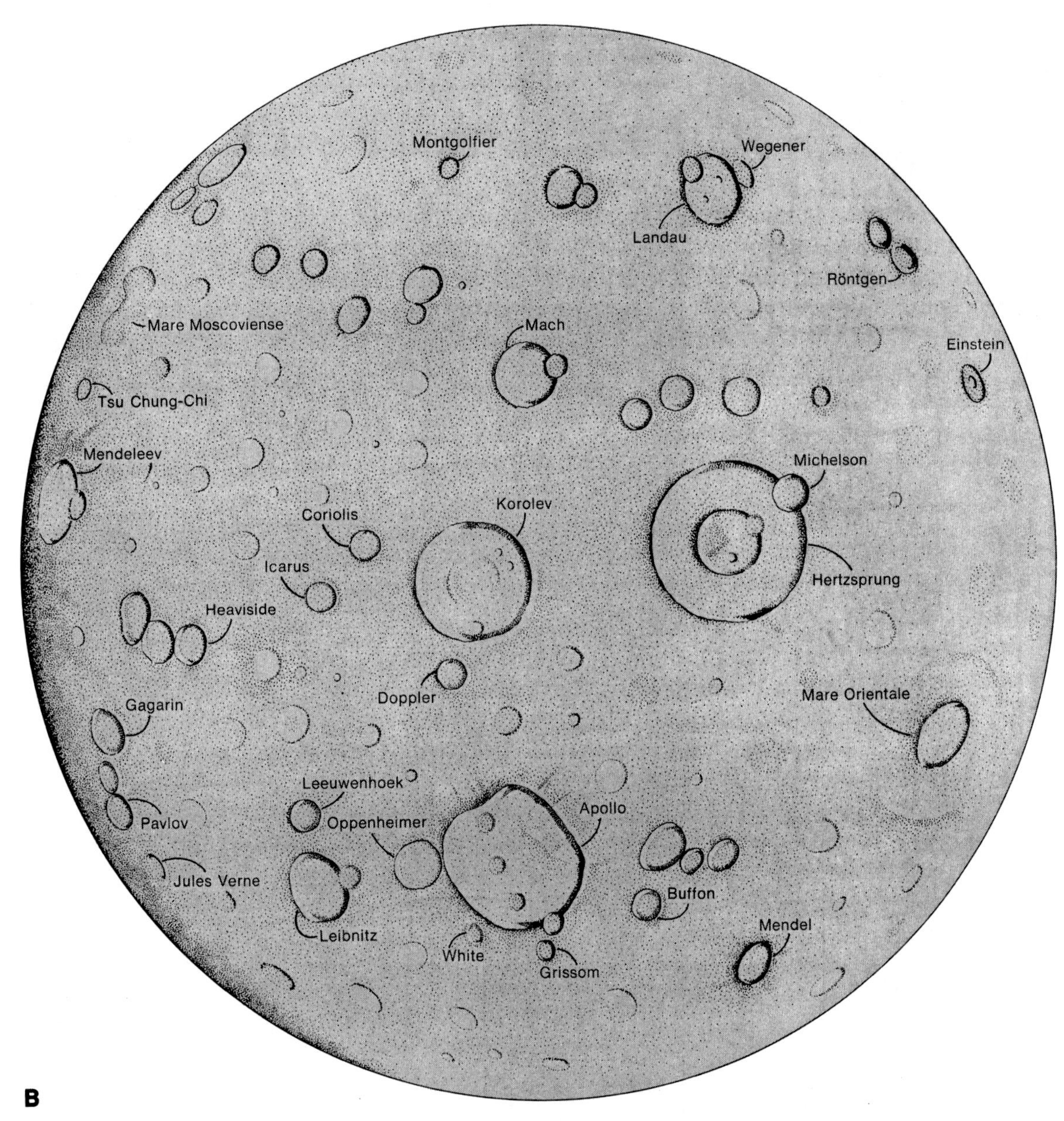

Figure 5/8B *The farside (B) of the moon was never seen until a spacecraft orbited past it and astronauts looked down on it.*

Stargazing with Binoculars and Telescopes

Perhaps by now you have been able to go outside and do some stargazing. Maybe you were even lucky enough to look at the sky through binoculars or a small telescope. If you were, then you know how much more you can see if your eyes have a little help. For example, the features of the moon stand out more clearly. You can easily see mountain peaks and mountain ranges, craters, rays, and even the Straight Wall.

If the planet Jupiter is visible in the night sky, a good pair of binoculars will enable you to see its larger moons, as in the photograph in A. A small telescope will bring out Jupiter's atmospheric features clearly and perhaps Saturn's mysterious rings. (See Figure 4/19.) You might see the moonlike crescent of Venus (B) and the red planet Mars. Points of light reflected from artificial satellites are also clearly seen with binoculars or a small telescope. Amateur astronomers sometimes assist scientists in observing the brightening and darkening of variable stars, as well as in tracking artificial satellites.

Several features of the sky described or pictured in previous chapters are very striking when seen through a telescope. A few examples are the Great Nebula of Orion, the Andromeda Galaxy, and the globular cluster in Hercules.

Using binoculars or a small telescope can make your stargazing more exciting. You might want to try making photographs of objects in the night sky with a telescope. Perhaps you'll even want to make your own telescope. One of the books listed at the end of this chapter will show you how to make and use a simple telescope.

Did You Get the Point?

The changing shapes, or phases, of the moon as seen from earth are caused by the moon's changing positions in relation to the sun and the earth.

Solar eclipses happen at new moon, when the moon comes between earth and sun. Eclipses of the moon come at full moon, when the earth moves between moon and sun.

The moon's path around the earth is elliptical, as seen from space above the earth. The moon's path as it moves with the earth around the sun is a wavy line, as seen from above the system.

The moon is so large in relation to the earth and so close to the earth that they are called the double planet.

Earth's gravitational attraction keeps the moon almost facing the earth.

The moon's attraction helps make the earth wobble on its axis of rotation.

Check Yourself

Imagine you are a pioneer astronaut helping others make the first settlement on the moon. You find you have the following questions to answer:

1. How long would your daytime and nighttime last on the moon?

2. Could you see an eclipse of the sun from the moon?

3. Could you see half the earth all the time?

4. Would you see any meteor flashes or trails on the moon?

5. If you went to the farside of the moon would you ever see the earth?

6. Would it always be night on the farside of the moon?

Answers (students may use sun-earth-moon model of Section 5.1 to help them in answering these questions):

1. Lunar day and night are each 14.5 earth days long, caused by a single rotation of the moon on its axis in one complete revolution of earth.
2. You could see an eclipse of the sun by the earth from the moon.
3. You could see half the earth only at "full earth," at time of new moon, and partial phases of earth the rest of the time.
4. You would never see flashes of meteors on moon, which has no detectable atmosphere to heat the meteors.
5. You would never see the earth from farside of the moon, unless you were very close to nearside and glimpsed the earth during librations.
6. Farside of moon has same daytimes and nighttimes as nearside. Farside's daytime noon occurs at new moon and nighttime midnight at full moon.

Earth and Moon Interact

5.4 How the moon makes tides.

The sun and moon together make tides, raising and lowering the water on earth. The moon's gravitational pull is about 2.4 times that of the sun, because the moon is so much closer than the sun to the earth. At high tide, the

Figure 5/9
Two high tides (A and B) and two low tides (C and D) occur at the same times on opposite sides of the earth. What happens as the earth rotates?

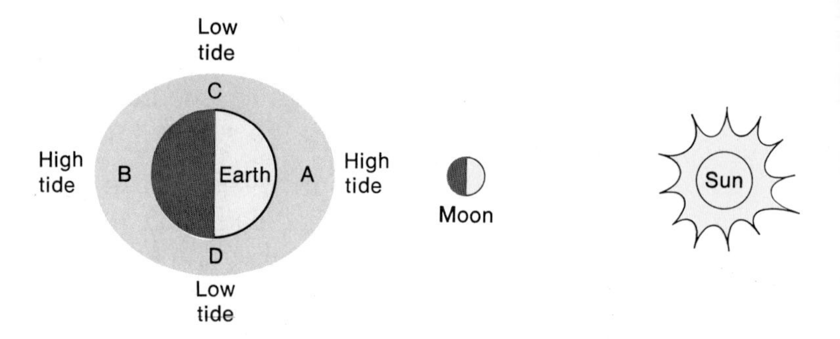

Figure 5/10
The different positions of the moon in relation to the earth and sun cause neap and spring tides. Why does the spring tide bulge out farther than the neap tide?

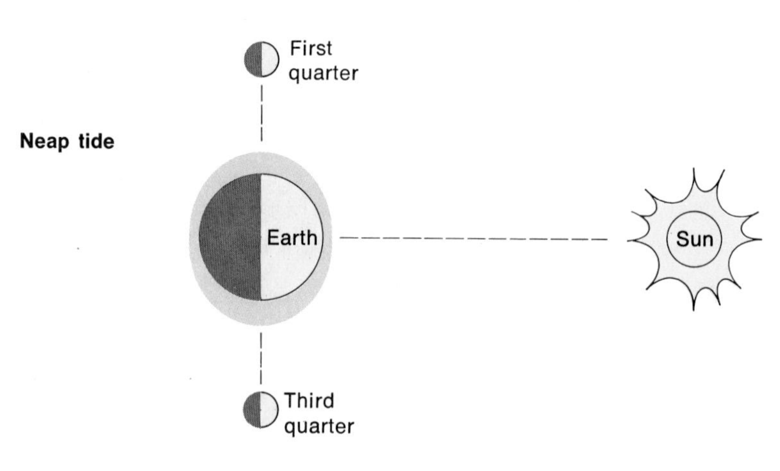

Moon raises bulge in oceans facing it to cause one high tide, and at the same time moon's smaller gravitational force on the other side of earth allows water to rise with centrifugal force of rotation, producing second high tide. **Concept: the interaction of matter**

moon makes the water bulge up in the oceans on the side of the earth facing the moon. Notice the bulge in Figure 5/9. You see the result when the water rises on the coastline. The moon's pull is less on the opposite side of the earth, so it allows the water to bulge up there with the spin of the earth in another high tide at the same time. At low tide, the moon has let the water sink where the moon is at right angles to the water.

Two high tides and two low tides a day (24 hours) sweep around the earth as it rotates in relation to the moon. At **spring tides** with each new and full moon, the sun and moon combine their attractions to make higher and lower tides, as in Figure 5/10. But at the quarter moons, sun and moon

Figure 5/11 *Low tide and high tide in the Bay of Fundy.*

work against each other, so the high tides, called **neap tides,** are lower.

The shape of a coastline plays a part in determining how high the tides will be along that coast. If the rising tide flows into a narrow channel or bay, the water may rise very high. A good example of this is the Bay of Fundy, shown in Figure 5/11, the part of the ocean between New Brunswick and Nova Scotia. Because of this bay's funnel shape, the tides there rise higher than anywhere else in the world.

5.5 What do lines on seashells tell?

Process: observing

Shellfish like clams and mussels live in marshes and bays along coastlines. If you've ever looked closely at a clamshell, you may have noticed lines around its edges. As tides flow in and out, the shellfish grow out along their edges. They live to be 8 to 12 years old. Yearly bands on the shells tell how old they are. The bands along the edges of shells show growth, somewhat like the growth rings you see on tree stumps. Similar rings of hard material are formed by some kinds of single-celled algae.

Scientists have found shells so old they have become **fossils,** the remains of living things made into rock. Some of these are millions and hundreds of millions of years old. These shells grew and made fine rings or bands with every high tide. At the higher spring tides twice a month, the shellfish made even bigger bands as their shells grew. With a microscope to see them, these bands can be counted to tell the number of days (or tides) in a lunar month, during

Figure 5/12
In this microscope photograph of a fossil, the thick dark bands were formed by algae twice a month at the spring tides. One light layer and one thin dark layer were deposited each day. The fossil is about 510 million years old and the lunar month at the time was 31.5 days.

which the moon goes through all its phases. You can count growth bands in fossil shells and in the fossilized deposits of algae. Count the growth bands in Figure 5/12.

Materials
graph paper

★ Use the number of days in a lunar month in Figure 5/13 to make a graph showing how these days changed over millions of years. ★

Figure 5/13
The decrease of days per lunar month over millions of years.

Millions of years ago	Days per lunar month
20	29.50
50	29.82
80	29.92
210	29.68
290	30.07
340	30.37
380	30.53
500	31.56

Students can best set up graph from Figure 5/13 data with millions of years in past horizontally and days per lunar month vertically. Completed graph appears in Teacher's Manual.

As your graph shows, there were about 31.5 days per lunar month 500 million years ago. By about 300 million years ago the lunar month had come down to only 30 days. Little change took place then until about 80 million years ago, when again the lunar month decreased to its present 29.5 days. The bands on the shells showed this.

Astronomical measurements confirm the slowing of earth's rate of rotation adding about 2 seconds per day in 100,000 years. You might ask students how many fewer minutes there may have been in a day in the early stages of man's evolution about 3 million years ago.

Answer: $\frac{3,000,000}{100,000} \times 2$ seconds

$= \frac{30}{1} \times 2$ seconds $= 60$ seconds,

or 1 minute, less.
Process: inferring

The earth must be spinning more slowly on its axis, to make fewer tides and days in a lunar month. Scientists think the pulling of the moon with the tides gradually makes the earth spin more slowly. Changes in the depth and size of the oceans make the tidal effect of the moon weaker or stronger.

As the rate of earth's spinning slows, with fewer days in a lunar month, the moon speeds up a little in its path and moves farther away from the earth. The moon now may be moving about one inch farther from the earth each year, scientists believe.

So the moon causes tides in the earth's oceans. The tides affect the growth of clams. The width of the bands on their shells changes with the height of the tides. And counting these bands tells how earth and moon were moving hundreds of millions of years ago, compared with the present.

5.6 How did the moon form?

If the moon is moving an inch farther from the earth each year, where was the moon a million years ago? a billion years ago? Much closer to the earth, of course. Some scientists think perhaps the moon actually split off from the

$\frac{1,000,000 \text{ years}}{12 \text{ inches}} = 83,333$ feet $= 16$ miles per million years or 16,000 miles per billion years.

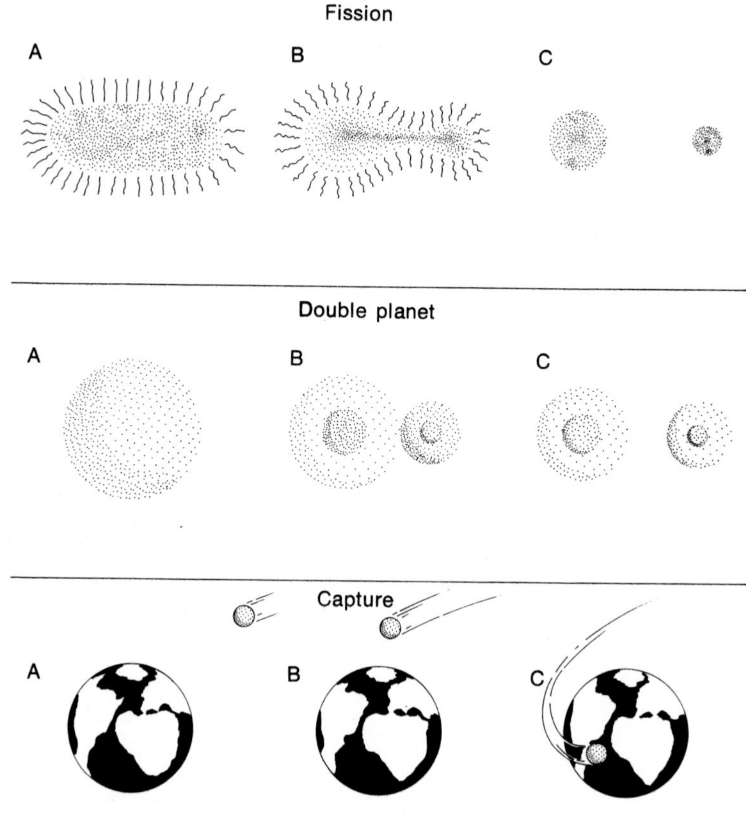

Figure 5/14 *Three different theories of the stages (A, B, and C) by which the moon became a satellite of the earth.*

earth at a time when the earth was spinning much faster and was much hotter. This theory is pictured in Figure 5/14.

The splitting of anything is called fission. So this is called the **fission** theory of the moon's origin. The moon may have split off from the Pacific region, according to this view, where the greatest ocean depths occur. According to another recent suggestion, Mars broke off from an earth originally much larger. Then the moon formed from droplets left between Mars and the earth.

According to another idea, the **double-planet** theory, the moon was formed separately, during the same time the earth was formed. Both bodies grew from the vast dust cloud in which the solar system began. Many scientists favor this view.

Capture theory does not explain *origin* of the moon, as do the other two theories.

Another popular idea is the **capture** theory. According to this idea, the moon formed elsewhere in the solar system, or even zoomed in from space between the stars. Nearing the earth, the moon was drawn close enough by gravitational attraction to be captured by the earth, but not close enough to strike it. So the moon went into orbit around the earth. This theory and the double-planet theory are also pictured in Figure 5/14.

People who disagree with the capture theory say that during capture the moon would have raised giant tides on earth as high as four miles. And possibly this capture would have heated the earth so much that the oceans would have evaporated and the earth's crust would have melted. However, the effects might not have been so great if the moon had been captured at a considerable distance from the earth.

More studies of the moon and of earth rocks and fossils are needed. Then there may be enough evidence to decide which of these ideas is the most satisfactory.

Did You Get the Point?

The gravitation of the moon, with some help from that of the sun, causes the tides in the earth's oceans.

Two low and two high tides a day sweep around the earth as it spins under the moon.

Spring tides at new and full moon are higher, with the sun's pull added; neap tides at the quarter moons are lower.

Growth lines on shells change with the tides, becoming wider at high tides. Counting these lines has shown that the earth's rotation has been slowing down gradually.

As the earth spins more slowly, the moon moves farther from the earth. Long ago the moon may have been much closer to the earth than it is now.

Perhaps the moon broke off from the earth, or perhaps it formed at the same time as the earth. Perhaps the moon was captured long ago when it passed too close to the earth.

Answers:

1. Earth completes full rotation in 24 hours. So moon is directly above a given place (first high tide) and directly above the opposite side of the earth (second high tide) during the 24 hours.
2. Two moons would double the number of high and low tides per day. Moons in orbit would sometimes pull with each other and the sun to make the high tides higher and the low tides lower, and sometimes pull against each other to make the high tides lower and the low tides higher.

Check Yourself

1. Explain why there are two high tides and two low tides in a day of 24 hours.

2. What would happen to the tides if the earth had another moon like its present one?

3. Use your imagination in describing the events that happened to the moon and earth according to the theories of the moon's early history.

Lunar Landscapes

3. Vast events assumed from the three theories of moon's early history can be described, drawn, or modeled by students.

5.7 Make some lunar landscapes.

A **landscape** is the combination of all the surface features of an area of land. When you observe the many forms of the land around you—plains, hills, ridges, valleys—this is your landscape. Landscapes on the moon are very different from those on earth, as you can see in Figure 5/15. You can make real-looking models of these lunar landscapes.

Believe it or not, regular dry cement powder is very similar to the moon's surface and makes excellent "moonstuff" for your model landscapes. Some of the moon's craters were caused by the striking of small and large meteorites from space. Other lunar craters may have been caused by gas and melted rock blown out from inside the moon through volcanoes. You can make both kinds of craters with cement powder.

Materials

cement powder (15–20 cups), deep pan or dish, marbles, nuts, washers, pencil, sticks

Activity Time: 25 to 30 minutes. Groups of 4–6 students model lunar landscapes by impacting "meteorites" and by outgassing processes. Half of groups can work on each process and then switch. Remarkably realistic models can be made in both ways. See photograph in Teacher's Manual.

★ To model with meteorites, pour moonstuff into a deep pan or dish to a depth of 2 to 3 inches. Stir it with a pencil until it is level. Smooth the surface lightly. Drop a couple of marbles into the surface from a few inches above. Drop some larger washers or nuts. Take them out. Smooth the surface, and try again.

Try various ways to make features like those shown in Figure 5/15. A pencil can make a deep crack or rill. What will circular stirring with a pencil do? How can you make a mountain or a mountain range? Can you make a peak

Figure 5/15 *Craters, rills, and the Apennine Mountains form a varied lunar landscape. Part of the Mare Imbrium is at the upper left.*

in the middle of a crater? The **rays** fanning out from some craters can be made, too. Try adding a **mare** (MAH-ray) or flat plain. The plural of mare is maria (MAH-ree-ah).

After you've practiced, try creating the whole lunar landscape shown in Figure 5/15. Or find your own photographs of the moon and make models of them. Or make up an entirely imaginary lunar landscape. ★

Have damp cloths handy to wipe off powder spills on floor or clothes. Tell students to keep powder away from clothes or eyes.

Figure 5/16
Setup for making blowout craters in moonstuff.

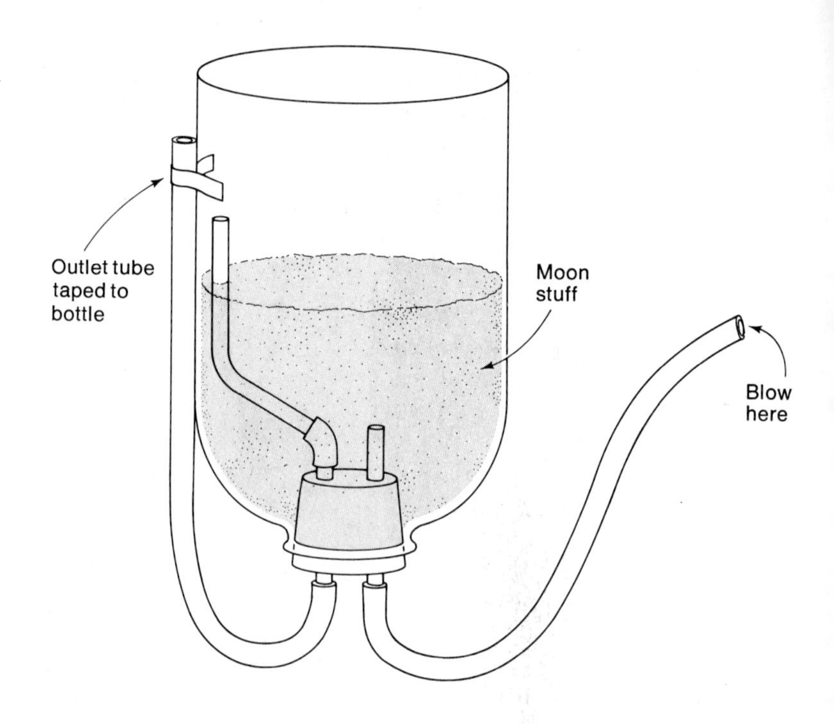

Here is another way to manage your moonstuff, in order to model the volcanic blowout way in which some lunar features may have formed.

Materials
quart glass bottle, stopper (2-hole), plastic tubing (2 feet), cement powder

Caution students not to suck in on either tube and not to blow in air-outlet tube, which makes a stream of moonstuff squirt out of other tube.

★ Fill a quart glass bottle about one-third full of moonstuff, as shown in Figure 5/16. Arrange two pieces of tubing in a stopper, insert the stopper in the bottle, and turn the bottle bottomside up. Be sure the ends of both the air-outlet tube and the blowing tube are above the lunar surface.

Blow gently, like a small volcano, into the tube opening into the lower neck of the bottle. (Don't suck in on any tubes.) Note how air bubbles burst up through the lunar surface. They raise clouds of moonstuff that settle around the forming craters. If you tilt the bottle as you blow, two or three craters may form, with one on the rim of another. Or you can tie up the end of the blowing tube and make holes near the end, if you wish to spread the ''gas'' more widely and make a number of craters. Holes in a piece of cardboard taped over the end of the tube will also work.

Compare your blowout craters with the real moon crater shown in Figure 5/17. Did you notice that fine particles lifted in air begin to flow like water? This same process is

Figure 5/17 *Craters on the moon photographed from a lunar orbiting satellite. Compare your blowout craters with some of these.*

used to blow finely powdered coal through pipelines or to blow powdered iron ore into blast furnaces. Blowouts or larger volcanic actions may have formed many of the lunar craters. ★

5.8 How were the moon's features formed?

Now that Apollo astronauts have explored the moon's surface and brought back lunar rocks, a better comparison can be made between lunar landscapes, such as the one in Figure 5/18, and those on earth.

Figure 5/18
The surface of the moon is covered with material that looks something like loose soil.

Figure 5/19
Pieces of hardened lava in lunar soil, with several glass balls, and brown and yellow glass chunks.

Astronauts can walk easily across the lunar soil, which is much like fine soil on earth, although moon soil contains no living things. It is a mixture of rocky and glassy particles and fragments. The soil is loose on the surface and packed more tightly down to solid rocks at depths of 10 to 20 feet or more.

Lunar soil also contains a few metal particles, like iron and nickel, which may be the remains of meteorites. Most fragments and solid rocks are of the **igneous** kind, rocks that have been heated and melted like the lava rocks from volcanoes on earth. (See Figure 5/19.) Lunar rocks and

Comparison of moon with earth throughout this section is a means of transferring attention from moon to earth, its landscapes, and in particular its atmosphere, to which Unit II turns.

Lunar soil and rocks are largely basaltic and anorthositic (like gabbro) igneous rocks. Such rocks are described in Chapter 9.

Directly under the sun at lunar noon, temperatures over the boiling point of water (212°F) have been measured.

If outgassings are large, the moon may on occasion have local hazes and clouds. These would soon escape from the low lunar gravity and be blown away in the solar wind. Observers from earth have often claimed they have seen hazes obscuring surface features in such craters as Alphonsus and Clavius, and such ringed plains as Plato and Grimaldi.
Process: inferring

soils are over 3.5 billion years old, much older than those found on the earth's surface.

The moon's surface is covered with craters, from small ones a few feet across to great holes tens of miles in diameter. Most of these must have been blasted out by meteorites, large and small. These meteorites blasted holes, melted rocks to glass, and showered chunks of rocks and clouds of dust around. Even little fragments of rocks in lunar soil are pitted with holes from tiny meteorites.

The large areas of melted, or igneous, rock spread across the maria must have flowed from beneath the moon's crust, however. This rock could have been melted inside the moon and blown out of volcanoes. Also, it could have been melted by heat from large meteorites colliding with the moon.

You probably know that water and wind break up rocks on earth and carry rock pieces from one place to another. But there is no wind or water on the surface of the moon. So how do moon rocks break up and get moved from one place to another?

During the long lunar night, the moon's surface may cool to $-250°$ to $-300°$ Fahrenheit. During the day, the surface may be heated up to the boiling point of water. Rocks contracting in the cold and expanding in the heat gradually break up into smaller and smaller pieces.

Meteorites further pit and powder the fragments. Moonquakes shake and tumble the pieces around. Lunar gravity carries the dust down to fill in craters and smooth the surface. So even without wind and water, which wear down the earth, the lunar surface can be gradually smoothed. But the process is much slower than on earth. Unlike your footprints on earth, those left by the astronauts on the moon will not be erased for millions of years.

Many lunar mysteries remain. The strange rills look like old, dry riverbeds on earth, but without water what could have made them? What formed the lunar mountains, which look so different from earth mountains? Was the moon once hot, even melted? Some day, lunar explorations will answer these and many other questions, and may also tell us more about earth processes. The more we can learn about our seemingly changeless moon (Figure 5/20), the more we will know about our ever-changing earth (Figure 5/21).

Figure 5/20 *The astronauts on the Apollo 10 spacecraft saw this view of the moon after they had blasted off on their return to earth. The round dark spot is Mare Crisium.*

Figure 5/21 *The astronauts on the Apollo 11 spacecraft had this view of earth as they left it, outward bound to the moon. Parts of Africa, Asia, and Europe are visible.*

Did You Get the Point?

Landscapes combine all the surface features of an area.

Lunar landscapes have many different features from earth landscapes.

Most soil and rocks on the moon are igneous in origin, having been heated and melted at some point in the moon's history.

Large and small craters, broad round plains called maria, mountains, rills, and rays are prominent surface features of the moon.

Meteorites from space and gases and melted rock escaping from below its surface were the main factors in shaping the moon's features.

The moon lacks the wind and water that wear down the earth's surface.

Lunar rocks are gradually broken up by meteorites and by being heated and cooled over and over again.

The lunar surface is smoothed and craters filled in by the gravity of the moon itself and by moonquakes that occur when the moon is nearest the earth.

Check Yourself

1. How may meteorites and blowouts have produced craters on the lunar surface?

2. Describe what might happen to a square mile of the moon's surface compared with a square mile of the earth's surface.

3. How would people have to live and act on the moon compared with the ways we live and act on the earth?

Answers:
1. Students can draw on Section 5.7 and Section 5.8 to answer this question.
2. The solar radiations of all kinds, including particles, strike the moon directly, although the earth is shielded from many of these by its atmosphere. More meteorites hit the moon, which has no atmosphere to burn them up before they reach the surface. Wind, rain, and running water work on the earth, but not on the moon.
3. Encourage students to summarize forces and changes active on the moon by contrast with those on earth, all of which would affect living conditions: weaker gravitation, barely detectable gas and water, great temperature extremes, no detectable life, no protection from meteorites and solar particle and wave radiations, and long day and night.

What's Next?

The lonely magnificence of the lunar view contrasts with the beautiful variety of the earth as it is seen from space. As opposed to the stark clarity with which we see the moon, the earth is wrapped in a deep and cloudy atmosphere. Oceans lie vast across much of its surface, and its landscapes vary endlessly in color and shape. Even more, the earth's surface supports life in highlands and lowlands, tropics and ice caps, oceans and atmosphere. This earth is our own home planet. Let's take a closer look at it.

Skullduggery

Answers: (1) Straight Wall; (2) crescent; (3) craters; (4) maria; (5) landscapes; (6) phase; (7) solar eclipse; (8) terminator; (9) farside; (10) tides; (11) meteorites; (12) fission; (13) perigee, apogee; (14) capture; (15) igneous.

Directions: The answer to each of the questions is given above the question, but you have to unscramble it. Read the question *before* trying to unscramble the answer. It will help!

1. thaigrts lawl
Astronomers were wrong in thinking that the ______ ______ was built by people on the moon.

2. tensercc
This shape of the moon comes just before and just after new moon.

3. sacrret
Meteorites have blasted these out all over the moon.

4. iraam
Both the name of large round features on the moon and a girl's name.

5. spldacnase
Painters paint them, you look at them every day, and both moon and earth have many.

6. spaeh
The ______ of the moon as seen from earth depends on its position in relation to both sun and earth.

7. lorsa piescle
What can happen only when the moon is on a direct line between earth and sun?

8. rrttnioaem
The earth has one, the moon has another, and each is constantly moving and partly caused by the sun.

9. dafeirs
This part of the moon is all lighted by the sun at new moon.

10. sedti
Lakes have these as well as oceans, to say nothing of the earth's atmosphere and solid ground, which have them too.

11. tormteeeis
We have these ______ on earth, too, but their effects are more obvious on the moon.

12. snosiif
Uranium atoms do this, and some people think the moon started by this ______, too.

13. egeripe ageope
When the moon is at ______ with the earth, it is more likely to have moonquakes than at ______.

14. rupcate
If one theory is true, the moon's path changed a great deal after ______ by the earth.

15. songieu
An ______ rock is one that was formed by heating and melting.

For Further Reading

Alter, Dinsmore. *Pictorial Guide to the Moon.* New York, Thomas Y. Crowell Co., 1967. Full of excellent photographs and maps of the moon, with discussions of what made the moon look like it does.

Dwiggins, Don. *Eagle Has Landed.* San Carlos, Calif., Golden Gate Junior Books, 1970. This book describes the landing of Apollo XI and the exploration of the moon with photographs and diagrams. Also discusses the history of the moon as revealed by stones and NASA's cameras.

Sagan, Carl, and Jonathan Norton Leonard. *Planets.* New York, Time, Inc., 1966. An excellent chapter on the moon, as well as unusual information on the origin of the solar system.

Texereau, Jean. *How to Make a Telescope.* New York, Natural History Press, 1963.

For a fascinating history of the exploration of the moon, read the following publications of the National Aeronautics and Space Administration, Washington, D.C.:
Apollo 8/Man Around the Moon. Publication EP-66
Code-Name: Spider/Flight of Apollo 9. Publication EP-68
Mission Report/Apollo 10. Publication EP-70
Log of Apollo 11. Publication EP-72
Apollo 12/A New Vista for Lunar Science. Publication EP-74
Apollo 13/"Houston, we've got a problem." Publication EP-76
Apollo 14/Science at Fra Mauro. Publication EP-91
Apollo 15/At Hadley Base. Publication EP-94
Apollo 16/At Descartes. Publication EP-97
Apollo 17/At Taurus-Littrow. Publication EP-102

Unit Two

The Earth's Atmosphere

Chapter 6 Atmosphere

During the mid-1970's, NASA may launch several Skylabs with Apollo Saturn rockets. The Skylabs attached to each other will form a space station or observatory as cabins or "modules" are brought up by shuttle. They will be outfitted as orbital workshops, to be staffed by three crews of three astronauts each. The first Skylab launch is described here. The scientists in the Skylabs will make observations of the earth and perform experiments in space.

You are lying in your spacesuit in the command cabin at the top of a huge rocket. The countdown comes through the earphones of your space helmet. "Seven, six, five, four . . ." Watching the instruments you must read, you grip the sides of your couch. The oxygen pumps scream higher and higher. Here it comes! "Three, two, one, zero . . ." A shattering rumble shakes your couch as the rocket engines fire. The whole rocket shudders and shakes. Then you are gently lifted as in an elevator. You and your two companions are off!

The rocket tilts into its course. You are pressed harder and harder into your couch. The rocket's nose seems to sway a little as it plunges faster through the air currents. The air pressure needle drops rapidly. Down, down it goes. In a moment the needle rests at zero. The second engine drops away with a thud. You are in orbit. The streamlined covering of your command cabin falls away. Outside, on the night side of the earth, space is black, spotted with stars. It takes a few minutes to get used to the weightlessness. You let go of your pencil and pad and they just stay there in front of you.

You are one of the astronauts working in the first section of America's space observatory. Next month, a new spacecraft with astronauts will be brought up by a space shuttle, to be joined to yours and make the observatory bigger. You will return to earth in the shuttle.

Clouds and decreasing density of the atmosphere with height are observable. Encourage students to explain why colored bands appear with sunrise. Work out with students a method for finding the radius of the earth as pictured using a compass or a piece of string to the horizon. From this radius, the 100-mile scale length can be derived. Compare this then with the depth of visible atmosphere. Compass in Sky Lab is reacting to passage through disturbed portion of magnetic field (where Van Allen ring dips fairly close to earth).

Your first job is to observe everything you can of the earth's blanket of air above the horizon at sunrise. A colorful space sunrise is brightening in the east as you turn to this assignment.

What can you see of the earth's air, lighted from behind by the sun? Do you notice anything that might be clouds along the horizon, or near it? Where is the air the thickest or hardest to see through? What do you notice about the colored bands in the air? Compare the size of the band of air in the picture with the size of the earth. Can you think of any way to find how long (in inches) 100 miles is in the photograph? Describe all of your observations about the air at sunrise as seen from the spacecraft.

You finish observing the sunrise. As you relax, you notice the pointer on a large compass in front of you beginning to move across the dial. "Look, we're already coming into the first Van Allen belt," you tell your fellow astronauts.

Space Meets Atmosphere

6.1 A magnet makes its own field.

You have seen in the photograph that many features of the earth's air can be observed from space. You can see clouds and dust layers in the air. And you can see rainbow colors as the light waves are separated by the air. But what you observe is only the earth's **atmosphere,** the blanket of gases that surrounds the earth above its surface. These gases and dust particles are made visible by light. But there are many invisible things in and around the earth, too. These include magnetic fields, gravitational fields, and charged atomic particles. And they can have great effects, as you will see.

What is a field? "A place where animals feed, crops grow, or people play football," you may answer. But in addition to these everyday meanings, "field" has another meaning in science. A **field** is a region or area throughout which an effect or a force exists and can be measured. For example, a sound field is the area in which a sound can be heard and measured. The region in which a clock's ticking can be heard is a sound field. The earth's magnetic field is the whole region around the earth in which magnetic effects can be observed and measured. This is often called the **magnetosphere** (mag-NEE-toh-sphere). Let's look into this magnetic field first.

★ Look over a bar magnet, an iron bar that has been magnetized. Even if you smell and taste the magnet, it will still be just a piece of iron to you. People have no senses with which to experience magnetic fields directly. They can only observe the effects of the magnetic fields.

Opportunity to open a discussion of atmospheric features, rare or common, that students and you find interesting. Ask students about other forces we cannot experience directly, but only through their effects: atomic particles (protons and electrons), X-rays (which rats can sense somehow), cosmic rays (constantly passing right through us), gas molecules (large masses of which are felt as wind), and smoke (directly observed, but not the particles in it).

Other examples of common fields are the field of the intensity of light from a light bulb and the field of the intensity of smell around a skunk. Earth's gravitational field, air temperature field, and air pressure field, and other fields are discussed in this chapter. The "field of view" of our eyes, of microscopes, of telescopes, and of cameras is another meaning of "field," namely, the whole area covered by our senses or by instrumental observations. Fields are covered more generally in physical science texts.

Concept: levels of organization

Activity Time: 10 minutes. Individual students or groups of two. Overhead projector can be used to demonstrate field forming, if desired. Place magnet on projector table, cover with glass or plastic plate, and sprinkle the iron filings. Students can make drawings of the best defined magnetic fields produced in this way.

Materials

bar magnet, cardboard, iron filings or powder, pencil and paper

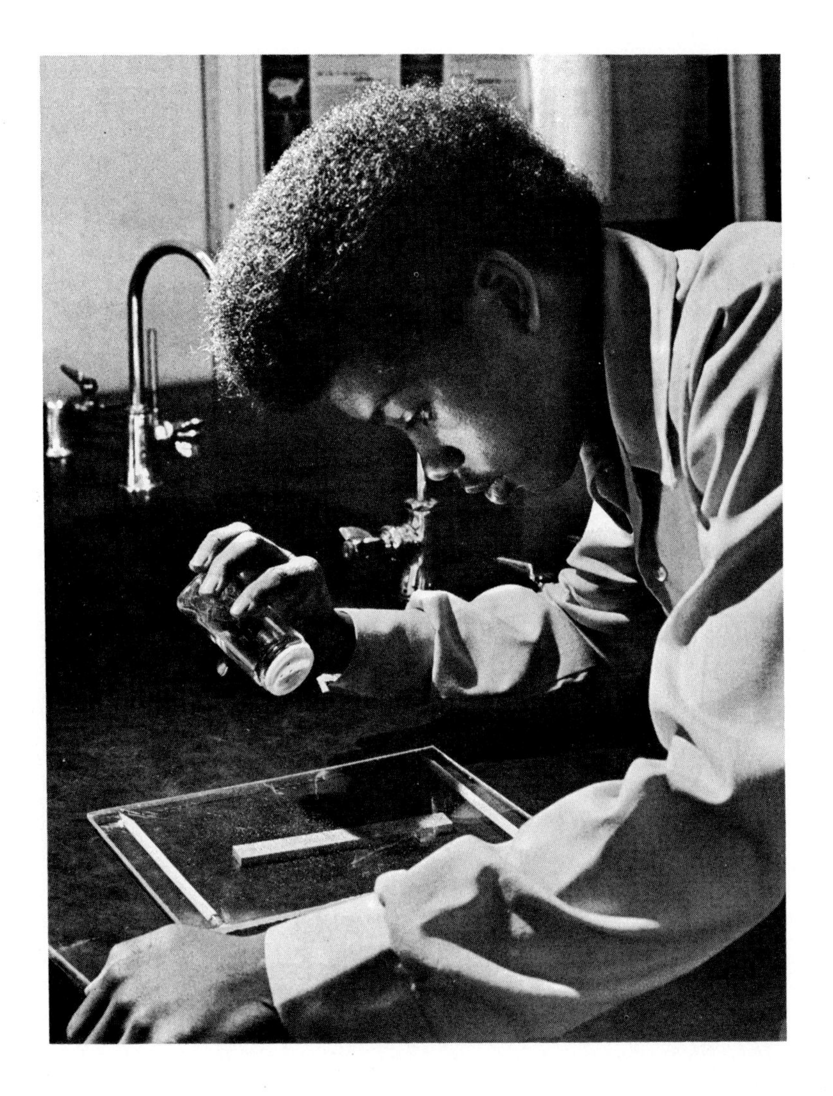

Figure 6/1
You can make a magnetic field and see its effects in an area.

As you know from using magnets, magnetism affects iron by attracting it. See how a bar magnet affects iron powder or iron filings. Place the magnet on a table. Put a sheet of cardboard over the magnet on two supports, such as pencils, as shown in Figure 6/1. Slowly sprinkle the iron filings through your fingers (or sprinkle iron powder through a shaker) all over the cardboard, directly over the magnet and around it. The effects of the magnet which you see traced on the cardboard by the iron filings are called **lines of force.** Tap the cardboard to bring them out more clearly. Make a drawing for your notebook of these lines of force, which curve around the magnet from pole to pole. ★

Are There Tides in the Atmosphere?

Chapman's measurement of the lunar air tide at Greenwich opened up the field for other scientists. Forty years later, some 85 measurements of the change in pressure with the lunar air tides had been made at stations all over the earth, as shown in the map in the Teacher's Manual. You can display this with overhead projector. You can describe how a barometer works here, if you wish. (See also Section 8.3.)

In 1917 a young man named Sydney Chapman was working at the Greenwich Royal Observatory in England. He was fascinated by anything and everything about the atmosphere. Chapman had read that a lunar air tide had been measured on the island of St. Helena. This island is in the Atlantic Ocean near the equator. A lunar air tide is a tide in the air made by the moon, like the tides the moon makes in the ocean. The air around the earth moves up and down as the water does in the oceans. The lunar tides cause changes in air pressure that can be measured by barometers. Since the moon orbits the earth fairly near the equator, the effect of the moon's gravitational attraction is strongest there. Chapman wondered whether a lunar air tide could be found anywhere else on earth.

The Royal Observatory had records of exact measurements of changes in air pressure for the last 64 years, hour by hour and day by day. Chapman decided that he would look over all these changes in air pressure covering about 18,000 days. Perhaps he would find high and low air pressure in England indicating high and low air tides.

Lo and behold, Chapman did find a small regular change in air pressure of about 0.001 inch of mercury. The mercury in the barometer went up and down twice a day with the changing position of the moon. He found two lunar high air tides each day, just like the tides in the oceans.

Chapman made many other discoveries about the earth's atmosphere, the magnetic field, the magnetosphere, and auroras, with the same patience and originality. He became a world famous authority on the earth's atmosphere, and he inspired many young people to make the study of it their life's work. Chapman died in 1970, the same year that a crater on the moon was named for him.

Slicing an apple shows a cross section similar to that of the magnetic field students made. The section does not show the whole apple or full shape of the core. Next activity lets students look further into the whole magnetic field.

Activity Time: 15 minutes. Individual students or groups of two. The iron filings clearly show the whole shape of lines of force around the magnet's pole and down the sides of the ball. Students draw the field with lines of force as they see them, not in cross section from above or from the side. For a more permanent model, submerge the ball-magnet in corn syrup, corn oil, mineral oil, or motor oil in a glass jar. Sprinkle the powder in slowly. The field will show as long as the jar is not shaken vigorously.

Process: observing, inferring

6.2 Model the earth's magnetic field.

When you cut through the trunk of a tree or slice an apple, you see rings of the tree or inner parts of the apple. In the same way, you made what looks like a slice or cut of a magnetic field by sprinkling iron filings on the cardboard. How does the whole magnetic field look? You can find out by making a model of the earth's magnetic field.

Materials

modeling clay or dough, scissors, compass, cardboard, paper and pencil, bar magnet

★ Enclose the bar magnet in a ball of modeling clay or dough that stands for the earth. (Keep track of where the magnet is under the clay or dough.) Place the ball on a table so that the magnet is upright, with one of its poles pointing straight up. Cut out a circle in a piece of cardboard to go around the ball's equator, as shown in Figure 6/2.

Now sprinkle iron powder or filings down on the model earth. Continue this gently as you see lines beginning to build up around the poles at the top half of the ball. In your notebook, draw these lines of force just as you see them. Use one color of pencil or pen. With another color, extend the lines as you imagine they must go around the whole ball, including its other pole on the bottom half. Scientists often need to use their imagination this way to fill in gaps in their observations.

A **compass** is an instrument that tells us about magnetic fields. The needle of a compass points along the lines of force around the earth leading to the north and south magnetic poles. Move a small compass around your model

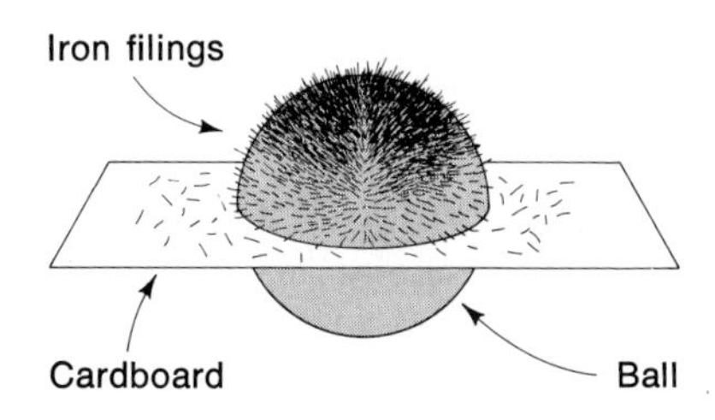

Figure 6/2
Use your model and your imagination to draw magnetic lines of force going around a ball like the earth.

Students might like to report on: How William Gilbert (1540–1603) studied the shape of the earth's magnetic field with a "terrella," a naturally magnetized lodestone ball, and small compasses; how compasses work; how like poles of magnets repel and unlike poles attract; or what magnetic declination is and how it varies across the United States. A map of magnetic variations in declination is given in the Teacher's Manual.

Get students to offer as many suggestions as they can about what is wrong with their drawings of the model fields. Some may well know the answer: the solar wind changes the shape of the earth's magnetic field.

Materials

large tray or dish, water, pencil

Activity Time: 15 minutes. Groups of two students. Students can do this activity as a simple exercise with emphasis on observation and drawing, or you can demonstrate in a glass tray of water on an overhead projector or in a ripple tank. The enlarged streamlined image reveals the details nicely. Shadows on the bottom of the pan reveal the streamlined shape.

Curved arc of streamlined shape forms in front of pencil in direction in which it is moving; shock wave moves out on each side of pencil toward side of dish; long, disturbed streaming tail forms behind pencil, all more clearly defined as pencil is moved more rapidly through water. Eddies like those behind a very fast-moving pencil also form in earth's geomagnetic tail. Streamlined effect in fluid is the same whether the object is moving or the fluid itself is moving, or both. Comets are a good example of this shape, with gases forming about their heads as they near the sun and long streamers extending far out behind them as they plunge through the solar wind.

Jupiter is the only other planet for which a magnetosphere has been identified. It is speculated that the whole solar system may have a magnetosphere as it moves through outer stellar space. The moon has a weak magnetosphere.

Concept: interaction of matter

earth. Try to locate the north and south magnetic poles. See if you can trace the lines of force along the model's magnetic field with the compass. Are the magnetic poles on the earth the same as the geographic poles?

You have made a full drawing of the magnetic field around the ball. Your model ball stands for the earth. You have drawn the shape of the magnetic field around the earth as scientists imagined it to be for hundreds of years. But farther away from the earth, this shape has been found to be different. The earth's magnetic field is only partly like your drawing. Do you have any idea what is wrong about your field? ★

6.3 The solar wind shapes the magnetic field.

★ You can find the shape of the earth's magnetic field far away from the earth by moving the end of a pencil through water in a dish or tray. The moving pencil end represents the earth in its swift passage through space. The water stands for the solar wind. (See Section 1.11 and 4.6.) In this model the effects around the pencil represent the earth's magnetic field.

Fill the tray or dish with about an inch of water. Dip either end of an unsharpened pencil a quarter of an inch into the water and move the pencil end slowly across the dish. Nothing much happens, does it?

Perhaps your earth isn't moving fast enough. Move the pencil end gradually faster and faster through the water. Watch closely what happens. What kind of shape begins to form in the water in front of the pencil in the direction in which it is moving? At the sides of the pencil? Behind the pencil?

Draw the whole shape that forms in the water around the pencil earth as it moves. Improve as much as you can on your drawing by watching the pencil move rapidly some more. Doesn't the shape look somewhat like the shape of a fish? Or the shape of the cross section of an airplane wing, made to move swiftly through the air? Compare your drawing with those shapes shown in Figures 6/3 and 6/4.

The shape that the earth's magnetosphere forms around the earth is very similar to the streamlined shape you have seen and drawn. The earth forms this magnetosphere in its rapid plunge through the solar wind. ★

Figure 6/3 and 6/4 *Water flows past a shark and air moves by an airplane wing the way the magnetosphere streams around the earth moving through the solar wind.*

The magnetosphere has many effects on our lives on earth. It shields us from damaging radiations from the sun. Recently scientists have noticed that climates have changed with changes in the strength of the earth's magnetic field.

6.4 The earth has streamlined "doughnuts." Until spacecraft came along very little was known about the magnetosphere. Then man-made satellites circling the earth and craft shooting out into space carried instruments called **magnetometers** (mag-neh-TOM-eh-ters) to measure

Concept: particulate nature of matter

the magnetic field. They carried other instruments to measure the charged particles in the space around the earth. (See Figure 6/5.) From these measurements, scientists gradually built up a picture of how the earth's magnetosphere is shaped. A cross section of this picture is shown in Figure 6/6.

Doesn't the magnetosphere look something like your drawing of the water moving past the pencil? Where the solar wind strikes the magnetic field of the earth you see a shock front. A **shock front** is made when a speeding body compresses the water, or air, or space particles it is moving through so much that they move in all directions. Aircraft breaking the sound barrier make shock fronts.

Concept: interaction of matter

Magnetic lines of force and charged particles are jumbled up within the shock front. Your compass would spin wildly in this region. From the shock front a wide wave spreads out, like the wave you saw in the water. Behind the shock front is the magnetosphere.

A student may like to report on the discovery of Van Allen radiation belts with the first American satellite (Explorer I, 2-1-58) and their exploration by later American and Russian space probes and satellites.

Deep within the earth's magnetosphere are the magnetic "doughnuts" called the **Van Allen belts.** Charged particles bouncing back and forth along the magnetic lines of force make the Van Allen belts. The earth is in the holes of these

Figure 6/5
The instrument pointed toward the sun on this Mariner spacecraft bound for Mars measured the charged particles in the magnetosphere and solar wind.

Figure 6/6
The solar wind is strong enough to push the earth's magnetic field, or magnetosphere, into a streamlined shape.

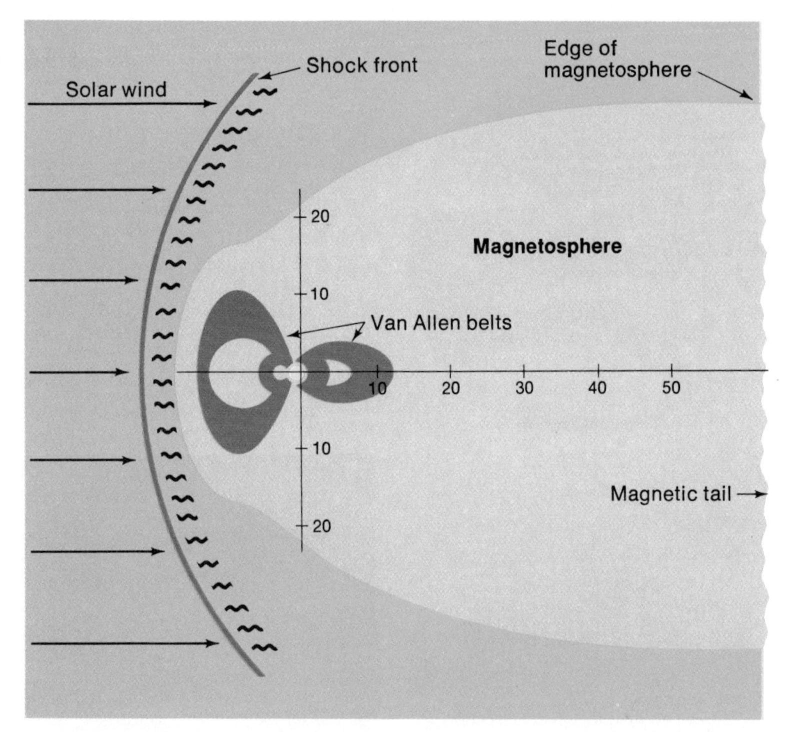

doughnuts. Notice how they are flattened out by the solar wind on the side of the earth facing the sun, just like ripples before the pencil. On the other side of the earth, the magnetic field is pushed far out from the earth, forming a long tail.

This tail of magnetic lines of force and charged particles goes out at least as far as the moon. Probably farther. Remember the positions of sun, earth, and moon when the moon is full? Check Section 5.1 if you don't. Can you see why the earth's magnetic tail whips across the moon's surface at full moon? Instruments left by astronauts report extra amounts of particle radiations on the moon's surface at full moon.

The gases of the earth's atmosphere are also contained within this magnetosphere. But the lighter gases in the atmosphere, like hydrogen and helium, rise through the air like a balloon. They may escape from the earth along the magnetic tail that streams out behind the earth, away from the sun. On its trip toward Mars, the Russian spacecraft Mars 3 reported signs of the earth's magnetic tail 12 million miles out in space.

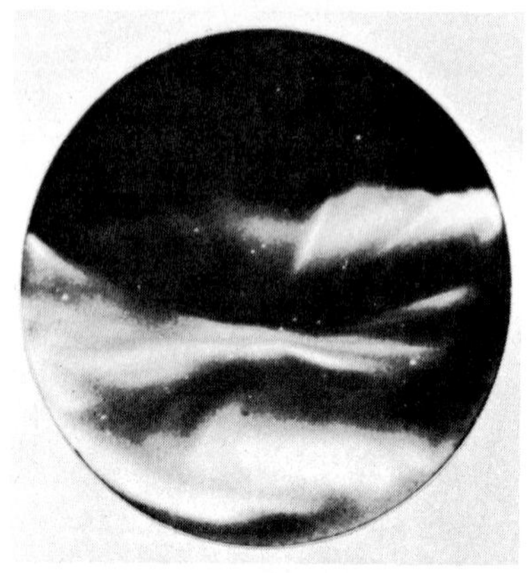

Some of the charged particles in the solar wind get into the act, too. They plunge down into the magnetosphere and are trapped in the stronger magnetic field close to the earth. This is how the Van Allen belts are formed.

Sometimes particles pour out of the openings of the Van Allen belts down through the air near the earth's magnetic poles. These streams of particles make beautiful patches and sheets of light in the sky that are called **auroras,** or northern and southern lights. Perhaps you have seen auroras like those in Figures 6/7 and 6/8.

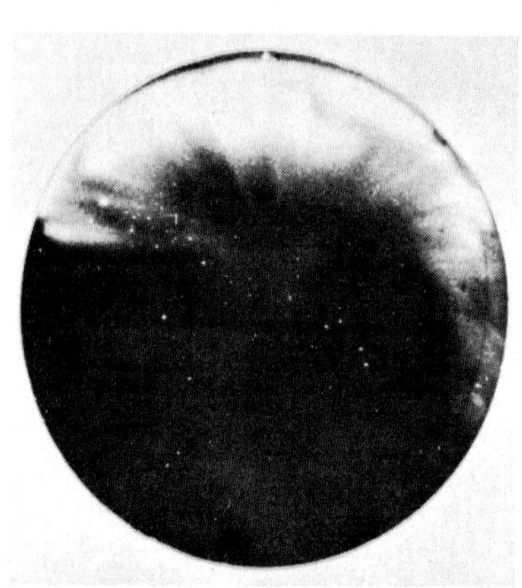

Figure 6/7
Photographs of auroras taken at the same time in the Arctic (above) and the Antarctic (below). Why don't auroras occur over the equator?

Figure 6/8 *Particles pouring out of the ends of the Van Allen belts cause auroras like this one photographed in January northwest of Fairbanks, Alaska.*

Did You Get the Point?

A field is a region through which an effect exists and can be measured.

The earth's whole magnetic field, or magnetosphere, consists of effects like those of a bar magnet, that form lines of force far out around the earth into space.

A shock front is formed where the magnetosphere meets the charged atomic particles in the solar wind from the sun.

The solar wind flattens the magnetosphere toward the sun and pushes it out in a tail away from the sun.

Van Allen belts are formed of charged solar particles trapped in the earth's magnetosphere.

Auroras brighten the sky when the charged particles fall out of the Van Allen belts and pour down through the atmosphere.

Check Yourself

1. Name some things that have shapes similar to the shape of the earth's magnetosphere.

2. A spacecraft approaches the earth from the direction of the sun. What regions of the magnetosphere will the craft cross as it comes in for a landing?

3. A manned spacecraft is launched from earth orbit directly toward the moon at a speed of 25,000 miles per hour. How long will the craft take to pass the inner Van Allen belt? The outer belt?

Answers:

1. The water around a rapidly moving boat, water streaming past a rock in a road or brook, and so on.
2. Shock wave, outer magnetosphere, and outer and inner Van Allen belts.
3. Students can use scale on Figure 6/6. Inner belt, about 4,000 miles:

$\frac{4000}{25000} = 0.16$ hour $= 9.6$ minutes.

Outer belt, about 16,000 miles:

$\frac{16000}{25000} = 0.64$ hour $= 38.4$ minutes.

What Makes Up the Atmosphere?

6.5 What goes on in the atmosphere?

Suppose someone has bored a large hole right down through 1500 miles of atmosphere to the earth's surface like that in Figure 6/9. Imagine you are looking down through this hole, as if you were looking down a deep well or an elevator shaft. You can see land and water on the

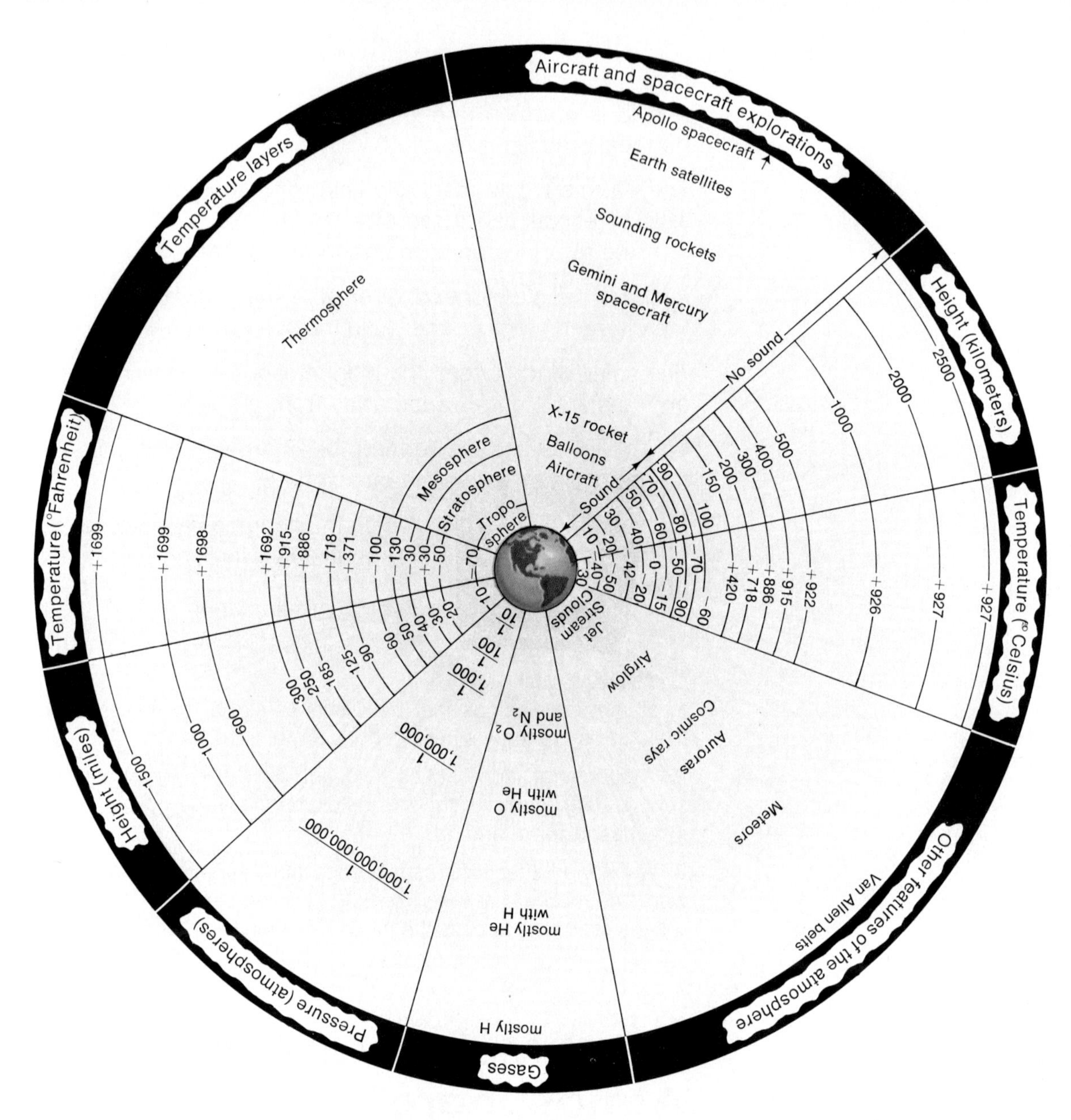

Figure 6/9 *An imaginary hole in the atmosphere shows how things change as you go down to the earth's surface.*

The scales shown are logarithmic to display lower levels of the atmosphere adequately. This means that the scale or length for a mile constantly decreases from the center of the chart outward to its circumference.

earth's surface at the bottom of the hole. Each section around the hole shows something that is known about the atmosphere, or some events going on in it. Distances up from the earth's surface are marked in miles (and kilometers) so you can see where events take place in relation to each other in the atmosphere. Each of these sections shows a different way of studying the atmosphere.

Venus has a very heavy atmosphere, believed to consist largely of carbon dioxide. Atmospheric pressure at the surface of Venus is about 95 atmospheres, with surface temperature of about 900 degrees Fahrenheit (482° Celsius). Possibilities for life as we know it on Venus are slim.

One section of the hole gives the pressure of the air on or above the earth. The mass of air above each square inch at the earth's surface, pulled down by the earth's gravitational field, weighs 14.7 pounds (6.7 kilograms). The weight of the air at the earth's surface is known as **air pressure.** Air pressure is measured in atmospheres, and one atmosphere equals 14.7 pounds for every square inch.

Answers: One-hundredth of an atmosphere at 20 miles and one-thousandth of an atmosphere at 30 miles.

The view "down the hole" in Figure 6/9 shows that the pressure is only one-tenth of one atmosphere at 10 miles above the earth's surface. What is the air pressure at 20 miles altitude? At 30 miles? You can see that air pressure decreases very rapidly as you go higher.

If you had a magic space belt that would lift you to any height in the atmosphere just by turning a dial on the belt, you could easily go up and explore the lower clouds. But above them, the air pressure gets very low. If you went any higher, you'd have to wear a helmet and carry air with you in tanks, like an underwater scuba diver.

Materials

graph paper, pencil

Activity Time: 20 minutes. Individual students.

Process: observing

★ How about the temperature of the air? What kind of a spacesuit would you have to wear to keep comfortable as you explored higher and higher in the air? The temperature section in Figure 6/9 will help you figure out the answer. Make a grid like that shown in Figure 6/10 to plot the temperature changes with height in the air.

Behind each of the zigs and zags on your temperature graph was a scientific discovery. The first ones were made with balloons, the later ones with rockets. The section of Figure 6/9 next to the temperatures shows the temperature layers in the atmosphere around the earth. Mark these temperature layers on your graph and name them. ★

Absolute zero is −273.16° Celsius, or −459.7° Fahrenheit. High temperatures in the thermosphere represent rapid movement of atoms or molecules of the gases there, but have little effect on spacecraft since gases have such low densities. Since temperature measures the mean kinetic energy (or energy of motion) of atoms or molecules, low-density gases may have very high temperatures, but very low total energy.

See how rapidly the temperature goes down as you go up through the layer nearest the earth, the **troposphere,** and into the **stratosphere,** the next layer of the atmosphere. Much of the earth's weather is made in the troposphere. Scientists used to believe that the temperature kept going down as you went higher, until you came to the coldness of space. But they know now that the temperature rises in the stratosphere. Then the temperature goes down again

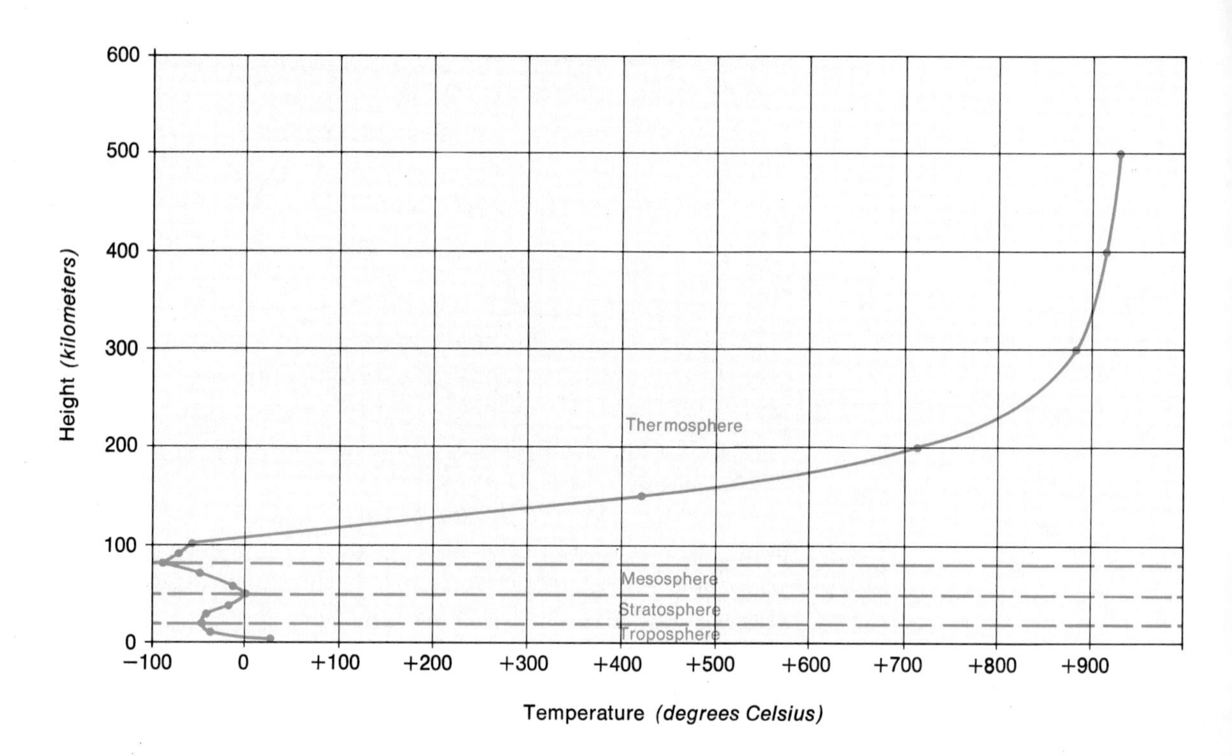

Figure 6/10 *Make a grid like this to see what happens to air temperature with increasing height above the earth. Use the data in Figure 6/9.*

in the **mesosphere,** or "middle," sphere, just above the stratosphere. Finally, the air temperature climbs steadily in the thermosphere, the next higher layer.

Other sections in Figure 6/9 tell about the clouds and other things and events in the air. The figure shows how man has been exploring higher and higher in the air with balloons, aircraft, and finally spacecraft.

6.6 How much oxygen is in the air?

Up to about 50 miles (80 kilometers) above the earth, there is a mixture of many gases in the atmosphere. Molecules of oxygen (O_2) and nitrogen (N_2) are the two main gases.

Higher than this, the gases are more separated out in layers or shells, as shown in Figure 6/9. The layers do not have sharp borders, but blend into each other.

This process is, of course, photosynthesis, which you can discuss in more detail if you wish.

You may know that the molecules of oxygen gas in the air come from plants. Plants make their own food with the energy in sunlight, water, and carbon dioxide in the air. During this process they give off oxygen gas. Perhaps half the oxygen in the air comes from tiny plants in the oceans and half from plants on land.

How much oxygen is in the air? In other words, what part of the whole volume of the air is the volume of the oxygen in it? **Volume** is the amount of space occupied by the air. And by part is meant the **percentage** of the whole air or hundredths of the whole air. So this question means for each 100 parts of air, how many parts (hundredths) are oxygen?

Materials

beaker, millimeter scale or ruler, cellophane tape, tray or deep dish, candle stub, water, jar

Activity Time: 20 minutes. Groups of three students.

Process: experimenting

★ One way to answer this question is to take the oxygen out of a volume of air and see what percentage of the whole volume is gone. When a fire burns it uses oxygen from the air. So why not burn a candle in a volume of air and see how much of the air is gone when the oxygen is all used up?

Prepare the equipment and set up the activity as shown in Figure 6/11. Light a candle, place a jar over it with its rim just under the surface of the water, and let the candle burn out. Then measure the height that the water has moved up in the jar as the oxygen was taken out of the air. Divide this height by the total height of the jar. Multiply by 100. You have the percentage of oxygen of the air in the jar. ★

Activity will show 15 to 20 percent of oxygen in air by volume. Candle burning yields carbon dioxide, which gradually and almost completely dissolves in the water, forming carbonic acid. This partly explains why the water level gradually rises a little after the candle flame has gone out. The level rises later also because the gas is cooling to room temperature. The presence of carbonic acid can be shown by placing blue litmus paper in the water near or under the mouth of the jar, where it will slowly turn red. If students set the jars directly on the tray bottoms, make sure they do not seal tightly. A tight seal prevents the water from rising.

Did you find about 20 percent of the air is oxygen? This means that in every 100 parts of air, 20 parts are oxygen. The most accurate methods for measuring the percentage of oxygen place it close to 21 percent, or 21 parts of oxygen to 100 parts of air. Nitrogen makes up about 78 percent of the air. What percent of the air is left that is neither oxygen nor nitrogen?

Close to one percent of air left.

Figure 6 / 11
Setup of equipment to find out the volume of oxygen in the air.

6.7 Discoveries are made in leftovers.

It was early in the eighteenth century (1700–1800) that scientists found out that air consists mostly of oxygen and nitrogen. Then in 1785 the British scientist who had discovered nitrogen found a little bubble, or leftover, of air after he had taken out all the oxygen and nitrogen. The bubble was a very small part of the air, only about one percent. And for over 100 years everyone forgot all about this little leftover bubble.

Henry Cavendish (1731–1810).

In 1894, two other British scientists began to wonder about what could have been in this bubble. They were so eager to answer the question that they camped in the laboratory day and night studying the leftover bubble. By this

Lord Rayleigh and Sir William Ramsay. The discovery of "noble" gases is a fascinating story, illustrating scientific method, about which some students might like to report.
Process: observing

time, the spectroscope had been developed, as you saw in Section 1.7. The two scientists looked at the spectrogram of the bubble of gas. They saw not only the bands of nitrogen, with which they were familiar, but also groups of red and green lines never before seen in the spectrum of any gas. After more study, they announced that they had discovered a new gas in the air. They named it argon. Argon, with a chemical symbol Ar, makes up a little less than one percent of the air.

But one new gas was not enough. The scientists cooled argon down to where it became a liquid. In this liquid they discovered two more new gases. They named one krypton (Kr) and the other neon (Ne). Later, the gas xenon (Xe) was discovered as a leftover in krypton. Still later, another gas, helium, was discovered by its spectrum, first on the sun and then in the gas from gas wells on earth. Then traces of helium were discovered in the air.

6.8 What else is in the atmosphere?

The air is a mixture of many other gases besides oxygen, nitrogen, and the "leftover" gases, as you can see in Figure 6/12. Some of the oxygen is combined with other elements to form larger molecules. Unless specially dried, the air contains a little water vapor (H_2O) which is water as a gas. Notice that the percents of gases listed in the table in Figure 6/12 are for dry air. Air also contains carbon dioxide (CO_2).

Although the percentages of other gases in the atmosphere are very small, they often become greater when the air is polluted. Fires and automobiles give off carbon monoxide (CO) as well as carbon dioxide. The burning of coal with sulfur in it makes sulfur dioxide (SO_2).

Short wave solar radiation splits oxygen molecules (O_2) into atomic oxygen (O), some of which then reunites with oxygen molecules to form ozone (O_3).

Electric motors, short circuits, and lightning make ozone (O_3). So does the sun's radiation when it strikes oxygen in the upper atmosphere, as you can see in Figure 6/9. The band of airglow, which is a faint glow in the night sky like a weak aurora without shape, is made mostly by radiations from atoms of oxygen (O) and oxygen combined with hydrogen (OH) in the air at heights of 50 to 100 miles.

Nitrogen combines with oxygen to form various oxides that are also gases. Oxides are compounds of oxygen with other elements. Nitrous oxide (N_2O) is the gas dentists

Gas (from largest to smallest amount)	Dry air (percent of total amount)	Molecular mass (sum of masses of atoms in molecules)
Nitrogen	78.1	28.02
Oxygen	20.9	32.00
Argon	0.9	39.94
Carbon dioxide	Trace	44.01
Neon	Trace	20.18
Helium	Trace	4.00
Methane	Trace	16.05
Krypton	Trace	83.70
Carbon monoxide	Trace	28.01
Sulfur dioxide	Trace	64.06
Hydrogen	Trace	2.02
Nitrous oxide	Trace	44.02
Ozone	Trace	48.00
Xenon	Trace	131.30
Nitric dioxide	Trace	46.01
Radon	Trace	222.00
Nitric oxide	Trace	30.01
Water vapor	0.1–2.8	18.02

Figure 6/12 *Percentages and molecular masses of gases in the atmosphere.*

Students can learn a lot from the table in Figure 6/12. It introduces many common compounds—gas molecules—and accustoms students to variation in molecular, as well as atomic, masses. Finally, it is a key to some of the major air pollutants.

Gravitational attraction gathers the gases more or less in layers. Gases with higher molecular masses are held closer (on the average) to the earth's surface than lighter gases.

sometimes give people so they will not feel their teeth being pulled.

The percentages of oxides and other gases in the earth's atmosphere are given in Figure 6/12. This table also shows how these gases differ from each other in terms of the mass of each kind of molecule. The **molecular mass** of a molecule is the sum of the masses of all its atoms.

Examine the layers or shells of gases as you go up through the mesosphere and thermosphere as they are drawn in Figure 6/9. Compare the gases in these shells with their molecular masses in Figure 6/12. Can you explain what makes these gases gather in layers in the upper atmosphere?

Dust is also present in the air. Much dust is blown up into the air by volcanoes, and man's activities add to the

Figure 6/13 *Sunrises photographed from a high-altitude balloon over Australia, (A) shortly after the eruption of the Agung volcano, and (B) seven years later after the dust had settled.*

dust, too. Dust lets the long, red wavelengths of light through the air and cuts out most other colors. As you can see in Figure 6/13 this makes our sunrises and sunsets colorful. But it does not improve the air we breathe!

Note that the table in Figure 6/12 gives percentages of the main gases in the atmosphere in dry air, but that normally some water vapor is present in air also. The fact that water vapor is lighter than most other gases in the atmosphere plays an important role in the weather.

Water vapor present in the air is invisible. But when this gas forms liquid drops (rain) or solid crystals (snow), the water becomes visible in the form of clouds. In whatever form, water in the air plays a very important part in the winds and the weather, as you will see in Chapters 7 and 8.

Did You Get the Point?

The pressure of the atmosphere increases down to the surface of the earth, where it is 14.7 pounds per square inch at sea level.

The temperature of the atmosphere varies with height above the earth, making a number of different temperature layers.

The air near the earth is a mixture of gases, mainly nitrogen and oxygen.

The air at greater heights above the earth is separated into layers of different gases, like oxygen, helium, and hydrogen.

Oxygen makes up about 21 percent by volume of the air near the earth, nitrogen 78 percent.

Varying traces of many other gases produced naturally or by human activities are found in the atmosphere.

Answers:

1. Gases would assemble by mass, making a thick layer of heavy gases at the earth's surface. In this case, life might not have developed, at least in the forms we know, most of which depend on oxygen.
2. Helium and hydrogen gases are the lightest of all, and atomic hydrogen atoms or protons are even lighter.
3. See finished graph, Figure 6/16. This graph shows that the masses of these gases are the cause of this situation. The "molecular mass' column of Figure 6/13 gives these masses. The mass of atomic oxygen is extrapolated as one-half of that of molecular oxygen.

Check Yourself

1. Suppose the gases in the air near the surface of the earth were not mixed. What would be the effects on the atmosphere? How would this affect life on earth?

2. Use Figure 6/12 to explain why helium and hydrogen are the principal gases escaping from the earth's atmosphere.

Materials

graph paper

3. Draw a graph showing the changes in percentage of the three gases listed in the table in Figure 6/14. How would you explain these changes in percentage as you go higher and higher in the atmosphere? Set up your graph as shown in Figure 6/15.

Working from chart (Figure 6/9) and/or from other charts, articles, or books, students can give "Tourist Guides" orally, or write them up in notebooks, or bulletin board displays can also be developed by groups or by whole class.

4. Make a sort of "Tourist's Guide" to the atmosphere. Start the trip from the earth's surface or from space. Describe the conditions and the events "air tourists" will experience at several levels on their trip up or down through the air. You might stop over at 10 miles up, 50 miles, 100 miles, and 1000 miles. Or you might start from 1000 miles up like astronauts, then, unlike them, come

Figure 6/14
Percentages of the main gases in the lower atmosphere at increasing heights.

Height (kilometers)	Percentage composition: Molecular oxygen (O_2)	Atomic oxygen (O)	Molecular nitrogen (N_2)
97	20	1	78
145	17	8	74
201	11	17	71
250	8	26	65
298	6	37	56
403	3	60	36
499	2	80	17

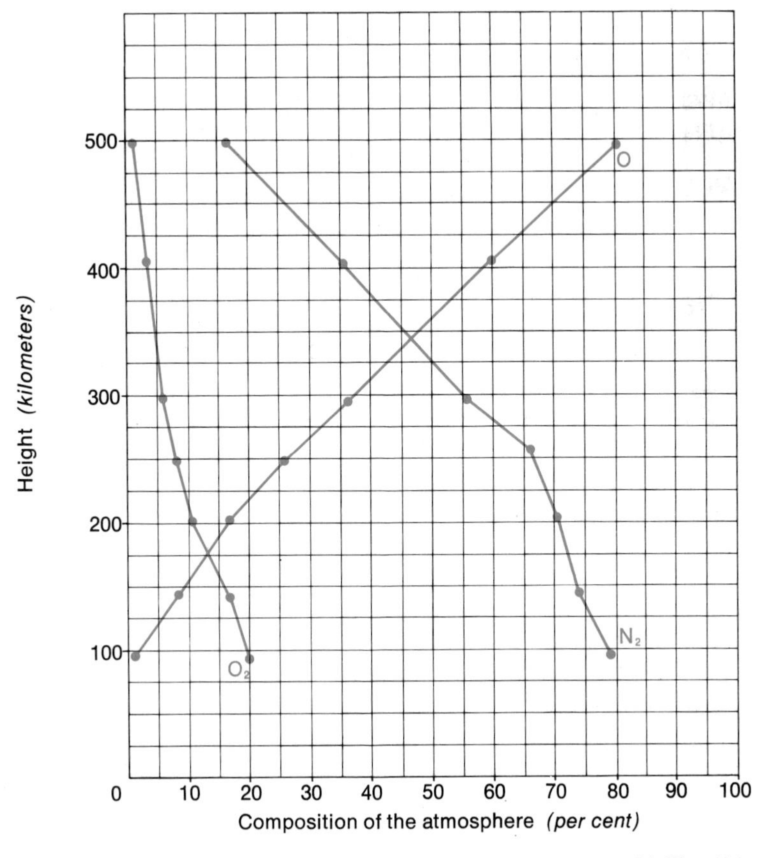

Figure 6/15 *Plot the data in Figure 6/14 on a grid like this to see how the composition of the atmosphere changes with height.*

down with several stopovers. Base your "Guide" on the information in Figure 6/9, and add whatever else you know about the atmosphere. If you go right around a circle on the chart in Figure 6/9, you will see what the atmosphere is like at the height of that circle.

The class could make a bulletin-board display of a trip through the air. Use yarn to show the tour route. Label and illustrate several stopping-off places.

What's Next?

People have skin, where they stop and other things begin. In the same way, the earth has its atmosphere, where it is separated from, but also joins with, space. The solar wind in space creates a shock wave where it meets earth's magnetic field. This makes the streamlined magnetosphere around the earth. Within the magnetosphere are the doughnut-shaped Van Allen belts. Charged particles from these belts cause auroras that brighten and fade.

People walk. If they trip, their falling is caused by gravitational attraction. Gravity affects all things in the air, from spacecraft to meteors to every single molecule of gas. Many gases, light and heavy, are found in the atmosphere, all reacting to the pull of gravity.

People are warmed and tanned as they are exposed to sunlight. As you will see in the next chapter, radiations from the sun, such as light and heat waves, flood into and through the atmosphere. In reaction the air stirs and moves. Winds begin to blow. How do these winds move around the earth? What effects do the sun's radiations have on the whole world around us?

Skullduggery

A rough, blocked-out drawing of the magnetosphere is made.

Questions A through P are multiple choice. Only one of the choices for each answer is correct. The teacher will give you a sheet with numbered squares on it like that shown in Figure 6/16. For each question, blacken all of the squares that have the same number as the correct answer. The blackened squares will trace out something you know from this chapter. What is it?

Example: The answer to question A is choice number 2. All the squares in the grid with the number 2 in them should be filled in.

1	1	1	1	1	1	1	11	11	11	11	32	32	15	8	3	3	3	3	3	3
3	3	3	5	5	5	5	11	5	5	5	3	3	3	15	3	5	5	5	5	5
6	6	6	6	6	2	2	6	6	47	8	47	47	6	6	15	1	1	1	1	1
7	7	7	7	4	4	7	7	7	7	5	5	5	5	6	11	11	6	6	7	7
8	8	9	4	4	9	9	4	9	47	8	47	9	25	9	9	9	32	15	9	9
10	4	4	4	10	10	4	4	10	10	10	10	10	25	9	9	9	9	44	44	9
32	32	15	13	12	17	36	36	10	10	2	10	10	8	25	10	10	12	12	2	2
47	15	13	12	13	40	17	14	14	13	14	14	14	8	8	25	13	12	4	4	44
47	3	12	13	13	40	40	14	13	47	8	47	14	14	25	8	12	13	14	14	44
16	12	13	16	40	40	40	13	16	16	18	18	14	18	8	25	18	14	13	18	18
12	19	19	8	8	8	19	19	8	8	17	47	47	12	25	8	30	19	19	13	19
23	23	23	32	32	32	23	23	23	23	23	23	12	23	32	32	8	23	23	23	13
22	14	26	29	29	29	26	26	22	8	8	8	22	26	8	8	8	8	26	26	26
53	22	53	36	36	36	53	22	27	24	24	24	27	22	27	17	36	11	27	27	27
28	28	22	11	40	40	22	53	28	32	20	24	28	28	22	29	17	11	28	28	28
29	32	32	20	11	44	32	32	53	32	24	20	31	31	31	17	29	11	32	32	32
33	29	33	44	20	11	47	33	33	24	20	44	33	33	15	36	11	15	33	33	33
34	34	29	44	44	20	11	34	34	44	32	32	53	31	11	11	15	29	31	31	31
34	34	34	31	44	11	20	31	31	20	20	17	31	53	20	15	20	34	22	22	22
22	22	22	22	11	47	47	35	35	35	17	36	31	35	15	20	29	35	35	35	35
37	37	37	37	20	11	20	37	37	37	2	37	31	37	20	32	29	25	37	37	37
39	39	39	39	44	47	11	39	39	39	2	29	41	15	29	29	32	41	25	41	41
41	41	41	41	44	47	17	43	43	43	2	43	43	29	17	32	45	45	45	53	45
41	41	41	41	40	17	36	43	43	31	11	43	43	47	32	17	45	45	45	45	53
42	42	42	42	17	36	40	42	29	42	15	42	42	11	44	44	42	42	42	53	42
50	50	50	50	36	17	40	31	48	48	24	37	37	36	11	36	37	37	53	37	37
48	48	47	48	48	40	17	48	48	48	8	48	44	44	44	11	39	53	39	39	39
49	50	49	49	31	40	15	17	42	42	8	42	8	8	8	39	53	45	49	51	45
50	51	49	31	51	15	40	24	42	42	29	41	47	47	24	53	45	49	51	50	52
51	50	29	52	52	40	24	29	54	54	17	49	24	24	47	45	49	51	47	52	54
51	51	50	52	52	52	29	38	54	54	38	49	17	38	45	49	51	50	52	54	51
51	51	50	53	53	53	38	38	53	53	38	53	38	17	45	45	50	52	54	51	54

Figure 6/16 *Use a grid like this for the Skullduggery.*

A. What is the scientific meaning of the word "field?" (1) an area that has a fence around it (2) an area in which an effect exists or changes (3) all the space around the earth

B. What are magnetometers used to measure? (4) the strength of magnetic fields (5) the direction in which compasses point (6) places where iron is located

C. What is the magnetosphere? (7) the charged particles thrown out by the sun (8) the shape formed by the earth's magnetic field (9) the sphere around the earth where there is no magnetism

D. What are lines of force? (10) the lines that show air pressures (11) lines around a magnet, indicated by a compass or iron filings (12) the lines along which bodies fall toward the earth

E. What are the Van Allen belts composed of? (13) particles of dust in the air, blown up high above the earth (14) molecules of gases like nitrogen and oxygen (15) charged atomic particles from the solar wind

F. What is the temperature of the air 50 miles above the earth's surface? (16) about +32 degrees Fahrenheit (17) about −150 degrees Fahrenheit (18) about +150 degrees Fahrenheit

G. What gas is most likely to escape from the earth's air into space? (19) helium (20) hydrogen (21) argon

H. At what height above the earth's surface is the airglow found? (22) 500 miles (23) 10 miles (24) 75 miles

I. What is the pressure of the air in inches of mercury at sea level? (25) 30 inches of mercury (26) 15 inches of mercury (27) 3 inches of mercury

J. In what layer of the atmosphere are meteors seen? (28) stratosphere (29) thermosphere (30) mesosphere

K. What percent of nitrogen is found in air in the lower atmosphere? (31) 65 percent (32) 78 percent (33) 80 percent

L. How is ozone made? (34) naturally only (35) by man only (36) both naturally and by man

M. What gas was discovered in the sun before it was found on earth? (37) argon (38) helium (39) hydrogen

N. What is the molecular mass of a gas? (40) the total of the masses of the gases' atoms (41) the total of the mass of the gas at sea level (42) the attraction of the atoms in the gas molecule for each other

O. How does the molecular mass of water vapor compare with that of oxygen? (43) water vapor has more mass than oxygen (44) water vapor has less mass than oxygen (45) water vapor has the same mass as oxygen

P. How do the molecular masses of carbon dioxide and carbon monoxide compare with that of oxygen? (46) both have more mass than oxygen (47) one has more mass and one has less mass than oxygen (48) both have less mass than oxygen

For Further Reading

Asimov, Isaac. *The Noble Gases.* New York, Basic Books, Inc., 1966. A science and science fiction writer describes the composition and sources of the six rare gases, the recent discovery that they can form compounds, and the uses of these gases in science and industry.

Craig, Richard A. *The Edge of Space: Exploring the Upper Atmosphere.* Garden City, New York, Doubleday & Co., Inc. 1968. A research weatherman tells how the air at great heights is explored with rockets and satellites. He describes what goes on in the various layers of the air and in the magnetosphere.

Chapter 7 Sun, Water, and Wind

Many of the topics discussed in the chapter are exemplified in the scene, which, with the text and illustrations accompanying it, can lead to questions and discussion.

Other misunderstandings and myths can be discussed, or drawn from students, such as "wet weather is coming because the crescent moon is pouring out water" and "ground-hog" day predicts the spring. But weather is complicated, determined by many variables, and the principles of sun-air-earth energy interchanges must first be understood.

Answers to all but the last of these questions will be found in the chapter. Questions are raised here to establish attitude of "Let's look and see."

"Boy, does that smell good!" Bob called to the others from the campfire, where he was busy cooking supper.

"Wood smoke, meat frying. Super!" Maria agreed. She came over to the fire from the large, flat rock where she had been putting out the picnic plates and cups. "Can't you hurry it up, Bob? I'm starving!"

"Gee, that hike sure gave me an appetite," Diaz said, sniffing the sizzling hamburgers. "I could eat just about anything, even Bob's cooking."

"Am I freezing after that swim in the lake!" Kathy said. "The water was warm enough, but the air sure was cold! And on top of it, the sun's gone down behind those clouds near the horizon." She pointed across the lake.

"Say look, the sun's drawing water!" Bob said. "Anyway, that's what my Dad calls it."

"That's just a nutty old saying," Diaz said. "It just *looks* as if the sun was pulling ribbons of water up into the sky. But it really isn't. Water's everywhere in the air. The sun's always evaporating it."

"Anyway, I know all I need to know about the sun and clouds and air and fires," Bob said. "We come closer to the fire when the sun goes down, don't we? The sun warms us up, we get a tan from it. Fires cook our suppers. So what else is new?"

"I bet we all have a lot more kooky ideas about the sun and the air and heat . . . and about the wind and weather, too," Maria said. "Like a watched kettle never boils."

"Go on, see if we do," Kathy said. "Go ahead. Let's ask some questions and see if we can answer them."

"O.K.," Maria said. "Here goes . . . Why is it hotter at some times of year than at others? Why is it hotter around the equator than around the poles? Why . . ."

". . . does the wind blow? Does the earth's spinning affect the circulation of the air?" Bob asked.

"Does a black iron frying pan cook faster than a shiny one?" Maria pointed to the fire.

"Hey, let's look and see," Bob said. "I think maybe it's time to eat!"

Heat Comes and Goes

7.1 Model the sun and atmosphere.

Heat energy enters the earth's atmosphere as infrared radiation from the sun. (See Sections 1.2 and 2.6.) Some of these radiations are taken up, or absorbed, in the atmosphere. Most of them strike the oceans and land.

Materials

aquarium or earth box with lid and opaque shield, 300-watt light bulb on extension cord, ringstand and clamp, 5 small thermometers, soft wire (12 inches), piece of plastic foam or stopper, graph paper, watch or clock with second hand, black construction paper

Details on the earth box and its use in several chapters will be found in the Teacher's Manual. Students can make their own boxes at home.

★ Let's look at the different ways in which heat energy travels. Think of the air in the aquarium or earth box as a sample of the atmosphere. Set up the box as shown in Figure 7/1. Put a square of black construction paper at the end where the light will shine.

Wrap one end of a piece of soft wire around the bulb of a thermometer. Make a spiral at the other end, which

Figure 7/1
This setup will show that heat energy can be passed along in three ways. Make sure your thermometers are placed with their bulbs facing the way they do here.

Activity Time: 25 minutes, groups of 4 to 6 students. Groups can be divided between activities in Sections 7.1 and 7.3 to save time. Use 300 watt photoflood bulbs.

will rest on the black paper. Place the thermometer and wire at position E in the figure. Arrange the other four thermometers. Thermometer C should be insulated from the bottom by a piece of plastic foam or a stopper. Place the lid on the box and cover it with the opaque shield except for the end where the light will shine through.

Typical temperature readings (Celsius) for the five thermometers appear in your copy of Figure 7/2 only.
Process: measuring

Hang the light bulb directly over the uncovered end of the box top. In your notebook, set up a data table like that shown in Figure 7/2 for recording your temperature readings.

Figure 7/2
Make a data table like this to record your thermometer readings.

Light turned on (minutes)	Thermometer readings (degrees Celsius)				
	A	B	C	D	E
0	21	21	21	21	21
1	25	21	21	22	22
2	29	21.5	21	23	24
3	33	22	21	24	26
	Radiation	Convection	Convection	Radiation	Conduction, radiation

Give the thermometers a few minutes to adjust to the temperature inside the earth box. Then record the reading of each thermometer as the original row of temperatures in the table. Turn on the light and record the readings of each thermometer each minute for five minutes, or a longer time if possible. CAUTION: Turn off the light if the fluid in any thermometer comes close to the top.

Caution students to turn off lights if fluid in any thermometer, particularly thermometer A, is rising close to top of the scale.

See how much you can find out from your table. Which thermometer showed the greatest increase? Which showed the smallest? Why do you suppose you got the results you did?

Thermometer C shows the smallest increase in temperature, with larger increases in thermometers B and D, and the greatest increase in thermometer A.

Now here is another way to present your data. Make up a grid like the one in Figure 7/3. Show the time of temperature readings horizontally and the degrees of temperature vertically. Enter the points for each thermometer with

Figure 7 / 3
Copy this grid for plotting the temperature changes.

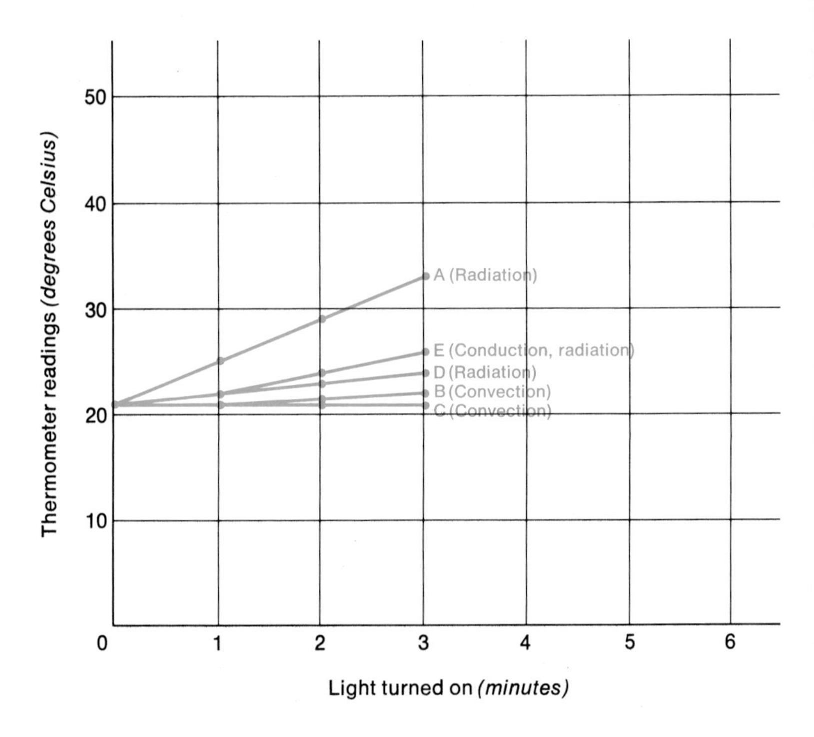

Completed graph for readings in table given appears only in your copy of Figure 7/3. Thermometer A registers largely radiation, B gives largely convection, C shows largely convection, D gives some radiation, and E gives radiation and conduction. Such combined effects are typical of heat-transfer conditions on earth. Section 7.2 will help students in interpreting graph.

a different-colored pencil. Connect the points for each thermometer with a line. Label each line with the letter of the thermometer it stands for. Does the graph give you a clearer idea about the changes in the temperature of the different thermometers than the table does? ★

7.2 Heat is transferred in several ways.

Questions here are answered briefly in text, but you may wish to discuss further the ways heat energy travels, pointing out some of the combinations.

How did heat from your light bulb affect the different thermometers in your earth box? The heat energy traveled directly by infrared radiation to thermometer A. (See Section 2.6.) Infrared waves traveling from the light bulb transferred their energy to the thermometer, making the fluid expand and rise. But the infrared waves did not directly reach thermometers B and C. A few infrared waves may have reached thermometers D and E.

Concept: particulate nature of matter

Why did thermometer E go up, even though it received little direct radiation? Why did it go up more than D? The heat energy in the infrared waves made the molecules of the metal in the wire directly under the lamp vibrate faster than usual. This vibration was passed along the wire from molecule to molecule until it got to the thermometer. When molecules or particles transfer heat energy by vibration from one to another, the transfer is called **conduction** (kon-DUCK-shun).

Heat energy traveled in another way to thermometer B, under the cover of the box, and to thermometer C, near the bottom of the box. This way is called **convection.** In convection (kon-VEK-shun), heat energy is transferred differently because the material is a fluid, not a solid. The fluid may be a gas (like the air) or a liquid (like the water in the oceans).

Here's how convection works. The air molecules receiving heat energy move faster, and spread and bounce farther apart. Thus, there are fewer molecules in a gallon of hot air than there are in a gallon of cold air. So hot air is lighter than cold air. As warmer air rises, colder air flows into the space the warmer air left behind. As a result, the air begins to circle, as you will see in Section 7.5. The black paper and the bottom of the aquarium or earth box were warmed by the radiation from the light, and they warmed the air above them. And the warmer air makes the temperature go up in thermometers B and C. Can you explain why the temperature in C goes up less than the temperature in B? Study the diagram in Figure 7/4. Infrared waves from the

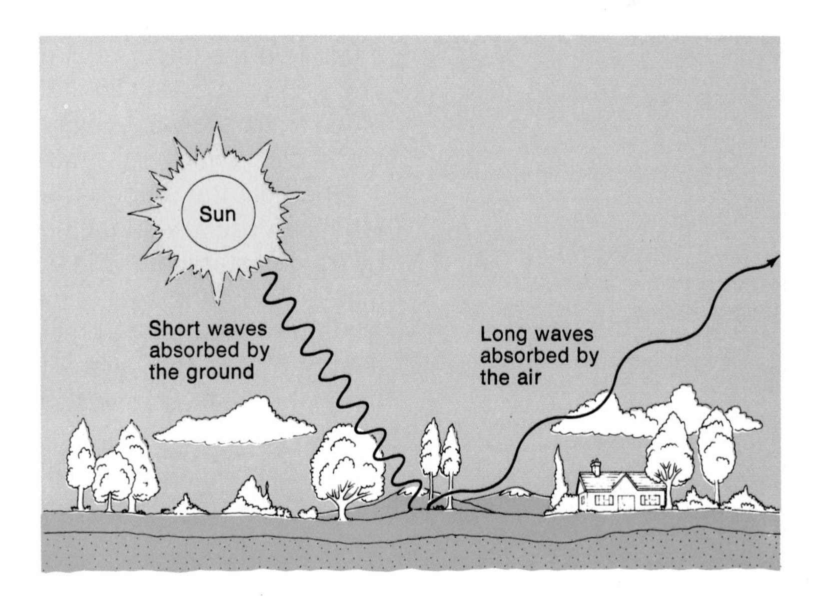

Figure 7/4
The sun does not heat the air directly. The sun heats the earth and the earth heats the air.

sun enter the atmosphere. They go through it the way visible light waves do. When they strike the lands and oceans, they are absorbed by soil, rocks, and water. Heat which is absorbed by the earth then radiates back into the air, but it is different from the heat that came from the sun. The reason is that if one heat source is hotter than another, it radiates shorter infrared waves than the other does. Since the sun is hotter than the earth, the heat from the sun radiates as shorter waves than the heat from the earth. The longer waves from the earth are absorbed by the air.

Why is it that sunlight does not warm glass, but the light from a light bulb does?

The light bulb radiates longer waves. You can introduce the "greenhouse effect" here.

7.3 Heating of the earth varies.

Activity Time: 10 minutes, groups of five. An earth box shields wax drops from stirring air, but use of a box is optional. Can also be used as a demonstration, as described here or with wax drops on a 4" to 6" plastic ball.

You know that the earth is warmer near the equator than north or south of it. You know, too, that it is coldest at the poles. Maybe you've never stopped to think why this is so. Doesn't the same amount of heat radiation from the sun fall on every part of the earth?

Materials

construction paper (strip about 2" × 12"), candle, light bulb (150- to 300-watt) on extension cord, ringstand and clamp

★ You can answer this question by doing a simple experiment. Cut a strip of construction paper. Fold it as shown in Figure 7/5, and then flatten it out. Light a candle and drop three spots of wax, all the same size, onto the paper in the places shown in the figure. When the wax has hardened, fold the paper again. Then put the paper on a table and turn on a strong light directly above it. Watch the spots of wax. As the wax melts, it spreads in a dark ring around the spots. Which spot melts first? second? last?

Melting occurs around the edges of the drops, with dark rings spreading out. Watch carefully for first signs of dark rims appearing. The drops directly perpendicular to the heat source melt first, with little difference in time, even though one is slightly farther from the source than the other. The drop at an angle melts last. Point out that the upper edge of the drop at an angle melts first, with the lower edge shaded by the drop.

When did the drop that was at an angle to the light melt? What can you say about the amount of heat energy received by slanted surfaces? Did both drops directly facing the light melt at about the same time? What can you say about the amount of heat energy received by surfaces directly facing the light? Did the different distances of these drops from the light change the amount of heat energy they received? ★

Heat received by the drop at an angle to the source is less than that received by the drops at right angles to the source. Nearly the same amount of heat radiation came in toward the three drops as shown in Figure 7/9. But the same amount of radiation received by the drop at an angle was spread over a larger area, so less radiation was received by it per unit area. Therefore, this drop melted last.

Do you know now why it is cold around the earth's poles and warm around the equator? The distances of drops from

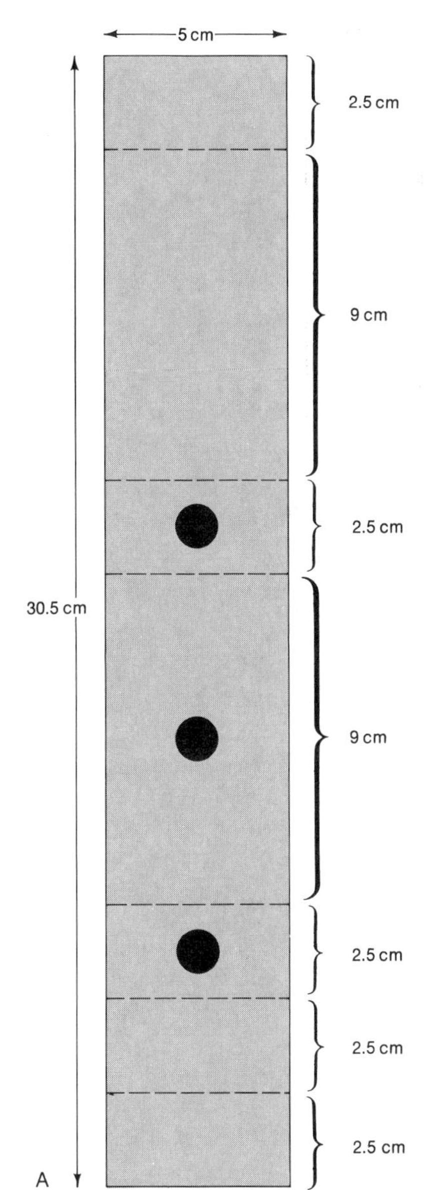

Figure 7/5 *Cut and fold a paper strip like the one drawn here and place it under a lamp, to show where the radiation has the most effect on wax.*

the light did not make much difference, did they? The angle of the drops to the light did make a difference. Look at Figure 7/6, which compares the heat energy received from the sun at the equator with that at a pole. At the equator, in the drawing, the earth's surface more directly faces the sun. At the poles, the earth's surface is at more of an angle to the sun, and a "ray" of light spreads over more area. Therefore, there is less heat at a single point. At the equator, the same amount of light shines on a smaller area, so a single point gets more heat.

★ You can check out this difference by letting the light you used to melt the wax represent the sun shining. Use a globe for the earth. Hold the globe five to ten feet away from the light. Now look at the areas around the poles of the globe. Doesn't it look quite a little darker there than by the equator? But not as dark as on the side of the globe away from the sun? ★

Materials

ringstand and clamp, light with extension cord, globe

Activity Time: 10 minutes, one setup viewed by all students. Variations in the light radiation to equator and poles are noticeable, to which the variations in heat radiations will correspond.

Figure 7/6
The sun's radiation is weaker at the poles than at the equator because each unit of sunlight spreads over more area.

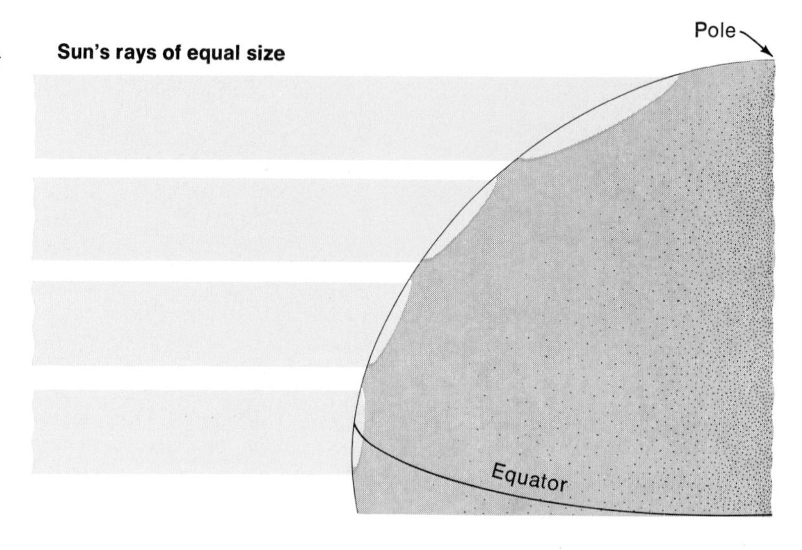

With globe, and flashlight to stand for the sun, you can have students demonstrate seasonal radiation changes resulting from the parallelism of earth's tilted axis in its orbit of the sun. Student standing at center with flashlight follows the globe with flashlight as another student moves around flashlight-sun carrying globe with tilted axis pointing in same direction at all times. Have students identify the four seasons.

The earth turns around an imaginary line joining its poles, as you saw in Section 3.3. This line is known as the earth's axis of rotation. It points in the same direction (near the North Star) all year, as the earth moves around the sun. As shown in Figure 7/7, the earth's axis is tilted at an angle. This results in different parts of the earth's surface getting varying amounts of heat energy from the sun at different times of the year.

It is winter in the northern hemisphere when the North Pole is tilted away from the sun. Areas in this hemisphere are then tilted at a greater angle to the sun's radiation and get less heat energy. At the same time, in the southern hemisphere, areas face the sun more directly, so it is summer there.

Six months later, the North Pole is tilted toward the sun. Areas in the northern hemisphere then face the sun more directly. It's our turn to get the swim suits out! Do you see from Figure 7/7 why the sun is higher in the sky at noon during the northern summer than it is at noon in the winter? This variation in heat energy received from the sun at different places on the earth through the year gives us the seasons.

Materials

globe, light bulb (300- or 150-watts), ringstand and clamp, thermometers (2) (optional)

★ You can show with a light and a globe how the amount of heat from the sun changes with the seasons. You will need a setup like that shown in Figure 7/8. For a rough

Figure 7/7
Because the earth's axis of rotation is tilted, the seasons change as the earth moves around the sun. Can you see why the summer sun never sets at the North Pole?

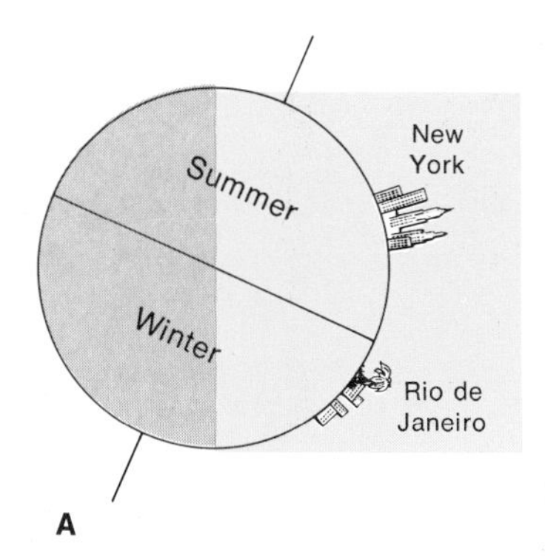

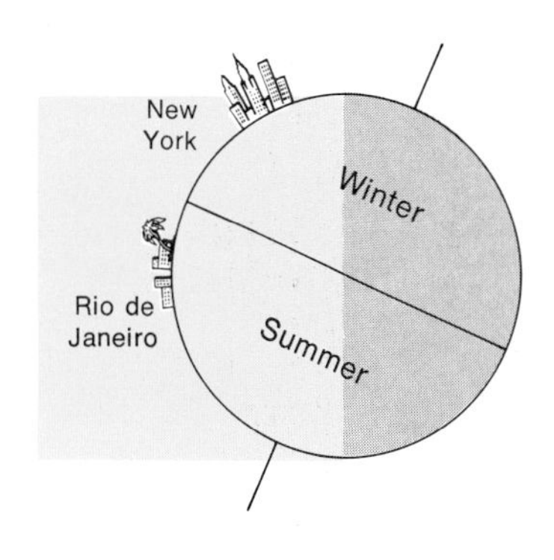

A

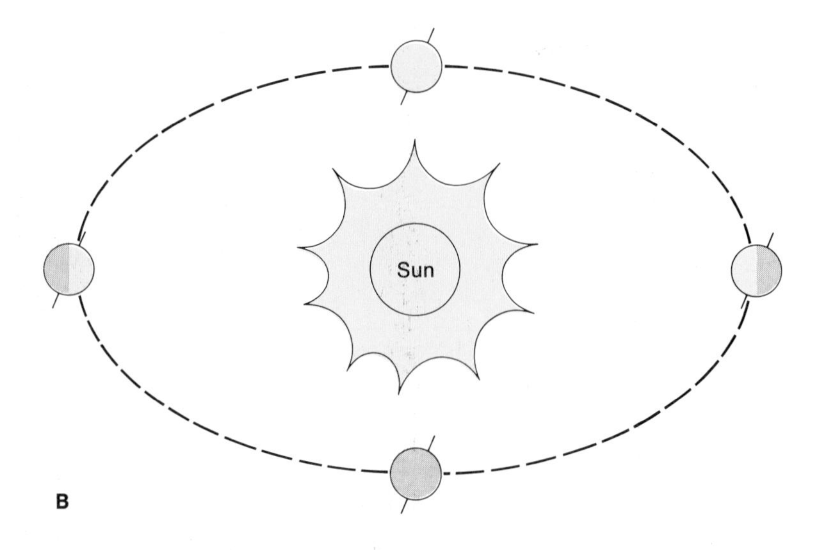

B

Figure 7/8
You can feel the temperature differences at high and low latitudes on a globe. Test your own latitude in its summer and winter positions.

measure of temperature, you can just use your finger on the globe across the region where you live. Or you might use a thermometer to measure the heat more precisely. Now run through the seasons with the light and the globe. Compare your results with those of others. ★

Process: observing

7.4 Different radiation angles cause different climates.

As you can guess, the sun's radiation is just as important in your climate and weather as it is in the seasons.

Weather is the state of the air at a given time, hot or cold, fair or rainy, and so on. But **climate** is the average state of the air, or the weather, in a place or region over many years. Climates remain the same or change very slowly over hundreds of years.

The broadest zones or bands of climate around the earth are based on the amount of yearly radiation from the sun. They are shown in Figure 7/9. A zone is a fairly large area or region within which certain similar conditions are found. These broad climate zones are the hot tropics on either side of the equator, the cold polar caps, and the temperate zones in between. **Temperate** (TEM-purr-ett) means neither very hot nor very cold.

The zones shown here are based on the way the climates affect plant life. In the polar zones the average tempera-

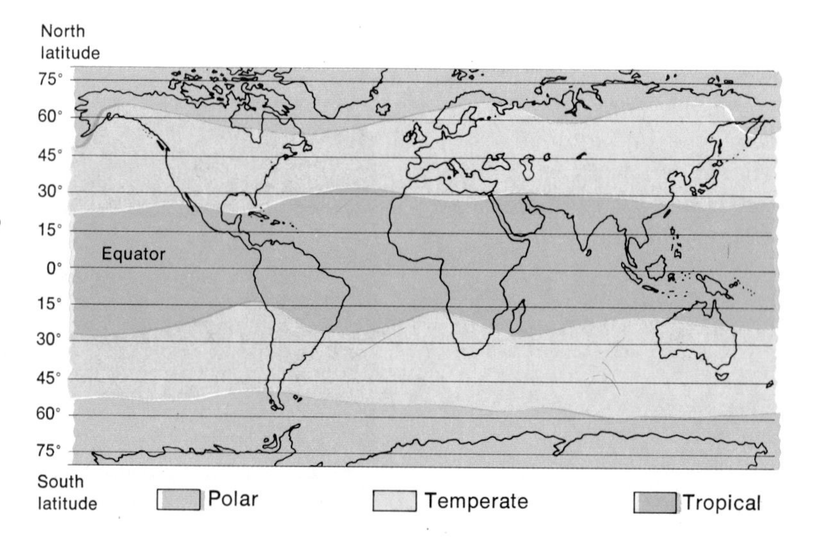

Figure 7/9
The basic climate zones of the earth. The fact that their borders are not as straight as the latitude parallels suggests that climate is due to other things beside radiation from the sun.

tures through the year are 50 degrees Fahrenheit (10 degrees Celsius) or less. The poles receive large amounts of radiation during their summers. Remember, the summer days are longer at the poles than they are in the temperate zones. When the days are longest in the United States, the sun never sets at the North Pole. On the other hand, the amount of heat the poles gain in their summers is lost in their winters. When the days are shortest in the United States, the sun never rises at the North Pole. So in polar regions, few trees grow. It is too cold for anything more than low bushes and smaller plants.

In the tropics, there is a good supply of solar radiation all year round, and the average yearly temperatures are 65 degrees Fahrenheit (18 degrees Celsius) and above. Jungle plants need the warmth to thrive. Between these polar and tropic zones lie the bands of the north and south temperate zones.

So you can see how the different angles of solar radiation do result in the various climates in which people live.

Answers:
1. Expect a variety of answers. While the sun is principal source of heat, students will realize that there are many other sources, including artificial ones, and that heat is continually being transferred in many ways.
2. Angle of pyramid to the sun remains the same from bottom to top, so temperature would be same. Since there is no breeze near the top, there is no chance for heat energy to be carried off faster by convection. Pyramids are not high enough to make a difference in temperature caused by altitude in air, but heat close to the ground might be conducted into stones at the base.
3. Season of the year will be: (A) winter in Miami; (B) summer in Capetown; (C) summer in Sydney; (D) winter in Honolulu; (E) winter in Anchorage; and (F) winter at home in U.S.
4. Tropical conditions in shower room, temperate in homes, polar in walk-in freezers in stores. Answers for outdoors will vary, depending on local conditions.

Did You Get the Point?

Heat energy is transferred by radiation in infrared waves, by conduction in the passing along of molecular vibrations, and by convection in the motion of gases and liquids.

When heat radiation is absorbed by materials, heat is radiated back.

Bodies at an angle to heat radiations receive less heat per unit area than bodies directly facing the heat radiations.

Heat radiations from the sun strike the earth's surface at different angles through the year, producing the seasons.

The amount of heat a region receives from the sun over a year's time causes its climate to be warm, cold, or temperate.

Check Yourself

1. Where is heat energy radiated, conducted, and carried by convection in your home? At school? On your trip to and from school?

2. Imagine you are climbing one of the stone sides of an Egyptian pyramid. The sun is directly overhead and there is no breeze. If you measured the temperature of the stone near the bottom and then near the top of the pyramid, would one temperature be higher than the other, or would they both be the same? Why?

3. Imagine you are taking a January around-the-world trip by plane. What season of the year will it be: (A) in Miami, Florida, at 26° north latitude, your first stop? (B) At your next stop in Capetown, Africa, at 34° south latitude? (C) At Sydney, Australia, your next stop, at 33° south latitude? (D) At Honolulu, Hawaii, at 21° north latitude, on your way home? (E) On a side trip to Anchorage, Alaska on the spur of the moment, at 62° north latitude? (F) When you get back home to the continental United States, 30 days later?

4. Name some places around where you live that have tropical, temperate, or polar climate conditions.

Water Moves in the Atmosphere

7.5 Air moves by convection.

How does air move from one place to another? Why does it move the way it does? How do winds get started, and why do they blow in different directions? You have seen that convection is one way heat is transferred from one place to another. You may not realize, though, that convection has a lot to do with the answer to these questions.

Materials

punk stick, wood or clay base

Activity Time: A few minutes, either with a number of pieces of punk for students to light or as a class demonstration. Punk must flame for a minute to develop a long-burning coal.

★ Light a stick of punk. Punk is pressed woody material that burns very slowly, with a great deal of smoke. Set the stick upright in a holder and observe the smoke. Notice how it goes straight up at first and then tumbles about. The smoke is acting as a tracer, marking the way the air is moving.

Hold your finger near (but not too near) the glowing end of the punk. Heat is radiating from the end. In other words, the lighted punk is a source of heat. ★

Heat sources are places that are warmer than the things around them and from which heat moves. Places cooler

Figure 7/10
With this setup, you can mark a convection current with punk smoke.

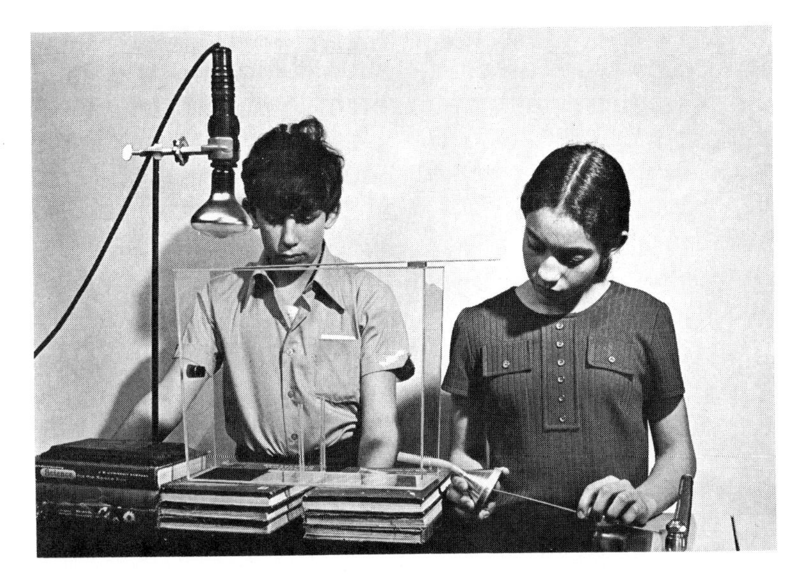

than the things around them, to which heat moves, are called **heat sinks.** Heat always moves from warmer places toward cooler places. Heat moves, for example, from a hot lighted stove burner (heat source) into a cooler teakettle (heat sink).

Materials

aquarium or earth box and lid with opaque shield, punk stick, wood or clay base for punk, ringstand and clamp, light bulb, funnel and tube

Activity Time: 20 minutes, groups of 4 to 6. This activity and the next, in Section 7.6, which takes perhaps 25 minutes, can be done at the same time by different groups in class, observing each others' results. The basic equipment is the same, only the materials vary.

★ To see how air moves from a heat source in calm air, set up an aquarium or earth box as shown in Figure 7/10. Light the punk, place it near the end of the box bottom where the light will shine and put the lid on the box. Cover most of the lid with the opaque shield. If you have an earth box, hold the punk so the smoke rises into the funnel. Turn on the light. Watch the smoke from the punk as it travels in the box. Draw the way or ways the smoke moves and how it looks. Whenever the smoke gets too thick, lift the lid of the box a moment to clear it out.

Now place the lighted punk in the middle of the box bottom. Put on the lid and repeat your observations and drawings. Turn on the light over the middle of the box for a minute or two if you wish. ★

The nearly round motions you see in the air are called **convection currents.** Convection currents are formed when the heated air around a heat source expands and rises, as described in Section 7.2.

With the punk and light at one end of the box, one convection current is formed. When you put the punk and light in the middle of the box, you see two convection currents. When the rising air reaches the lid of the box, it tends to go toward both cooler ends, which are the heat sinks.

7.6 You can move water through the air.

Heat energy always flows from hotter places to cooler places. Here you use a light bulb as a heat source, representing the sun. Water on one side of the aquarium or earth box stands for the ocean. A hill of cold, dry sand at the other side of the box stands for land. It will be the heat sink. Punk smoke traces the moving air currents. With the air, water, and sand in the box, you have small models of the oceans, the air, and the land.

Materials

aquarium or earth box with lid and opaque shield, light bulb on extension cord, ringstand and clamp, 4 thermometers, dry, cold sand, water, punk stick, wood base, wood partition for box, clay, putty or play-dough

After the convection current has been demonstrated, the punk can be removed for the remainder of the activity.

Answers: Punk smoke traces formation of a convection current. Motion is up and over the land, and back down and across to heated water again, but soon most of smoke (with water vapor) settles down on land. Droplets form on the glass sides and end of box, close to sand, then rise above sand and spread. Sand becomes wet. Droplets grow to drops and begin to smear the glass and run down. Along the line of sand on glass a wet darkened line of moisture appears. Pebbles become shiny with moisture.

★ Place a watertight partition in the middle of the aquarium or earth box, as shown in Figure 7/11. Then set up the box with water and cold dry sand. A few pebbles in the sand will help to show what happens as your experiment continues. Place the light bulb as shown. Put the lid on the box and cover two-thirds of the lid with the opaque shield to stand for clouds. What would happen on the earth when water and the air above it are heated by the sun?

Float a piece of lighted punk on a wooden block in the water close to the end of the box. What is the punk for? Turn on the light bulb. Watch carefully what happens. Notice the way the smoky air drops over the cold sand. Draw the whole process and its effects for your notebook. Whenever the smoke gets too thick, open the lid of the box briefly to let it out.

Does a convection current of air form? How does it move? Notice where water droplets are forming on the sides of the box.

Can you explain what has been taking place in the box? Do you see any evidence that the sand is getting wet? Do you agree that water has moved through the air in the box? ★

Figure 7/11 *What will happen to the model ocean when the light is turned on over the earth box?*

7.7 What changes happen in the water cycle?

A cycle is a series of events that keeps repeating. The **water cycle** of the earth is shown in Figure 7/12. In this cycle, water moves from ocean or lake or land to atmosphere and back to land or lake or ocean. Can you find

Figure 7/12 *This diagram of the water cycle shows how many trips a molecule of water can take from the time it leaves the ocean until it returns to the ocean.*

Smaller cycles, parts of larger cycles, are often called subcycles.
Concept: levels of organization

one small cycle that takes in only ocean or lake and atmosphere? Another that includes only land and atmosphere? Let's look more closely at what happens during the water cycle.

Heat radiations from the sun enter the water and warm it. They add energy to the water molecules so that they move faster. Some move fast enough to leave the liquid water and enter the air as water vapor. This change from a liquid (like water) to a gas (like water vapor) is called **evaporation** (ee-vap-oh-RAY-shun). The opposite change, from a gas to a liquid, is called **condensation** (kon-den-SAY-shun). Water molecules condense to form drops of water in clouds.

The air is said to be humid when it contains a lot of water vapor. At any given temperature, the air can hold only a certain amount of water and no more. Warm air can hold more water vapor than cold air can. When the air holds all the water vapor it can, it is said to be **saturated** (SAT-your-ay-ted). The amount of water vapor the air *does* hold, divided by all it *can* hold is its **relative humidity** (hue-MID-ih-tee). This is given as a percentage.

At a relative humidity of 100 percent, water vapor usually condenses. The temperature at which water vapor condenses is called the **dewpoint.** In your water cycle activity, the dewpoint was reached when you saw droplets form on the colder sides of the box above the sand.

As air rises with convection, it cools. When it has cooled to the dewpoint, the water vapor in the air condenses into clouds. It condenses around tiny particles of dust or salt particles from the ocean or other materials. Without these particles, the vapor does not condense in the air. When the drops of water or particles of ice in a cloud get so large and heavy that they can no longer float in the air, they fall as rain or snow. For clouds to form and rain to fall, a steady supply of moisture must be evaporated into the atmosphere for a period of several hours.

Tiny droplets of sea water are thrown into the air by waves. The water evaporates, and salt particles are carried by winds.

A little water in a half-gallon or gallon glass jug will demonstrate condensation. Shake the jug for plenty of vapor (evaporation), blow into the jug, sealing the mouth with your lips to increase the pressure, then let the pressure down and the air expand. No fog. Shake the jug again and throw in a lighted match whose smoke will provide condensation nuclei or particles. Repeat the blowing and release pressure: fog, condensed water vapor.

Did You Get the Point?

Heat sources are places from which heat energy moves. These places are warmer than the places around them.

Heat sinks are cooler places to which heat energy moves. Heat energy always moves from warmer to cooler places or materials.

Convection currents are circular motions of fluids, such as air and water. These currents are formed when warm material expands and moves from a heat source and heavier cooler material moves in to take the place of the warm material.

The water cycle is the movement of water from ocean and land to air to the land and ocean, to begin all over again.

Evaporation is the change of a liquid into a vapor. Condensation is the change of a vapor into a liquid.

Water vapor in the air condenses into clouds when the air is saturated with the vapor at 100 percent relative humidity.

How Warm Is the Earth Today?

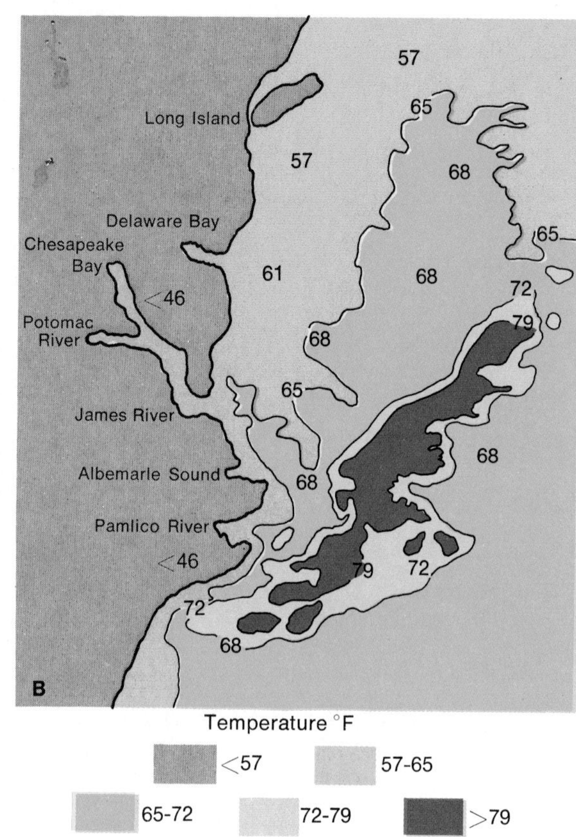

Do you recognize the region of the earth shown in the picture in A? This is an unusual picture. It was made from an environmental satellite circling the earth at a distance of 910 miles (1465 kilometers). The satellite orbits across the poles, so it can picture the whole earth every 12 hours. The picture shows the land in the eastern United States and Canada and the water in the Great Lakes and the western Atlantic Ocean. The picture also tells how warm the water and the land are. How?

The picture was taken at night, 4 A.M., in October. The satellite carried a scanning camera that reacted only to infrared radiations coming from the land and the water. Where the area is warm, intense infrared waves make the picture shades darker. The light areas like the clouds you see near the bottom of the picture are cold. The water is warmer (darker) than the land. The darkest wavy ribbon of water running up the Atlantic is the Gulf Stream, coming from the Gulf of Mexico.

The map of part of the region included in the picture shows how the actual temperatures can be read from the shades in the picture. The water in the Gulf Stream is warm enough for swimming, over 79 degrees Fahrenheit. But along the coast, where the water is only 55 to 60 degrees, it would be rather cold for a dip.

Temperature pictures and maps like this from satellites are valuable for fishermen, weathermen, and ship captains. Maps can be made from them of the surface currents in which water circulates in the oceans.

Answers:

1. Cycle moves from water to air (steam vapor) and back to water, with some condensation of vapor on lid of pot because it is cooler.
2. Heat source—cylinders of engine where gas ignition takes place. Heat sink—radiator. Heat energy is conducted and radiated into air, after conduction through radiator. Air stream through radiator carries off heat energy. Cooled water in radiator passes cylinders again, takes heat energy conducted through cylinder walls, and carries it (with help of water pump) back to radiator again. Engine can overheat if cycle is broken.
3. Balloon filled with hot air rises in air, as does balloon filled with helium or hydrogen. Molecules in hot air have moved faster, and expanded, giving hot air less density (mass per unit volume) than normal-temperature air. When air cools, balloon comes down.
4. Hot air from candle, collecting under one bag, is lighter than the air under the other bag, so bag containing hot air rises.
5. Single water molecule would trace the path of the whole water cycle more or less as shown in Figure 7/12.

Check Yourself

1. Describe the water cycle for water boiling in a tightly covered pot on a stove.

2. What is the heat source, the heat sink (the cooler place the hot water goes to), and the water cycle for the water in the cooling system of a car? What can happen if the cycle is broken?

3. What happens to a balloon if it is filled with hot air rather than with helium or hydrogen? Why? What happens when the air cools?

4. Hang two open paper bags from a meter stick hung from the ceiling, as shown in Figure 7/13. Move the bags until the stick balances. Hold a candle under one bag. What happens? Why?

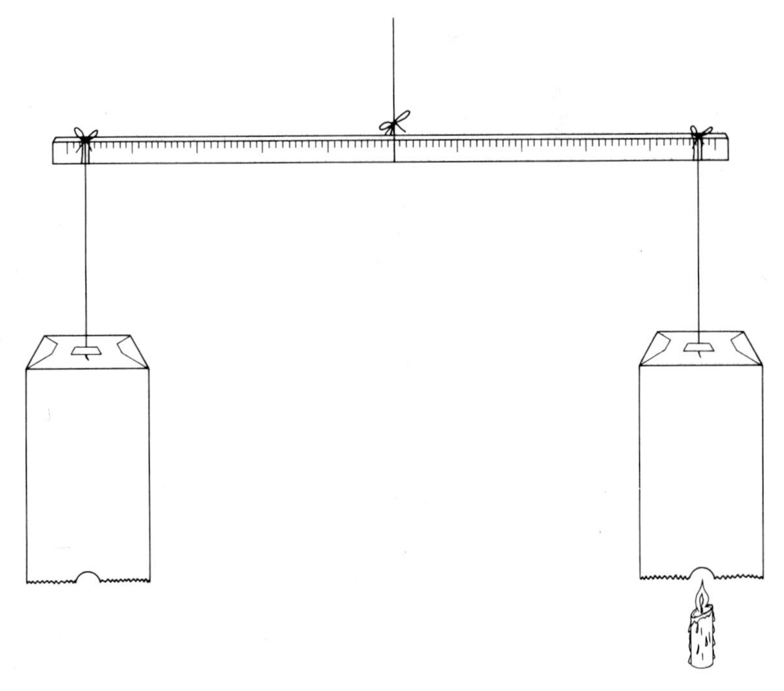

Figure 7/13
Will the meter stick stay balanced when there is a burning candle under one bag?

5. Where would a single water molecule go if it passed through all possible paths of the water cycle?

Air and Water Motions

7.8 Warm and cool air mix around the earth.

You have probably guessed by now that winds can be caused by convection. Cooler air flowing in to replace warmer rising air will form a wind.

Suppose you made a map of wind directions all around the world on a single day, and then looked for general patterns. You would find something like what is shown in Figure 7/14. North and south of the equator you see the easterlies, a broad band of winds coming from the east.

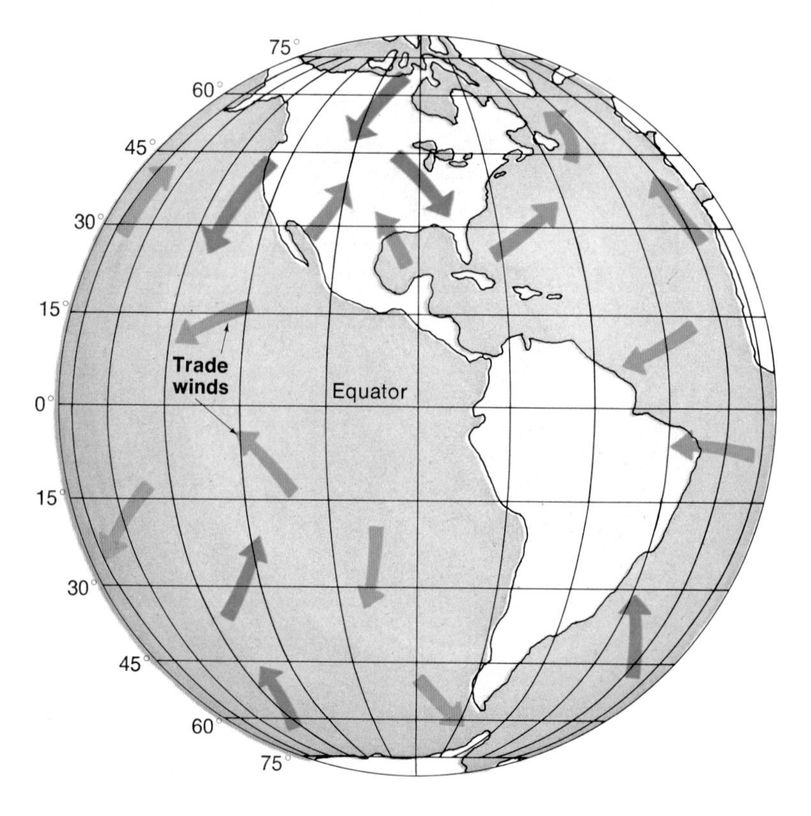

Figure 7/14
A sample of general wind patterns. On any day, trade winds will be blowing from the east in the tropic zone, while major winds at higher latitudes blow toward the poles or the tropics.

Winds are named by the direction from which they come. The easterlies are called **trade winds.** This is because they blow so regularly that sailing ships carrying on trade across the seas tried to get into their paths. Where the trade winds blow, the weather does not change much from day to day.

North and south of the trade winds, you can see that the patterns get more complicated. Warmer air from the tropic zone flows toward the poles, and colder air from the poles flows toward the tropics. The mixing of polar and tropical air must have something to do with the changeable weather that is found outside of the tropics.

7.9 Earth's turning affects the winds.

See note on Hadley cell concept in Teacher's Manual.

From what you know about convection, you might think that the warm tropical air would rise, move toward the poles, and be replaced by cooler polar air. You might expect to find a big convection current north of the equator and another to the south. But the motion of the air is not that simple, and the reason is that the motion of the earth complicates the air flow.

Process: observing

Figure 7/15, which was simplified from the actual northern hemisphere weather map of a January day, shows the result. The air to the north and south of the tropic zones is thrown into **eddies.** Eddies are circular motions in the air similar to whirlpools in water. Eddies are never exactly the same two days in a row. They form, grow in size, and move eastward and toward the equator or the poles. In a few days they disappear and new ones take their places.

There are two kinds of eddies in Figure 7/15. The ones marked "H" are areas of high air pressure, where colder, heavier air is found. These areas are called "highs." The ones marked "L" are areas of lower air pressure, where warmer, lighter air is found. These are called "lows." Figure 7/16 is a diagram of wind directions inside a high and a low of the northern hemisphere. The winds are curving clockwise in the high and counterclockwise in the low.

If nothing else is happening to a molecule of air in the northern hemisphere, it is traveling along in a counterclockwise direction around the earth's axis of rotation just as fast as the earth underneath is going. There can be another pull on the molecule, though. You remember from the

Figure 7/15
High and low pressure eddies in the northern hemisphere air. High pressure eddies occur often over the central and eastern United States and northeastern Asia during the winter. Figure 7/16 illustrates how winds in the eddies combine to form the major flows drawn here.

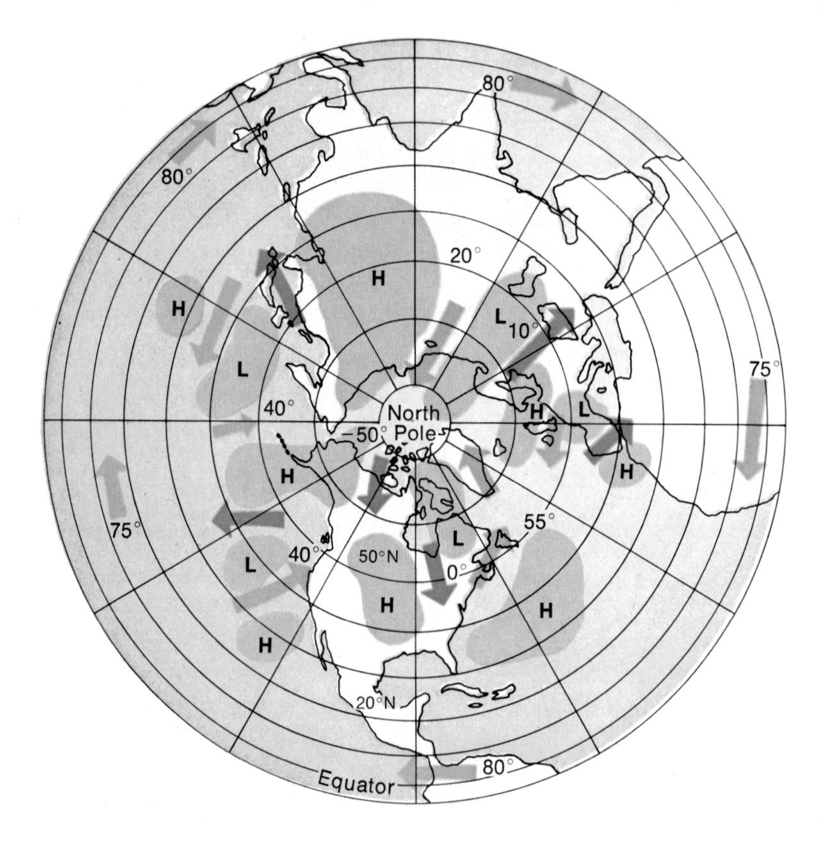

convection activity that air will flow from where it is cooler (heavier) to where it is warmer (lighter). This pull will move the molecule from high pressure (heavy air) to low pressure (light air).

See notes on Coriolis effect and its illustration in Teacher's Manual.

The air molecules in a high pressure eddy flow outward in all directions, toward areas of lower pressure. But the rotation of the earth and its atmosphere causes the air flow to bend toward the right. The result is shown by the arrows in Figure 7/16. The air flow is toward the left in a southern hemisphere high.

In a low pressure eddy, the air flows toward the center, where the pressure is the lowest. In the northern hemisphere, the rotation of earth and atmosphere causes the flow to bend as the arrows do in the figure.

If you look back at Figures 7/15 and 7/16 you can see that the largest air flows occur between highs and lows. These winds help move polar air toward the equator and tropic air toward the poles. In Figure 7/15, notice the temperatures around the 50th parallel of latitude. Can you see

Figure 7/16
Typical wind directions in and around a high pressure eddy and a low pressure eddy. The largest continuous winds flow between highs and lows.

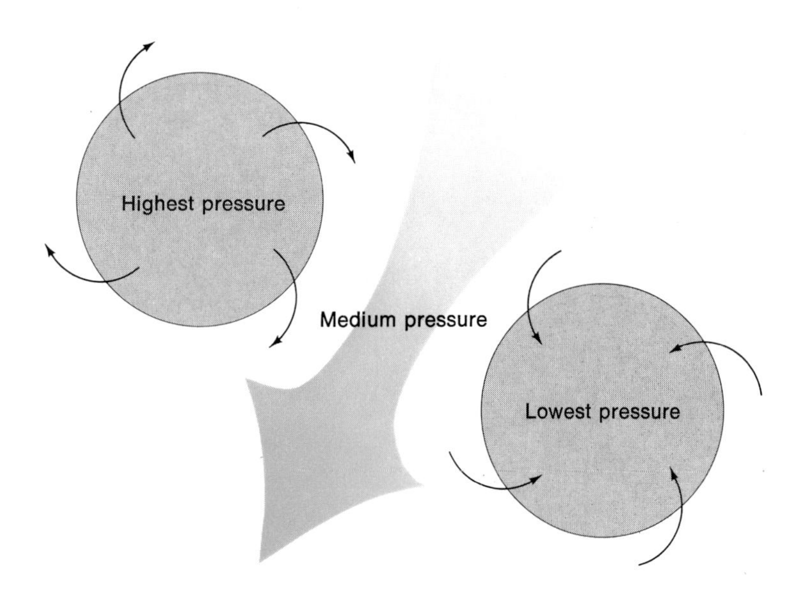

why some places were warmer than others, even though they are all at the same latitude?

7.10 There is more to climate than temperature.

You know from your study of the water cycle that air picks up moisture from both land and sea. It can pick up more in some places than in others. Air can be dry or moist as well as warm or cold. Dry medium temperature conditions can produce deserts or dry grasslands called steppes. Wet warm conditions can produce humid tropical climate and rain forests.

See in Figure 7/17 how many different climates there are *across* a single continent. How can this be when the sun's radiation should be the same all across it? Some other things must affect the earth's climates. How about the landscape?

Materials
aquarium or earth box, dish or cup, blocks, powdered calcium chloride (1 cup), black construction paper

See notes on this activity in the Teacher's Manual.

★ Let's look further into the climate of a continent like North America. Do mountain ranges affect climates?

Cut a piece of black construction paper and fold it in half, so it will make a "mountain" like the one illustrated in Figure 7/18. A strip that is the length of the sheet of paper and about 9 centimeters wide will do. Put a block at each end of the aquarium or earth box, so the mountain won't collapse. Rub calcium chloride dust on the surface of the

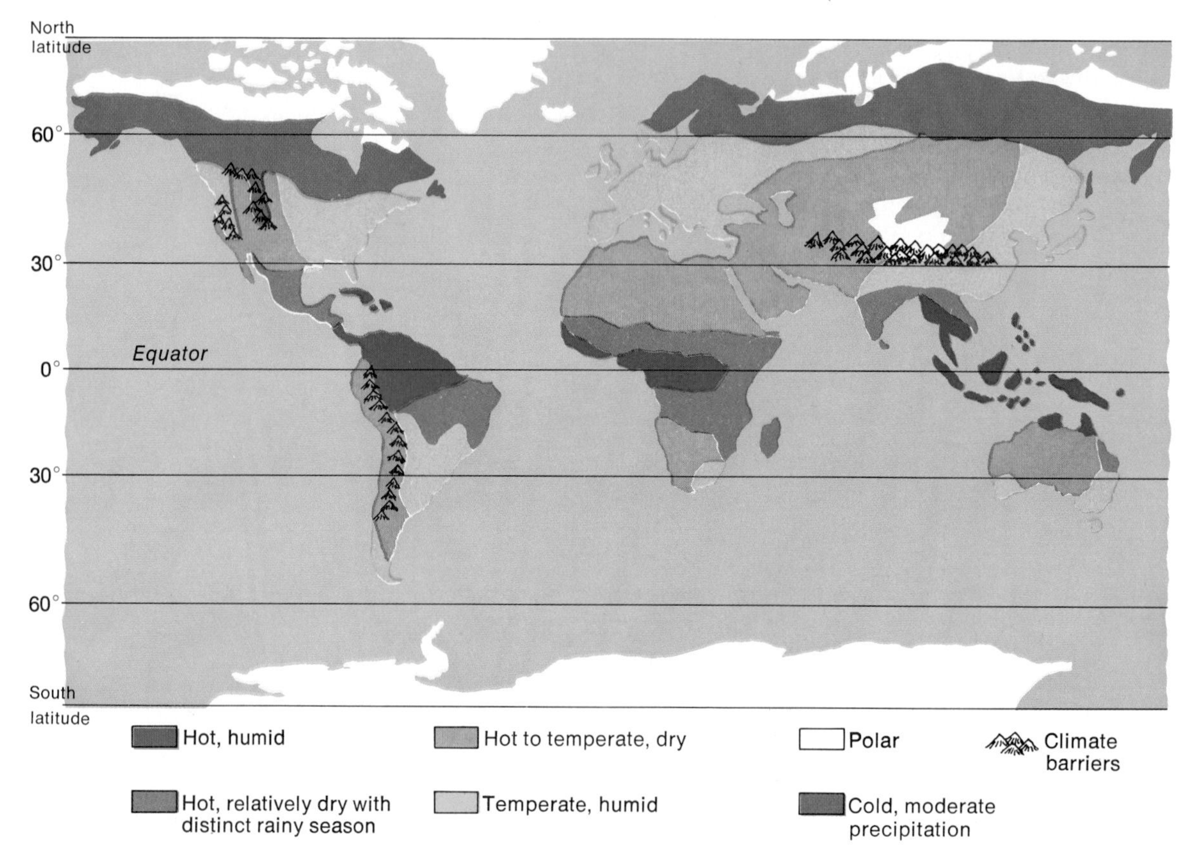

Figure 7/17 *A more detailed map of the earth's climate zones.*

paper strip that will be the mountainsides. Be sure to spread it evenly from one end of the strip to the other. Put the mountain into the box and place a container of hot water on one of the blocks.

Put the top on the box and watch the paper carefully for a few minutes. Calcium chloride absorbs moisture, and as it does, it darkens and becomes invisible. Which side darkens first? How much difference is there between the two sides after a few minutes? Do you think there is a convection current in the box?

If the water represents the Pacific Ocean and the paper represents the Sierra Nevada mountain range, what can you say about climates in northern California and in Nevada? Would you expect the Rocky Mountains to have a similar effect in Colorado?

Point out that the western Great Plains are relatively arid, yet there is a range in the Colorado Rockies called the Wet Mountains.

Figure 7/18
With this setup you can show how mountain ranges affect climates.

Repeat the activity with a piece of black paper parallel to the bottom of the box and at the level of the hot water. Where on the paper is the water absorbed? What do you conclude about the effects of mountains on climates? ★

Did You Get the Point?

Easterly winds blow in the tropic zones.

Air north and south of the tropic zones forms eddies, because of the earth's rotation.

There are two kinds of eddies, areas of high pressure with clockwise winds and areas of low pressure with counterclockwise winds.

Mountain ranges have a lot to do with whether a climate is dry or moist.

Answers:

1. Answers follow text and illustrations. Identify heat source and sink in room. Ocean currents on the surface move from warmer regions to cooler regions. Trace some on map or globe.
2. Equator and poles. Others to consider are volcanoes, hot springs, desert areas, cool water, glaciers.
3. Seattle—cool and moist; Phoenix—warm and dry; Denver—cool and dry; Memphis—warm and moist. Seattle is on windward side of mountains, near sea. Denver is on leeward side of mountains. Phoenix receives dry air from west and south. Memphis receives moist air from Gulf of Mexico. (Temperature and rainfall data are in the Teacher's Manual.)

Check Yourself

1. Describe convection currents in the aquarium or earth box, in a room, in the earth's atmosphere, and in the oceans.

2. Where are the best heat sources and heat sinks in the earth's atmosphere? Where are the others?

3. Explain how and why the climates are different in Seattle, Phoenix, Denver, and Memphis.

What's Next?

Energy from the sun pours over the earth in heat radiation. Much of it is absorbed and then moved around to provide energy for the earth's activities. All these processes affect the earth's surface and its atmosphere.

Solar energy powers the water cycle, which links air, water, and land. Water vapor passing through the air from land or oceans combines with prevailing movements of air to produce our weather. Clear days, stormy nights, humid summers, and cold winters all come from the sun, water and wind. How do they combine to make these different kinds of weather?

Skullduggery

Answers: To riddle—circulation. To central column of words on grid—The answer is circulation. To questions: (1) heat-2; (2) radiation-1; (3) radiates-9; (4) water-7; (5) seasons-3; (6) conduction-5; (7) convection-12; (8) climate-10; (9) axis-4; (10) source-15; (11) condensation-11; (12) dewpoint-8; (13) humidity-19; (14) evaporation-13; (15) saturation-14; (16) trade-18; (17) cloud-16; (18) wind-22; (19) vapor-21; (20) circulation-20; (21) absorb-6; (22) cycle-17.

Riddle: What is very important for your health, the earth's atmosphere, and newspapers?

Directions: Maybe you have solved this riddle right away. Maybe not. Even if you think you have, check your solution to the riddle by filling in the vertical column of word blanks in a grid like the one in Figure 7/19. Your teacher will give you one, or you can make one yourself. When you get the words that answer the following questions, fill in the words on the grid in the row of squares with the same numbers as those following the questions in parentheses. If the letters of the word do not fit the row, you know your answer is incorrect. (The answer to one of these questions is also the answer to the riddle.)

To get you started, the answer to question A is "heat." Since the question is followed with a (2), fill in "HEAT" in the squares in row 2 of the grid.

A. What are both your body and the sun a source for? (2)

B. What is one of the ways in which heat moves? (1)

C. The earth ______ some of the heat from the sun right back. (9)

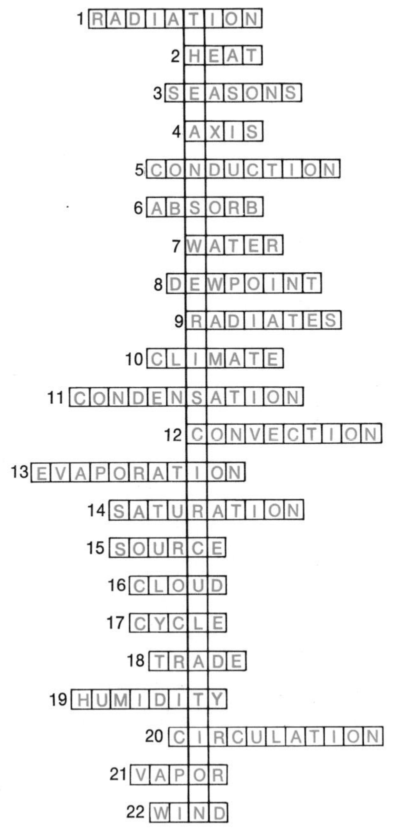

Figure 7/19
Solve the Skullduggery riddle on a copy of this grid.

D. What condenses when the dewpoint is reached? (7)

E. Changes in heat energy received by the earth's northern and southern hemispheres make the ______. (3)

F. What is the passage of heat along a spoon in hot soup? (5)

G. Heat moves through the air by ______. (12)

H. The yearly solar radiation in an area helps determine its ______. (10)

I. This is just imaginary, but the angle of it affects the seasons. (4)

J. Heat always moves to a heat sink from a heat ______. (15)

K. ______ of water vapor in the air usually makes clouds. (11)

L. The ______ is the temperature at which water vapor condenses. (8)

M. Relative ______ measures how much water vapor is in the air compared to what the air could hold. (19)

N. What is taking place when water changes from a liquid to a gas? (13)

O. The air reaches ______ with water vapor when the air can hold no more. (14)

P. ______ winds blow toward the west both north and south of the equator. (18)

Q. What does not form without particles for water to condense on? (16)

R. A westerly is a ______ coming from the west. (22)

S. Evaporation of water makes water ______. (21)

T. ______ of water in ocean currents is helped by convection, as is the circulation of air. (20)

U. Water and land ______ heat energy. (6)

V. When one ______ is completed, another just like it begins. (17)

For Further Reading

Adler, Irving. *Hot and Cold.* New York, The John Day Co., 1959. More about how heat is distributed and how the highest steady temperatures have been obtained, in a plasma jet.

Chapter 8 Weather

Chapter opener raises amateur weather questions about forecasting, rule-of-thumb weather prediction, weather superstitions, and the National Weather Service (formerly U.S. Weather Bureau). Discussion might follow questions in opening about what factors actually make the weather and what factors to consider when predicting weather. This leads into first section, in which students examine weather maps, compare forecasts with actual weather, and conclude how good the forecasts were.

''What a drippy, messy rain!'' Jeri said to Dan as they burst through the main school doors into the hall. She unbuttoned her soggy raincoat and then screeched, ''Hey, Dan, don't shake it all over me!''

Brad pushed through the door to join them, followed by Pam, panting from the run up the walk from the bus. They stamped their feet and began to peel off wet coats.

''Phew!'' Brad exclaimed. ''I don't mind a little rain and wind, but this is unbelievable! I sure hope it won't last through tomorrow until Saturday for the game.''

''Shows no sign of stopping,'' Pam said gloomily. ''How can I lead cheers in my raincoat? You'll have a hard time winning this one, Dan.''

''You said it! It's tough to hold the ball in the rain,'' Dan said. ''It's like a slippery fish. And forward passes—impossible!''

''It can't rain Saturday,'' Jeri said. ''I saw one of those extended forecasts yesterday, and I think it said fair for the weekend.''

''Can't believe the forecasts,'' Brad grumbled. ''On the radio this morning the prediction was rain tapering off tomorrow but nothing about Saturday.''

Discussion will draw out many of these presumed connections that people often use to predict the weather.

''My Dad said he heard the trains to the west last night,'' Pam offered. ''He always says that west winds are a sure sign of bad weather.''

"All these amateur weather prophets!" Dan scoffed. "Think they know more than the weather bureau with stations all over America and satellites whizzing around the earth!"

Chance to discuss probability of predictions with students. Mom's statement means she believes Weather Service forecasts are 100 percent incorrect; even forecasting at random, the Service would do better than this. Students use probabilities in Weather Prediction Chart activity of Section 8.4.

"Mom says if you prepare for just the opposite of the forecasts you never get caught," Jeri offered.

"Predictions aren't as bad as all that," Dan said. "I'm only afraid it'll snow on Saturday, it's so late in the season. Then we'll be skidding around in frozen slush."

"How about looking at some weather maps in the newspaper?" Pam suggested. "If we look at quite a few maps for last week and compare them with the weather that came, we can see how good the predictions are."

"That makes sense," Jeri agreed. "Look, I saw Mr. Lorenzo with a bunch of weather maps in earth science yesterday. Let's get them from him."

A student might like to look up recent reports on the successes and failures of cloud seeding.

"Good idea. They better give us fair weather for the game Saturday," Brad said. "Maybe they could do some cloud seeding and dump the rain somewhere else."

"Come on, let's get going. We don't have too much time before first bell," Pam said.

"O.K., let's go," Brad shouted. He whirled his raincoat around, showering the others, as they ran off toward earth science.

Weather Conditions

The comparison of forecast and actual weather introduces students to variables that determine weather. In the process, they become more familiar with weather maps, symbols, and simple definitions of weather events. They need not memorize these definitions. Encourage them to study the maps enough to decide how good the forecasts were.

Process: predicting

8.1 Compare weather forecasts with actual weather.

People are always wondering what the weather will be like. Really they are wondering about several things: What will the temperature of the atmosphere be? Will it be cloudy or fair? And people worry about how windy it will be and how humid.

Perhaps most of all, people want to know if it will rain or snow; that is, will there be any precipitation? **Precipitation** is the word used for any and all forms of water, liquid or solid, falling from the atmosphere to the ground. Precipitation can come down in the form of drizzle, rain, snow, sleet, or hail. The reason everyone wonders about the weather, of course, is that weather conditions have a lot to do with our physical comfort, our moods, and our activities.

Meteorologists are the scientists who study conditions in the atmosphere in order to understand and predict the changes that affect weather. From their observations, meteorologists make weather maps of actual weather or forecasts of it that indicate these changes.

Students study and compare forecasts and actual weather individually. **Activity Time:** 20 minutes for the map set. This includes questions and discussion of symbols on maps and of photo compared with maps. Activity can also be done as homework, with differences between forecast and actual weather written out.

★ The first map in Figure 8/1 shows the forecast made by the National Weather Service for the next day's weather across the whole United States. Note how many different kinds of conditions are given on this map: the temperature, pressure highs and lows, and precipitation. The heavy black lines made of bumps or points show various kinds of fronts. **Fronts** are the boundaries between air flows of different temperatures. All these conditions help to make the weather.

Notice that the weather map and the forecast in Figure 8/1 are for the same day. The day's actual weather is given in more detail than the conditions shown in the forecast. The actual weather map shows the degree of cloudiness, wind speeds and directions, and temperatures at various places. The degree of cloudiness circles on the report map stand for weather stations.

Forecasts and actual weather always vary considerably in their details on predictions for the whole country. Have students back up their judgments of forecasts with specific evidence. Because of the general nature of country-wide predictions, local, short-term predictions still have a useful role to play. Further activities using weather maps from local papers might be interesting for students. They could keep individual cards of forecasts versus actual weather from papers, TV, or radio to see for themselves how accurate forecasting is. The class will do one or more weeks of forecasting in Section 8.4, using the Weather Prediction Chart as adapted to your local area.

Process: observing

The satellite photograph in Figure 8/2 shows you what the clouds across the United States looked like on the same day as that of the maps. Examine these clouds and compare them with the weather conditions shown in Figure 8/1. See if you can match the clouds and the clear air shown in Figure 8/2 with conditions in the atmosphere shown in Figure 8/1.

Note how closely the weather forecast map matches the actual weather as shown in the two maps of Figure 8/1. Check the temperatures in a number of places, for example. Check where precipitation was forecast and where it actually fell. Check whether the lows and highs were actually in the positions predicted. Were the fronts of the same kind, and were they in the places where they were forecast? List the differences between the forecasts and the actual weather.

How good was the forecast? ★

C

Fronts

Cold

Occluded

Warm

Stationary

Wind speed

Wind direction

Degree of cloudiness

Wind speed (miles per hour)

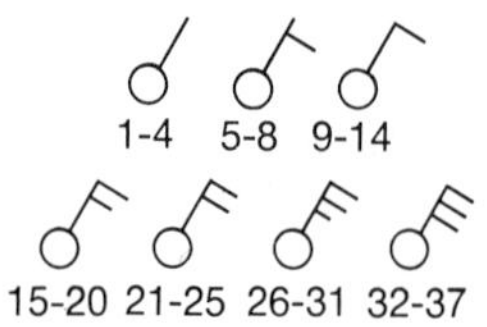

Degree of cloudiness

Clear

Scattered clouds

Sky ¼ covered

Sky ½ covered

Sky ¾ covered

Sky 9/10 covered

Completely overcast

Sky not visible

Other symbols

High pressure

Low pressure

Showers

Steady rain

Snow flurries

Steady snow

Thunderstorm

29.97 Atmospheric pressure (inches of mercury)

Direction and speed (mph) of high or low

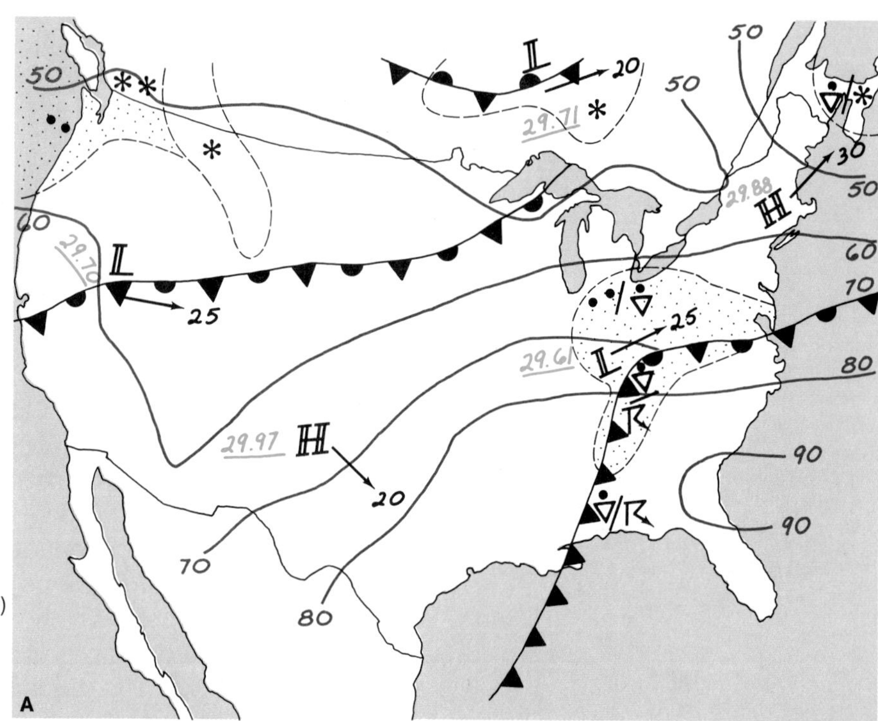

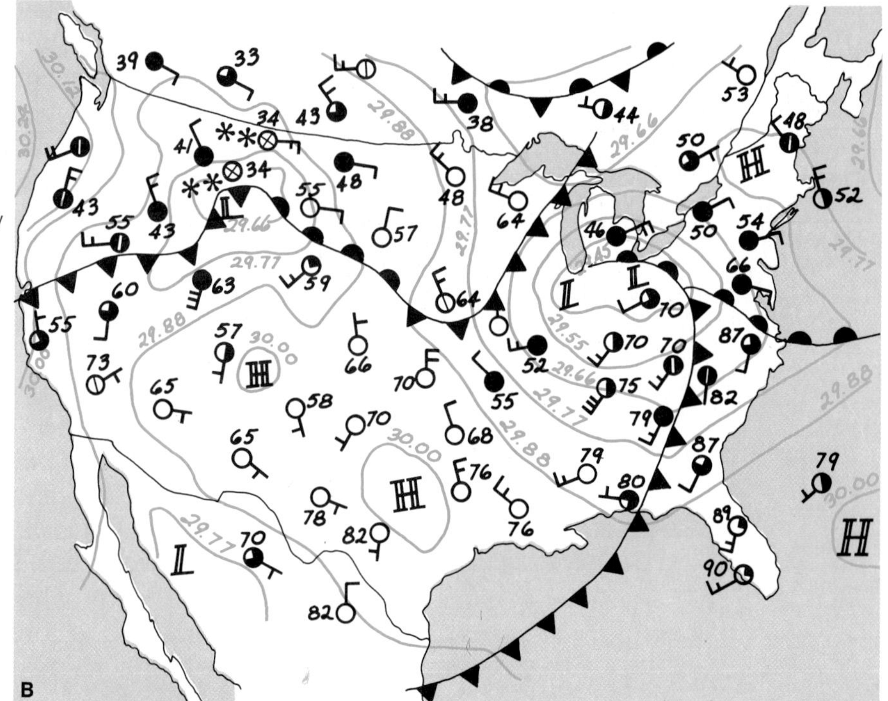

Figure 8/1 *Map A shows the National Weather Service 24 hour prediction for April 16. The red isotherm lines are lines of equal predicted temperatures. The temperatures are given at the ends of the isotherms. The prediction—Northeast, partly sunny; east of lower Mississippi River, showers and thunderstorms; Great Lakes to Ohio Valley, showers or steady rain; Pacific Northwest, showers and snow flurries; elsewhere, sunny and mild. Map B shows the actual weather for April 16. The blue isobars are lines of equal atmospheric pressure, with barometer readings in inches of mercury. Study the key to the symbols on the map carefully. You can tell from the map how accurate the prediction was.*

Figure 8/2 *A satellite photograph of the cloud patterns over North America taken on April 16, the same day as the actual weather map (B) was made. You can see how the conditions marked on the map resulted in the clouds in the photograph.*

8.2 What is the weather?

Every day, right around you, without any instruments at all, you can observe most things people call the weather. You feel the temperature rise as the sun climbs higher, and fall as the sun sets. You notice whether the air is damp or dry

Actual thermometer reading (degrees Fahrenheit)		How cold it feels when the wind is blowing (degrees Fahrenheit)							
		5 mph	10 mph	15 mph	20 mph	25 mph	35 mph	45 mph	50 mph
60	Little	56	50	48	46	44	42	41	40
50	danger	45	37	33	30	28	24	23	22
40	for	34	23	20	15	11	7	5	4
30	person	24	12	3	−1	−5	−13	−14	−16
20	warmly	12	−4	−13	−18	−23	−28	−32	−33
10	dressed.	0	−17	−27	−33	−38	−45	−48	−50
0		−13	−29	−40	−46	−53	−61	−64	−65
−10	Danger of	−23	−42	−53	−60	−68	−77	−80	−83
−20	freezing ex-	−33	−55	−67	−76	−85	−93	−98	−101
−30	posed flesh.	−45	−67	−81	−92	−100	−109	−114	−120

Figure 8/3 *A wind chill chart. If the temperature is 50° and the wind is blowing at 20 miles per hour (mph), the chart tells you that you would feel as if the temperature were 30° with no wind. See how much colder it feels if the wind speed increases.*

or if there is any precipitation. Anyone can tell if the sky is cloudy or clear, how windy it is, and what direction the wind is from. Another part of the weather, visibility, is especially important for drivers and pilots. **Visibility** is the distance at which you can clearly see and name sizeable objects.

The weather consists, then, mainly of the temperature, humidity, precipitation, wind, and visibility of the atmosphere.

This basic common sense definition is the set of factors that people usually mean by weather, according to the American Meteorological Society.

You may wish to work with students on the use of the wind-chill chart. If convenient, you can introduce them to wind-speed and humidity measurement, or you can wait until these are described in Section 8.3 and used in Section 8.4.

You know a great deal about how the weather affects you. When you say, "It's chilly out," you are often talking about cold temperature and wind. The combined effect of the temperature and the wind is shown in the wind chill chart in Figure 8/3.

You have probably also noticed how the **humidity**, or the amount of water vapor in the air, affects you, both in the summer and in the winter. High temperatures are especially uncomfortable on humid days. Low temperatures feel much colder when the air is humid.

By contrast, on a hot, humid summer day, isn't a breeze a great relief? This is because wind speeds up the evaporation of moisture from your skin, so you cool down.

Figure 8/4
The Nimbus weather satellite can record temperature data on the ground, and temperature and humidity in the upper atmosphere. It photographs clouds where it is day by visible light and where it is night in infrared light. It even detects ozone.

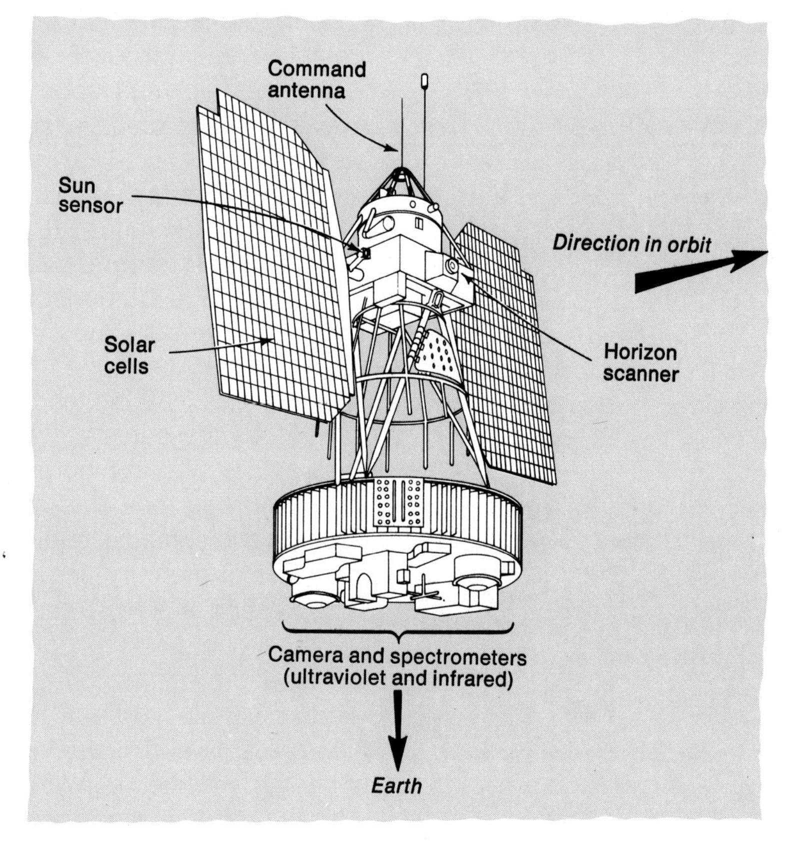

Scientific observations of all the conditions that affect the weather are collected at weather stations across the United States and around the world. Low- and high-altitude weather balloons are very important collectors of weather data, too.

British and American weather scientists have cooperated with experiments with Nimbus IV satellites making temperature measurements from polar orbits.

Now artificial satellites like the one in Figure 8/4 orbit the earth and are being used to make weather observations. These observations give meteorologists a more complete picture of weather conditions around the globe.

8.3 Many conditions determine weather.

Process: measuring

You have seen in Section 6.6 how many different ways there are to study the gases of the atmosphere. Most of the conditions of the atmosphere change, or vary. These variables are measured with weather instruments. You can use some of these instruments yourself.

Some conditions that play an active part in shaping the weather are air pressure, wind, air temperature, and relative humidity. You read about some of these in Section 7.7. How do meteorologists study and measure these conditions?

Describe and/or demonstrate use of various weather instruments. Go into various scales in temperature, inches and millimeters of mercury, and so on.

Air pressure is the weight of the air above a given place on the earth due to the attraction of gravity on the air above that place. **Barometers** are the instruments that measure air pressure. A barometer shows how high (in inches) a column of mercury must be to balance the weight of the air on the surface of the mercury. Air pressure varies from a low of about 29 inches to a high of about 31 inches.

Some students may benefit from identifying and following the isobars of pressure and the isotherms of temperature in weather maps in Section 8.1. Some weather maps show pressure in millibars rather than inches of mercury.

What air pressures are shown in the highs and lows on the weather map in Figure 8/1? The lines along which the air pressure is the same are called **isobars.** "Iso" means equal or the same. The air pressure in Figure 8/1 is given in inches of mercury across the isobars.

In Section 7.9 you saw that winds swirl around low-pressure centers in a counterclockwise direction. These low pressure eddies are called **cyclones.** By contrast, the winds blow clockwise around high pressure centers. The high pressure eddies are called **anticyclones.** They usually bring fair weather conditions. You can see examples of cyclones and anticyclones in the weather map in Figure 8/5. Look back at the weather map in Figure 8/1. Are the winds blowing as you would expect them to in the highs and lows shown on that map?

Usually winds will be following described motions around low- and high-pressure areas, but some winds will be running in various directions in relation to fronts. Point out that the described motions are only rough generalizations.

The speeds of the winds vary a lot. Wind speed is measured by instruments called **anemometers** like the one shown in Figure 8/6. Their rotating cups whirl with the varying winds. Notice how the flags on the wind lines on the weather map in Figure 8/1 stand for varying wind speeds. The wind direction is down the wind lines from the flags to the circles.

The air temperature is shown by the single numbers scattered over the weather maps at places where there are weather stations. Air temperature is measured by **thermometers** in degrees Fahrenheit. The red lines on the weather forecast map of Figure 8/1 show where it is predicted the temperatures will be the same. These lines are called **isotherms.**

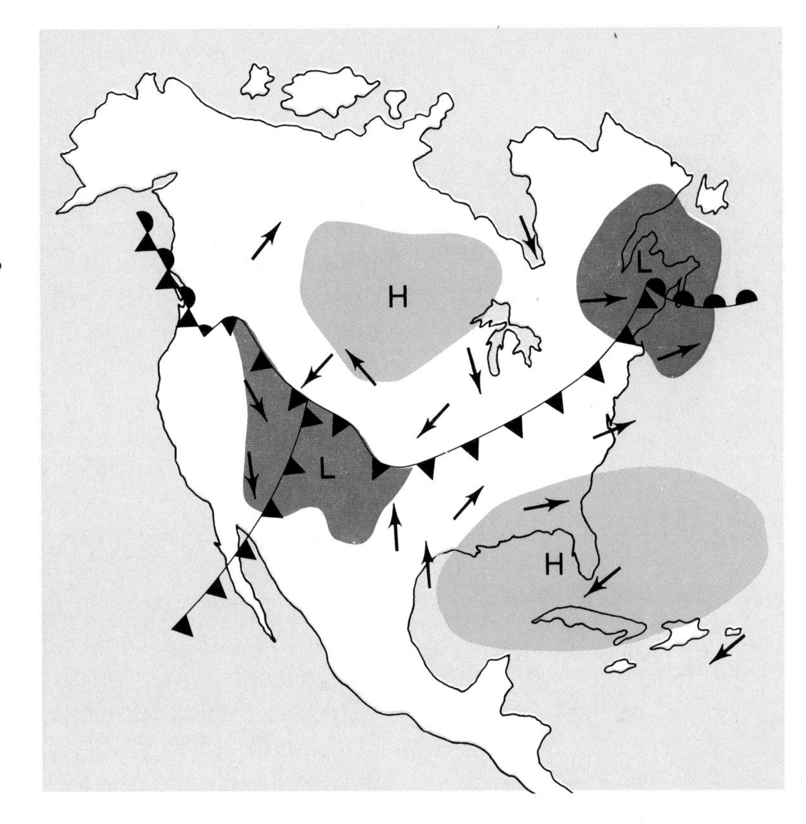

Figure 8/5
A winter weather map (December 26) with two lows and two highs. Wind directions from stations at various places in North America illustrate the clockwise air flow around the highs and the counterclockwise flow around the lows.

Figure 8/6
Workmen repairing the anemometer and wind direction indicator on top of Mt. Washington in New Hampshire. The world's record wind speed of 231 mph was recorded there in 1934.

The amount of water in the air varies also, making the air dry or humid. In Section 7.7 you saw how the condensation of this water vapor takes place at the dewpoint. You can discover how relative humidity can be measured with a wet-bulb and dry-bulb thermometer. The drier the air, the more water evaporates from the wet-bulb thermometer, and the lower the wet-bulb temperature becomes. Using the dry-bulb temperature and the difference between the dry-bulb and wet-bulb readings in the table in Appendix R, you can find the relative humidity of the air.

When the dewpoint is reached, clouds form like those shown in Figure 8/2. Cloud systems like these condense in the atmosphere at low, medium, or high altitudes above the earth.

Look at the five main kinds of clouds in Figure 8/7. These cloud types can tell you a lot about the coming weather. The names of these clouds indicate their shape and sometimes tell their altitude in the atmosphere. "Cirrus" means curly, which cirrus clouds are, as you see. "Stratus" means spread out or stretched in layers. "Cumulus" means piled or heaped up. "Nimbus" means rainy. So nimbus clouds are most likely to bring a downpour.

The natural place to introduce a classification of cloud types in conjunction with Figure 8/7. If possible, go outside with students to observe cloud types on a day with some variety of clouds. Encourage students to identify cloud types while reading this chapter, particularly with Section 8.4.

The portion or percentage of the whole sky in which there are clouds is called **cloud cover.** Cloud cover can be judged by looking at the sky and seeing how much of it

Figure 8/7
Cloud types: (A) cirrus, the kind that appears before a front comes along; (B) stratus, or layered; (C) fair weather cumulus, the kind that often develops in a dry, clear summer sky; (D) large cumulus or thunderhead; and (E) nimbus, ordinary rain clouds.

When cloud cover is sparse to medium, estimates can be made more precise by judging cover in 4 quadrants of the sky, north to east, east to south, and so on, and then combining the 4 estimates. Meteorologists often use tenths of cloud cover per quadrant, which can then be added for the 4 quadrants and divided by 4 to arrive at the overall cloud cover.

is clear and how much filled in with clouds. If the sky is half clouded, then the cloud cover is 50 percent. Cloud cover runs from 0 percent (clear) to 100 percent (entirely cloudy). Symbols for cloud cover, as well as for precipitation, appear on the weather map in Figure 8/1.

8.4 Predict the weather from changing conditions.

Many people claim they can predict the weather by watching the moon, listening to crickets chirp, or judging the pain in aching joints. Strange as it may seem, they are often right.

Although they are often not aware of it, these people are estimating weather changes. All their calculations are unconscious. People are predicting tomorrow's weather from what has happened in the past.

Activity Time: About 20 minutes for discussion of "No" and "Yes" columns and "Probability" percentages when activity is started.

Materials

thermometer (Fahrenheit), barometer, anemometer (or some type of wind speed indicator), sling psychrometer (or wet- and dry-bulb thermometer), wind direction indicator

This is a weather watch. You should plan to run it for at least one or two weeks—even after you have started the next chapter. It takes only a few minutes out of each class to get it done, if it is well organized. Students who want to continue the weather watch beyond one or two weeks should be encouraged to do so.

See Teacher's Manual for sample observations and notes on what students should observe to fill in the chart. Try to get students to deduce the information you already have. Encourage them to draw on their accumulated experience with weather.

With a little practice and a little help from the rest of the class you can become at least as accurate as the old-time cricket and ache studiers. At the same time you will get some idea of how the daily weather predictions are made.

★ Figure 8/8 shows a partially blank chart. Use a page of your notebook to make a larger chart that you can fill in. When each member of the class has done this, decide who is going to measure the variables or conditions listed on the chart each day. One student can be appointed to assemble all the daily data on the chalkboard for the entire class to copy. Discuss which variables you think will be the most important.

After the class has gathered data for a week or more, discuss the data and decide which measurements of each variable are likely to indicate that rain or snow *is* on the way. Then decide which measurements of each variable are likely to indicate that *no* snow or rain is on the way. Record these observations that favor precipitation in the YES column and those that do not favor precipitation in the NO column.

For instance, increasing wind speed often means a change in the weather. Some changes in the weather variables often bring precipitation. You would write "high or

Probability	**Variable**	**Observations favor precipitation?** No	Yes
	Present temperature		
	Temperature change		
	Present pressure		
	Pressure change		
	Cloud type		
	Cloud cover	*Do not*	*write*
	Direction wind comes from	*in this*	*book.*
	Wind speed		
	Difference between dewpoint and temperature		
	Conditions upwind		

Figure 8/8 *A weather prediction chart for predicting precipitation.*

Feel free to change the variables or substitute others as students learn more about weather and what causes it. You can also change "Yes" and "No" factors and probability percentages. Students will make discoveries as the weather watch proceeds. Keep the excitement of "trying to be right" alive.

Check some weather maps with students to identify what conditions lead to precipitation or fair weather.

Stress the fact that computers can be fed much more data than your class can collect. Also, you are not using weather balloons or satellites.

increasing wind speed" in the YES column and "low or no wind" in the NO column opposite "Wind Speed."

Now decide which of the ten variables are the surest, or most probable, signs of precipitation. First put a small check by these. Then try rating each of the variables with a number. Put this number in the "Probability" column. Assign the highest number rating to the variables that most often mean rain or snow. Adjust the numbers so that all together they total 100.

A weatherman would refer to your ratings as a "percent probability" that rain will follow a YES observation of this variable. The more measurements found to favor precipitation, the higher the percent probability of precipitation the next day.

Each day for a week or two weeks, just before class, help measure the variables listed on the chart. Decide which of the conditions fall under the NO column and which fall under the YES column. Keep a percent probability record of YES's and NO's. The main reason for putting observations on paper is to be able to make predictions. On the chalkboard write the class' percent probability prediction that *it will rain (or snow) tomorrow.* A day's wait will tell you how accurate your prediction was. After a few days of observations, you may find that you need to change some of the YES and NO conditions or the percent probability numbers. They have just not been accurate enough. Weathermen use observations made over many years to make their predictions more accurate. Besides this, they can measure more complicated air conditions than you can measure in just one place. Among these conditions are the fronts that you will make in Section 8.6. ★

Did You Get the Point?

Meteorologists study atmospheric conditions in order to explain and predict the weather.

Weather consists mostly of the temperature, humidity, wind, and visibility of the atmosphere, and of the clouds and precipitation in the atmosphere at a particular time.

Weather fronts are the boundaries between air flows of different temperatures.

The wind-chill chart shows the effect on people of the wind and temperature together.

At the dewpoint, water vapor condenses into clouds in the atmosphere to produce the cloud cover.

Answers:

1. Good place for a discussion of the environment and the need for a balance between these and other factors. "Ideal" conditions will vary greatly among students. Unchanging "ideal" conditions over a long period would raise havoc with the environment, as students can point out.
2. Examples of occupations affected by the weather would be: mowing lawns, shoveling snow, outdoor work of all kinds, farming, fishing, piloting aircraft, all types of transportation, construction, and manufacture of all items exposed to weather.

Check Yourself

1. List the weather conditions, such as air temperature, air pressure, wind, humidity, precipitation, and cloud cover that you would call ideal. What would happen to your environment if your ideal conditions should be kept up for a long time? Give reasons for your answers.

2. What occupations in your part of the country are affected a great deal by the weather? Describe some of the effects of the weather on each of these occupations.

Storms and Fair Weather

See notes on "air mass" concept in Teacher's Manual.
Concept: energy in motion

Maps of worldwide prevailing surface winds in January and July and surface ocean currents appear in Teacher's Manual. Point out that prevailing winds are statistical averages, and on a given day might not look like those on the prevailing wind map. With overhead projection, these can be discussed with students, revealing general circulation in both hemispheres. NASA's *Weather in Motion* shows clouds in satellite photos moving across continents (See Additional References in Teacher's Manual).

8.5 Moving air creates weather.

You have seen how the air moves in convection currents and how the turning of the earth is believed to affect moving air. You learned in Section 7.9 that air to the north or south of the tropic zones moves in circular patterns called eddies. Eddies are either high pressure anticyclones or low pressure cyclones. How does this moving air create weather?

Weather conditions are mainly determined by the temperature and pressure of the air and the amount of moisture in the air. The temperature of the air and the amount of moisture it contains depend on where the air comes from. On the map in Figure 8/9, you can see the directions from which winds often approach the continental United States as they form around highs and lows. Air from the poles tends to be cold. Air from the equator or tropic zones tends to be warm. Air from over the land tends to be drier than air from over the ocean. Where a cold eddy of air with high pressure is close to a warm eddy with low pressure, a front may form. The **front** is the narrow region where the eddies meet. Weathermen can usually identify a front by checking the temperature differences in the air where the eddies

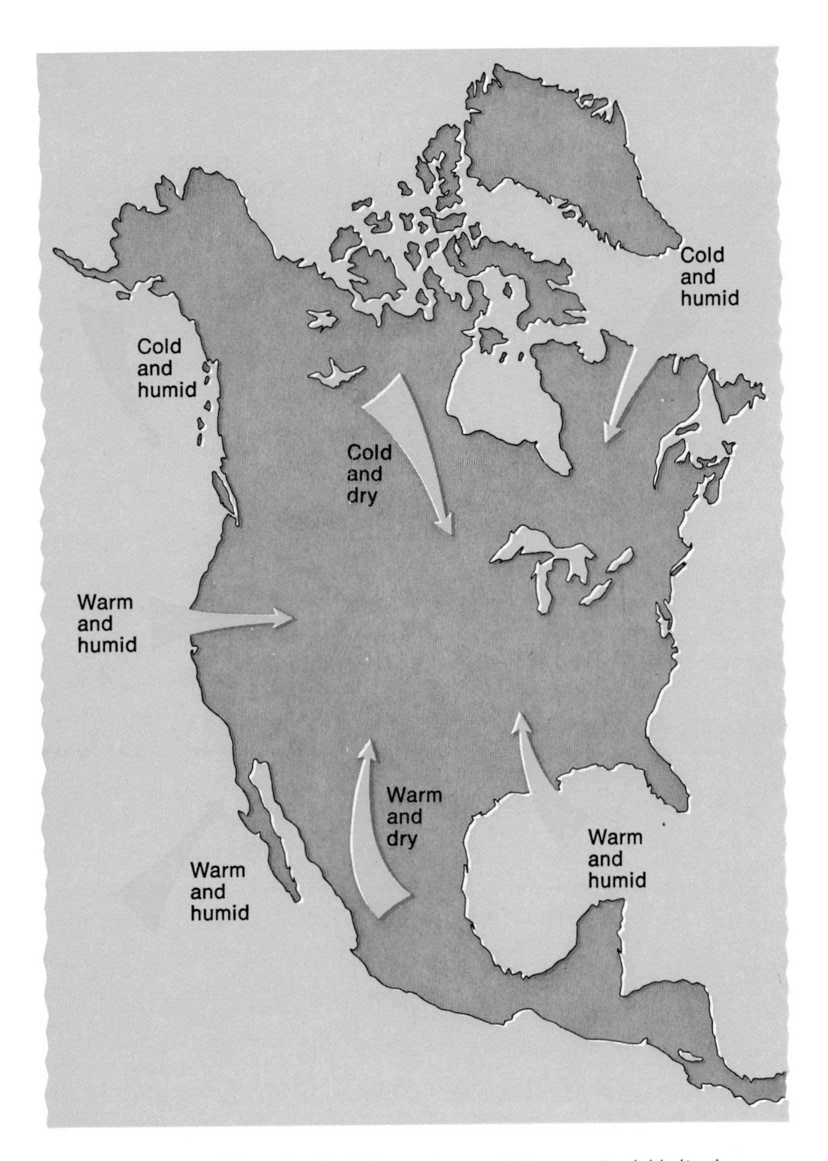

Figure 8/9 *The air that flows toward the central United States comes from many directions and varies greatly in temperature and humidity. Fronts form where differing flows come together.*

come together. For example, if air at 70 degrees Fahrenheit is close to air at 50 degrees Fahrenheit, a front is probably forming.

Along the front the weather is likely to be unsettled because of the different conditions of pressure, temperature, and moisture in the eddies of air. A front passing over the place where you live might produce weather conditions ranging from cloudiness to a violent storm.

Arizona air comes over land. Great Lakes area has a mixture of cold and dry, and warm and humid, promoting precipitation.

Study the map in Figure 8/9 again. Can you tell from it why Arizona has so much clear weather? Can you tell why the Great Lakes region has many cloudy days and much more precipitation than Arizona? Look back at the weather map in Figure 8/1. Can you see how weathermen predict tomorrow's weather by studying the kinds of eddies that are moving into their region? Wouldn't your predictions be better if you knew the conditions of the air that was heading your way?

8.6 Make your own weather fronts.

The map key in Figure 8/1 defines various kinds of fronts and gives the symbols for them. The four main types are warm fronts, cold fronts, occluded fronts, and stationary fronts (not moving), which could be either cold or warm fronts.

A **warm front** is a boundary between warm air and colder air, both moving in the same direction. As they move, the warm air rises over the cold air. You can see what happens in Figure 8/10.

The earth box is a good place to make models of weather fronts and watch how they behave. Use plastic bags of cold sand to cool the air. A pan or cup of hot water will form warm, moist air.

Materials

aquarium or earth box, stoppers, container of hot water, cold sand bags

Activity Time: 30 minutes. Groups of 4–6 students with each earth box. Groups can be assigned different portions of Section 8.6 at the same time, sharing results. Narrow box permits weather effects to form quickly and precisely. Bags of refrigerated sand for cold air are better than water, which cannot be molded when frozen. Discussion of student weather observations over a week or two should clearly indicate the passage of one or more fronts.

★ Make a cold front. A **cold front** is a boundary along which cold air moves under and into warm air. Set up cold sandbags and hot water in an aquarium or earth box as shown in Figure 8/11. Because of convection, cold air moves toward the air over the hot water. That makes a model of a cold front. Watch the pattern of "clouds" that forms. Where does most of the moisture appear on the sides of the box? Make a drawing of what you observe and then compare it with the front in Figure 8/12. ★

Have you noticed any warm or cold fronts in the weather conditions you have been observing?

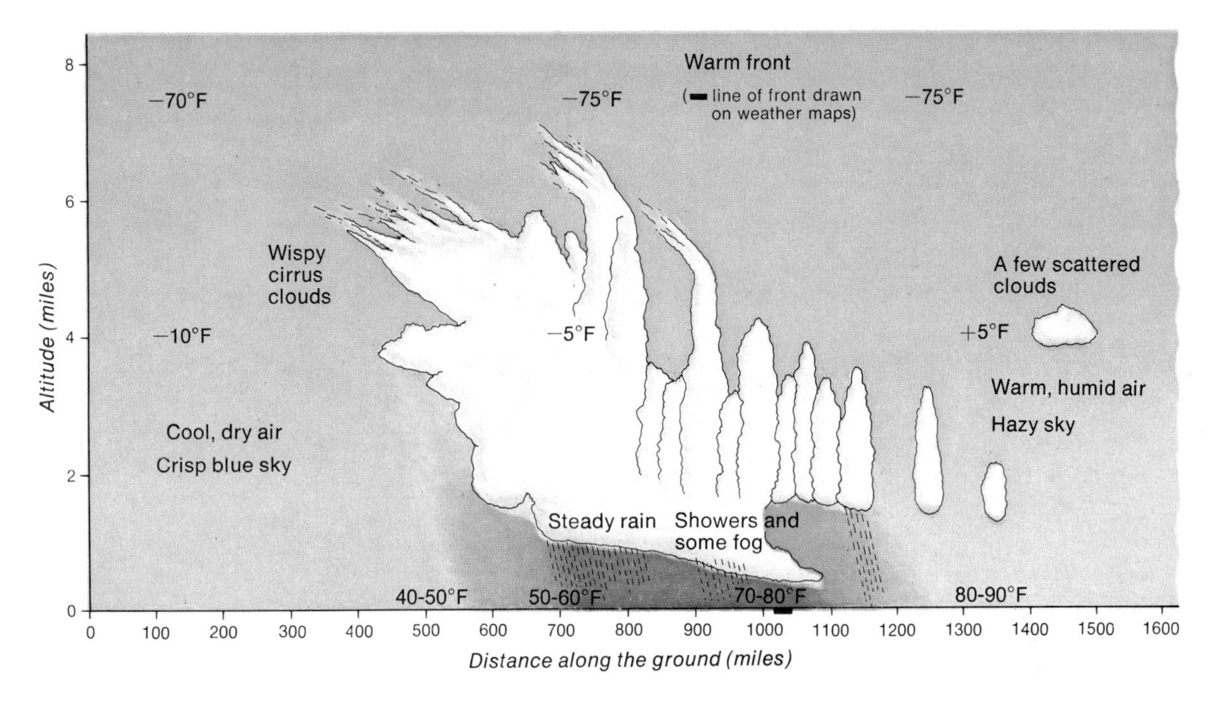

Figure 8/10 *A cross section of a warm front. The vertical scale has been greatly stretched out so that you can see details of the clouds and weather more clearly.*

Figure 8/11
Model a cold front with this kind of an arrangement.

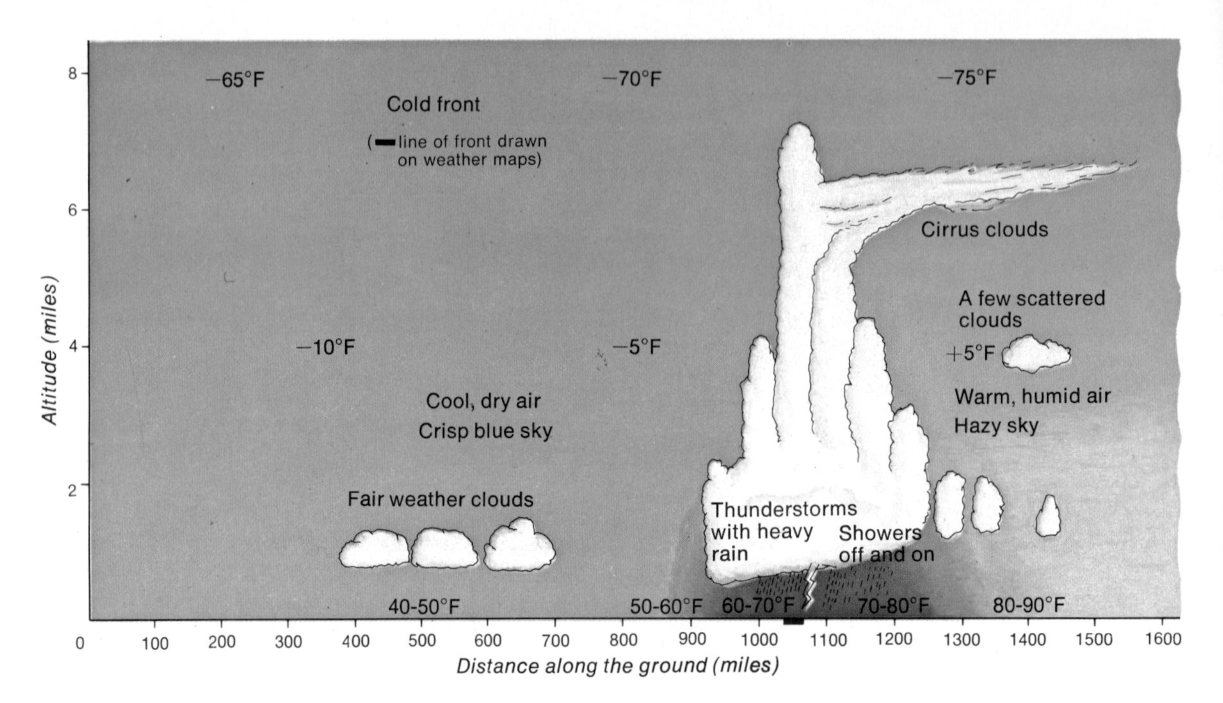

Figure 8/12 *A cross section of a cold front. As in Figure 8/10 the vertical scale has been stretched out.*

Figure 8/13 *With this setup in your earth box you can demonstrate the effect of cool air flows on both sides of warm moist air. This models an occluded front.*

An **occluded front** is made when cold air moves under warm air, lifts it up, and then meets more cold air. The cold air holds the warm air up, so it is cut off from the ground. "Occluded" means cut off or closed off.

Materials
same as previous activity

★ To make an occluded front, set up the aquarium or earth box as shown in Figure 8/13. Use a cup of hot water as a heat source. Notice where the "clouds" form in the air and where water first condenses on the sides of the box. Make a drawing of the occluded front as you see it in the box. ★

8.7 Some cyclones cause violent storms.

Concept: energy in motion

Cyclones, as you know, are eddies with winds that blow into and around a center of low atmospheric pressure. (See Section 8.3) They rotate counterclockwise in the northern hemisphere and clockwise in the southern hemisphere.

Where warm and cold air meet along a front, storms are often produced. These storms may be mild, with only light winds and heavy rains, or they may be violent. One such violent storm that nearly everyone has experienced is a thunderstorm. **Thunderstorms** are local weather disturbances in which strong convection currents rise rapidly to great heights. Because of this, water vapor in the cloud is carried upward, condenses, and freezes very quickly. Lightning strokes zip from cloud to cloud or from cloud to earth. Heated by lightning, the air expands and explodes in bursts of thunder. Columns of rain or hail pour down on the earth. Nowadays radar photographs like those in Figure 8/14 are often used to spot heavy rainstorms or thunderstorms that are 100 to 200 miles away. This gives weathermen time to warn people that a storm is coming.

Another kind of local storm, a **tornado,** is a violently spinning column of air hanging below a thunder cloud. The funnel-shaped columns of tornadoes contain winds that may be whirling at more than 300 miles an hour. Currents of air may stream upwards at more than 150 miles an hour. **Waterspouts** are like tornadoes, but they form over bodies of water. Both types of tornado are shown in Figure 8/15.

Some violent storms form near the equator. In photographs they look like cyclones, but they have more energy. People call them **hurricanes** when they form over the

Figure 8/14
Varying amounts of rainfall recorded on a radar scope by radio signals bounced back from storms. The bright areas mean heavy rain, and the gray means lighter rain.

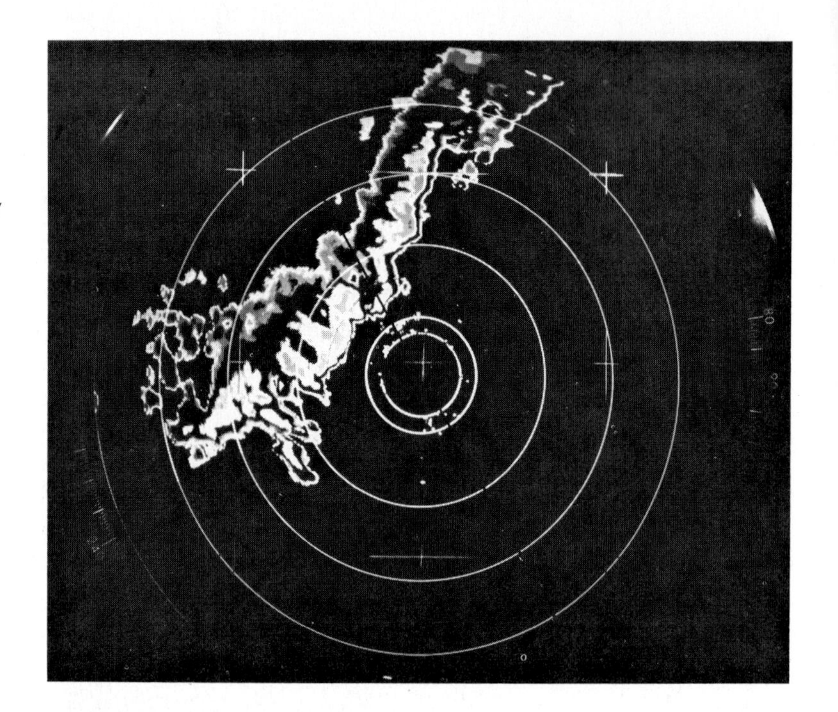

Figure 8/15 *A tornado is much smaller than a hurricane, but it is extremely destructive. When a tornado forms over water, it is called a water spout. Weathermen can now predict tornados better, but cannot control them.*

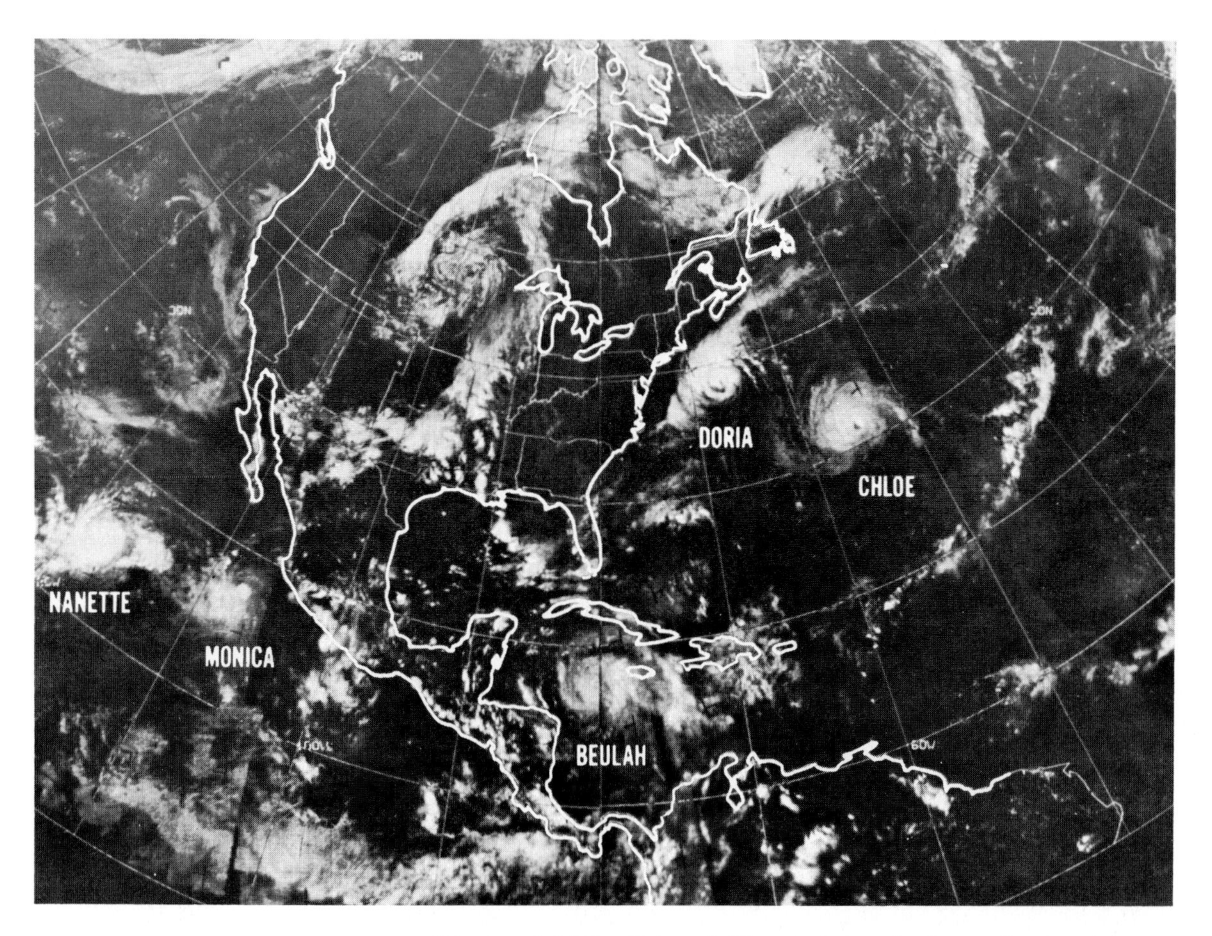

Figure 8/16 *This satellite photograph shows five hurricanes in action at one time on September 14, 1967.*

Atlantic and **typhoons** when they form over the Pacific. They travel in westward or northward paths across the Atlantic Ocean, sometimes devastating the land with their fury. A typical hurricane like those shown in Figure 8/16 is about 600 miles in diameter. In its center is an open "eye" of clear calm air, about 25 miles across. The strongest winds in a hurricane surround the eye and may be as high as 135 miles an hour.

Is there a way to control violent storms? Planes carrying instruments fly through hurricanes to record what is happening within them. Meteorologists study conditions that produce violent storms in the laboratory as well. Perhaps what they learn will enable them to find a way to decrease the intensity of violent storms, if not to control them.

The Moon and the Weather

Down through the centuries, farmers have claimed that the moon and the weather are connected. They have said, for example, that the second full moon after the beginning of spring is the time to plant. Other people will say that the weather is likely to be wet when the crescent moon is tipped so that it can hold (or not hold) water! You can imagine how modern scientists react to ideas like these.

A few years ago, however, a scientist studied 50 years of weather records and found that there may be a connection between the position of the moon in relation to the earth and fair and stormy weather.

The scientist found that the greatest amount of precipitation during the month is most likely to come shortly after full moon and new moon. This is not true every single month, but it is true for most months over many years. The moon is in full phase, of course, when it is on the opposite side of the earth from the sun and at new-moon phase when it is between the earth and the sun. (Look back at Figure 5/10.)

Not all meteorologists agree that the position of the moon affects the weather on the earth in this way. Those who think it does, wonder if the tides in our atmosphere that are caused by the moon's gravity (Chapter 6, Are There Tides in the Atmosphere?) influence the amount of precipitation on earth.

This connection between the moon and weather has not been proven. Maybe there really is a connection. Maybe it is pure chance that the weather records suggest a connection. What do you think?

Did You Get the Point?

The temperature and amount of moisture the air contains depend on the region the air comes from.

The weather along a front is generally unsettled.

The four main types of fronts are warm, cold, occluded, and stationary.

Strong cyclones rotating counterclockwise around an area of low pressure may produce violent storms, such as thunderstorms and tornadoes.

Hurricanes and typhoons are violent storms that form in the tropic zones and move into temperate regions.

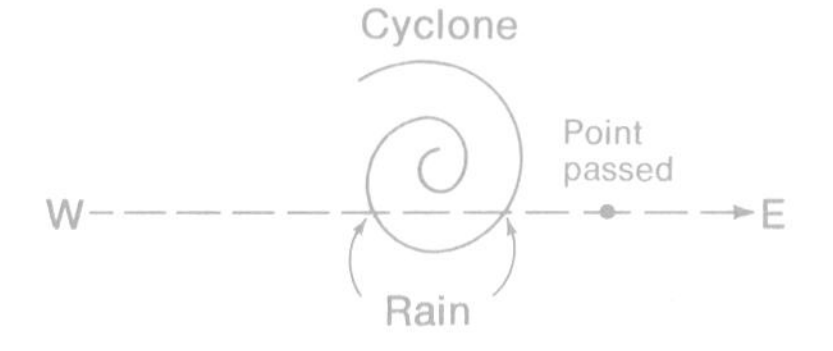

Check Yourself

1. Explain why you might get several rainstorms with lulls between them when a cyclone passes over. The photograph in Figure 8/16 may help to answer, if you think of the cyclones in it as moving from left (west) to right (east).

Students may realize that some cyclones have several distinct weather zones. As the cyclones move from west to east one zone may bring precipitation, followed by clearing, then by precipitation with the next zone.

What's Next?

Like a ship at sea, the earth plows its way through the solar wind around the sun. Gravity hugs a blanket of air close to the earth's surface. The sun pours light and heat into the earth's atmosphere. This energy radiating from the sun evaporates water into the air and helps to move air in sweeping paths around the earth.

The motions of air and water make our weather. Fronts are formed. Cyclones and anticyclones develop, bringing storms or fair weather.

The next unit moves down from the atmosphere to the land on which we live. Unlike the weather, the land seems solid and permanent. Yet it is constantly being changed, too, not only by different weather forces, but also by powerful forces on and within the earth that destroy and create.

Skullduggery

On July 4, 1985, many people across America saw a huge silver flying saucer darting about in the sky. It was trailing a long cable, from which hung the following words, which seemed to be in some kind of code. A junior high student in Cirrus, Illinois, finally managed to break the code and read the message. Can you?

Message reads: ASK PERMISSION TO LAND TO BUY ALL EARTH'S URANIUM FOR ROCKET FUEL FOR RETURN TRIP TO OUR PLANET. FOR PAYMENT WILL GIVE YOU KNOWLEDGE TO CONTROL HURRICANES, TORNADOES, AND PRECIPITATION.

Message Trailed across the Sky by a Flying Saucer

IXK MOBCPXXPEA GE RIAD GE FUZ IRR
OIBGSX UBIAPUC NEB BEHKOG NUOR NEB
BOGUBA GBPM GE EUB MRIAOG. NEB
MIZCOAG TPRR VPLO ZEU KAETRODVO GE
HEAGBER SUBBPHIAOX GEBAIDEOX IAD
MBOHPMPGIGPEA.

Code runs: (1) A = I; (2) B = F; (3) C = H; (4) D = D; (5) E = O; (6) F = N; (7) G = V; (8) H = S; (9) I = P; (10) J = J; (11) K = K; (12) L = R; (13) M = Ç; (14) N = A; (15) O = E; (16) P = M; (17) Q = Q; (18) R = B; (19) S = X; (20) T = G; (21) U = U; (22) V = L; (23) W = T; (24) X = W; (25) Y = Z; (26) Z = Y.

A summit conference of world leaders was called immediately to discuss whether to give permission or not. Would you?

Directions for Breaking Code of Message

The answers to the questions and the missing words in the statements below are given in the code. When you have the first few answers, the ones you *know* are correct, you will have enough of the code to help you figure out the answers to the other questions. The code is simple. The letters of the alphabet have been rearranged. Z is now Y and so on. Since answers to *all* the questions give you the alphabet, you can use the code for your own messages. Answer the questions and away you go!

1. An PXEFIB shows equal pressures and an PXEGSOBC shows equal temperatures.

2. When the LPXPFPRPGZ is zero, it is much safer not to try to drive anywhere.

3. What is a scientist who studies the weather called? COGOEBEREVPXG

4. Fog is not MBOHPMPGIGPEA because it does not fall from the atmosphere to the ground.

5. A NBEAG forms when cool and warm air flows meet.

6. Weathermen often call their predictions NEBOHIXGX.

7. If the earth had no atmosphere, then we would have none of this. TOIGSOB

8. Cyclones rotate HEUAGOBHREHKTPXO, but anticyclones rotate in the opposite direction.

9. What kind of an instrument is used to measure wind speed? IAOCECOGOB

10. Humidity is due to OLIMEBIGPEA evaporation.

11. With what kind of clouds are thunderstorms associated? HUCURUX

12. What is a very severe tropical storm in the Atlantic Ocean called? SUBBPHIAO

13. A modern way of making weather observations is to use APCFUX XIGORRPGOX.

14. XGBIGUX clouds look pulled out.

15. Precipitation that is NBEYOA is called sleet or snow.

16. At the DOTMEPAG, relative humidity is 100 percent.

17. Air pressure can be found with an instrument called a FIBECOGOB.

18. Where warm air is held up above cold air flows, an EHHRUDOD NBEAG occurs.

19. Air from over the EHOIA is likely to be filled with moisture.

20. Fair weather is related to IAGPHZHREAOX.

For Further Reading

Buck, Robert N. *Weather Flying.* New York, Macmillan, 1970. An aircraft pilot tells about the effects of air temperature, turbulence, and thunderstorms.

Lehr, Paul E., R. Will Burnett, and Herbert S. Zim. *Weather—A Guide to Phenomena and Forecasts.* Golden Science Guide, Racine, Wis., Western Publishing Co., 1965. A description of the earth's ocean of air. You can see how to identify the various weather happenings, and learn weather forecasting and reporting.

Spilhaus, Athelstan F., *Weathercraft.* New York, Viking, 1951. How to build your own home weather station with anemometers, nephoscopes, hygrometers—the works.

Unit Three
The Earth's Crust

Paul's house

Chapter 9 Rocks

Paul sat at the dinner table eating as fast as he dared with his mother watching. Every once in a while he would look at his watch.

"Anything happen at school today?" Paul's father asked at dinner.

"Yeah. School wasn't half bad, for once." He paused just long enough to jab a large forkful of food into his mouth. "Had an assembly. Some guy gave a talk on the history of this place."

"History? I thought you didn't like history." His mother's attention shifted from his table manners to his sudden interest in history.

"This wasn't history like in a book. This was the history of this town before there were any people around anywhere." The food on his plate continued to disappear.

"If there were no people around, who wrote the history?" his mother asked.

"He did," answered Paul. "He's one of those guys who studies rocks. He collected all kinds of rocks from this area and put the story together himself."

"How do you put a story together from rocks?" said his father.

According to his watch, Paul didn't really have the time for all these questions. His answer was short and to the point.

"There was some of it that didn't make sense, but most of it was okay. See, he could tell by studying the igneous rock that this place was probably buried at one time. Real deep." He pointed to the floor with his fork. "The crystal sizes told him that. Then there's the sedimentary rock on top of that. We must have been under water at one time, too. I'm not sure how he knew just when. He also said

Technical terms and words used here will be defined later in this chapter. Their purpose is to preview material to be covered in the chapter.

The cartoon was redrawn from an original by Scott Stewart, Grade 9, of Churchville, New York.

there's a bunch of metamorphic rock somewhere near here."

"So?" his mother added.

"It has something to do with being close to some melted rock. Mr. Powell, our earth science teacher, said he'd explain that part if we brought some of it in." He glanced at his watch again and gobbled down the rest of his supper. "Oh-oh. I'm late. I have to meet Terry. He's going with me to get some of that rock. Can I leave now?"

"Good hunting," said his mother. "Let me see the rocks when you get back." Paul left and she turned to her husband.

"Does he mean this area was near a volcano? And under water?"

You can see how Paul's mother pictured the history of the area. If Paul had not left out so much, his mother might not have been so confused.

What Is Rock?

9.1 Rocks are made of minerals.

Paul never did get around to saying what a rock is. Everyone knows that there is a hard material called "rock" which is found in some road cuts, hillsides, and riverbanks. You know that it is covered up in most places by dirt and plants. You know that it is a natural material. **Rock** is the solid part of the earth's crust, not part of any living organism, and not made by man. Smaller pieces which have broken off from the main mass of rock are usually called "rocks."

Materials
magnifying glass

★ Look at the earth materials in Figure 9/1. They are all pieces of the solid part of the earth. None of them was made by man. Would you call them rocks? Certainly!

Now study the rocks shown in Figure 9/1 more closely. A magnifying glass will help. You can't see the molecules that make them up no matter how strong a magnifying glass you use. But you can see other things. Write a description of one of the rocks, but do not mention color. Give your description to a classmate. Can your classmate pick out the rock you chose, using only your description? ★

Figure 9/1
The names of these rocks are: (A) rhyolite, (B) granite, (C) porphyry, (D) basalt, and (E) gabbro. How would you describe each rock without mentioning color?

Figure 9/2
The minerals (A) plagioclase feldspar, (B) orthoclase feldspar, (C) quartz, (D) calcite, and (E) biotite mica. How are they different from the rocks in Figure 9/1?

Concept: levels of organization
Process: observing

Did you find different things in your rock? Most rocks have more than one kind of material in them. Sometimes a rock is made of just one kind of material all the way through. The different materials that make up rocks are solid chemical compounds called **minerals.** Figure 9/2 shows some of the common minerals that make up rocks.

References and notes for the study of minerals are given in the Teacher's Manual.

A rock, then, can be a mineral or a mixture of minerals. Both types as seen in Figures 9/1 and 9/2 are parts of the hard portion of the earth. Here is where Paul's explanation should have begun.

Display pictures or actual rocks that can be used to tell of events that have happened in the past. Fossils provide obvious clues. Show students limestone and explain that it tells of shallow seas flooding an area. Show them schist and explain that it is an indication of great heat and pressure in the area in which it was found. Indicate results of research on moon rocks.
Process: inferring

9.2 Rocks tell us about the history of the earth.

The rocks in Figure 9/1 or the different minerals in Figure 9/2 may not always have been the way we see them now. For over its long history the earth has changed. Muddy ocean bottoms have been changed to solid rock and raised 20,000 feet into the air. Flat rock layers have been cracked and forced upward to form mountains, and rocks have been changed from one kind to another in the process.

Most of these changes happened before there were people to write about them. But rocks that formed during these changes are still around. If you know what makes a rock look the way it does, you can tell a lot about what has happened at the place where the rock was formed.

Did You Get the Point?

A rock is a solid piece of the earth that was not made by man. It is not and never was alive.

A rock may be made up of only one mineral.

Rocks may be mixtures of minerals.

Rocks can be used as clues in putting together the history of an area.

Check Yourself

Answers:
1. Examples might be concrete, cinders, glass, rusted metal, plaster, plastic, hard rubber.
2. a) limestone or white marble, quartzite, sandstone (if it has a quartz cement), coal, hornblende schist
   b) some granite (syenite), some gabbro, some schist, diorite
   c) granite, granite gneiss, some gabbro
   d) granite (granodiorite)

1. Many things look like rocks but are not really rocks because they were made by man. List some of these things. Try to bring in examples.

2. If it is possible to do so in your area, collect and label four different rocks, one of each kind listed below:

a. a rock made of one mineral
b. a rock containing two different minerals
c. a rock containing three different minerals
d. a rock containing four different minerals.

Igneous Rocks

9.3 Very high heat causes minerals to melt.

Have rock samples in the classroom: some very hard (quartzite); some very soft (soft sandstone or shale); rocks that soak up water (porous sandstone); rocks that break into flat chips (shale); rocks that shatter (obsidian); colorful rocks; patterned rocks; banded rocks; rocks with ripple marks.
Process: observing

Rocks can differ from one another in several ways. Some rocks are harder than others. Some rocks are so full of small holes called **pores** that they can actually soak up water. Some rocks break into flat pieces of different sizes and shapes. Rocks come in many colors. Figure 9/3 shows a rock that has a pattern on it. **Geologists** (jee-AHL-uh-jists), who are scientists who study what makes up the

Figure 9/3
Some rocks come with patterns in them.

earth and how the earth changes, try to explain these differences.

Molecules in rocks may line up in regular patterns called **crystals.** Crystals can form when some liquids turn to solids. For example, ice crystals and snowflakes form when liquid water freezes. So mineral crystals may tell us that some rocks were once melted and part of a "mineral soup." Compare the crystals shown in Figure 9/4.

There has always been plenty of heat in the earth to melt rocks. For instance, a lot of heat may have been produced when the earth was very young. This heat might have left large pools of melted rock below the surface of the earth. Slow cooling of the melted part would form the rocks containing crystals.

But there is evidence that the earth may have formed billions of years ago. The hot, melted material produced when the earth was forming must have become the first rocks, the billions-of-years-old ones. The rocks that have been melted since the first rocks cooled have to be explained another way.

The material below the surface of the earth is still very hot, as a glance at Appendix S will show you. Scientists

Figure 9/4 *The orderly structure of the quartz crystal (A) suggests that it cooled from a liquid as the snowflake (B) did.*

believe that there are two reasons for this great heat. In the first place, some of the original heat is still left over from the time when the earth was formed. The rest of the heat comes from **radioactivity.** Radioactivity is a process by which some elements naturally break apart and change into other elements. Since this process resembles rotting, or decaying, of wood or meat, radioactive elements are said to "decay."

When the elements decay, they give off heat, light, and pieces of atoms. The outer part of the earth keeps most of the heat in. This heat trapped inside the earth is more than enough to melt every kind of mineral.

Activity Time: one hour. If a double class period is not available, materials can be melted one day, cooled overnight, and be examined next day. A single hot plate (or Bunsen burner) is sufficient, and a one-liter beaker will hold enough boiling water for the entire class. You may prefer to use several 250 ml beakers. Since the test tubes are to be broken, try to get stained and discarded test tubes from the chemistry lab. If a microscope is available, try to make thickness comparisons of crystals in the three different models.

9.4 Make rocks from melted minerals.

The best way to see how the hot "mineral soup" turns into rock is to make a few model rocks of your own. According to the definition of a rock, all you really need is one mineral. But two will make it more interesting. Let crayon be one of your model minerals and mothballs the other.

Materials

hammer, moth balls, crayon, test tubes, corks, hot plate, beaker, pan, sand, ice, towel, microscope, magnifying glass

★ Put two whole mothballs in each of three test tubes. If the mothballs will not fit, crush six mothballs with a hammer and put one-third of the pieces into each of the test tubes. Add one-third of a crayon to each portion of mothballs, and your model rock mixture is complete. Be sure to identify your test tubes with a crayon mark. Now to turn the samples into mineral soup!

Stand the three test tubes in a beaker of boiling water until the model minerals have melted. There's the mineral soup. Seal each of the samples with a cork. Now, put one sample in a cold water bath or ice to cool. Push the second test tube into a pan of dry sand to cool. Leave the third sample in the beaker to cool as the water cools. Figure 9/5 shows the three tubes cooling.

Demonstration Time: 15 minutes. While the students are busy with the activity, you can set up the following demonstration: heat 17 g of sodium acetate ($NaC_2H_3O_2$) with 10 ml of water in a bottle or in a shallow, transparent dish. If the dish is used, students can watch crystals in the process of growing on an overhead projector.
Process: experimenting

When the model mineral soup has completely hardened, wrap each test tube in a towel and hit it with a hammer. Remove the broken glass and examine the rocks you have made. Use a low-power microscope to look for things you may not be able to see with an ordinary magnifying glass. In your notebook keep track of how each sample cooled. Write down your observations about the size of crystals in

The students should note that the faster the rock cools, the smaller the crystals. The more slowly the rock cools, the larger the crystals.
Process: interpreting data

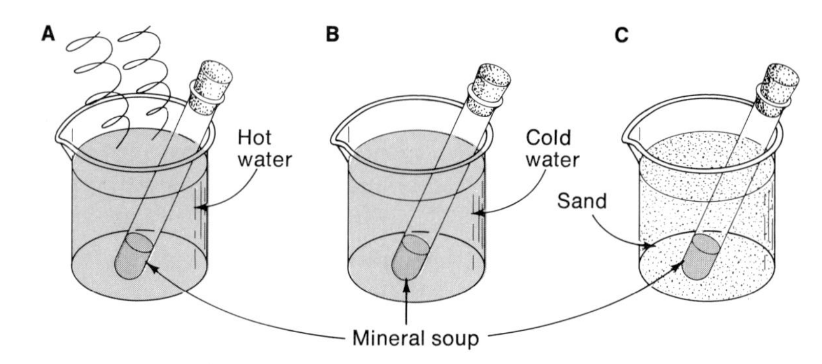

Figure 9/5 *The model mineral soup will cool slowly in the hot water and quickly in the cold water. It will cool at a medium rate in the sand.*

the rocks. What can you say about crystal size and how fast the rock cools? ★

Geologists call the real buried mineral soup **magma** (MAG-muh). To turn into rock, magma only has to cool. Rocks that form from magma are called **igneous** (IG-nee-us) rocks. The word igneous means "from fire."

9.5 Cooling rate affects the size of crystals in a rock.

Materials

obsidian, basalt, granite

Obsidian cooled fastest, basalt more slowly, granite most slowly.

★ Think about the three igneous rocks you made and how the cooling rate affects the crystal size in a rock. Then examine a specimen of the real rock called **obsidian** (ub-SID-ee-un). Did it cool fast, slowly, or in between? What about the rock called **basalt** (ba-SALT)? How fast did it cool? And **granite** (GRAN-it). Did it cool faster or slower than the obsidian and basalt? Use Figure 9/6 if you don't have any real samples.

If the magma comes to the surface where it can cool quickly, it can form either obsidian or basalt. Obsidian looks like glass. Basalt has crystals that are so small you will probably need a magnifying glass to see most of them. If the magma doesn't get quite to the surface, it will cool slowly enough for small visible crystals to form. Magma that stays trapped deep beneath the surface, but far from the source of heat, will cool even more slowly, and larger crystals will form. Try examining some other igneous rocks.

There are exceptions to this pattern. If there is a large amount of water in the magma, ions can move more freely, and large crystals can form at shallow depths and low temperatures. The result is the rock called pegmatite, whose crystals are much larger than the usual granite ones.
Process: observing

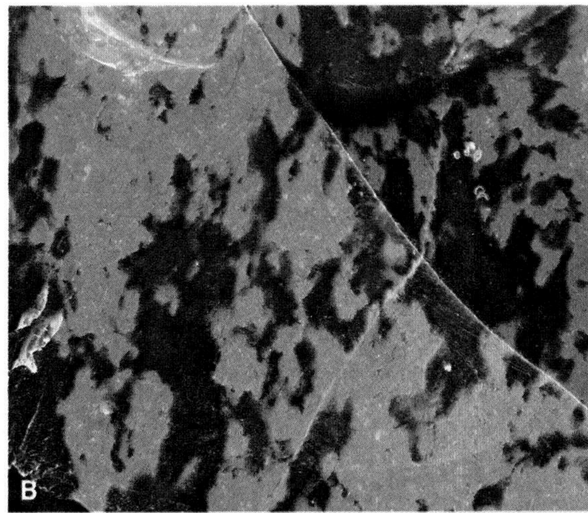

Figure 9/6
All of these rocks cooled from magma. Which one cooled fastest, the granite (A), the obsidian (B), or the basalt (C)?

See if you can tell how and where each rock in Figure 9/6 might have cooled. ★

Did You Get the Point?

Magma is a mixture of melted minerals.

Rocks that form from magma are called igneous rocks.

The heat to melt minerals comes from the original heat inside the earth and from radioactive decay.

Fast cooling may produce no crystals at all, medium-fast cooling usually produces medium-sized crystals, and slow cooling produces large crystals.

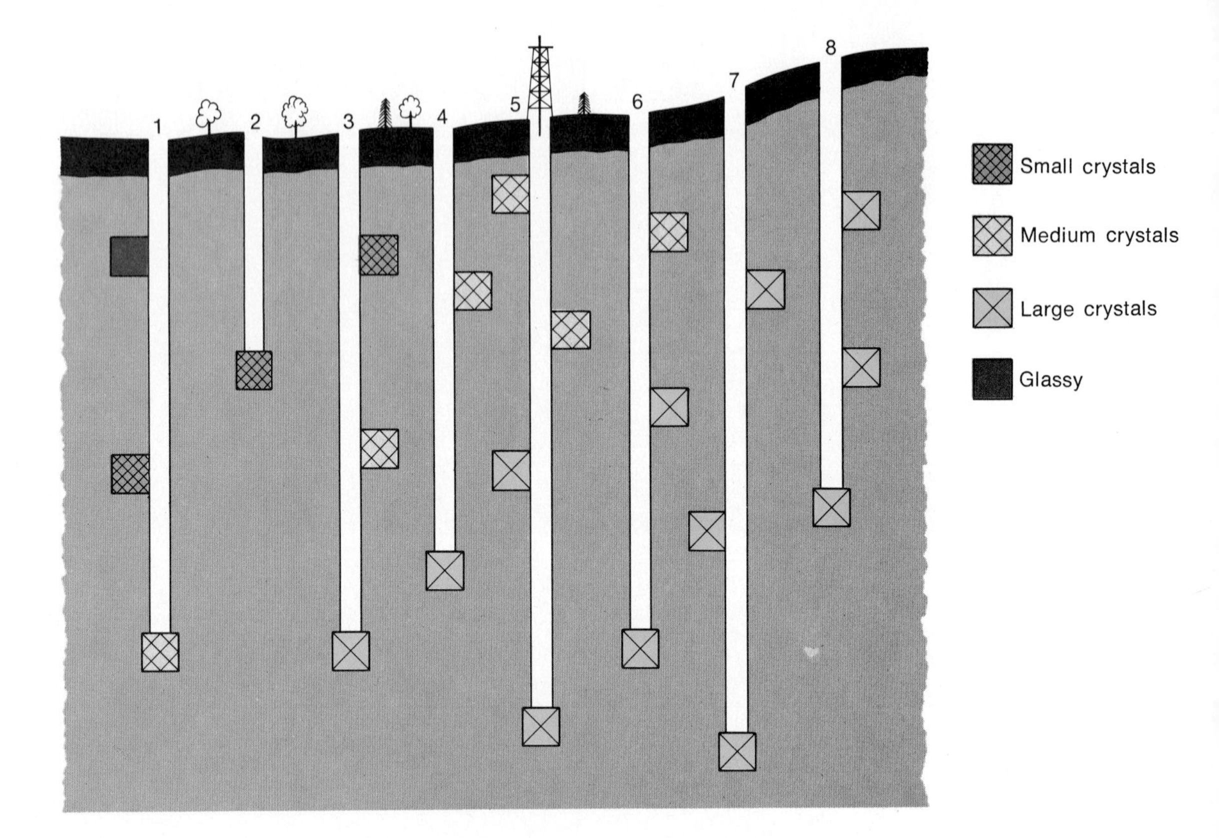

Figure 9/7 *This map shows information found by drilling into the solid rock. How were the different rock types formed?*

Figure 9/7 shows that rock has been removed from the area. Well 8 has rock with large crystals at the surface. The magma must have been far below the surface when it cooled, in order to cool slowly enough to form large crystals. So, the original surface rock must have been eroded from the surface at the right. The zone of medium crystals indicates faster cooling near the edge of the mass of magma. The zone of small crystals could be the edge of the magma mass, very close to the surface; it could also be a lava flow series, poured on top of the medium crystal zone and then tilted. The glassy rock was definitely formed at the surface and then tilted.

Check Yourself

1. Figure 9/7 shows information found by drilling into the solid rock of an area. The symbols on the diagram tell what crystal size was found at each place. The symbols are explained beside the diagram. How did the rocks get to be the way they are?

Sedimentary Rocks

9.6 Rocks are broken down and carried away.

Rock at the surface of the earth naturally breaks up. The weather won't leave it alone. Running water won't leave it alone. Chemicals won't leave it alone. Man won't leave it alone. It is chipped at, splashed on, eaten by acids, and carried away a piece at a time by streams. The pieces, called **sediment** (SED-ih-ment), eventually collect in piles. Follow a stream some time and see if you can find one of these sediment dumps.

Materials

quartz sand

Activity Time: 30 minutes. Some sands contain little but quartz. Others contain the minerals of granite. Sand from a glaciated region will have many rock fragments among the mineral grains. Sand from the Bahamas consists of calcareous animal fragments.

★ Examine a handful of dry **sand.** Geologists define sand as a material made up of grains that are from $\frac{1}{16}$ millimeter to 2 millimeters across. The grains can be pieces of single minerals or pieces of very fine grained rock, like basalt. How many different minerals can you find in your handful? How many minerals in the sand match the minerals that make up a piece of granite? Could sand be made when a rock like granite is broken down? Or could rocks other than granite have been involved in the making of sand? Check your answers with your teacher. ★

9.7 Cemented sediment makes sedimentary rocks.

Wherever the sediment is deposited, at the end of a river as in Figure 9/8, or at the bottom of a large body of water, the rock pieces will probably some day be cemented together to form rock again. Rocks that have been made by cementing sediment together are called **sedimentary** (sed-ih-MEN-tree) rocks.

Figure 9/8

Sediment dumped by a river running through the V-shaped cut in the mountain. Where did the sediment come from?

Materials

paper cups, gravel, glue-water mixture, pipe cleaner, eye dropper, sandstone

Activity Time: 30 minutes to set up. Cups must be left overnight. Plastic cups will not work. The smaller the pebbles in the cup, the longer the drying time.

★ You can make a model to observe the cementing process. Figure 9/9 will help you prepare your rock. Mix two parts of white glue to one part water. Let the glue stand for chemicals dissolved in the water that flows through sediments. Punch a paper cup full of holes. Put a handful of pebbles or crushed stone from a driveway into the paper cup.

Pour the water-glue mixture through the cup and collect what runs out the bottom in an extra cup. Wait a few

Figure 9/9
When the glue hardens, the pebbles will be cemented together into a model sedimentary rock.

minutes. Then pour the run-off back through the cup. Pour the run-off through the cup a few more times. Wait a few minutes between pourings. Stop when you are sure that the glue has touched every particle. Then set the cup and its contents in a warm place overnight to dry.

The next day, tear away the paper cup. Examine the hardened stuff. How is this model of rock different from the models you made of igneous rocks? Are the empty spaces between the cemented particles connected? Check by pushing a fine wire or a pipe cleaner into one of the holes.

The water squeezed onto the sandstone soaked in. Stone must have pores like the ones the fine wire revealed in the model. The model could be of sandstone.

Study a sample of a rock called **sandstone.** Can you see why the rock is called sandstone? With an eyedropper, put a few drops of water on it. What happens to the water? Could the model you made be a model of sandstone? ★

Other kinds of rocks are made of cemented sediment. If the particles are like those of sandstone but so small they can hardly be seen without a magnifying glass, the rock is called **shale.** Many of the rocks called **limestone** consist of sand-size or smaller pieces of sea shells and bits of other ocean animals, which have been cemented together.

9.8 Dissolved animal shells can cement sediment.

Materials

sandstone, clamshell, eyedropper, dilute hydrochloric acid

Activity Time: 10 minutes. Dilute hydrochloric acid (HCl) is not extremely dangerous, but you should warn students not to spill acid on skin or clothing. If they do, they should wash with plenty of water and soap, if it is also available. (You also might make a solution of baking soda for washing.) If the acid is too strong, dilute it by pouring acid into distilled water while stirring with a glass rod. *Never pour water into acid.* Water could splatter, throwing acid as it does. If HCl is not available, vinegar will work, but more slowly. Be sure the sandstone has a calcite cement, which will react with the acid. The acid on a clamshell will effervesce. **Process: experimenting**

★ Many minerals that dissolve in water can act as cement. Some of the cements are hard to recognize. But a simple test using dilute hydrochloric acid will help you identify one cement quite easily. CAUTION: Be very careful not to spill the acid.

Put a few drops of acid on the sandstone you examined earlier. What happens? What is the acid working on? the sand? a cement? What makes you think so?

Put a few drops of acid on a dry clamshell. What do you see? ★

Clamshells will dissolve in water if you use enough water and give them enough time. This is because they are made mostly of a chemical called **calcium carbonate** (KAL-see-um KAR-buh-nayt). When water carrying calcium carbonate dries up, the chemical from the shells is left behind.

The hardened calcium carbonate is the mineral called **calcite** (KAL-site). If the calcite is left between particles of the sediment, it forms a hard crust that holds the particles together. Such cementing is shown in Figure 9/10. If you used enough acid, it would dissolve all of the cement. If the sediment particles did not dissolve in the acid, there would be only a pile of sand left.

Figure 9/10
The dark spots and lines in this magnified view of a piece of sandstone are the "cement" between the sand grains.

9.9 Some sedimentary rocks need no cement.

Materials
pie tin, salt, sugar, baking soda, food coloring

Activity Time: 10 minutes to set up. Do this activity in groups, sharing pans. Leave overnight.

★ Use table salt, sugar, baking soda, and food coloring to stand for minerals. Dissolve a few grams of each solid in a pan of clean water. Add a drop of food coloring and mix well. Set the solution aside to evaporate overnight.

When chemicals drop out of solution, they become sediment. Test the hardness of your model sediment that was left after the water in the pan evaporated. If your model minerals had come from the hard part of the earth, could what is left over be called rock? If your model minerals had come from dead plants or animals, could the sediment still be called rock? Can you recognize the example of such a rock in Figure 9/11? Could the material left when the water evaporated be called sedimentary rock? Geologists say yes. ★

Gypsum is hydrated calcium sulfate ($CaSO_4 \cdot 2H_2O$).

Figure 9/11 shows rock salt and **gypsum** (JIP-sum), which originally formed from evaporating sea water.

Figure 9/11 *These rocks are sedimentary but they do not contain cement. The coal (A) is a deposit of plant remains. The halite (B), gypsum (C), and tufa (D) are minerals deposited when water evaporated.*

Demonstration Time: 20 minutes. As a demonstration, dissolve enough washing soda in boiling water to make a nearly saturated solution. Then let the solution cool. When soda precipitates out of solution on cooling, ask your students whether this material dissolves better in hot or cold water. You can point out that cooling is what causes previously dissolved sugar to be left in the bottom of coffee cup when drinker has loaded up hot coffee with too much sugar to begin with.

You can also point out that water containing CO_2 will dissolve more calcite than water without CO_2.

If warm water carrying dissolved material is suddenly cooled, the materials that dissolve easily in warm water will drop out. This helps form rock deposits around hot springs. Evaporation does the rest. The mineral calcite drops out of cold water from springs to form a kind of limestone called **tufa** (TOO-fa). A sample of tufa is also shown in Figure 9/11.

9.10 Sedimentary rocks contain records of the past.

As sediment is piled up and changed to sedimentary rock, many things can happen to it. Records of these events can stay pressed in the rock for millions of years.

For example, you can tell if a rock was under water before it became a rock. Shallow water moving across sediment may leave ripple marks that become part of the rock that eventually forms. Ripple marks in sediment and rock are shown in Figure 9/12. Or suppose that soggy sediment begins to dry up. Mud cracks could form when mud dries and become locked into the rock as it hardens. Figure 9/13 shows mud cracks both before and after they have become recorded history. Which is which?

Figure 9/13A shows dried mud and 9/13B shows mud cracks that have become rock.
Process: inferring

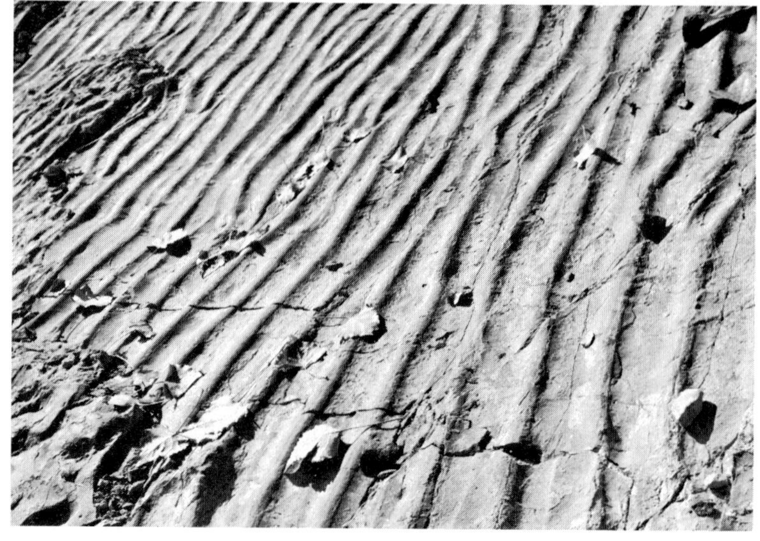

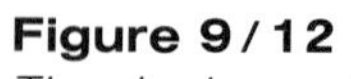

Figure 9/12
The ripple marks above were formed by water moving across soft sand along the shore. The ripple marks on the right were formed in shallow water near the shore and have hardened into rock.

Figure 9/13
One set of mud cracks is in soft sediment and the other set has become hard rock. Can you tell which is which?

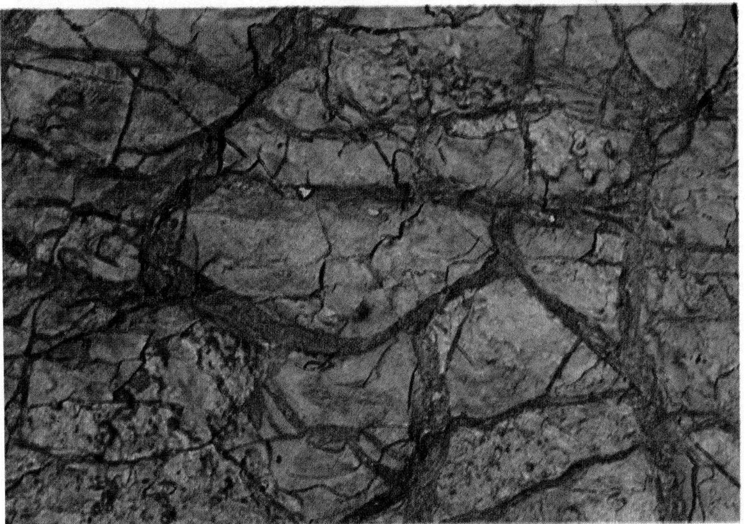

Then, too, the conditions where the sediment is building up could change greatly. An entirely different kind of sediment might be added to the first kind. A flood could dump very large particles on top of smaller ones already there. Or sediment of a different color might be added. The result would be a pile of rocks with easy-to-see layers. Because many sedimentary rocks contain such layers, they break into flat pieces. Figure 9/14A shows layers of sediment and Figure 9/14B shows layers of rock formed from sediment.

Figure 9/14
The photograph above shows different colored layers of sediment deposited on top of each other. The sediment layers below have become rock over millions of years and are now part of the Grand Canyon.

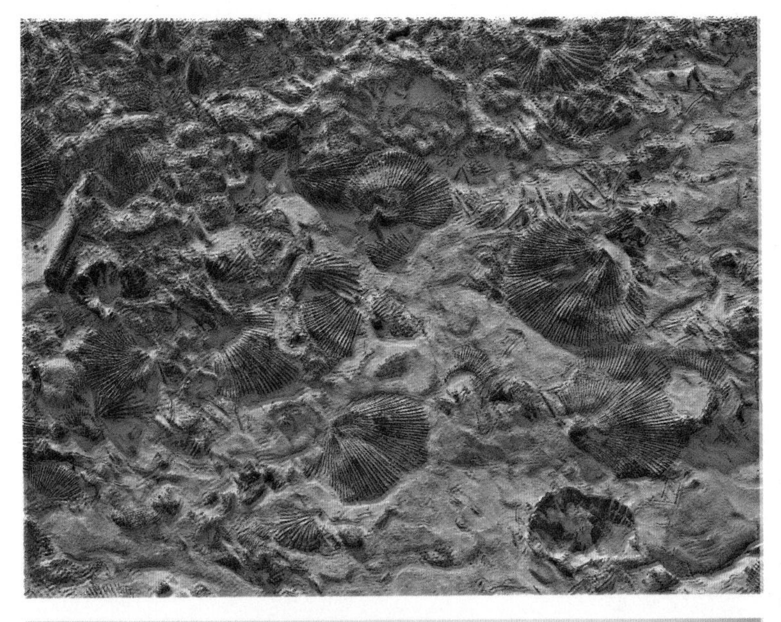

Figure 9/15
The shells of sea animals may be replaced by minerals to form fossils like these.

Figure 9/16
Empty spaces in the rock are the impressions left by ancient sea shells when they are dissolved. This kind of fossil is called a fossil mold.

In some cases, outlines of whole plants or animals become part of a cemented rock. Records of this sort are **fossils**. Sometimes pieces of the plant or animal are dissolved out of the rock and replaced by minerals as in Figure 9/15. Sometimes only an empty space having the shape of the plant or animal is left to show that it was once there. Such a fossil is illustrated in Figure 9/16.

Figures 9/17 and 9/18 *Water flowing through the sand in the cup will dissolve the sugar. The empty space left by the dissolved sugar is a mold.*

Materials

cup, sand, sugar cube, tissue, food coloring

Activity Time: 10 minutes to set up. Students take them home. If possible, have each student make a model. Drying after leaching process is not strictly necessary, but a few days on the window sill should dry the sand enough to show signs of cementing. Remind students to dump the dried sand carefully. The cementing is quite delicate. Students can pour a glue-water mixture through the sand to make the model a permanent one.

See supplementary activity in Teacher's Manual.

★ You can make a model fossil of your own. Get a clear plastic cup with holes in the bottom. Put a tissue across the bottom. Then fill the cup with damp sand. The tissue will keep the sand from falling through the holes. Bury one or two sugar cubes, colored with food coloring, in the sand so that you can see them through the sides of the cup. The sugar will represent small animals that have died and been buried in sediment.

Take the cup home and put it under a dripping faucet as in Figure 9/17. Make the faucet drip fast enough so that as much water goes into the cup as drips out through the bottom. After the dripping starts, be careful not to shake the sand. The next day, turn off the water. Look at the remains of the sugar cubes. Let the sand dry for about three days before bringing it back to school. Then hold a piece of paper over the cup. Carefully turn the cup and paper upside down. Lift the cup off the sand. Look at the sand. What has happened?

Even though most of the sugar cube has been dissolved and carried away, you can tell its size and shape. You have made a model fossil like the one in Figure 9/18. If water containing dissolved minerals were allowed to flow through the sand, the empty space, called the **fossil mold,** might become filled again. Then you would have what is called a **fossil cast.** Some of the original cube might even remain unchanged and become part of the rock. Cementing has already begun. Look carefully at the sand for some cemented sand grains near the fossil mold of the sugar cube. ★

Possible records of past events include ripple marks, mud cracks, layering, and fossils. The lack of fossils in a particular area can also tell a story.

Records of past events are everywhere. Collect rocks from your own area. See if you can spot some of these records. Perhaps you can begin to put together a rock history of your own town.

Did You Get the Point?

Pieces of rocks can become cemented together to form new rocks.

The dissolved shells of sea animals can act as a cement.

The rock formed by the cementing of small pieces of sediment is called sedimentary rock.

Sedimentary rocks may contain records of past events and past life.

Minerals or other compounds that have been left behind as water evaporated or that have dropped out of a water solution because of temperature changes also make sedimentary rocks.

You may want to have a class field trip.

Check Yourself

1. Examine several mud puddles. Even dried-up ones will do. If there are none to examine, make your own. See how many records of present-day events are being written in the sediment. You should be able to spot such records as evidence for lowering of the water level, ripple marks, layering, buried plants and animals, mud cracks, and splash marks from raindrops.

2. Make a list of differences that you can observe between igneous and sedimentary rocks.

3. If there are sedimentary and igneous rocks in your area, collect five different specimens of each. Label each rock "sedimentary" or "igneous." If you know the name of a rock, like granite, write that on the label too.

Answers:

2. Igneous rocks are harder than some sedimentary rocks. Igneous rocks often contain larger crystals, and a greater variety of minerals. Igneous rocks seldom have a layered appearance, unless they were formed as lava flows. Have samples of these kinds of rocks to show the class.
3. Sedimentary rocks to be expected are sandstone, shale, limestone, conglomerate, and possibly coal. Igneous rocks will probably be granite, diorite, or basalt; gabbro and rhyolite may be collected, too. Remind students to look in stream beds as well as outcrops.

Collecting and Naming Rocks

Collecting rocks is an enjoyable hobby and it's a good way to learn some geology. All you need is a hammer, a bag to carry your rock samples, and maybe a book on rocks and minerals to help you identify your rocks.

The naming of rocks may be confusing at first. Usually the kinds and amounts of different minerals in the rock determine what the rock will be called. The chart in A will help you name igneous rocks. Run your finger from the top X to the bottom X on the chart and look at the names of the minerals in the sections you touch. According to the labels on the chart, a rock that has some biotite mica and a lot of quartz and orthoclase feldspar will be called granite or rhyolite, depending on how big the crystals are. How would you describe the rock called gabbro? Appendix T also has a chart that will help you name some minerals often found in rocks.

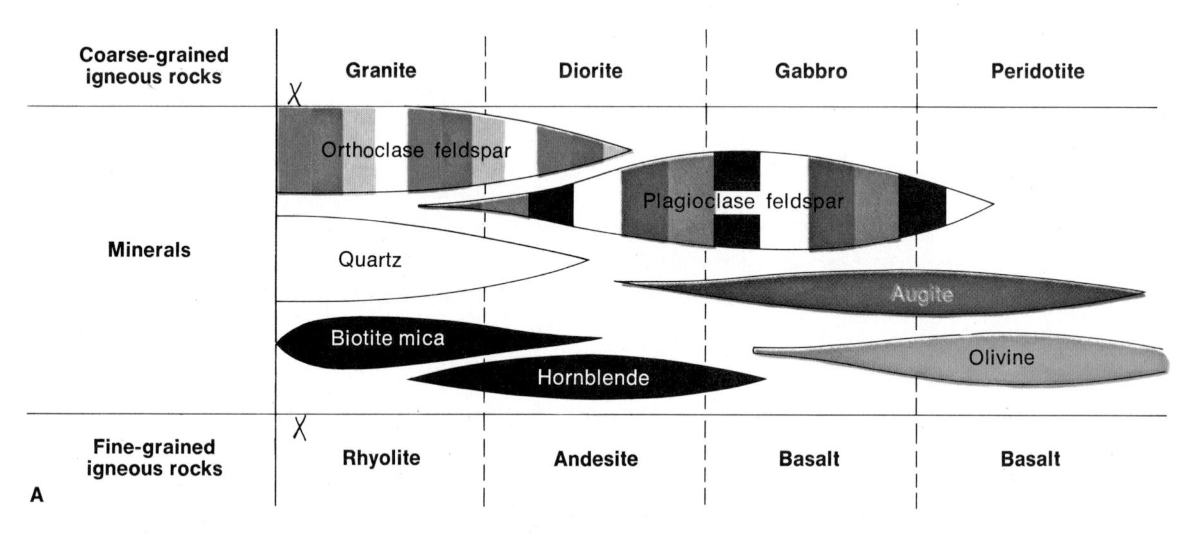

Metamorphic Rocks

You might wish to represent schematically the changes rock can undergo by drawing circles and lines on the chalkboard. The drawing would be the beginning of what is called the rock cycle.

9.11 Sedimentary rock may be turned to magma.

What happens to the rock pieces after they have been cemented together? Rain, wind, chemicals, and ice still work on them. Streams still carry off the chipped-away pieces—that is, if the rock gets to the surface. But what happens to rock that doesn't get to the surface? Suppose more and more sediment keeps piling on top of it. Will it stay sedimentary rock forever?

Maybe the sedimentary rock will be buried so deeply that the heat of the earth begins to raise its temperature. The sedimentary rock could get hot enough to turn into magma. The new magma might even reach the surface to start the entire process over again: magma to igneous rock, igneous rock to sediment, sediment to sedimentary rock, sedimentary rock to magma.

Geologists have found that this can and does happen to some sedimentary rock. But they have also found that sometimes sedimentary rock melts and cools again without turning to magma. Sometimes great pressure squeezes the rock until the minerals actually flow, without melting.

If possible, have a few sedimentary-metamorphic rock pairs, such as sandstone and quartzite, limestone and marble, shale and slate, or shale and mica schist.
Process: observing

Materials

sandstone, limestone, shale, quartzite, marble, slate

When a rock is metamorphosed, it may
a. become harder (shale to slate).
b. become crystalline (sandstone to quartzite and limestone to granular marble).
c. take on a banded appearance (granite to gneiss).
d. change color (limestone to marble).

9.12 Sedimentary rock may become another kind of rock.

Samples of the three sedimentary rocks—sandstone, limestone, and shale—are shown with samples of **quartzite** (KWORT-zyt), **marble,** and **slate** in Figure 9/19.

★ Examine the rocks themselves if they are available. Quartzite, marble, and slate are examples of the three sedimentary types after they have been changed by great pressure, or by great pressure and great heat together without turning into magma. Match the changed rocks with the sedimentary rocks from which they came. How have the original rocks changed? List any changes you can see. ★

Rock that has been changed by heat or pressure or both together is called **metamorphic** (met-uh-MOR-fik) rock. The changing process sometimes seems to make the metamorphic rock harder than the original rock. Spaces between the particles in a sedimentary rock fill in as the separate crystals flow together or are pushed together. The cement and the particles blend. Sometimes even the color changes and new crystals appear, producing a rock that looks very different from its sedimentary parent.

Figure 9/19
Sedimentary rocks and their metamorphic twins. Sandstone (A) changes to quartzite (B), limestone (C) to marble (D), and shale (E) to slate (F).

9.13 Igneous rock can also be changed.

Materials

clay, washers, gneiss, magnifying glass, microscope

Activity Time: 15 minutes. You can use chips of mica in place of washers. Make the substitution if at all possible, for using mica makes a model composed entirely of earth materials.

★ Push as many washers or pennies as you can into a ball of soft modeling clay. The clay will stand for igneous rock that is squeezed by a force in the earth's crust that puts enormous pressure on it. The washers will stand for flat crystals in the igneous rock.

Put your whole weight on the clay. When the pressure of your weight has flattened the clay ball, break it in half. How have the washers arranged themselves? Does your mixture of clay and washers look like Figure 9/20?

When an igneous rock is squeezed tightly enough, flat or long crystals tend to turn and arrange themselves just as the washers did. Examine a sample of the metamorphic

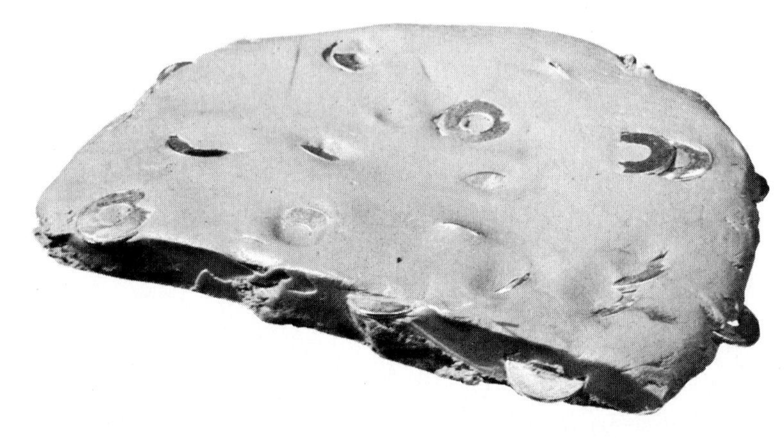

Figure 9/20 *Crystals in metamorphic rocks arrange themselves in one direction like the washers in this piece of squeezed clay.*

rock **gneiss** (nice) in Figure 9/21. Do you see any flat crystals in it? Take a close look at several crystals. Can you see the washer effect? Gneiss is a rock that forms when granite is squeezed and heated. The pressure and heat cause some of its crystals to change position. Using a low-power microscope, compare the dark crystals in gneiss with those in granite. ★

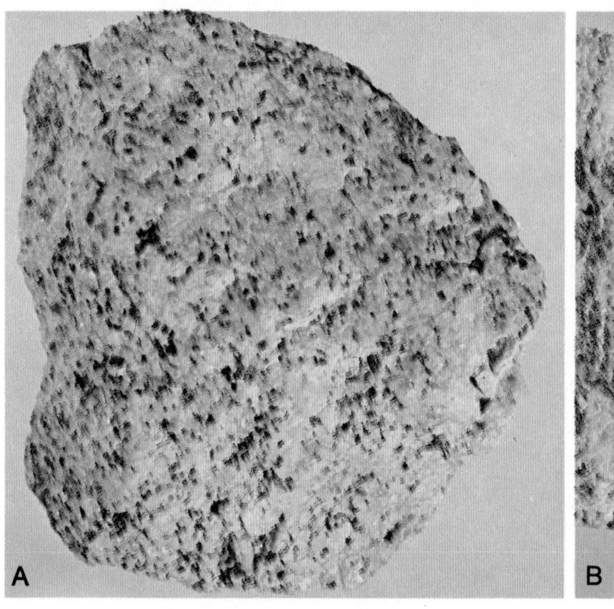

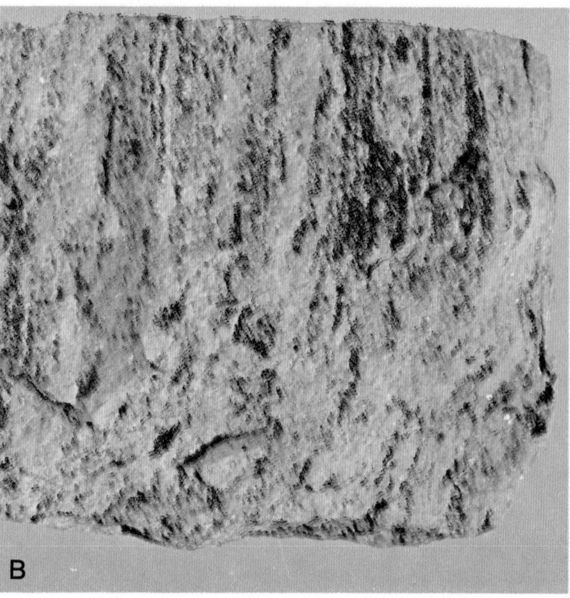

Figure 9/21 *Igneous rocks and their metamorphic twins: Granite (A) changes to granite gneiss (B), diorite (C) to diorite gneiss (D), gabbro (E) to hornblende schist (F), and pyroxenite (G) to serpentine (H).*

Process: observing

Ask your teacher to give you some igneous-metamorphic rock pairs such as those in Figure 9/21. Mix them up and see how many of them you can match. Watch for crystals turned by pressure and for bands formed by the settling of the crystals as they flowed under pressure. You might find an example of a rock with two or more minerals becoming a rock with only one mineral.

Did You Get the Point?

Rocks can be buried deeply enough to melt again. A new igneous rock may then form.

Rocks can be changed by heat and pressure. These rocks are called metamorphic rocks.

While igneous rocks are in the softened form because of heat and pressure, some of their crystals may turn and flow.

When sedimentary rocks are changed, the new rock is sometimes harder and sometimes contains more kinds of crystals than the original rock.

Check Yourself

Answers:

1. Expect mica schist and granite gneiss, and marble where it is locally abundant.
2. See chart in Teacher's Manual.
3. Paul could have explained just what a geologist is and what he does. He could also have explained the differences between the different kinds of rock. To prevent his mother's confusion over his explanation he might have explained that all the changes did not take place at the same time.

1. If there are metamorphic rocks in your area, collect five samples. Add them to your collection of sedimentary and igneous rocks. Label each rock with the word "metamorphic" and its name if you know it. Try to find metamorphic rocks that are the changed forms of the igneous and sedimentary rocks you already have. Label any sedimentary-metamorphic, metamorphic-igneous, or igneous-sedimentary rock pairs you have.

2. Make a list of the differences that you can see between igneous, metamorphic, and sedimentary rocks.

3. Read again the conversation Paul had with his mother and father, in the introduction to this chapter. How could his explanation be improved, and what should he have said?

What's Next?

Any rock you find may have once looked completely different, because there is no real end to the story of a rock. A rock may change in many different ways, and even repeat the changes over and over again. Geologists interpret each change as part of a cycle—the rock cycle.

Figure 9/22
Follow the arrows to see how rock of one kind can be changed into rock of another kind.

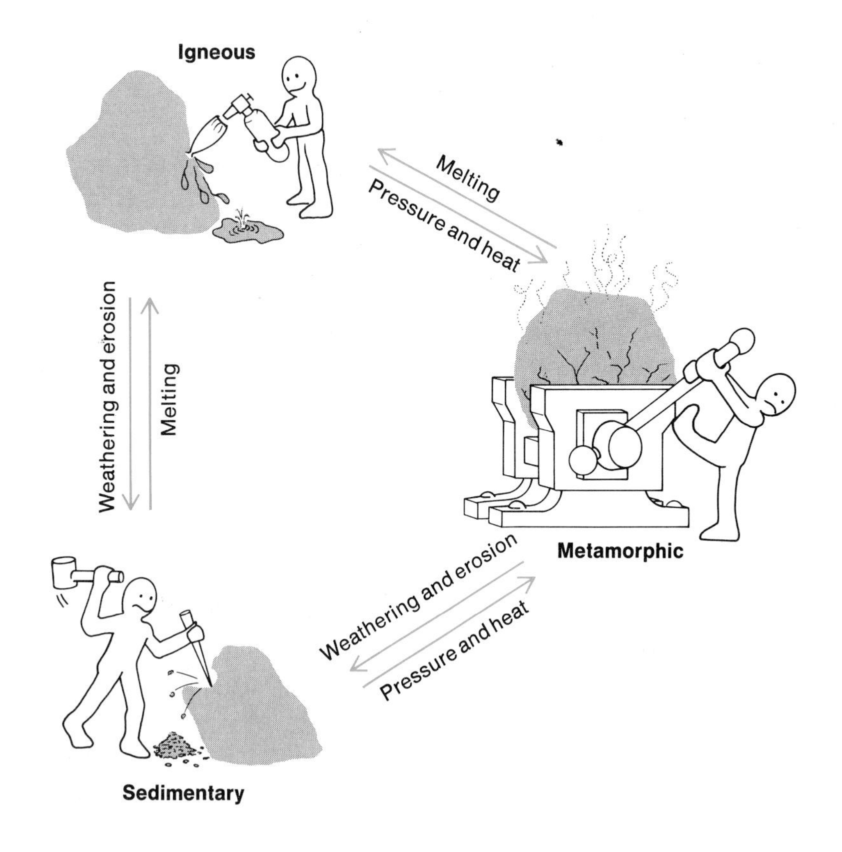

It's a good idea to have a large permanent drawing of the rock cycle. Real samples of rock can be attached to it. Thumb tacks can be glued to the rocks with epoxy so that they can be stuck to a cork board. The drawing can be made by the students as a class project.

Unlike the change of seasons and other cycles, however, changes in a rock's form do not have to take place in any special order. This can be shown by a kind of map. Look at Figure 9/22. The arrows show the paths a rock can take to get to each other type of rock. Can the paths be traveled in all directions? For example, can a metamorphic rock turn into either a sedimentary or an igneous rock? Can an igneous rock turn into either a sedimentary or a metamorphic rock?

The map says yes. Any rock can change into any kind of rock in any order. But be careful, the map of the rock cycle can be a little misleading. The arrows do not show the many in-between steps which can take millions of years. The change from one type of rock to another is very slow and complicated. In fact, it will take the rest of Unit Three in this book to explain what the arrows do not show.

You will learn how volcanoes can turn flat land into mountain ranges with new igneous rock making up all solid earth materials beneath the mountain peaks. You will see that earthquakes are related to volcanoes, and how they show that new rock is being made. You will study the many ways that the new rock is broken down and carried off. You will follow the sediment as it becomes part of the sea floor. Finally, you will study the processes that can turn the sediment back into rock and make it part of a mountain again. Keep the sketches and arrows in mind, but be watching for what goes in between.

Skullduggery

See Teacher's Manual for directions for the game.

Answers to Skullduggery questions are in Teacher's Manual.

Directions: Your teacher is going to divide the class into two teams for a game. Below are the questions to be answered in the game. You will not know ahead of time which questions will be asked so you had better know the answers to all of them.

1. Why can't concrete be called a rock?

2. Of what value to the geologist is the fact that a rock has been around a long time?

3. Why can't radioactive decay on the surface of the earth melt rock?

4. Why isn't the original heat of the earth the only source of heat for making new rocks?

5. How can you usually tell how deeply a rock was buried when it hardened from magma?

6. Where might a glassy-looking rock have hardened?

7. How can a sedimentary rock form where there are no shelled animals around?

8. How are fossil molds and fossil casts formed?

9. Why don't fossils form in igneous rock?

10. How is it possible for water to flow through some sedimentary rocks?

11. How might a sedimentary rock that is held together by natural cement differ from one that formed when water evaporated?

12. What things show up in sedimentary rocks to indicate that they might have formed in a body of water?

13. How can melting or softening change the appearance of a sedimentary rock?

14. How can softening change the appearance of an igneous rock?

15. What might happen to fossils in a rock when it is melted, softened, or squeezed?

16. What is meant by the term "rock cycle?"

17. How is the rock cycle different from the cycle of the changing seasons or the changes in phase of the moon?

For Further Reading

Zim, Herbert S. and Paul R. Shaffer. *Rocks and Minerals—A Guide to Minerals, Gems, and Rocks.* New York, Golden Press, 1957. A good place to start. This book shows you how to identify rocks and minerals and tells you how to collect and study them.

Harland, W. B. *The Earth: Rocks, Minerals, and Fossils.* New York, Franklin Watts, Inc., 1960. You will enjoy this history of the earth, its rocks, and its climate.

Chapter 10 Volcanoes

In 1783, two unexpected and violent explosions, one in Japan and another in Iceland, threw so much dust into the air that the next morning it got light an hour later than usual. Around the year 1814 an explosion in the Philippines and another near Java filled the air with great amounts of dust. It was said that there was total darkness for three days, even 300 miles away. The dust blocked out enough of the energy coming from the sun to the earth to lower average temperatures all over the earth for an entire year. The year 1815 was called "the year without a summer."

In the year A.D. 79, Pompeii (pom-PAY), a city in Italy, was completely buried by volcanic pebbles and ash following one of these surprise explosions. Thousands of people choked to death in poison gases and fine dust.

One of the most famous of these natural explosions happened in 1883 on the island of Krakatoa (krak-uh-TOH-uh) in Sumatra. Part of the island was blown right off the map. The explosion caused a giant wave to flood the island, drowning more than 36,000 people. The noise was heard 3000 miles away. Barometers all over the world recorded a change in air pressure caused by the sound waves.

All of the explosions were part of the rock-building process. Hot melted minerals reached the surface of the earth to cool and harden into new rock. And all construction jobs bring noise and dust with them.

Causes of Volcanic Eruptions

10.1 Melted minerals take up space.

An opening where melted minerals reach the earth's surface with great force is called a **volcano** (vol-KAY-noh). When a volcano explodes, or erupts, a violent mess can result. Ask someone, "What comes out of a volcano?" He or she will answer, "**Lava** (LAH-vuh), of course. Melted rock." But is the answer really that simple?

Read the newspaper clipping in Figure 10/1. It is dated June 5, 1929, and describes the effects of an eruption of Mount Vesuvius (veh-SOO-vee-us) in Italy. Does the article tell you something about flowing lava that you did not know before? Did you know that the flow could be 23 feet high? Could you pile water that high? Is 500 feet per hour a fast flow or a slow flow? How can lava explode? According to the newspaper, it did, and quite unexpectedly, too. Does lava behave like any liquid you know? Something drastic must happen to a mineral when it melts in order for magma to rise to the surface with tremendous force. Perhaps a close look at the melting of minerals will provide a clue to what happens.

Materials

crayons, graduated cylinder, paper towels, ringstand and clamp, Bunsen burner or hot plate, water container

Activity Time: 25 minutes. The water level rises 5 ml, because one crayon takes up 5 ml of space. When the same crayon is melted, it takes up 6.5 ml of space. The ratio of unmelted to melted should be about 10:13. Mothballs may be used instead of crayons. Three mothballs will take up 12 ml of space unmelted and 14 ml when melted. If wax solvent is not available, boil the graduated cylinders in water to remove crayon.
Process: measuring

★ Fill a graduated cylinder half full of water. Record the number of milliliters of water in the cylinder. Drop a crayon into the graduated cylinder. If the crayon is too big, break it up into pieces first. How much does the water level rise? How many milliliters of space does the crayon take up?

Remove the crayon and dry it with a paper towel. Put it into a second, empty graduated cylinder and melt it by standing the cylinder in boiling water over a hot plate. How much room does the melted crayon take up now? ★

10.2 Melted minerals build up pressure.

The crayon behaves as most minerals do when they are melted. They take up more room in liquid form than they do in solid form. What would happen if a crayon were melted in a closed space?

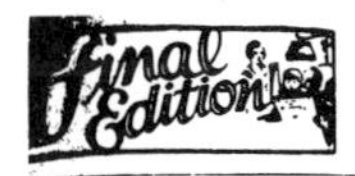

ROCHESTER TIMES-UNION

79,274
Average Daily Circulation for Week Ending June 1st

THE WEATHER: Mostly cloudy tonight and Thursday; slightly cooler tonight.

VOL. XII. NO. 73. 36 PAGES ROCHESTER, N. Y., WEDNESDAY EVENING, JUNE 5, 1929 Daily Entered as Second Class Matter, Postoffice, Rochester THREE CENTS

DEBENTURE PLAN STRICKEN FROM FARM BILL

Germany's Tremendous War Burden For Next 58 Years Finally Fixed

LAVA WAVES FROM VESUVIUS NOW SUBSIDE

BULLETIN

Naples, Italy—(AP)—Lava waves which have been threatening to engulf villages around Mt. Vesuvius were stated by the Vesuvian Observatory late today to have begun diminishing.

Students Trapped by Lava

Naples, Italy—(U.P.)—Twenty students were trapped by a flow of lava today while they were endeavoring to witness the latest eruption of Mt. Vesuvius at close hand. Several were injured by the bursting of the lava as they tried to flee.

Dwellings in Hamlet Destroyed

Torre Annuziata, Italy—(AP)—Molten lava from wrathful Vesuvius, moving on a front of 650 feet wide, invested the hamlet of Pagani. Terzigno township, at 3 a. m., destroying 40 farm dwellings.

The fiery torrent, 23 feet high, then proceeded westward from Terzigno toward the railway station at the rate of nearly 500 feet an hour. One branch of the flow invested the woods at Campitello, the change in direction arousing hopes the rich agricultural section around Terzigno, supporting 7,000 people, might be spared, in large part.

The director of special volcanic services of the ministry of public works, Commendatore Romano, spent the night at Terzigno, conferring at length with military and engineering authorities.

MOLTEN ROCK CRUSHES ONE TINY HAMLET

Students Trapped, Several Injured, Dwellings Destroyed by Fiery Torrent Before Ebb Comes.

BULLETIN

Naples, Italy—(AP)—Lava waves which have been threatening to engulf villages around Mt. Vesuvius were stated by the Vesuvian Observatory late today to have begun diminishing.

Students Trapped by Lava

Naples, Italy—(U.P.)—Twenty students were trapped by a flow of lava today while they were endeavoring to witness the latest eruption of Mt. Vesuvius at close hand. Several were injured by the bursting of the lava as they tried to flee.

Dwellings in Hamlet Destroyed

Torre Annuziata, Italy—(AP)—Molten lava from wrathful Vesuvius, moving on a front of 650 feet wide, invested the hamlet of Pagani. Terzigno township, at 3 a. m., destroying 40 farm dwellings.

The fiery torrent, 23 feet high, then proceeded westward from Terzigno toward the railway station at the rate of nearly 500 feet an hour. One branch of the flow invested the woods at Campitello, the change in direction arousing hopes the rich agricultural section around Terzigno, supporting 7,000 people, might be spared, in large part.

The director of special volcanic services of the ministry of public works, Commendatore Romano, spent the night at Terzigno, conferring at length with military and engineering authorities.

Pall Over Naples

Naples, Italy—(U.P.)—Huge clouds of smoke, laden with ashes, from the belching throat of Mount Vesuvius, hung over Naples today and turned the brightness of noon into a kind of twilight.

Neapolitans were unable to see their familiar landmark, which was clouded by the screen of smoke issuing from its crater.

The Associated Press correspondent visited Terzigno this morning and found conditions desperate. Families were withdrawing with all their belongings and cattle. They flocked into the churches on the line of retreat, invoking Saint Gennaro and praying before shrines.

Troops took possession of villages to prevent marauding after the inhabitants had departed.

250,000 UNITE IN LUTHERANS' SYNOD MERGER

Albany—(U.P.)—A union of 250,000 Lutherans in New York and New England states was effected today with the filing of a merger resolution by three Evangelical Lutheran synods at the secretary of state's

Not a Reel Wedding

Here they are—married. Douglas Fairbanks Jr., and Joan Crawford, youthful screen stars, slipped away from Hollywood to New York and were wed in St. Malachy's Roman Catholic Church after telling newspaper men that the ceremony wouldn't take place till Fall. Doug Sr., back in Hollywood, knew about it, though, and wired them his blessing.

Copper Shares Lead Continued Recovery On Stock Exchange

By George T. Hughes
By Consolidated Press Leased Wire

Wall Street, New York—The recovery in the stock market was carried still further today notwithstanding the fact that the averages were now back over half the ground since the high of the year, made the first week in May. The leaders today were the copper, the utility and to a less degree the motor stocks.

The improvement in the copper shares was in evidence early in the day. It har the foundation of a stronger market for the metal in London and an increase in export sales. The price was firm around 18 cents and producers were confident that another domestic buying movement was at hand.

Then, too, copper stocks had been well deflated and so were in a position to respond to any fresh demand. Anaconda selling at 108 before noon was up three points net. Greene Cananea at 149 was up over four points and gains by Nevada, Anes, Arizona, Kennecott and others in the group varied between these two extremes. The

Life Prison Sentence For Mail Order Bride

Sonora, Calif.—(AP)—Mrs. Eva Rablen, "mail order bride" from Texas, is under a sentence of life imprisonment today following her plea of guilty to the murder of her deaf husband, Carroll Rablen.

Mrs. Rablen told Judge J. B. T. Warne she put poison in a cup of coffee and gave it to her husband as part of a suicide pact concluded April 26. She said she gave him the coffee as he sat in a motor car in

House To Get Revised Farm Bill Tomorrow

Three Senators Vote Against Elimination of Debentures—Agreement Reached After Many Conferences—Measure Held Acceptable to Hoover.

Washington—(AP)—A farm relief bill with the export debenture plan eliminated was agreed to today by the Senate and House conference committee on the measure.

The revised relief measure, composed from the two farm bills passed separately by the Senate and House, will be formally engrossed by the committee tomorrow and immediately submitted to the House for consideration.

After the House has acted the measure must go back to the Senate for consideration.

In the conference committee all of the House members voted to eliminate the debenture plan.

Thre of the five Senate conferes, McNary of Orgon, and Capper of Kansas, Republicans and Ransdell of Louisiana, Democrat, voted to eliminate. Senators Norris, Republican, Nebraska, and Smith, Democrat, South Carolina, voted against elimination.

The measure, as finally agreed to after more than a week of conferences, is held by conference committee members to be acceptable to President Hoover. There were no material differences except for the debenture plan in the Senate and House Bills but the measure framed by the conference group follows more nearly the House measure.

LABOR PARTY HEAD ACCEPTS CABINET OFFER

Macdonald Confers With King and Agrees To Organize New Government—To Replace Baldwin.

London—(AP)—Ramsay Macdonald, leader of the Labor party, victorious in Thursday's polls, at noon today accepted King George's invitation to form a cabinet to replace that of Stanley Baldwin, Conservative premier, who resigned yesterday.

Official announcement of the acceptance was made at Windsor Castle, where King George is ill, a few minutes after the Labor leader had called and left to go back to London.

He arrived at 11:04 a. m. with his son, Alastair Macdonald, and Lord Arnold. The King's equerry, Colonel Seymour, escorted him inside the castle. He remained slightly more than an hour leaving at 12:18.

It was the second cabinet Mr. Mucdonald has been called on to form. The first was that of more than five years ago when the Labor

LIGHTNING BOLT TERRIFIES 19TH WD. RESIDENTS

Bolt Struck Near Mayor's Home—Shock L. Felt—Tree Bark Scaled—Dog Hysterical.

A blinding flash of lightning striking with deafening report at 11 a. m. today terrified residents of the Wellington and Post Avenues and Trafalgar Street neighborhood in the 19th ward.

Pupils in School 16 jumped from their seats, housewives were momentarily stricken with terror. Some left their homes.

The bolt so far as could be learned caused no property damage. It struck near the home of Mayor Joseph C. Wilson, 405 Wellington Avenue. The mayor's daughter-in-law, Mrs. T. M. Crandle, was home at the time but was not injured.

Mrs. Charles C. Eshelman, wife of a vice-president of the Union Trust Company, reported lights in her home at 425 Wellington Avenue were put out of commission. She reported a sulphur-like odor was noticeable in the basement after the detonation and that bark was torn from a maple tree in front of the house.

Mrs. A. H. Dale of 419 Wellington Avenue was ironing when the bolt hit, and her arms was paralyzed for a time from shock.

Residents in describing the phenomenon declared there was no warning except a sudden sizzling, followed by a blinding flash and a terrific detonation. No thunder preceded it, nor was any heard afterward, they said.

A fox terrier at Wellington Avenue and Trafalgar Street became temporarily hysterical.

Meteorologist Jesse L. Vanderpool said a dark cloud hovered over the city at the time the lightning struck. He said it was an ordinary bolt caused by discharge of electricity between a cloud charged

Figure 10/1 *An actual newspaper story written after the eruption of Mount Vesuvius in June, 1929.*

Materials

scissors, paper cups, plaster of Paris, spoons, yarn, crayons

★ In a paper cup, mix plaster of Paris and water. The mixture should be about as thick as a thick milk shake. Break a crayon up into two or three pieces. Tie a 15 cm length of knitting yarn around each piece and push the pieces, one at a time, into the wet plaster. Be sure that

Important notes on this activity are in the Teacher's Manual.

each piece of crayon is completely surrounded by plaster and does not touch the sides or bottom of the cup. Let the ends of the yarn hang over the sides of the cup as in Figure 10/2. The plaster will harden overnight.

After the plaster has hardened, cut the yarn off close to the surface of the plaster and tear away the cup. Then put the piece of plaster directly over an open flame. Leave it

Figure 10/2
Your plaster volcano should look like this.

there until the crayon inside melts. What happens at the surface of the model? The same sort of thing happens when minerals melt in a closed space. ★

Process: inferring

When rock is deeply buried, it may be melted by heat from radioactivity and the original heat of the earth. The resulting increase in volume causes pressure to build up. You can feel this kind of pressure by cupping a balloon in your hands while blowing it up. The yarn in the model is like the spaces between your fingers or weak spots in the earth. When those weak spots are scattered around, the stream of magma under pressure simply divides and takes many different paths toward the surface. Look at the magma paths in Figure 10/3. It's a little like mud being squeezed up between your toes when you walk barefoot through a mud puddle.

Places where the magma comes out at the surface are called **vents.** Sometimes the magma never reaches the surface. When this happens, it cools off slowly to form an igneous rock like granite.

Demonstration Time: 10 minutes. You may wish to display a few chemical bottles and jars containing hydrated compounds. The formulas will show an "n H_2O," but examination of the material will not reveal the water.

You can put about 10 grams of hydrated (blue) copper sulfate crystals in a long test tube and heat them. As heat drives off the water, the crystals will turn white. The steam will condense into droplets of water visible near the top of the test tube.

Concept: particulate nature of matter

10.3 Gases and steam change the melted rock.

Some mineral molecules have water molecules hooked to them. As long as the water molecules stay hooked to the minerals, they cannot be recognized as being water. How-

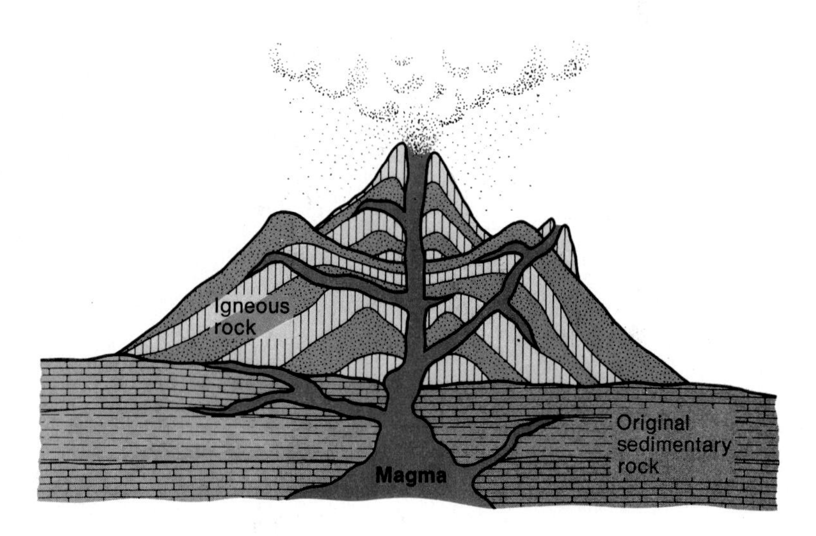

Figure 10/3
The melted rock feeding a volcano can take many paths to the surface. The easiest path is often between two rock layers.

ever, deep within the earth, where the magma starts, there is great heat. The water molecules there are not hooked to the mineral molecules but are in the form of steam. The heating also causes some other minerals to produce poisonous gases. But under the tremendous weight of the rocks above these minerals, the steam and gases are trapped. In fact, they are forced to become part of the melted rock just as the gas in soda pop is part of the soft drink. The gases dissolve in the melted rock.

As the melted rock moves up toward the surface of the earth, the dissolved steam and other gases can come out of solution. In the same way, when you shake up a bottle of soda pop, bubbles form in the pop and pressure builds up. The bottle of shaken-up soda pop is like a volcano ready to go off.

Did You Get the Point?

When most minerals melt, they expand.

The expansion of melting minerals builds up pressure. The pressure forces magma to the surface through weak spots in the rock above it.

Gases and steam in melted rock build up pressure like gas in a shaken-up pop bottle.

Figure 10/4
Fiery lava flowing across a highway in Hawaii. Notice how much gas and steam are given off.

You may send home a small test tube for this simple experiment. Suitable containers may be hard to come by in some kitchens. The surface will be smooth and almost flat as long as the wax is still melted. When the wax cools, the surface becomes noticeably concave. Sometimes there is a hole in the center extending to the bottom of the container. This occurs because the wax takes up less space in solid than in liquid form. Refer to the crayon experiment.

Check Yourself

1. Melt candle wax, paraffin, or crayons by placing wax in a container in boiling water. Observe the surface of the melted wax. Allow the wax to cool slowly. Look at the surface of the hardened wax. How has the shape of the surface changed? Why?

What Comes Out of a Volcano?

Try to get pictures of the Surtsey eruption from sources like *Life Magazine*. Try to get pictures of other volcanoes. Display them and discuss them. Note permanent changes in the face of the earth. Students are fascinated with the subject.

10.4 Volcano eruptions come in many forms. When the hot melted magma containing the bubbles of steam and gases reaches the surface of the earth, lava is formed. The lava comes out through an opening in the surface called a vent. In Figure 10/4 you see steam and gases escaping from a flow of molten lava. If the lava contains a lot of water, it is thin and runny, and the gases can easily escape from the melted rock. Gases bubble off as the gas does from a bottle of soda pop that has been sitting open. Lava flows quietly out over the land. The volcanoes in Hawaii usually erupt this way. Figure 10/5 shows the flat-shaped volcano that forms when the lava flows out.

Figure 10/5
Mauna Loa, a volcano in Hawaii, has the usual flat shape of a lava cone. Part B shows how the layers build up when lava flows over the land.

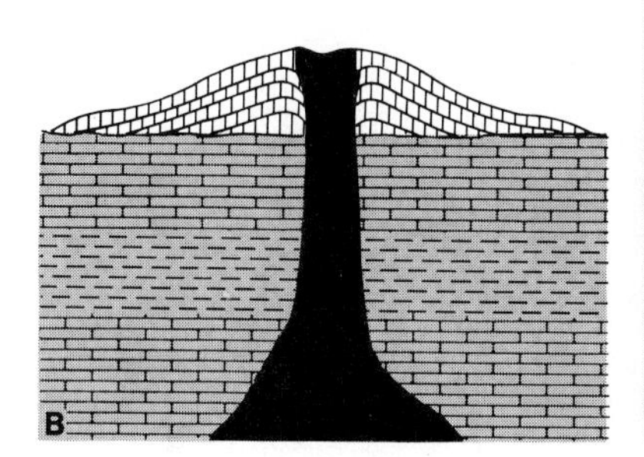

The Birth of a Volcano

Señor Pulido was cultivating his field near Parícutin, Mexico, on February 20, 1943, when the ground began to tremble. The soil under his bare feet got hot. White smoke rose through cracks in the ground and from one small hole in particular. The frightened farmer ran to town to tell the villagers.

When the crowd returned an hour later, the hole in Señor Pulido's field was 30 feet deep. Black smoke, hot rocks, and dust were coming from it. The next morning, the hole was covered by a 25-foot volcanic cone. A week later the volcano was 550 feet high. Ten weeks later it had reached 1100 feet. In six months the volcano was 1500 feet high. Meanwhile, loud rumblings continued.

Fifteen weeks after steam had first escaped from the ground, lava began to flow from the cone. "The fire monster," as the people called the volcano, threw out 2700 tons of material every minute during these first months. It completely covered the village of Parícutin. Black ash drifted as far as Mexico City, 200 miles away.

After nine years of activity, the volcano Parícutin stopped erupting. It is now considered to be dead. But as the first new volcano to appear in North or South America in 200 years, Parícutin gave scientists a firsthand view of the life cycle of a volcano.

But if the lava is thick and pasty, gases are trapped in the melted rock and pressure builds up. When the pressure is released suddenly at the surface of the earth, the lava explodes into the air as small chunks called **cinders** and fine dust called **ash.** In Figure 10/6 volcanic ash is being sprayed into the air during an eruption. The eruption of Mount Vesuvius was an explosion like this. The cinders and ash may pile up around the opening in the ground as a mound called a **cinder cone.** Notice the steep sides of the

Figure 10/6
Volcanic ash being sprayed into the air. Notice what the hot ash can do to plant life in the area.

cinder cone in northern New Mexico called Capulin Mountain, shown in Figure 10/7.

After the volcano has exploded and the pressure has been released, magma may harden in the vent. When this happens, the flow stops for a while, and the volcano is quiet or dormant. But the pressure continues to build up. Eventually, the gases in the magma build up enough pressure to blast the plug out of the way. Kerpow! Krakatoa all over again. The same thing can happen to toothpaste that has a hard lump at the mouth of the tube. Look at Figure 10/8. In a volcano, pieces of the plug may be scattered for miles. The noise can be deafening. The total effect can be deadly.

Some volcanoes alternate between eruptions of ashes and cinders and quieter periods of lava flows. These volcanoes build up in layers as shown in Figure 10/9. This type of cone may grow to the size of a large mountain high enough to have snow on its peak. In the Cascade Mountains of the western United States are good examples of layered volcanoes.

Sometimes a large depression or hole called a **caldera** (kol-DARE-uh) forms at the top of a volcano. Figure 10/10 shows the three different ways a caldera may form. An explosion may blow off the top of a cone. Or after a lot of magma has flowed or been blown out of a volcano, the top of the cone may collapse into the empty cave left down below. Or, finally, if the magma in a passageway of the

Figure 10/7
Mount Capulin in New Mexico is made up of layers of cinders.

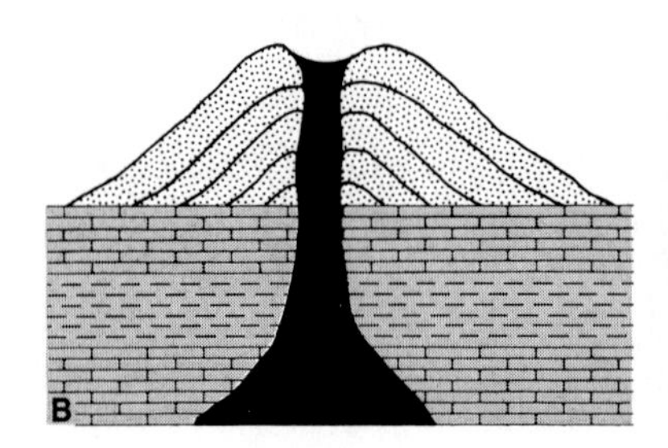

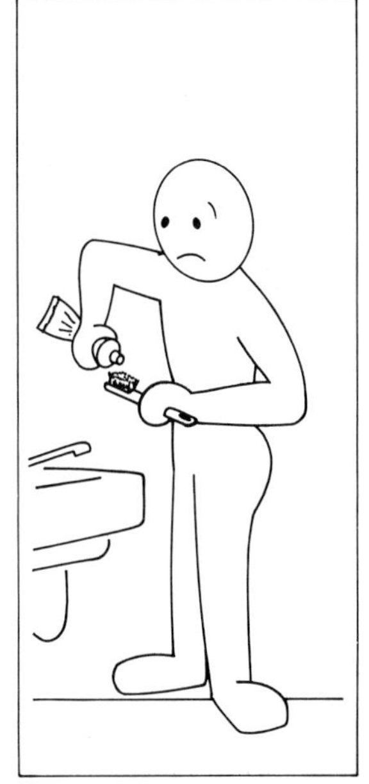
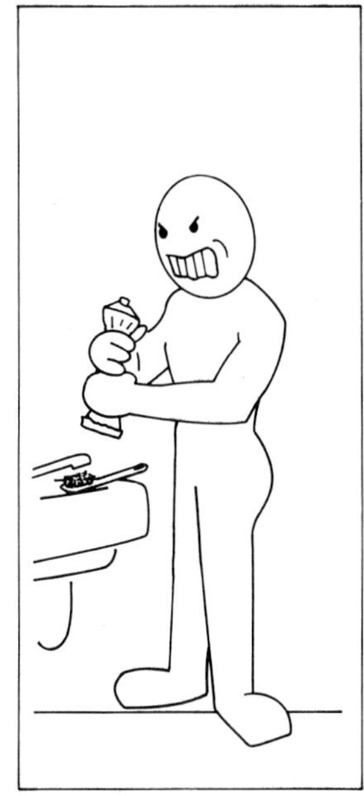

Figure 10/8 *A tube of toothpaste can become plugged like a volcano. Pressure builds up until the plug is blown clear.*

volcano sinks downward, a piece of the top can sink with it. The calderas of Hawaiian volcanoes usually form in this way. In Figure 10/11 you see a caldera that has filled with water to become a lake.

Figure 10/9
Mount Fujiyama in Japan is a large layered volcano. The diagram shows how the layers of lava and cinders build up.

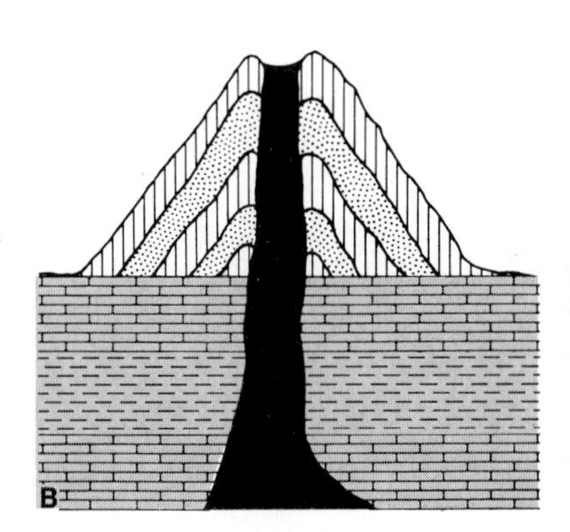

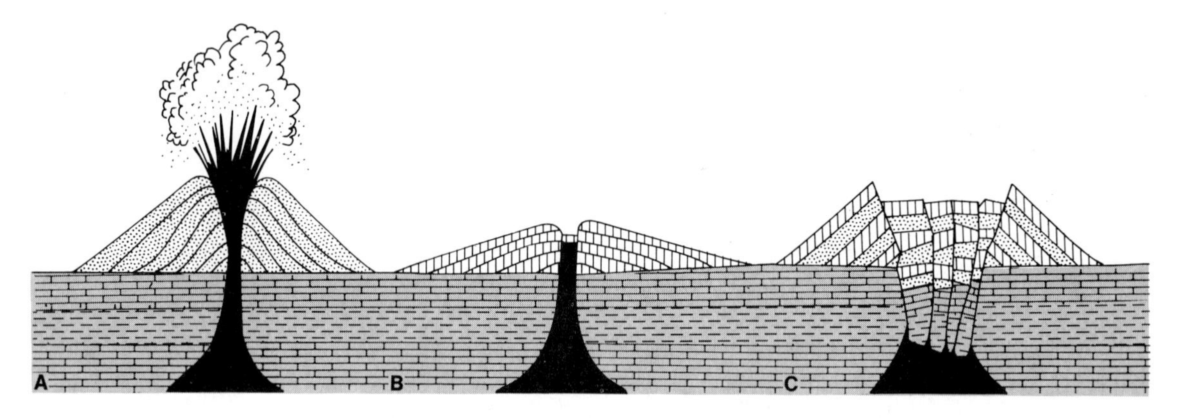

Figure 10/10 *A caldera may form when a) the volcano explodes, b) the cooling magma sinks into the passageway, or c) when the cone collapses into the empty space left down below.*

Figure 10/11 *Crater Lake in Oregon is a caldera filled with water. The small cone sticking above the rim is Wizard Island.*

Some volcanoes start on the floor of the ocean. If they reach the surface, they become islands like those that make up Hawaii. If they do not reach the surface of the ocean, they are thought to become underwater mountains called **seamounts.** Figure 10/12 shows the difference between a volcanic island and a seamount.

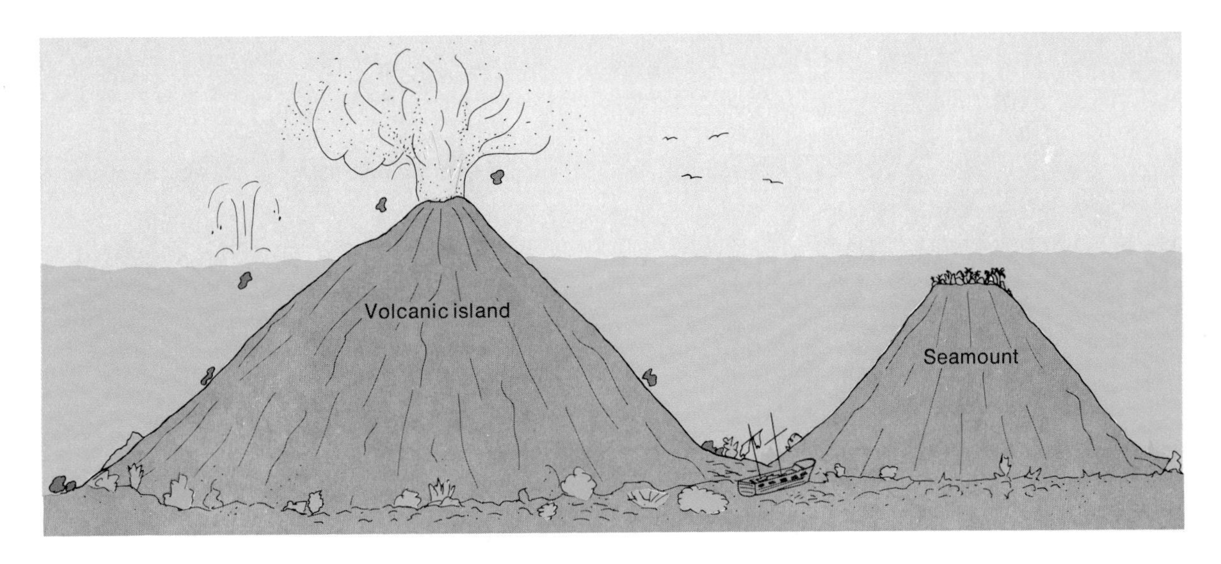

Figure 10/12 *The cone of an underwater volcano may form a mountain called a seamount. If the cone reaches the surface, it becomes an island.*

Concept: interaction of matter

10.5 Gases in magma leave holes.

Imagine what very bubbly soda pop would look like if it were suddenly frozen. Now try to imagine how a glob of foam from a bottle of shaken-up soda pop would look if it were suddenly frozen. Wouldn't it look a little like a chunk of swiss cheese?

The hardening of a rock is the same kind of process as the freezing of a liquid. Figure 10/13 shows some bubbly and foamy lava that "froze," or hardened quickly. Do the rock samples look like your mental picture of the frozen soda pop? The scoria and pumice are the "frozen" foam from a volcano.

Activity Time: 10 minutes. You will need samples of volcanic rock, especially pumice. You can have students bring in slag if there are steel mills in your area or cinders from the school outdoor track to model foamy volcanic rock. Discuss how they might have been made. Obsidian cools so fast that no crystals can form.
Process: observing

If your school has some samples of real volcanic rock, your teacher will show them to you. Notice how light some of them are. Some can even float on water because of the holes and air spaces in them. Notice how small the crystals in the samples are. They must have cooled very fast. Look back at Figure 9/6. Notice the glassy appearance of the obsidian. Can you find any crystals in obsidian?

Figure 10/13 *Foamy lava that hardened quickly. The holes in the rocks show where gas bubbles used to be.*

10.6 The dissolved materials can kill.

Flying chunks of rock, hot lava, and violent explosions are not the only features of a volcano that make it dangerous. When the steam and poisonous gases come out of the melted rock, they too can destroy life and property.

Most of the poisonous gases that leave the lava are heavier than air. They sink and stay close to the ground, and sometimes they flow into valleys and fill them. The deadly gases replace breathable air. Since the gas flow cannot be seen, people and animals may be trapped and suffocated. Plants will die, too.

Concept: interaction of matter

The steam is lighter than air, so it rises. The dust particles thrown out by the volcano act as condensation nuclei. Remember these "cloud starters" from Section 7.7. In the same way, condensation nuclei from the volcano cause clouds to form when the steam cools. It may start to rain heavily. The loose ash and dust near the volcano can be wet by the rain and become mud. The mud can become thin enough to flow if enough water is added to it. Some mud flows have been known to move at 60 miles per hour. They can sweep huge boulders and trees along with them. Entire villages can become buried.

The fiery terror of a volcano doesn't last. Eventually the pressure that caused the eruption is relieved and the flow of lava stops. The poison gases are carried away by the wind. The steam forms rain clouds that drift away. The vent becomes clogged with the cooled magma left there after the explosion. The dust eventually settles. Soon everything is back to normal. Or is it? The land doesn't look the same, and no one knows for sure when the next eruption will be.

Did You Get the Point?

There are three different kinds of volcanoes.

Magma (melted rock) containing dissolved gases is called lava when it comes to the surface.

When the heavy, poisonous gases escape from the lava, they can collect in low spots and harm plants and animals. The steam can form rain clouds that provide water for mud flows.

Volcanoes build islands in the ocean and mountains under the sea.

Check Yourself

1. How can mud flows form from the ash and dust from a volcano even if there are no rain clouds in the area?

2. How can the material coming out of a volcano affect the weather in the area?

3. Imagine a glob of shaving cream sprayed onto a slanted surface. Spray it yourself if you can. How could the spraying process and the changes that take place afterward be used to model material that comes from a volcano? List any other man-made materials that could be used to model materials that come out of a volcano. Be sure to indicate what each item is supposed to be modeling.

4. Open a small bottle of a carbonated soft drink. Stick a cork in the top. Make sure the cork fits tightly. Take the bottle outside, point it away from people and windows, and shake it. What happens? What parts of an exploding volcano can you compare to the bottle, the soft drink, and the "pop?"

5. Have one person in your class write to the National Park Service to get information and travel folders about Yellowstone National Park. When they arrive, try to identify those features in the photos that might have been the result of a volcano.

Answers:

1. Water spilled from a crater lake. Ash and dust dumped directly into a stream.
2. The dust can cause a dense cloud cover to form and produce heavy rain. The dust cover can also cut down the amount of solar radiation reaching the earth, thereby lowering temperatures. This could in turn change the weather of that area until equilibrium is reestablished.
3. The shaving cream will flow down the surface as long as the gas remains in it. As soon as the gas escapes, the shaving cream stops flowing. Fireplace bricks, cinders on a school track, kitchen sponge, are all models of rocks with holes in them made by trapped gas.
4. You may have to supply corks for some students. The students should mention the "fizz" in the pop as being like the dissolved gases in magma coming out of solution. The neck of the bottle is the vent, and the cork is the plug of hardened rock. The noise is the "Kerpow" of an exploding volcano.

   These simple experiments are designed to be done at home. However, don't penalize the student who can come up with the answers without actually trying them.
5. Address:

   National Park Service
   U.S. Department of the Interior
   Washington, D.C. 20240

Predicting Volcanic Eruptions

10.7 The "where" for an eruption is no secret.

What if man could tell ahead of time when a volcano is going to erupt? He could satisfy both his interest in the rock-building process and his interest in self-preservation. He could measure lava flow, earth trembling, and all sorts of other things. He would be able to warn people living in the area in time for them to move their families and belongings out of the danger zone.

The first part of the prediction would have to be the "where." Perhaps plotting past and present volcanoes will give you an idea of how this could be done.

Materials
world map, thumbtacks, copy of a world map

Activity Time: 30 minutes. A wall map is best. It will give you a visual aid to refer to later. Pass out dittoed maps for students to record a copy of the pattern disclosed. Make thumbtack-pushing a class project. The most obvious pattern will be around the Pacific Ocean. A general line pattern will do.
Process: interpreting data

★ Appendix U is a chart of 39 volcanoes and their locations. Push a thumbtack into the world map at the location of each of the volcanoes. Do you see a pattern? Is a volcano likely to erupt where you live? Where would a new volcano be most likely to erupt?

Mark the general location of the tacks on the map your teacher gives you. On the back, write a brief description of the pattern shown by the tacks. Tape the map in your notebook. You will have to use it later. ★

10.8 The "when" for an eruption is difficult to predict.

Most volcanoes have been and are quite closely watched. The activity of Mount Vesuvius has been recorded since the year A.D. 63 when earthquakes rocked the area. The first record of the eruption of this volcano dates from A.D. 79. Figure 10/14 lists important eruptions that have occurred from that date to the present.

Materials
meter stick, scissors, tape, colored pencils, adding machine paper

★ Cut off a little over two meters of adding machine paper and tape it to the wall so that you can see all of it at one time. Draw a straight line across it near one end. Label this line 0. Draw a line for each eruption listed in Figure

Figure 10/14
Can you tell from this record of eruptions when Mount Vesuvius will erupt again?

Years when Mount Vesuvius erupted (all dates A.D.)			
79*	1631*	1794	1891
203	1660	1804	1895
472*	1682	1805	1900
512	1689	1822	1903
685	1694	1838	1904
993	1707	1850	1906
1036*	1737	1858	1913
1139	1760	1861	1926
1306	1767	1871–72	1929
1500	1779	1875	1944*

Activity Time: 30 minutes. One year equals 0.001 m or 1 mm. The 79 A.D. line will then be 79 mm from the beginning. You should make a time line of your own to display. The day after the assignment is given, yours can be taped to the front wall and be used in the discussion of the activity.

10/14. Space the lines according to dates by letting one meter stand for 1000 years. What distance will equal one year? How far will the line for the eruption in A.D. 79 be from the 0 line? Can you see why? In Figure 10/15 a student is already up to the year 1682.

See Teacher's Manual for notes on this activity and the eruption cycles of Vesuvius.
Process: interpreting data, formulating hypotheses

Examine the completed "time line." Can you see a pattern? Can you predict a future eruption of Mount Vesuvius by studying its behavior in the past? Could there have been eruptions before the year A.D. 79? Color the lines representing the eruptions that have been marked with asterisks in Figure 10/14. Those eruptions are considered more severe than the others. Do the colored lines help you find a pattern? ★

10.9 Modern methods predict eruptions.

Many *active* volcanoes, ones that have erupted since man began to record history, are being watched and studied at this very moment. Sensitive instruments measure and record any changes that might lead to an eruption. Thermometers that can measure very high temperatures up to 1000 degrees Celsius record temperature changes in pools of boiling lava. Satellites carry special equipment for measuring heat flow beneath the surface of the earth. They watch for changes that might indicate rising magma. **Tiltmeters,** like the one shown in Figure 10/16, continuously record any bulging of the earth's surface near volcanoes. Bulging might result from the buildup of pressure before an eruption.

Process: measuring

Figure 10/15 *The eruptions of Mount Vesuvius on a time line.*

Just which earthquakes belong to a series is determined by a plot of their focuses, which is demonstrated in the next chapter. All those quakes falling on a straight line plot of focuses are in the same series. You may want to review this section after completing Chapter 11.
Process: predicting

Since 1963, an entirely different method has been used to predict future eruptions in the southwestern Pacific. Scientists found that volcanic eruptions often happen after a *series* of earthquakes. The first earthquake in the series is deep below the earth's surface. The later earthquakes get closer to the surface. The eruption usually follows the last earthquake in the series, the one closest to the surface. Using earthquake data from the area around volcanoes, the scientists could predict eruptions three, four, and five months in advance. Can you see how such predictions might save lives and property?

Compare the depths and dates of the earthquakes in Figure 10/17. Using these data, scientists predicted that an eruption of the Gaua (GOW-uh) Volcano would take place some time in November of 1963. This volcano is located in the New Hebrides (HEB-ruh-deez) Islands, shown on the map in Figure 10/18. The actual eruption took place on November 3, 1963. Perhaps in your lifetime, live television

Figure 10/16
This tiltmeter can measure very slight changes in the slope of the land near a volcano.

Date of earthquake	Depth of earthquake
May 2, 1962	623 km
April 1, 1963	196 km
June 6, 1963	160 km
July 24, 1963	133 km
November 3, 1963	Eruption

Figure 10/17 *Some scientists believe that a series of earthquakes, each one closer to the surface than the one before it, means that a volcano will erupt soon.*

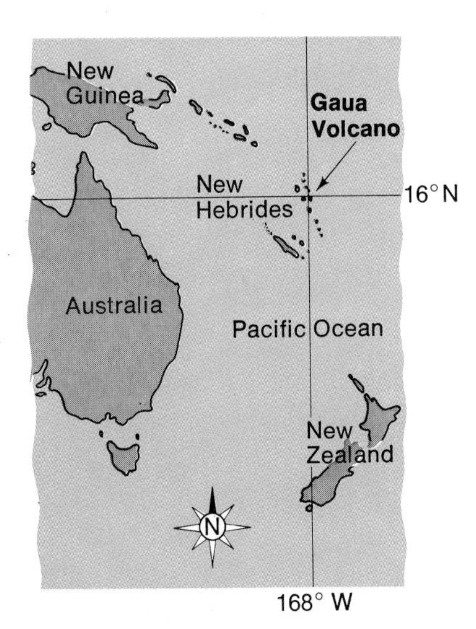

Figure 10/18
Scientists predicted the eruption of Gaua Volcano in November 1963 by studying earthquake activity in the area.

coverage of this part of the rock-building process will be as commonplace as coverage of a spaceship splashdown.

Did You Get the Point?

Volcanic eruptions occur in zones or beltlike patterns.

Past volcanic eruptions are little help in predicting future eruptions accurately.

Answers:

1. Students are only expected to notice that volcanic areas are mountainous.
2. It would not have a flammable roof or siding. There would be fireproof shutters to protect the windows. Gas and utilities could be turned off at a point away from the house so that breaks in a line would not feed a fire. A fire hose should be ready to go into action at any time. A sloping roof would shed dust and ash easily; perhaps roofing should be slippery.
3. a. Every family in the area would be warned of the impending eruption, what signals to look for, and what safety precautions to take. Then, instruments used to watch and measure the eruption would be put in position.

   b. Move important papers and records to a safe location. Arrange for food and shelter well away from the area. Plan escape routes. Plan to leave the area days before the actual eruption. Practice family safety drills. Plan to move to another area for good.

Sensitive instruments have been helpful in predicting future volcanic eruptions.

Earthquake data have been useful in predicting some volcanic eruptions as much as five months in advance.

Check Yourself

1. What landforms, events, weather patterns, or climate types, if any, also fit the pattern traced out by plotting volcanoes on a map?

2. Suppose you were given a piece of land near an active volcano. What "extras" would you build into your house?

3. Suppose you live in a town at the foot of an active volcano. Scientists predict an eruption in three months.

a. As a scientist, describe what preparations you would make right up to the day of the eruption.

b. As the head of a family, describe what preparations you would make right up to the day of the eruption.

What's Next?

Some volcanoes never seem to become totally inactive. Threats of eruptions can be seen in their craters. Boiling lava, rising steam, poison gases, and an occasional rumble

Figure 10/19
Hot steam rising from the ground in Yellowstone National Park in Wyoming. There might be magma close to the surface here.

Figure 10/20 *The volcano Surtsey began to erupt off the southwest coast of Iceland on November 14, 1963. Two years later, Surtsey was an island one square mile in area and 500 feet high, with plants growing on it.*

seem to indicate that magma is still trying to get to the surface, as you can see in Figure 10/19. Areas where this occurs are often made into national parks, like Yellowstone.

Other volcanoes stop erupting and become permanently quiet. The cones, hardened lava, steaming hot springs, and ash-covered islands on the earth's surface remain for what seems to be forever. But new volcanoes like the one in Figure 10/20 erupt every day, showing that this part of the rock-building process is still going on. The rumblings that may come before, during, and after these new eruptions will be discussed in the next chapter.

Skullduggery

Directions: Lay out a crossword puzzle grid to match the one in Figure 10/21. (Do not write in this book.) Use graph paper or grids your teacher will give you.

Across

2. Melted rock is called ______.

5. The visible part of a volcano is called a ______.

9. The bubbles in volcanic rock are formed when the gas in the lava ______.

10. The places where the magma escapes through the surface of a volcano are called ______.

12. Water and gases ______ in making magma liquid enough to flow.

14. The "coming alive" of a volcano is called an ______.

15. The path magma takes to the surface is seldom ______.

16. The sites of the rock buildup process seen on the surface of the earth are called ______.

18. The effects of a volcano reach ______ and wide.

19. Scientists are collecting ______ in an attempt to find a way to predict future volcanic eruptions.

21. As a general rule, a material takes up less space as a ______ than as a liquid.

22. Plotting locations of volcanoes, doing experiments, and making time lines of past eruptions are three ways to ______ about the rock-building process.

24. If the melting of minerals in a rock takes place in a closed-off area, the ______ increases.

Down

1. Magma that has reached the surface is called ______.

2. When a ______ melts, the amount of space it takes up changes.

3. Melted rock usually contains ______ which makes it flow better.

4. ______ can sometimes be evidence that there has been volcanic activity in the area.

5. When the vent of a volcano caves in, a ______ is formed.

6. The visible part of a volcano is smallest at the ______.

7. ______ volcanoes offer evidence that the earth is still changing.

8. Do the locations of known volcanoes form a pattern of some kind?

11. Some of the gases that are produced when minerals are heated are ______.

13. Some mineral molecules have ______ molecules hooked to them.

17. There are periods when an active volcano is ______.

18. The rock called pumice is hardened ______.

19. The approximate time it would take lava to flow two miles is ______.

20. Much of the material coming from volcanoes remains in the ______.

23. Is it possible at this time to predict just when a volcano will erupt?

Figure 10/21
Copy this crossword puzzle grid for the Skullduggery.

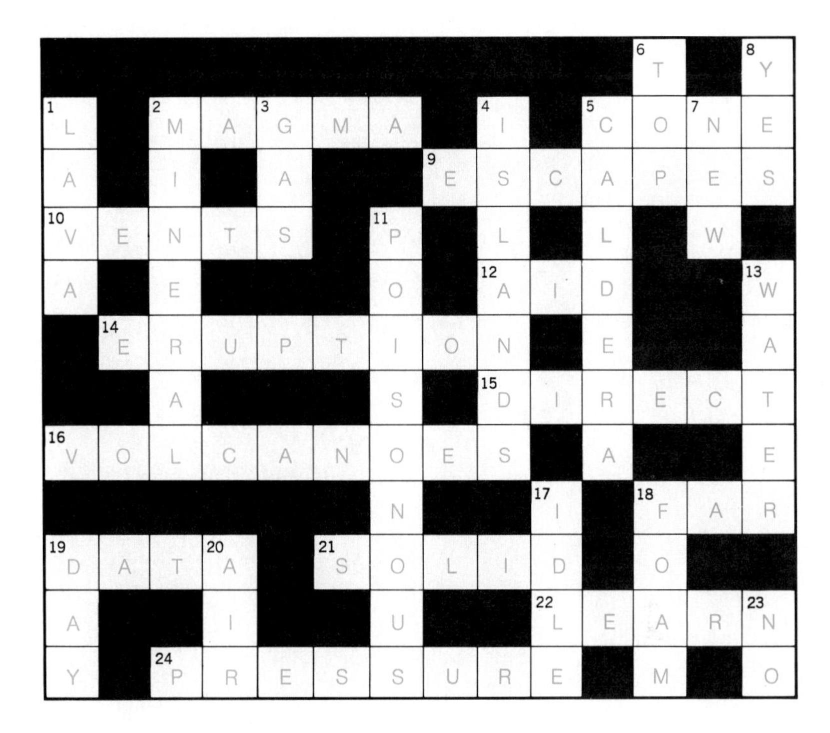

For Further Reading

Mather, Kirley F. *The Earth Beneath Us.* New York, Random House, 1964. This book not only tells you but also shows you how beautiful and exciting our earth is. It contains breathtaking photos of volcanoes and the effects of earthquakes.

Clayton, Keith. *The Crust of the Earth.* Garden City, New York, Natural History Press, 1967. An easy-to-understand book with interesting pictures. A good overall look at our earth with sections on earthquakes and volcanoes.

Matthews, William H., III. *The Story of Volcanoes and Earthquakes.* New York, Harvey House, 1969. This book describes both the destructive force and the beauty of volcanoes. The photographs are especially good.

28391
ALASKA

RISCO
CADE
CANDINAVIAN
LUB BAR
OPEN
24
HRS
PAWN SHOP
LOANS
OANS
SEMENTS
ANS
LOANS
WE SHIP
COD

Chapter 11 Earthquakes

When Karen entered the room, she was a little surprised to find that Garry was the only other student back early from lunch. ''Phew!'' she said as she sat up on the teacher's desk. ''This place still smells like burnt wax from that volcano business.''

Garry was so busy reading a book that he didn't even notice.

''Is that all you ever do, read?'' Karen asked.

''What?'' Garry looked up.

''I said, what are you reading now?'' She really didn't care, but he was the only one in the classroom to talk to.

''I found this book on Mr. Tracey's desk. He must be planning to teach earthquakes next. There's some stuff in here that's a little hard to believe.'' Garry flipped back a few pages from where he was reading.

''Like what?'' Karen asked.

''Wait a minute, listen to this. There's a report in here by a guy in Japan who lived through a real bad one. He says there were several rumblings to the same earthquake. Each one seemed to be a little worse than the one before it. Fires started up all over the place. People were crushed by falling buildings. There are pictures in here of damage done by earthquakes all over the world.''

Now Karen showed some interest. ''How can an earthquake start a fire? I thought the ground just shook a little.''

Garry was quick to answer. ''Well, the ground does shake. Sometimes like what you feel when a big truck goes by close to you. Sometimes worse. Buildings crack and cave in. Electric wires are pulled apart, and the sparks start fires.''

''Stoves could get tipped over, too.'' Karen added.

''And when the streets are lifted up, gas mains break, and that adds to it.''

''We've never had an earthquake here, have we?'' Karen asked, half as a comment and half as a question.

"Hmmmmmm. I don't remember any." Garry thought for a moment. "But I really don't know if we have them here or not."

"What causes them anyway? Where do they happen? Could we have one here someday?" Karen really wanted to know the answer to her last question.

"Well, I haven't read that far yet," Garry hedged. "There's a list here, though, of some of the places where earthquakes have happened." He pointed out the list, and both read a few of the locations to themselves.

"Hey," Karen said finally, "let's put some of these earthquakes on a map. Then maybe we can tell whether one will happen around here."

The Permanent Effects of Earthquakes

11.1 Earthquakes change the earth's surface.

Activity Time: 30 minutes. This could be a class project. Use thumbtacks or pins in a wall map, one color for volcanoes and another for earthquakes. Yes, the earthquakes follow a pattern like the volcanoes.

★ Karen and Garry were looking at a list of earthquakes like the one in Appendix V. You can find out about the chance of having earthquakes in your area, too. Plot the earthquake locations from the list on a blank map like the one you used to plot the volcano locations in Section 10.7. But use a different color this time. Do the earthquakes follow a pattern? ★

Just like volcanoes, earthquakes are reminders that the earth is still in the process of changing. In 1811, during an earthquake in Tennessee a large piece of ground dropped downward. The depression filled with water, and a lake 18 miles long was born. It is called Reelfoot Lake, and if you are a fisherman, you have probably heard of it. In the Owens Valley of California in 1872, a section of the earth jolted upward 23 feet and sideways 20 feet. The first shake was so violent that it threw fish right out of a river! During the famous San Francisco earthquake of 1906, a 200 mile strip of California moved and cracked. The quake, and the fire which followed, completely destroyed 490 city blocks of San Francisco itself. Because of a 1959 earthquake in Montana the rocks from a hillside slid down to form a ridge 300 feet high across a valley.

During the 1964 Alaska earthquake, an island offshore rose over 30 feet. People driving their cars in Anchorage were bounced as high as two feet above the road. Floors in large buildings tilted, rose, and fell while people were scrambling to safety. Tremendous ocean waves caused by the quake washed fishing boats into coastal towns. One skipper who managed to keep his boat off the land at Kodiak found himself with another problem. He claimed that a large building floated out and cracked the bow of his boat. Property damage amounted to more than $285 million in Anchorage alone.

Earthquakes can change the rocks themselves. While rocks are being pushed up in some places, they are being squeezed to the side in others. The squeezing causes some layers of rock to be folded and bent like ribbons.

Try folding the cigarette in class as shown in Figure 11/1. It works.

It is hard to imagine something as hard and brittle as a rock being bent without breaking. But certain conditions make it possible. For instance, have you ever seen a cigarette tied in a knot without breaking the paper? It can be done. Figure 11/1 shows how. As long as a constant pressure is exerted all around the cigarette, it will not break.

Figure 11/1 *A brittle cigarette can be rolled in cellophane, tied in a knot, and straightened again without breaking the paper. Rock buried deep within the earth is kept under pressure like the tightly wrapped cigarette.*

As long as constant pressure is exerted all around rock layers, they will not break either. Furthermore, the extreme pressure can help the folding process by causing the rock to soften.

Sudden but short-term changes can also happen along coastal areas during earthquakes. The first burst of energy from an earthquake can cause a gigantic wave to start across the ocean. If you hit a full glass of water with a spoon, or kick the side of an above-ground swimming pool, you will start a similar wave. When the wave breaks against the shore, millions of gallons of water may wash away whole beaches. The waves are called **tsunamis** (tsu-NAH-meez) and are greatly feared by people living along some coasts. The wave that hit Kodiak, Alaska, started from an earthquake on land. Tsunamis also may be started far from shore by earthquakes under the ocean. They have been known to travel at speeds of several hundred miles an hour toward the land.

Tsunamis, often incorrectly called "tidal waves," are caused by disturbances in the earth's crust. True tidal waves, by contrast, are simply changes in water level resulting from the motions of the sun and moon or a cyclonic storm:
Concept: energy in motion

11.2 Land in earthquake areas moves up and down.

The places that lie directly *over* the spots where each earthquake begins are called **earthquake epicenters** (EP-ih-sen-terz). The **focus** is the spot *beneath* the earth's surface where the disturbance causing the earthquake actually happened. Special instruments are used to record the depths of focuses. Figure 11/2 is a chart of the depths of focuses for several earthquakes that took place under Japan.

Materials

world map, graph paper

Activity Time: 20 minutes. Students will need help setting up the graph and plotting data. Perhaps a clay model will be useful. Cut a clay ball in half and show some exaggerated epicenters and focuses. The pattern of focuses clustering along the line could be due to some land rising on one side and some sinking on the other side.
Process: interpreting data, formulating hypotheses

★ Set up a graph like the one in Figure 11/3. Plot all of the information from Figure 11/2 on the graph. Then draw a straight line along the points so that about as many points are above the line as below the line. Do you see a pattern? What could be happening under Japan to make earthquake focuses cluster along the line you have drawn? ★

Plotting information from other earthquakes has produced lines similar to the one you found on your graph. Figure 11/4 shows a graph of earthquake focuses under South

Figure 11/2
Focuses of earthquakes near Japan.

Depth (km)	Distance east of Asia (km)	Depth (km)	Distance east of Asia (km)
55	600	125	650
300	490	280	520
375	425	410	400
405	350	75	625
240	625	60	675
500	60	300	300
305	375	100	700
150	625	40	650
305	400	410	425
75	625	50	825

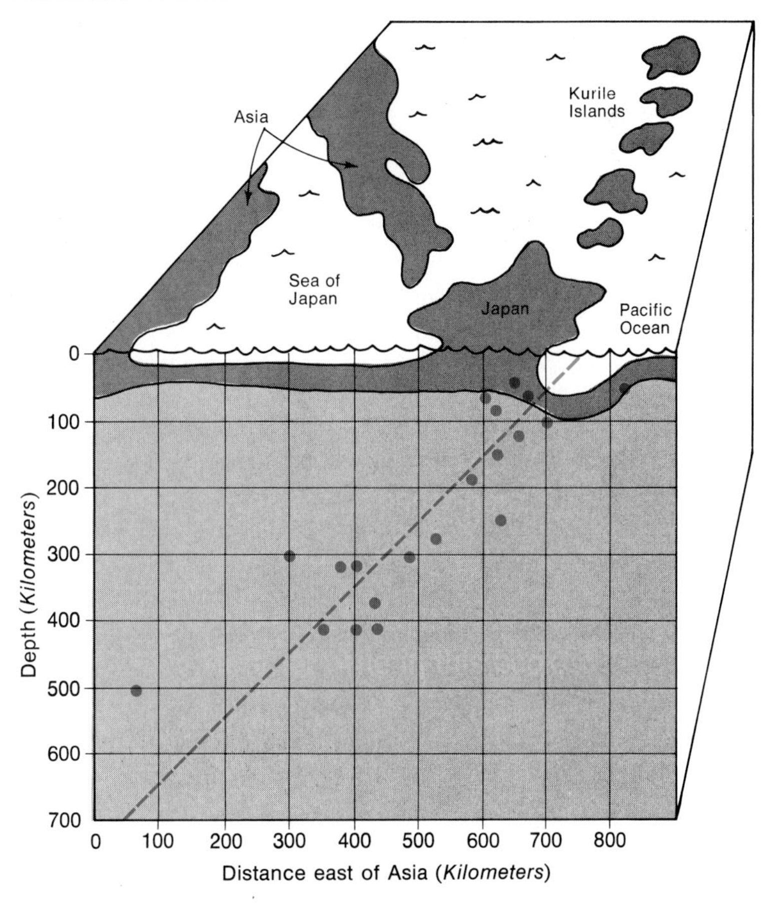

Figure 11/3 *Plot the earthquake focuses near Japan listed in Figure 11/2 on a grid like this one.*

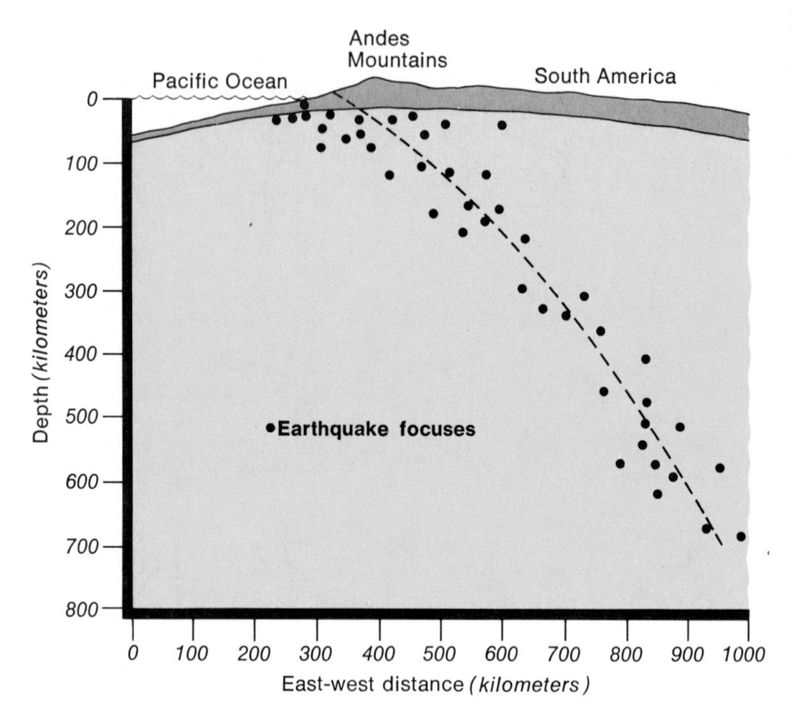

Figure 11/4 *The focuses of earthquakes under South America seem to trace out a line.*

Demonstration Time: 40 minutes, including setting it up. Do this demonstration if possible. It is beneficial to students to see the fault develop and to measure the angle on the front. If the angle measurement is not exact, discuss experimental error both in the operation of the fault box and in earthquake focus determination. The measurements should be close, however. See Teacher's Manual for directions for the construction of your own fault box.

The sand should be damp enough to stick together, but not so wet that water will seep out of it. Instead of dry sand, if you wish, iron powder, charcoal, or dark topsoil can be used as a marker between layers.

Process: experimenting

America. A simple experiment with the earth box will help explain these lines of focuses.

The earth box with an inflatable plastic bag placed at the bottom is shown in Figure 11/5A. The plastic bag has a pump to fill it with air. First an inch-deep layer of dark, damp sand is poured in and patted down. The sand has been dampened to make the grains stick together. Then to show what happens, a thin layer of dry topsoil is sprinkled on. Then another layer of damp sand and so on. (You can use light dry sand instead of topsoil in the classroom, but topsoil makes a clearer picture for a book.)

The plastic bag under part of the sand layers is slowly filled, a little at a time. Figure 11/5B shows the plastic bag partly filled. Notice in Figure 11/5B where a crack first cuts across the sand layers. Can you see in Figure 11/5C how a second crack develops after the first one on each side of the plastic bag? Such cracks at or near the earth's surface are called **faults**.

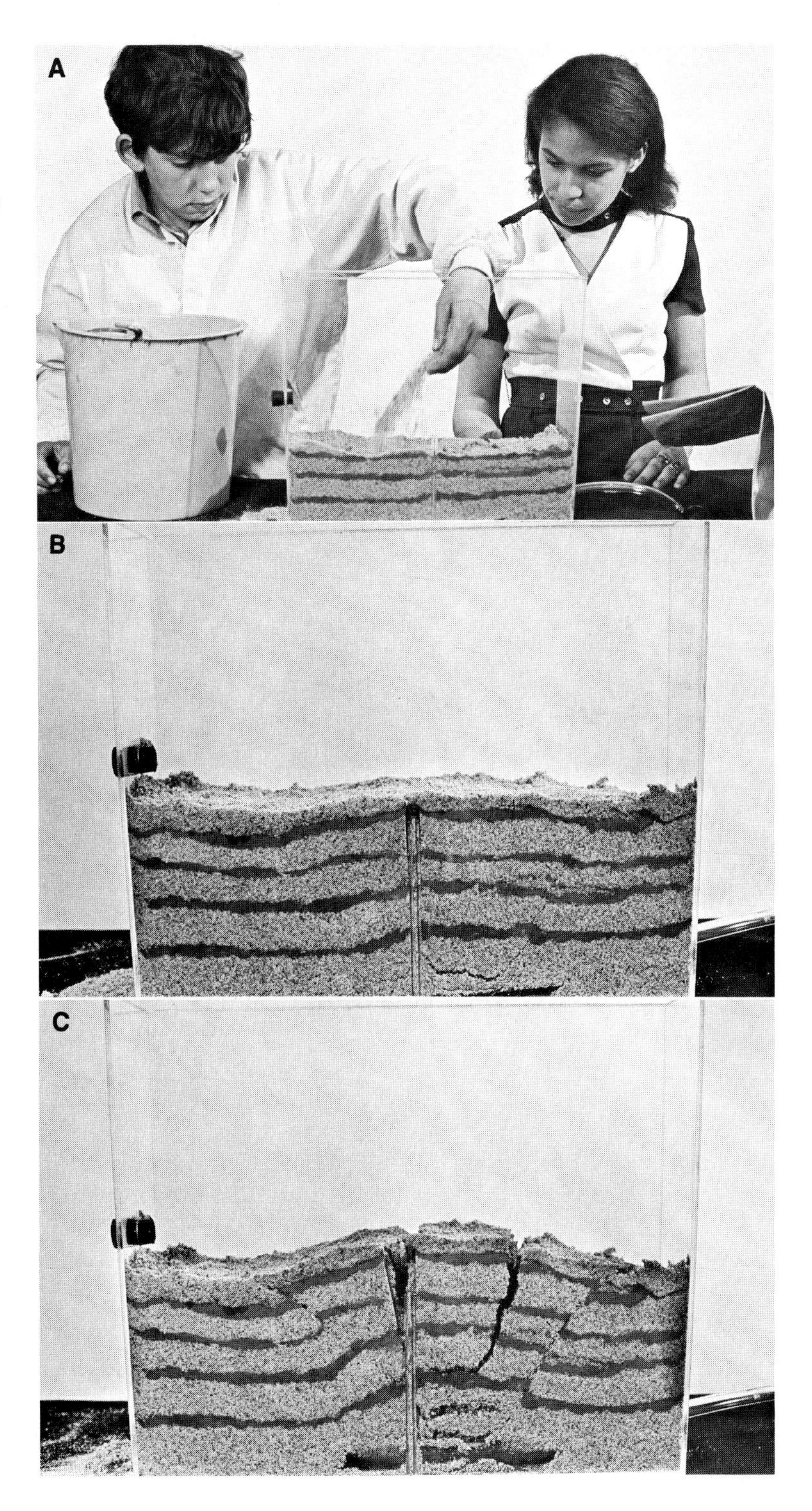

Figure 11/5
The alternating layers of sand and topsoil in A model rock layers in the earth. Pressure builds up in B as the small pillow inside the box is pumped full of air. In C, the sand layers crack like rock layers during an earthquake.

Use a protractor to measure the angle the cracks make with the surface of the sand in the box. Then measure the angle the earthquake pattern under Japan makes with the surface of the land in your graph. Could something be rising or sinking under Japan? Could two sections of the earth be sliding past each other?

The answer to both questions is "Yes."

11.3 Land in earthquake areas moves from side to side.

Try sliding two pieces of granite against one another to show the origin of what scientists call "tremors." Note the similarity to the nail file analogy used in the text.

There is a place in California where earthquakes occur along a line running north and south. Do you see anything "funny" in Figure 11/6? Can you see the line along which

Figure 11/6
The arrows show how the land moves on each side of the San Andreas fault in California. What has happened to the streams crossing the fault line?

the local earthquakes occur? This line is called the San Andreas (san an-DRAY-us) fault.

In this case, two land masses are sliding past each other a little more each year. Look at the farmer's fence that stretched across the line in Figure 11/7. Is there any doubt about what has happened? A very strong force must be acting to move western California northward and eastern California southward. The rocks on either side of the underground break slide past each other, like one nail file

Demonstration Time: 10 minutes. The various types of faulting can be illustrated with a 2″ × 4″ board that has been cut at an angle as shown. Try to use fault box in your explanation of faults. Ask questions like "How many faults can you see in the fault box (or in the photo)?"

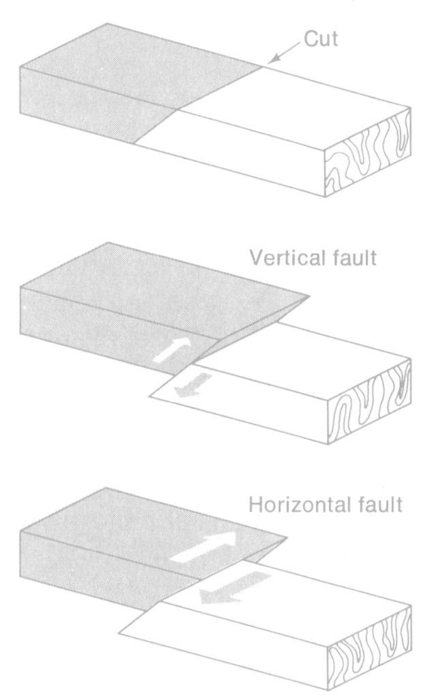

Figure 11/7 *This farmer's fence was built across the San Andreas fault. The moving land carried the fence with it.*

sliding across another. The earthquakes are felt on the surface.

Many other areas also have earthquakes along old breaks in rock layers. After the first break, there is likely to be further vertical or horizontal movement of the rock layers. Movement like that seen in Figure 11/7 is horizontal. When there is further movement it means that the fault is still active. The shaking that occurs when the rocks on either side of the fault are ground together can be just as destructive as the shaking caused by the first break.

11.4 Are earthquakes, volcanoes, and mountains all related?

Compare your map of earthquake locations with the map of volcano eruptions you plotted in Section 10.7. What do you find? It is easy to imagine that the great pressures forming volcanoes are also strong enough to shake the earth.

Many people who have seen volcanoes erupt have also felt the ground shake before, during, and after the eruptions. Sometimes the earthquakes started only a few minutes before the eruptions. However, on the island of Hawaii there were earthquakes for 11 months before lava started to pour out of Kilauea Volcano in February, 1955. Even without a map, people living near Kilauea could tell you that some earthquakes must be related to volcanoes.

You know that pressures in the earth can crack the outer layer, so that lava, ash, and steam can erupt from volcanoes. The same pressures can also bend thick layers of hard rock. They can move large areas of rock up and down and from side to side. Maybe all this has something to do with mountains.

Process: interpreting data

Accurate measurements have shown that the land in most of the areas where earthquakes occur is rising a few inches every ten years. In some places, as you have seen, sections of land have been known to rise or drop many feet during a single earthquake. Something must be happening deep within the earth in earthquake areas. Figure 11/8 shows how rising land, volcanoes, and earthquakes may be related.

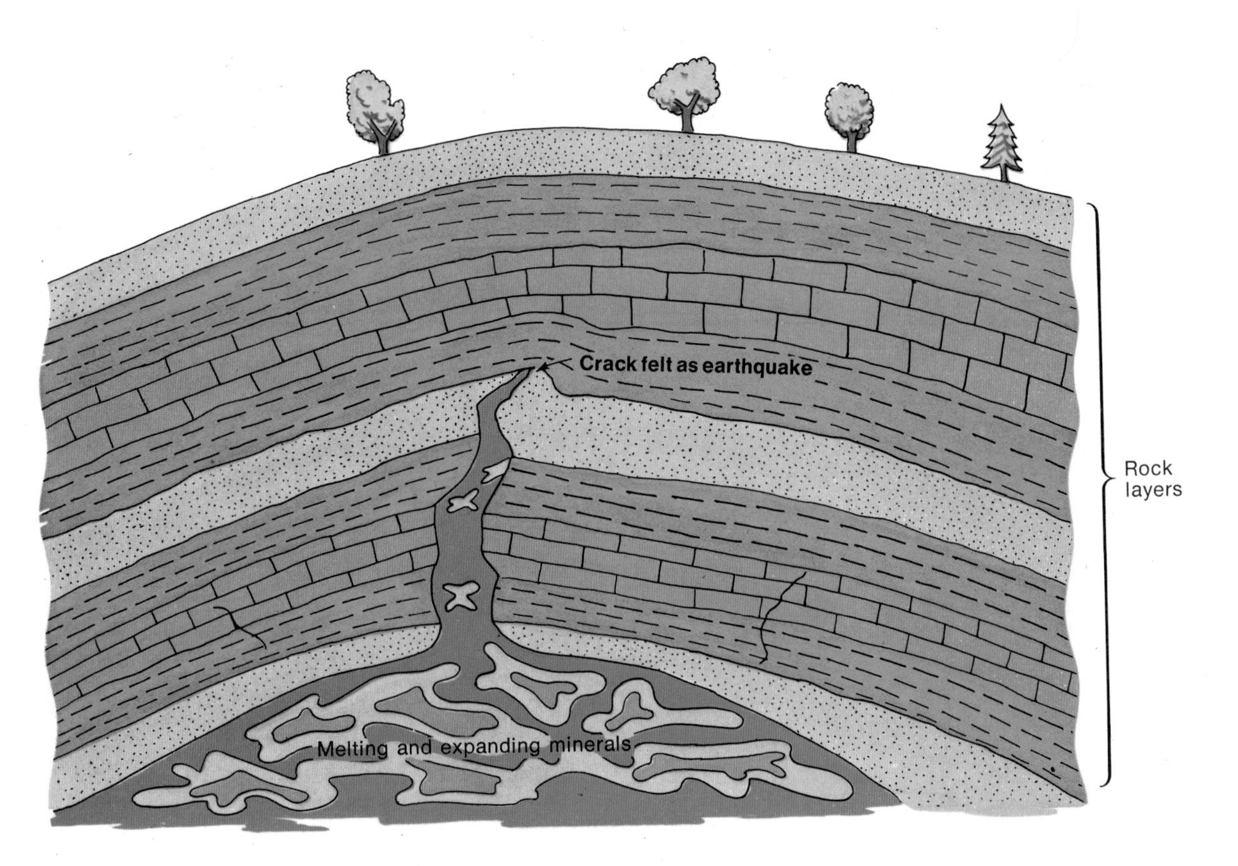

Figure 11/8 *Minerals melt deep within the earth and expand. The land above is pushed up and the pressure breaks rock.*

Can you see further evidence in Figure 11/9 that rising land, volcanoes, and earthquakes happen together? The heavy lines show areas where mountain building has gone on in the past or is still going on. Mountain building is land rising in a big way. Compare Figure 11/9 with your earthquake and volcano maps. What force could cause mountain building, volcanoes, and earthquakes in the same area?

Did You Get the Point?

Earthquakes have caused great damage in many places around the world.

Earthquakes occur when land masses break and slide past one another.

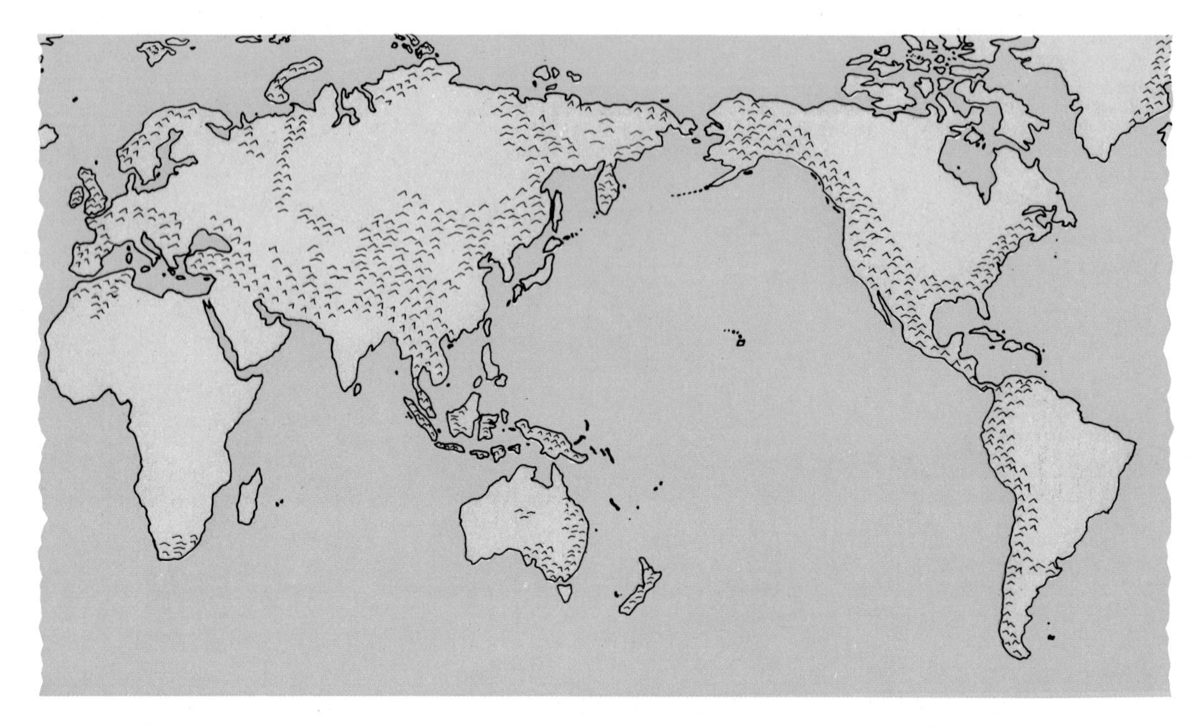

Figure 11/9 *Compare this map of mountains around the world with your maps of volcano and earthquake locations.*

During an earthquake, layers of rock may become bent without breaking.

Earthquakes may cause gigantic waves called tsunamis.

When earthquake focuses are plotted on a graph, they trace out a pattern. This pattern indicates that some force is acting beneath earthquake zones.

When many earthquake epicenters are plotted, they trace the same pattern as volcanoes.

Volcano and earthquake patterns match mountain-building patterns.

Answers:
1. See maps in Teacher's Manual.
2. The positions of moving fingers beneath the sand mark focuses. The places on the surface directly over the buried fingers mark epicenters. Faults will appear as areas of damp sand break. The pattern of the "mini" mountain ranges will reflect the zone of faults and earthquakes.
3. Berkeley Stadium in California is one example of a structure built across a fault. You might mention the recent interest in the fact that some Californian housing developments have been built across faults.

Check Yourself

1. Draw a map of an imaginary town in a major earthquake zone. Draw a second map showing the changes that might take place in the town as a result of a severe earthquake.

2. Imagine a hand buried in damp sand. The fingers are slowly raised one at a time. How can these moving fingers be used to model what happens in an earthquake zone? Can you find focuses? Can you find epicenters? Can you find a pattern? What about mountain building?

3. Write a story about someone who owns land or has built a house across a fault like the one shown in Figure 11/6. Your story can be funny. Be sure you mention an earthquake or two. Some true stories about earthquakes can be found in the school or local library.

Detecting and Recording Earthquake Disturbances

11.5 Sensitive instruments detect earthquakes.

Over 1 million earthquakes occur on the earth every year. Fortunately, only about 700 of them are strong enough to cause any damage or deaths. Even so, enough occur to make us worry about people who live in earthquake zones.

Naturally the towns, villages, and cities nearest the epicenter of an earthquake are shaken most strongly. People in places far from the epicenter may not feel the earthquake at all. For this reason, scientists depend on sensitive instruments, called **seismometers** (size-MOM-uh-terz), to tell them when an earthquake has occurred. These instruments are usually placed at the bottoms of old mines or wells and fastened securely to unbroken rock. This prevents the seismometer from mistakenly "detecting" trains and trucks.

A seismometer consists of a heavy weight, a spring to support the weight, a pointing needle, a measurement scale, and a sturdy base. The base and scale are fastened to the rock floor of the mine. The spring, the weight, and one end of the needle are free to move.

Materials

rubber bands, weight

Activity Time: 10 minutes. Allow students to get the movement just right. It will take some practice.

★ To get a better idea of how this arrangement can "feel" earthquakes, suspend a heavy weight from a rubber band held in your hand. Let the weight come to rest. Can you get the weight to stay perfectly still? Now, imagine that your hand is shaken by an earthquake. Move your hand up and then down a short distance just once. Watch the weight.

Notice that it takes a definite length of time for the weight to make one complete bounce. The time for one bounce is called a **period.** Compare your time with other people's results.

Now, move the hand holding the rubber band up and down faster than the period of the weight. Can you move your hand up and down fast enough so that the weight almost stops moving? This is what happens to the weight in a seismometer during an earthquake. ★

Yes. If the hand moves fast enough, the weight does not have a chance to react much in either direction.

When the ground shakes from an earthquake, the case of the seismometer moves up and down just as your hand did. Each up and down movement of the rock causes one up and down movement of the base that has been fastened to the rock. This movement takes less time than one period of the weight on the spring. As a result, the weight stays almost where it is. Since one end of the needle is fastened to the base, the needle will move like a lever. Figure 11/10 shows how. The moving end of the needle will point to some number on the scale.

Process: measuring

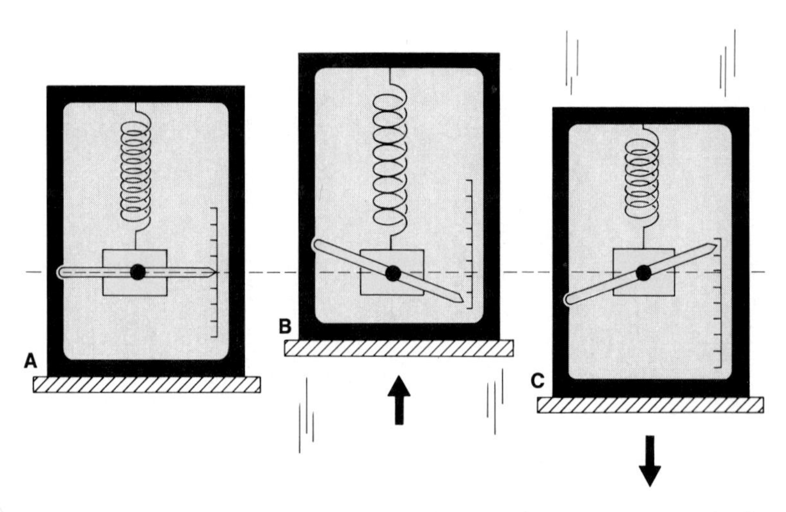

Figure 11/10 *Part A shows a simple seismometer at rest. In parts B and C, an earthquake is bouncing the box up and down but the heavy weight doesn't move. The pointer attached to the weight indicates the strength of the shaking.*

Electronic Seismometers and Seismographs

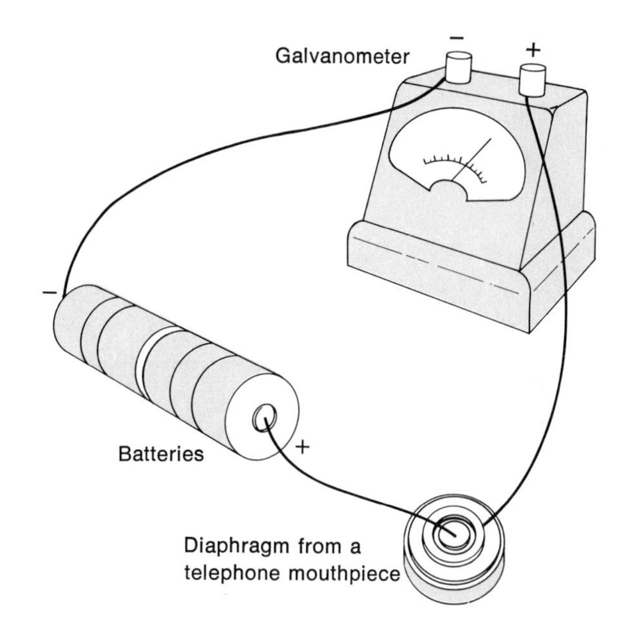

Most of today's seismometers and seismographs use electronic sensing devices instead of the mechanical spring and weight. The new devices are more sensitive because they change the earthquake shaking into bursts of electricity. Weak signals can then be amplified (made bigger) in the same way that the sound from the plucked string of an electric guitar is made louder. Only the sensing device of the electronic seismometer has to be placed in the mine shaft. Wires can connect it to the power source and meter many miles away.

You can make a working model of an electronic seismometer quite easily. All you need is two flashlight batteries, a galvanometer, and the mouthpiece from a discarded telephone. Wire them in series as shown in the diagram.

Fasten the mouthpiece securely to the floor or to the top of a heavy table. Weight it with a brick or pile of books or, if you put it on a table, hold it down with a clamp. Drop a heavy object or stamp your feet. Does this make the galvanometer needle move? The heavier the object or the closer it is to the galvanometer when it is dropped, the more the needle will move. See if your model seismometer will pick up a signal from the energy lost to the table in the activity in Section 11.8.

11.6 Sensitive instruments record earthquakes.

It might be helpful to compare a seismograph with a recording thermometer or barometer. Discuss the similarities and differences. If your school does not have a recording instrument, use a soup can and its paper label as a model of the drum and paper.

Most of the seismometers used today are designed so that scientists don't have to watch the machine to tell when an earthquake has taken place. A machine called a **seismograph** (SIZE-mo-graf) both detects earthquakes and makes a written record of them. It's all done with a pen, ink, a clock, and a metal can wrapped with paper. The clock gives the time when the earthquake occurs. Figure 11/11 shows a seismograph in action.

The strength of earthquakes is called their **magnitude.** It can be measured by the size of the waves they trace on seismographs. Magnitude runs from very weak (magnitude 1) to extremely powerful (magnitude 10). Newspapers often report this as the Richter magnitude of earthquakes. Another way to express the power of earthquakes is by the effects they have on people, buildings, and the land. This measure is called the **intensity** of earthquakes. It gives you an idea of what earthquakes actually do. Earthquake intensity levels run from very weak (intensity I) to very strong (intensity XII), as shown in Appendix W.

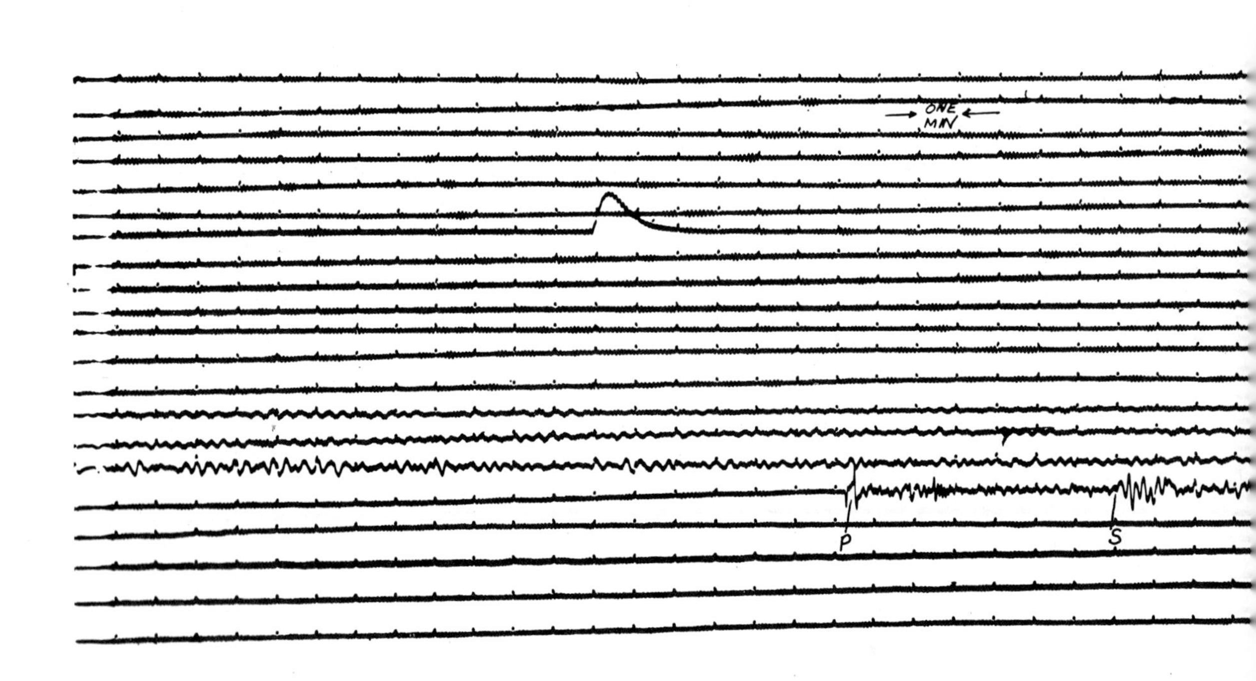

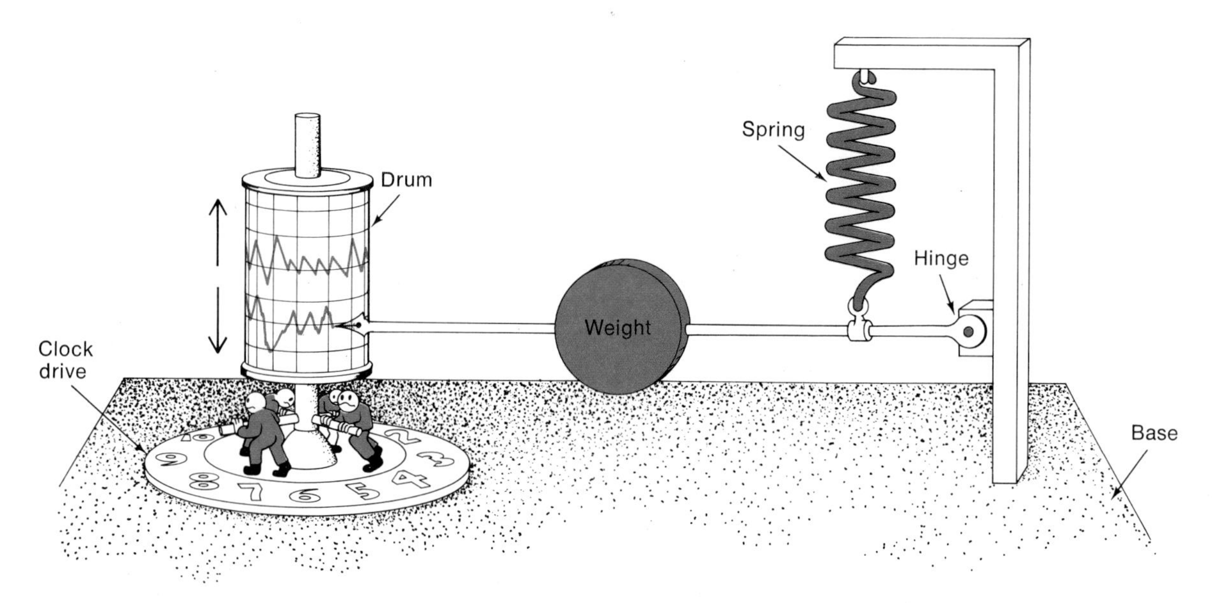

Figure 11/11 *The clock drive and the recording drum make this seismometer into a seismograph.*

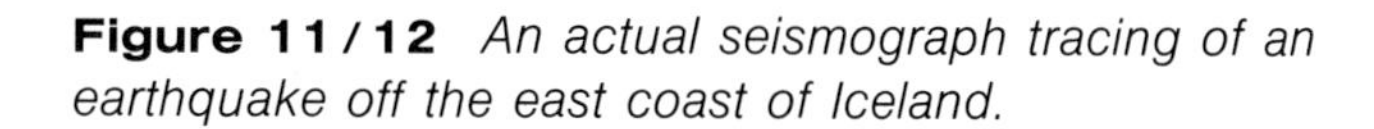

Figure 11/12 *An actual seismograph tracing of an earthquake off the east coast of Iceland.*

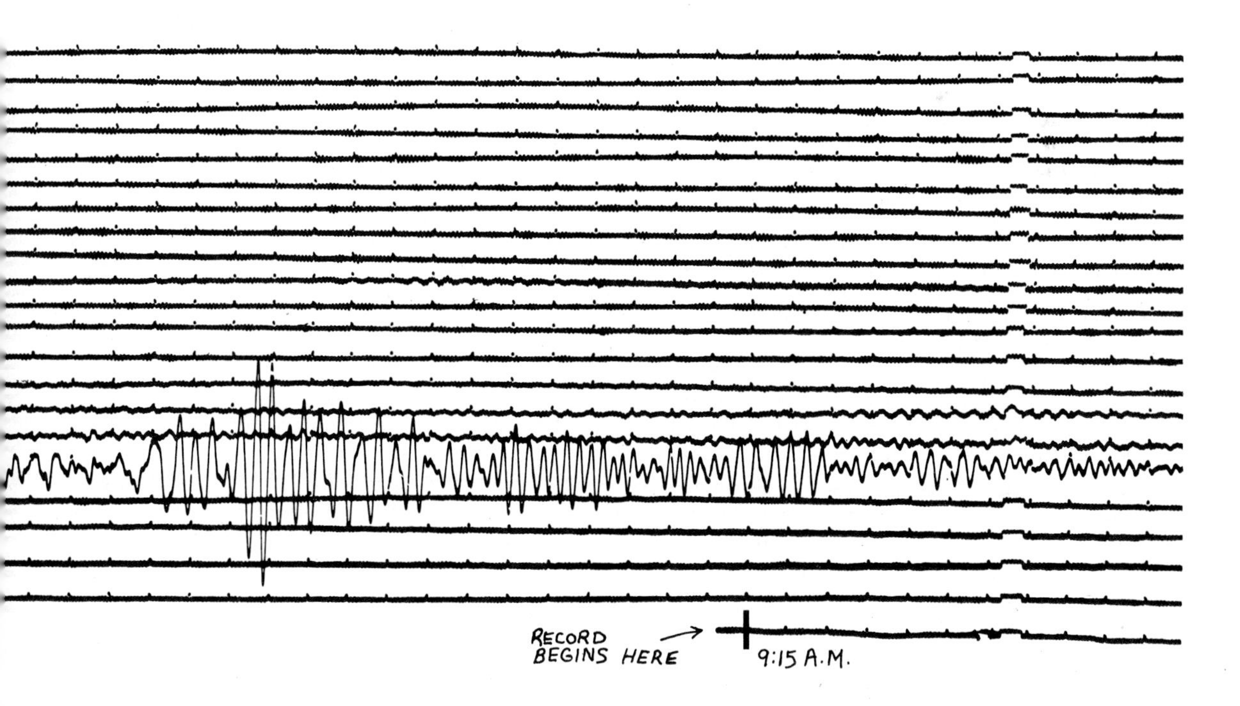

Figure 11/12 is the pen tracing taken from a real seismograph. Notice that the line the pen made is not smooth. The little bumps all along the line are simply a record of background vibrations going on all the time. These vibrations are like the shaking of your hand when you tried to hold the weight and rubber band perfectly still.

Process: measuring

At one place on the seismograph record, something out of the ordinary happened. Can you find the place on the line where an earthquake occurred? Suppose you were told that the record in Figure 11/12 started at 9:15 A.M. Can you figure out what time the earthquake waves reached the seismograph station? Try it! Each little up and down line is a minute mark.

Activity Time: 5 minutes. The quake occurred at 1:49 P.M. It is a lot of counting, but the variety of different answers will promote discussion.

Did You Get the Point?

Sensitive instruments are necessary to detect an earthquake at locations far from the epicenter.

The instrument used to detect earthquake disturbances is called a seismometer.

The instrument used to detect and record earthquake disturbances is called a seismograph.

Seismograph tracings can be used to tell the exact time waves from an earthquake disturbance reach the instrument.

Check Yourself

1. How are seismographs and seismometers different? How are they similar?

2. Make a model of a seismograph tracing as it would appear if it were still wrapped around the drum. Use it to show that the beginning of one line on the tracing continues where the end of the previous line leaves off. One way is to roll up a piece of paper into a cylinder.

Answers:

1. A seismometer can only detect earthquakes. A seismograph detects and records earthquake information. Consequently, a seismograph makes it possible to determine the exact time an earthquake occurred. They are similar in that both have quake detection devices.
2. Demonstrate the answer to the question by drawing a continuous line around a paper towel tube. Cut the tube lengthwise and open it. Lay it flat and compare it to Figure 11/12.

How Can an Earthquake Be Felt Far Away?

11.7 Energy is released when a rock layer breaks.

Any number of instruments can record the same earthquake, even though the instruments are at different distances and in different directions from the earthquake. The effects of an earthquake must travel out from the break. But how? Why?

Materials
wooden stick

Activity Time: 5 minutes.

★ A simple experiment will help you to see what happens when rock layers break. Ask your teacher to give you a wooden stick. It will be your model of a single layer of rock. Take one end of the stick in each hand and very carefully bend it. DO NOT LET THE STICK BREAK. Then let one end of the stick go.

You had to use a certain amount of energy to bend the stick. If you have ever bent a wire coat hanger and straightened it out again, you know that it takes as much energy to straighten it as it did to bend it. The hanger will remain bent until enough energy is applied to straighten it. But what happens when you let go of the stick? Where did the energy come from to do that? The stick stored the energy you put into it and used this stored energy to undo what you did. It would have stored your energy as long as you held it in the bent position. But there is a limit to how much energy the stick can store.

Activity Time: 5 minutes. Save all broken pieces. More energy was put into the stick than could be stored.

Take the stick and hold it again as before. This time, bend the stick until it breaks. In terms of energy and storage what happened?

Just as before, the stick stored up the energy you put into it to bend it, as a balloon will store up air you put into it. Eventually, you reached a point where the stick could hold no more, just like the balloon again. When you "overfilled" the stick with energy, it broke. All of the stored-up energy left at once. None of it was saved for straightening out the stick.

Where did all the energy go? ★

11.8 The released energy shows up elsewhere.

Now, fill another stick with energy, but this time keep a closer watch on what happens to the energy when it escapes.

Materials

salt, wooden stick

Activity Time: 10 minutes. You might point out that the table and the air absorbed some energy when the stick broke. Otherwise the breaking would not have been heard. The students should use safety goggles for this activity.

Concept: energy in motion

★ Lay a stick over the edge of a table as shown in Figure 11/13. Hold it in place while a classmate sprinkles some salt crystals on the end resting on the table. While holding the stick tightly to the table, push the free end down until it breaks. What happens to the salt crystals? What would have happened if you had used a whole pile of salt? Use another stick to find out. ★

This is the elastic rebound hypothesis that states that when rock bends to its elastic limit, it breaks. However, this explanation is only valid for earthquakes with shallow focuses (in solid material). Recent research has shown that deep focus earthquakes also take place in the plastic material of the core. Scientists theorize that the material ruptures or "jerks" every once in a while to cause earthquakes.

It takes energy to move a salt crystal. When you released the energy the stick was storing, it must have traveled down the stick to the salt. The energy made the crystals jump the way the case around a seismometer would. Maybe you felt some of the energy in your hands when you held the stick and broke it. The same thing happens when a baseball bat stings the hands of the batter, or when a rock striking one small corner of a window breaks the whole pane.

Rock layers are much bigger than your stick, and usually more than one rock layer is involved in an earthquake. The energy released when these rock layers break is millions of times greater than the energy released when you broke the stick. Sometimes when rock layers break, enough energy is released to send earthquake energy two or more times around the earth. When this happens, one seismograph will record the same earthquake twice. And when this much energy is released, much more than salt crystals will be moved!

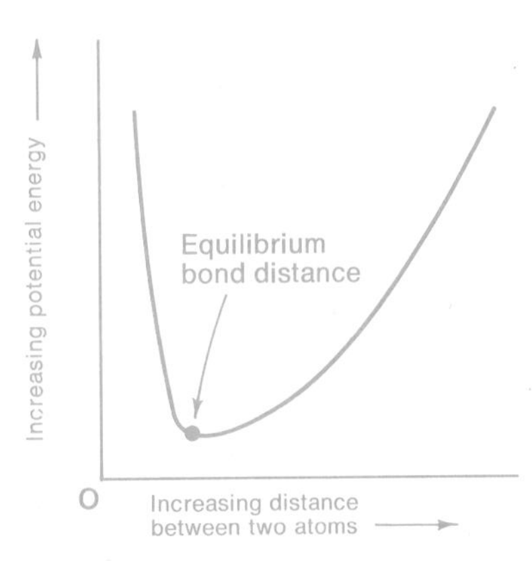

When the chemical bonds holding a mineral together were formed, a bond length was established. The bond is strongest when the potential energy of the atoms involved is least. The graph shows that squeezing or stretching the bond will increase the potential energy of the material.

Concept: particulate nature of matter

11.9 The energy is passed from one particle to another.

A rock is hard because the particles that make it up are fastened to one another very strongly. When enough pressure is applied to a rock to make it bend, the particles are pulled farther and farther apart. When the rock breaks, the particles on either side of the break snap back together again.

Figure 11/13
The salt will show what happens when the stick breaks and energy is released.

Demonstration Time: 2 minutes. Try this with real dominoes, matchboxes, blackboard erasers, or a row of student volunteers. If bodies are used, don't push too hard. The class may get the point without actually carrying out "the threat."

See diagrams of earthquake wave transmission in Teacher's Manual.

When you stretch a rubber band until it breaks, the two pieces snap back in the same way. The particles in the pieces of rubber band that snap back crash into particles already there. These particles then crash into particles behind them, and these into still others farther down the line. Think of a truck crashing into the last car of a long row of cars lined up bumper to bumper in a traffic jam. Even the car at the head of the line will feel the crash.

Figure 11/14 *This tree swings back past its upright position before coming to rest. Rows of rock particles do the same thing when an S wave passes through.*

The more particles pulled in the stretching, the more violent the snapback. The more violent the snapback, the more destructive the earthquake. In fact, some people in earthquake areas get a little scared when there hasn't been an earthquake in a while. To them it means that the energy that could have caused several small earthquakes is being stored up. When the break does come, the earthquake may be a big one because many more particles have been pulled and stretched.

11.10 Energy can be passed along by rows of particles.

The energy from an earthquake can get from place to place in another way. During the bending, whole rows of particles may be pulled upward or downward. Suddenly the rock breaks. But, like the small tree pulled down to one side and let go in Figure 11/14, the rows of particles next to the break zip right past their normal positions. And, just like the tree, the row of particles in the rock will keep swinging up and down before coming to rest. During this up-and-down movement, the strong forces between the rock particles have dragged rows of particles next to them up and down.

Materials
rope

Activity Time: 15 minutes. Send energy down the hall. *Each* student should have his turn at it. There will be no question that energy did move if each student feels the jerk the wave makes when it reaches the end of the rope. The "S" action can be illustrated by putting the rope on the floor and snapping it like a whip. This will show the same plucked wave in slow motion.
Point out that each time the wave makes a trip, it loses energy and becomes a little weaker. In the same way, the stick lost energy to the air and table. Similarly, an earthquake's energy is soon dissipated.

★ You can pass energy from one place to another in the same way with a rope. Ask two of your classmates to pull on opposite ends of a rope. The rope must be pulled tight. Then, pluck the rope downwards once, as you would pluck a guitar string. Can you see signs of the energy moving down the rope and back again? ★

When you allowed the rope under your finger to snap up, it snapped the section of rope next to it, too. This section snapped the section of rope farther along, and so on down the rope. Energy from an earthquake can be carried to seismographs all over the world in this way.

11.11 Moving earthquake energy is called a wave.

The energy released when a rock layer breaks travels in waves. You read about energy waves traveling through space in Section 2.6. However, earthquake energy can't travel through the vacuum of space. It must have a "touchable material" between the source and the place where it is felt.

The terms *longitudinal* and *transverse* or *compressional* and *shear* can be used instead of push-pull and shake.

There are two types of waves. When rock particles are running into each other, earthquake energy is traveling in push-pull waves. This is similar to what happens when a truck crashes into a row of cars in a traffic jam. When rows of rock particles are pulling up and down on neighboring rows of particles, the earthquake energy is traveling in shake waves. The shake waves are like the ones that travel down the stretched rope. The tree in Figure 11/14 passes on its energy in this way.

Shake and push-pull waves both travel through the interior of the earth. Compare these to Figure 2/17.

These waves have shorter names too. The push-pull wave is called a **P wave.** The shake wave is called an **S wave.** The initials *P* and *S* come from the words *P*rimary and *S*econdary. The P, or primary waves, involve the movement of fewer particles. So they are the first to arrive at a seismograph station. The S, or secondary, waves involve the movement of a greater number of particles. So they are second to arrive at a seismograph station.

Students can trace the seismograph tracing and mark the P and S waves on it themselves.

Look back at the seismograph tracing in Figure 11/12. Can you locate the P and S waves on it? How much slower was the S wave? Other types of waves travel on the surface of the earth where the forces between the particles are not very great (in soil, for example). Thus they always arrive after the S and P waves. Notice the pattern that each type of wave makes on the tracing.

The surface waves are sometimes called L waves (L for long) or Rayleigh (RAY-lee) waves.

Did You Get the Point?

Energy is released when a rock breaks.

The energy released during an earthquake can be passed from rock particle to rock particle.

A push-pull or P wave is energy moving from place to place as rock particles crash into one another.

The energy released during an earthquake can be passed from one row of rock particles to the next.

A shake or S wave is energy moving from place to place as rows of particles slide past each other.

P waves travel faster than S waves.

Seismograph tracings show the arrival of different types of waves at seismographs.

Check Yourself

1. If P waves travel faster than S waves, why didn't the salt on the stick in Section 11.8 jump twice, once when the P wave passed and again when the S wave passed?

2. Some materials will not allow S waves to pass through them. What kinds of materials are they? Why won't they allow S waves to travel through them?

Answers:

1. The stick was too short. The P wave didn't have time enough to get far enough ahead of the S wave for us to notice the difference. The S wave arrived while the salt was in the air.
2. Gases and liquids don't allow S waves to pass through them because the rows of particles involved are not held together tightly enough to pull each other. We know empirically that the rows are not held tightly together because we can walk through gases and liquids.
3. S and P waves travel best through hard material. Unbroken rock at the bottom of a mine would most likely be connected to the rock in the area of the quake and therefore best transmit the energy released.

3. Now that you know about waves of energy and how they travel, give another reason for locating seismographs at the bottom of old mine shafts and for fastening them to solid, unbroken rock.

Determining the Epicenter of an Earthquake

11.12 S and P arrival times change with distance.

If a boy on foot races a boy on a bike, it is obvious who will win. Suppose that three different races are run: a short race, a medium race, and a long race. Suppose that in one of the races the boy on foot loses by 10 feet. In another one he loses by a mile. In the third he loses by 200 yards. Can you match the three different results with the three different races?

Short race—lost by 10 feet. Medium race—lost by 200 yards. Long race—lost by a mile.

Process: using space/time relationships

This is how scientists tell how far away an earthquake recorded by a particular seismograph occurred. The P wave, since it travels faster, is like the boy on the bicycle. The S wave is like the boy on foot. The longer the race or the farther away the earthquake occurred, the farther ahead a P wave gets.

11.13 Three seismographs can determine the epicenter of an earthquake.

From Figure 11/15 you can tell the time it takes a P wave to get to a seismograph. The time depends on the distance of the seismograph from the place where the earthquake took place, the epicenter. Figure 11/15 shows how long it takes an S wave to get to a seismograph under the same conditions. Which kind of wave takes longer to travel a given distance, say 4000 kilometers? What is the difference between the time the P wave arrived and the time the S wave arrived after they had traveled the 4000 kilometers?

S wave

Materials
paper, compass

★ Suppose that a seismograph at position A on the map in Figure 11/16 recorded a P-wave arrival and an S-wave

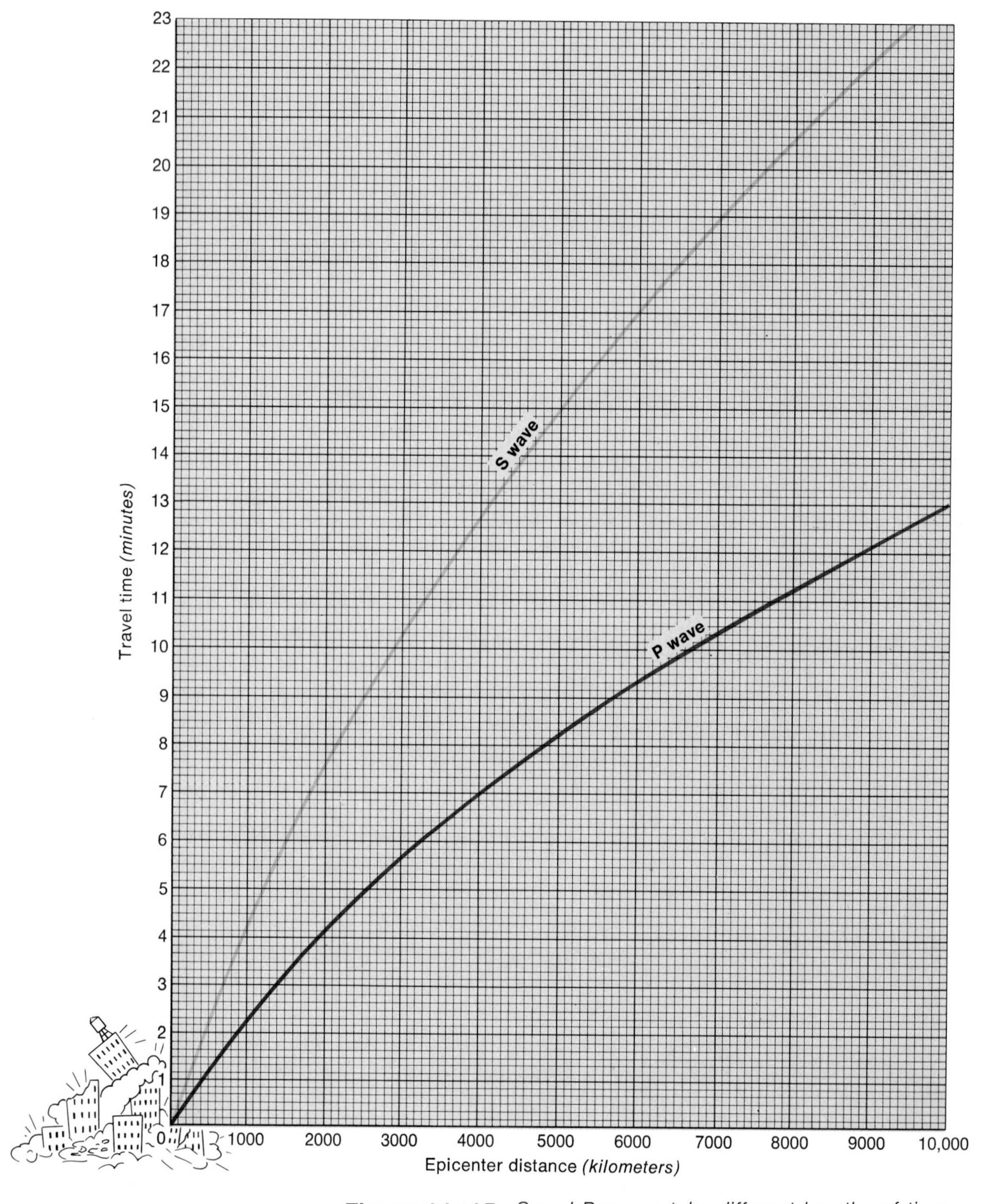

Figure 11/15 *S and P waves take different lengths of time to travel the same distance from an earthquake.*

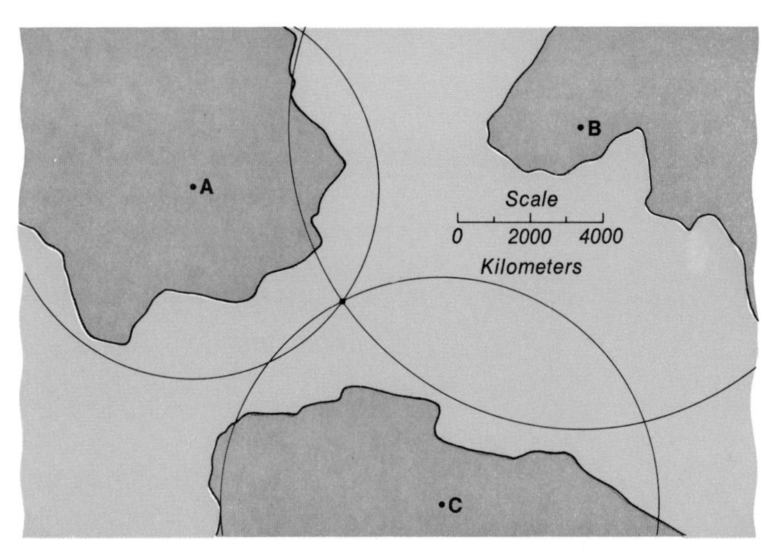

Figure 11/16 *Seismograph readings tell the people at A, B, and C that an earthquake occurred somewhere on the circle around each location. Where did the earthquake take place?*

Activity Time: 30 minutes. Expect some variation in students' numbers. Some students will be more careful than others. The map is small enough that errors won't show up when scaled.

arrival 6 minutes and 40 seconds apart. Using the graph in Figure 11/15, you can determine how far away the break in the rocks must have been. To do this, lay a piece of paper along the time divisions on the left edge of the graph. Along the edge of the paper, make two dots marking the difference between the zero-minute time mark and the 6-minute and 40-second time mark. Then find the points on the two curves on the graph where this distance on the paper just fits up and down. By following this vertical line to the bottom of the graph, you can read the distance the earthquake must have been from the seismograph. ★

5000 km

The circle around position A on the map locates all the places that are that number of kilometers from the seismograph. The earthquake must have occurred somewhere along that circle.

If two other seismograph stations also record arrival-time differences for S and P waves from the same earthquake, the epicenter of the earthquake can be determined. If the difference between the S wave and the P wave arrival times for station B was 9 minutes and 20 seconds, what was the distance of the S and P race? The circle around B in Figure 11/16 shows all of the places that many kilometers from

8000 km

6000 km

If students are interested in trying determinations on their own, make up a problem or two for them. Provide compasses. You may wish to try the investigation in *Investigating the Earth* (Houghton Mifflin); Chapter 16 of the 1967 edition and Chapter 13 of the 1973 edition.

You may wish to use Houghton Mifflin Co. Overhead Visuals—Earth Science #17 to illustrate locating an epicenter.

B. Notice that there are only two places both the right number of kilometers from A and the right number of kilometers from B. The earthquake must have occurred at one of these two places.

If a third station records a difference of 7 minutes and 40 seconds between the S wave and the P wave arrival times, how far does the distance of the earthquake epicenter from C have to be? The circle of all points that many kilometers from C hits only one of the two possible epicenter sites. This must be the place where the earthquake occurred. It is the only location that fits all three sets of data at once.

As you can see, there is no real trick to locating the epicenter of an earthquake. As long as you can determine a difference in arrival times of S and P waves for three different stations, you can determine the origin of an earthquake. Ask your teacher to give you information for an earthquake and see if you can find the epicenter on your own.

Did You Get the Point?

The greater the distance between the epicenter of an earthquake and the seismograph recording the earthquake, the greater the difference between the S wave and P wave arrival times.

The distance of the seismograph from the epicenter of an earthquake can be determined by figuring out the time difference between S wave and P wave arrival times. If the distance from a seismograph to the epicenter of an earthquake can be determined for three seismograph stations, the earthquake epicenter can be located.

Check Yourself

1. Why can't a single seismograph tracing locate the epicenter of an earthquake? Why can't two seismograph stations locate the epicenter of an earthquake?

2. What would have been the S wave and P wave arrival time difference at a seismograph located 3000 kilometers from the epicenter of the earthquake shown in Figure 11/16?

Answers:

1. A single seismograph tracing could represent many epicenter locations. Two tracings narrow the possible locations to two.
2. After the quake: 5 minutes 40 seconds for P, 10 minutes 10 seconds for S. Therefore, the difference is 4 minutes, 30 seconds.

What's Next?

The story of newly formed rocks and earthquakes and volcanoes is far from over. As soon as new rocks have been forced to the surface, shaking and scorching the earth's crust in the process, as soon as old rocks have been raised into mountains, other forces take over. The rocks begin to fall apart.

Some rocks break down fast. Others break down so slowly that only the results can be seen. Because much of the falling apart depends on the rain, snow, heat, and cold, the breakdown is sometimes called weathering. And, only the rocks that remain deep within the earth can escape it.

Skullduggery

Answers:
1. earthquake
2. epicenter
3. patterns
4. depth
5. mountains
6. horizontal
7. fault
8. weight
9. minute marks
10. period
11. energy
12. waves
13. push-pull
14. shake
15. seismograph
16. three
17. time difference
18. landslides, folded, streets

Directions: The answer to each of the questions is given on the line above the question, but you have to unscramble it. Read the question *before* trying to unscramble the answer. Write your answers on a separate piece of paper.

1. treqkahuea
The release of energy when a rock breaks is called an ______.

2. eienrctpe
The place at the earth's surface directly over the location of an earthquake is called the ______.

3. astptrne
Both volcanoes and earthquakes occur in ______.

4. etdph
Information about the ______ at which earthquakes occur helps to determine what causes them.

5. usamniotn
The formation of ______ is part of the earthquake-volcano pattern.

6. rothinaozl
The ______ movement of rocks can cause earthquakes too.

7. utfla
The break in the rocks along which movement occurs is called a ______.

8. egtwih
The part of the seismograph that does not move is the ______.

9. iumtne ramks
______ ______ are used to tell the exact time an earthquake took place.

10. ripoed
The amount of time for one swing of a weight is called a ______.

11. ngeery
______ is released when a rock snaps.

12. avswe
The energy released along a new fault travels in ______.

13. uhsplulp
______ waves are sometimes called P waves.

14. esakh
______ waves travel slower than P waves.

15. poissgaemrh
The instrument used to detect and record earthquakes is called a ______.

16. heetr
______ earthquake tracings can tell us where the epicenter of an earthquake is located.

17. meit fedfeircne
As the distance traveled increases, the ______ ______ between S and P wave arrival also increases.

18. ndlsialesd, oedfld, rseetts
______, ______ rocks, and broken ______ are all evidence of earthquakes.

Directions: Use Figure 11/15 to help you solve the following problems:

1. 4400 km
2. 1400 km
3. 3100 km
4. 9200 km
5. 40,000 years (400 centuries)

1. If it takes 7 minutes and 30 seconds for the P wave to reach a certain seismograph, how far from the seismograph must the earthquake have occurred?

2. If it takes 5 minutes and 20 seconds for the S wave to reach a certain seismograph, how far from the seismograph must the earthquake have occurred?

3. If the time lag between the arrival of the P and S wave is 4 minutes and 40 seconds, how far away is the epicenter of the earthquake?

4. Seismographs A and B are on the same line of latitude. An earthquake occurs between them. The time lag between the S and P wave at A was 5 minutes and 20 seconds. The time lag between the S and P wave at B was 7 minutes and 10 seconds. How far apart are A and B?

5. A certain railroad bed is rising at the rate of 2.5 meters each century. If the rate of rising remains constant, how long will it take the bed to rise one kilometer?

For Further Reading

Iacopi, Robert. *Earthquake Country.* Menlo Park, California, Lane Magazine and Book Company, 1969. You learn about the great fault areas in California and see excellent photographs of what earthquakes and faulting can do.

Matthews, William H., III. *The Story of Volcanoes and Earthquakes.* New York, Harvey House, Inc., 1969. Eyepopping photographs of the damage from earthquakes.

See also two articles about the Alaska earthquake in the July 1964 *National Geographic.*

Unit Four
The Changing Crust

Chapter 12 Weathering

Most people have to learn things the hard way. Take Janet, for example. She finally had to give up on her flower bed. Pieces of concrete from the wall next to the flower bed kept falling on the plants. A daisy would have to wear a helmet to take that for long.

On their way to a vacation spot along the coast, Steve and Peg had to drive through a desert area. Suddenly a violent sandstorm came up. When it finally died down, they found to their surprise that the windshield was pitted and scratched.

Mr. Morris, a local clergyman, has a real problem. Last winter a fire destroyed all the town records. He can only replace the oldest ones by copying information from old gravestones. But the oldest stones have been worn almost completely smooth. Minerals in the newer stones have rusted and run down over the letters. Mr. Morris will never be able to make out all the names and dates.

Even the mayor is in trouble. Many town sidewalks need to be replaced this year. He didn't plan on sidewalk repairs in the town budget. Guess the mayor thought the sidewalks would last forever.

You have to feel sorry for Tom, too. His mother has a small garden with stones around the edge. The moss she planted alongside the stones keeps growing up over them. So she makes Tom trim it back. You should see what the moss does to the stones—makes them all rough. You guessed it! He gets blamed for not taking care of the stones. "Polish those stones," his mother yells. "And water the moss." All in the same breath. What a grouch!

Causes of Weathering

12.1 Rocks must weather.

What do the falling cement, the pitted windshield, the aging gravestones, the rusting minerals, the flaking sidewalks, and the rough garden stones all have in common? The answer is weathering. **Weathering** is breaking down of earth materials into smaller pieces or changing them chemically. What results may be the same as the original material or may be changed into a new material. Without weathering there would be no soil. There would be no sandy beaches. All home cellars would have to be dug with dynamite.

So far you have learned about changes on the earth that build up the landscape. Volcanoes pile up new rock. Movements in the earth's crust raise up new mountains. But rocks and mountains must be broken down at nearly the same rate somewhere else. Otherwise the earth would become only new rock and all mountains.

12.2 Weathering results from natural disorder.

Institute a discussion of entropy here, if you wish.
Concept: the nature of equilibrium, the universality of change

When we look around us, many things seem to be in the process of becoming disordered. This "mixed-up-ness" happens all the time. In spite of the fact that weathering has helped to give us the soil we depend on for growing food, weathering is one of those things that annoys us. Like rain that brings an end to a picnic, watches that run down, pizza that gets cold before we can get it home, and rooms that have to be cleaned once a week, there seems to be no way to stop it from happening. And the reason is quite simple. Weathering is just another one of those things following the natural tendency for things in the universe to lose energy and become disordered.

Weathering causes the earth to become disordered. It happens naturally, without the aid of people. The things that do the actual weathering, called **weathering agents,** break down the hard rock. Weathering agents are usually water, wind, plants, animals (including man), and chemi-

Figure 12/1 *The pile of small rocks at the bottom of the hill is a talus.*

cals. Then a force like gravity moves the weathered particles so that fresh particles will be exposed to the agents. The work of gravity can easily be seen at the foot of a cliff. The pile of rock that collects there is called a **talus** (TAY-lus). Can you see the talus in Figure 12/1?

Anyway, Steve and Peg's windshield, Mr. Morris's gravestones, and the mayor's sidewalks were simply "doing what comes naturally."

Did You Get the Point?

Weathering breaks down earth materials and may change them into new materials.

Weathering happens naturally and causes the earth to become more disordered.

Weathering is caused by water, wind, chemicals, and plants and animals.

Check Yourself

1. Certain governments have put together "time capsules," containing information about our civilization. They hope that this information in the form of tape recordings, pictures, and diaries can be preserved for study by future civilizations, no matter what happens to us. How would you go about preparing and storing such a time capsule so that weathering could not destroy the information?

2. If a scientist wants a piece of rock that is unchanged since it was formed, he digs deep into the earth for it or breaks open other rocks. Why?

Answers:

1. Put capsules at the bottoms of mineshafts. There, the temperature is constant, and water is limited in most cases. Suggest salt mines. The items to be preserved are encased in plastic or treated with antiweathering chemicals. Such preparations constitute the energy spent to make and maintain order.
2. The scientist is looking for unweathered rock, so he has to get a piece that has not been exposed at the surface. The unweathered rock has been protected from entropy except where it may have cracked.

Mechanical Weathering

12.3 Mechanical weathering makes small pieces out of large ones.

Concept: universality of change

Mechanical weathering of rocks produces smaller pieces of rock, or pieces of the minerals that make up the rock. Sandstone weathers into sand grains. The sand grains are usually grains of the mineral quartz. Granite weathers into grains of at least three different minerals. Look ahead to Figure 12/10 to see the result. Obsidian changes into small pieces of obsidian. If the pieces of rock are moved out of

Concept: interaction of matter, energy of motion

the way when they break off, weathering agents can continue to work on the new surfaces that are exposed.

When weathering agents break rocks loose from cliffs and road cuts, gravity pulls the rocks down. They may crash into other rocks on the way down the cliff or break when they hit the bottom to become even smaller pieces. The various weathering agents then work on the newly broken pieces.

During windstorms most coarse sand grains remain within a foot or two of the ground. Therefore, wind abrasion is not very effective at greater heights.

Wind cannot do much weathering on its own. But wind can pick up sand to grind off a rock surface the way sandpaper smooths a piece of wood. Most sand grains are very hard. When the sand grains are thrown against a softer surface, they act like thousands of tiny bullets or knives. A little more rock may be chipped or scraped away every time one of the grains hits it.

Deep rocks are uncovered by being naturally uplifted and exposed by weathering and erosion. Processes that uplift rock masses are covered in Chapter 15.

You have probably noticed that masses of "solid" rock usually have cracks in them, even when the rock is just uncovered and not weathered yet. Geologists believe that over long periods of time, when rocks are uncovered, they expand a little and crack. Figure 12/2 is a picture of a mass of granite that has cracked into curved layers due to expansion. When rock cracks like this, it is said to **exfoliate** (ex-FOH-lee-ate).

Figure 12/2 *The surface of this granite is peeling off in layers by a kind of weathering called exfoliation.*

Figure 12/3 *How do you think this large tree split the boulder? Small plants can also act as powerful agents of mechanical weathering.*

Plants growing on the cliffs may help the pile of talus to grow. As the tree in Figure 12/3 grows in a rock crack, its roots act like wedges and force the rock to split. Eventually, large pieces are broken completely free. Gravity pulls them down, and the rock breaks into even smaller pieces.

Boulders can be split without dynamite. Wooden rods can be soaked in water and hammered into holes drilled in the rock. When the plugs freeze, the rock splits.

Water is the agent that formed much of the talus in Figure 12/1. Figure 12/4 shows what happened to a full bottle of water that was placed in a freezer. One hour of cold and presto! A weathered bottle! What happens when water freezes in a crack in a rock?

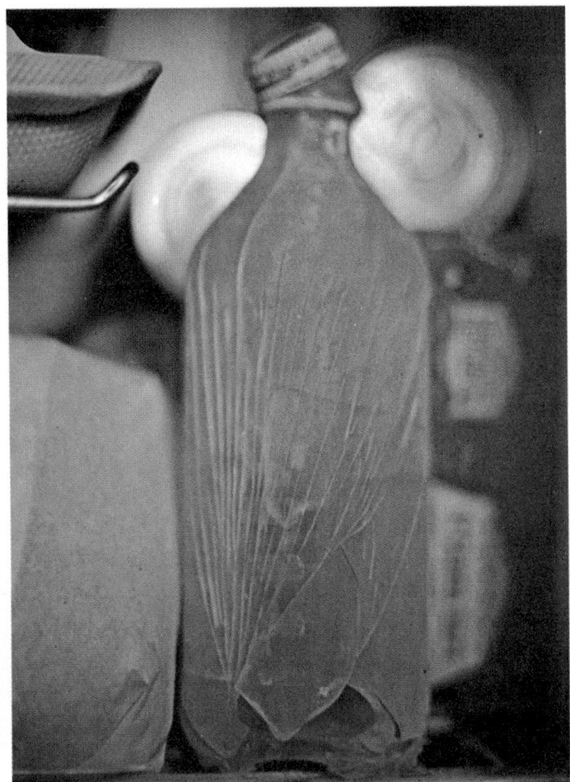

Figure 12/4 *Freezing water is strong enough to break rocks the way it broke this bottle.*

12.4 Mechanical weathering can happen under water.

Some weathering agents work quickly. Some work slowly. Some work every day. Others work only once in a while. How would you classify water as an agent of mechanical weathering? Does it work quickly? Can it work every day? How can it work on rocks only once in a while? You can try weathering with water yourself.

Activity Time: 40 minutes. Use fresh stones for each class. The edges must be sharp. *Investigating the Earth* (Houghton, Mifflin Co.) has a more sophisticated version of this activity. See notes in Teacher's Manual.

Materials

grease pencil, tape, limestone, bottle, cork, pill bottle

★ As a class project, select and wash enough small stones so that each person has 200 grams. All the rocks should be the same kind and about the same size. A box of crushed stone from a driveway will make your work easy.

Put most of the 200 grams of rock in a plastic bottle. Add water until the rocks are covered and stopper it. Save

You can bring in a discussion of controlling variables here.

the rest of the rock. When all the bottles have been filled and stoppered, your teacher will give each of you an even number from two to twenty. This is the number of minutes each person is to shake his bottle.

A record player and rock music work well.

Why is it important for each of you to shake your bottles at the same rate? See what you can come up with to help everyone keep the beat. Your teacher will signal you to start shaking and tell each shaker when it is time to drop out. Your class will look like the students in Figure 12/5.

You might suggest a control, to prove that the water alone did not do any visible weathering.

When your shaking time is up, swirl the water once more. Quickly pour enough of the water into a pill bottle to fill it. Label each pill bottle with the number of minutes it was shaken. Save the rocks that are left behind.

If possible, prop the bottles at an angle for the settling. This will make it easier to compare the amounts of sediment. Have students look at the stones that were shaken. Their edges should show some rounding. Compare them with fresh pieces of stone. Suggest that students look for pebbles that have been water worn in streams.

Arrange the pill bottles in order of the number of minutes they were shaken. Let them stand overnight. The next day your bottles should look like those in Figure 12/6. Which sample contains the most "stuff?" Has mechanical weathering taken place? How do you know? Compare the rock that was shaken to the rock that was not. ★

Process: inferring

12.5 Some weathering agents are more effective than others.

Think about the bottle shaking you did. Recall examples of weathering between school and home. Can you see that some mechanical weathering agents work better than others?

Look at the rocks in Figure 12/7. They are pebbles in a stream bottom. What would be the quickest way to turn your rocks into rounded pebbles? Would you put them in front of a fan to let the wind work on them? Would you put them in a flower pot and let the roots work on them? How long do you think it would take for plants and the wind to round the pebbles as much as your shaking did?

Rushing water in a stream or river does a fine job of mechanically weathering rocks, by banging them against each other and chipping off the corners and rough spots. Finally, they become more or less rounded. Sand in the water then polishes the pebbles.

If you have ever seen a river at flood stage, you can understand how water can break rocks. Even a polished

Figure 12/5
Shaking rocks in a bottle of water mechanically weathers the rock.

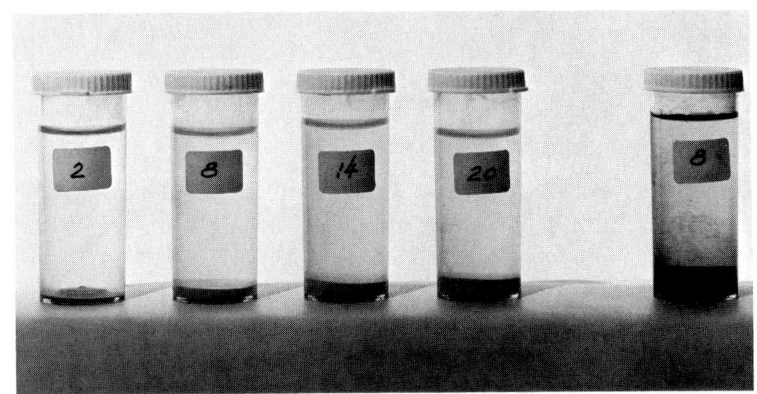

Figure 12/6
The four bottles on the left show the results of "classroom weathering" of limestone for 2, 8, 14, and 20 minutes. The bottle on the right contains weathered material from a soft shale.

Figure 12/7
These are pebbles in a stream bottom. What would be the quickest way to make other rocks look like them?

pebble might be cracked in two. Then the two pieces would go through the pounding process all over again.

Water does an even better job of weathering than you can see. The "stuff" on the bottom of the bottles was not all that was mechanically weathered from the larger pieces of rock. Some pieces dissolved in the water. Other pieces were too small to settle out. Evaporate some of the water from the twenty-minute bottle and see for yourself.

Mechanical weathering by plants can make small pieces out of large pieces, but the conditions must be just right. The wind is a good weathering agent only when it is carrying sand. But, water can work on rocks by falling as rain, and then freezing in cracks to split the rock. Water can roll and slide rocks against each other along a stream bottom. Water is a very good agent of mechanical weathering.

Did You Get the Point?

Mechanical weathering makes small pieces out of large pieces of the same material.

Moving or freezing water, wind, and plants are common mechanical weathering agents.

Some weathering agents are more effective than others.

Gravity helps mechanical weathering by causing rocks to fall. These rocks then chip and break other rocks.

Answers:
1. Examples are paint worn off, wood splitting, concrete floors and foundations cracking, concrete floors dusty (concrete breaking down), bricks cracking and chipping.
2. Leaves and trees will fall, wind will pick up soil, raindrops will wear soil away, freezing water in winter may crack logs and rocks, the tent will tear and fall apart if left.
3. Examples are sandblasting of buildings (wind carrying sand), sandpaper and files (rock ground against rock), tumbling barrels for rock polishing (streams carrying rocks), and a sledge hammer (rock falling on rock).

Check Yourself

1. List all the materials around your house that appear to be weathering mechanically.

2. Many of the things that will happen by themselves in the campsite shown in Figure 12/8 could be called mechanical weathering. List them.

3. Many of the tools and processes used by industry work like mechanical weathering agents. For example, a wedge will split wood just as roots split rocks. List some tools and processes that work like mechanical weathering agents and explain the agents they are like.

Chemical Weathering

Demonstration Time: 25 minutes. Heat the acid in an evaporation dish to show the powder. The powder (calcium chloride) will redissolve in the acid, but it will not bubble.
Concept: interaction of matter

12.6 Chemical weathering changes the rock itself.

If you put a piece of rock called limestone into some dilute hydrochloric acid, it will bubble as can be seen in Figure 12/9A. Soon the rock will disappear as shown in Figure 12/9B. It has weathered. If the acid is then evaporated, a white powder like that shown in Figure 12/9C will be left behind. Little pieces have been made from big ones, but these little pieces are quite different from the original rock. The new pieces may be a different color and will not bubble when put into dilute acid. Unlike mechanical weathering, chemical weathering changes the material itself, not just its size and shape.

12.7 Chemicals act as weathering agents.

Rainwater will join with some minerals in rocks by a process called **hydration**. The hydrated minerals, because they take up more space than the original minerals, cause the outer part of the rock to split into small pieces. That is exfoliation on a small scale.

Oxygen from the air or oxygen dissolved in water is another chemical weathering agent. Many rocks contain minerals that will combine with oxygen to produce compounds called **oxides** (ox-ides). When the original minerals are

Figure 12/8 *How will the different kinds of weathering affect this campsite?*

Figure 12/9 *Acid slowly dissolves limestone (A) until all of the rock is gone (B). When the acid is evaporated, a new material is left behind (C).*

changed to oxides and other compounds, they can become powdery or scaly. The oxide of iron is rust. Iron can be found in rocks, so it is possible for a rock to rust!

Many minerals can be broken down by acids. You may think that acids are hard to find in nature, but they are not. They are everywhere. Plant roots make acid while the plant is growing. Dead plants make another kind of acid as they decay. If you think of it the next time you are on a picnic, examine some moss-covered rocks or rocks with grass growing around the base. See if you can find any chemical weathering. If you push the moss back from the rock, you may see pitted places on the rock's surface. The pitted places can now be attacked by freezing water, and the rock will weather even faster.

See Background on calcium carbonate in Teacher's Manual for a more complete chemical explanation.

When a gas in the air, carbon dioxide, dissolves in water, it becomes carbonic acid. This acid is found in soda pop. The rocks containing the minerals called **carbonates** (CAR-buh-naytz) are easily weathered by acid. The acid changes the carbonate into a material that will dissolve in water. Limestone, which is mostly the mineral calcium carbonate, weathers this way. Mammoth Cave in Kentucky was formed by water weathering limestone underground.

Try to have some of the actual materials to display, such as feldspar, calcium carbonate, and quartz sand. Also put limestone in soda water.

The minerals in the feldspar family make up the largest part of igneous rocks. The feldspars can also be attacked by carbonic acid. The chemical reaction between the minerals and the acid forms special minerals called clay minerals, carbonates, and a solution of quartz. Figure 12/10 shows what happens when granite weathers.

Figure 12/10
Weathering of the minerals in granite produces sand, clay, and mica flakes.

Weathering—Natural and Artificial

You may think weathering of materials is a problem. Well, some people look for ways to weather woods for decorative reasons. Also, weathered woods for the outside of a house are often considered more practical than painting. Indoors, weathered wood is resistant to marks and scratches.

Nature will do the job within a matter of months or years, depending upon the particular wood. The process of natural weathering can be speeded up by hosing down the wood periodically during warm weather.

"Instant weathering" can be obtained with the aid of a wire-brush. This removes the soft spring wood and exposes hard summer grain. Or you can first burn the surface and then wire-brush to a deeper layer. This is most suitable for furniture or small wood objects.

If you want to weather large areas of wood you can paint a salt solution on them. This treatment must be repeated many times to give the desired appearance. Or you can bleach with a chlorine solution or with commercial bleaching oils. And finally there are weathering stains that imitate the effect of many years outdoors.

Sometimes man's artistic nature enables him to turn a destructive force of nature to his own advantage.

Janet's flower bed: mechanical and chemical. Gravestones: some hydration and attack by carbonic acid. Sidewalks: as with concrete wall above, some chemical and some mechanical. The last example is largely chemical.

Some of the statements at the beginning of this chapter described chemical weathering taking place. Can you pick out which ones are chemical weathering?

Did You Get the Point?

Chemical weathering changes the material itself, as well as making small pieces from large pieces.

Chemical weathering produces materials that are softer or more easily dissolved in water than the original rock.

Acids can act as chemical weathering agents.

Answers:
1. Examples are wood rotting, paint fading, chemical stains, rusting.
2. The trees will eventually rot, the boulder may weather chemically, the ax may rust.

Check Yourself

1. List all the materials around your house that appear to be weathering chemically.

2. Many of the things that will happen by themselves in the campsite shown in Figure 12/8 could be called chemical weathering. List them.

The Making of Soils

12.8 What does soil contain?

Mothers and soap manufacturers call it dirt. Poets call it earth. Men who have tried and failed to make it in the lab call it soil. All of them are talking about the "stuff" made by weathering.

Materials

soil, pill bottle

Activity Time: 5 minutes; let sit overnight. Try to provide a soil that has a wide variety of materials—some sand, some silt (clay), and some organic matter.
Process: observing, measuring

★ Look at some soil closely. Describe the different kinds of things that make up the soil. Can you find any pieces of plants? Is it sandy? Does it feel like clay? Put a few pinches of the soil into a bottle of water. Put your thumb over the top of the bottle and shake it up. The muddy cloud suspended in the water is called a **slurry.** Wait a few minutes and examine the bottle. Measure the thickness of each layer that settles out. Then allow the slurry to settle overnight.

How many layers have settled out now? How thick is each layer? Which layer is made up of clay? The sand should be easy to spot. Notice the layer of black material on top, if there is one. What do you think the black layer is? Is there anything floating on the water? If so, what is it? These layers are different materials that make up the soil. Soil looks like a simple mixture, doesn't it? ★

If the class is interested, see *Investigating the Earth* for photos of different soil profiles, as well as an investigation of the chemistry of a soil.

Actually, there are also chemicals in the soil that you cannot see just by looking into the bottle. Some of the chemicals dissolve in the water during the shaking process. But others can be found throughout the layers. The black layer alone contains thousands of different compounds, most of them from rotting plants. These hidden chemicals are probably the most important part of the soil because they are the food for growing plants.

12.9 Soil develops in steps.

How does a soil develop from a solid rock like that shown in Figure 12/11A? The first step is the breaking of the solid rock, sometimes called **parent material.** Continued weathering of the large pieces produces a mixture of smaller pieces of various sizes as in Figure 12/11B.

You may wish to introduce the term "horizon" for soil here. The topsoil layer is called the A horizon, the subsoil layer is the B horizon, and the weathered rock below is the C horizon.

Figure 12/11C is Figure 12/11B a hundred years later. Notice that the layer of weathered material is thicker. There are still large chunks of rock just above the solid rock. The rock chips at the top are finer than the chunks at the bottom because the weathering agents work more quickly near the surface, and they have been working longer. Also notice that many plants have taken root. As more and more plants grow and die, the decaying organic matter will darken the layer of finely weathered rock at the top. This layer is called **topsoil.**

After rainwater has had a chance to flow through the topsoil for a while, clay is carried down into the weathered rock layer, along with iron oxide. Clay is the part that contains the finest particles of soil. The **subsoil** shown in Figure 12/11D contains the clay and iron oxide. Many of

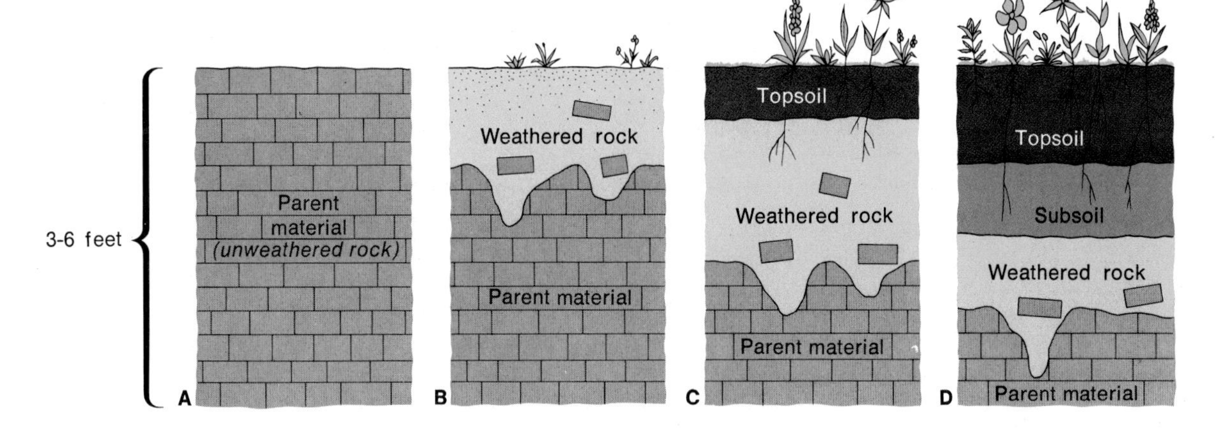

Figure 12/11 *A soil begins to develop when solid rock weathers into smaller and smaller pieces.*

the compounds at the top dissolve in the water as it runs down. This dissolving is called **leaching** (LEE-ching). If you dig a deep hole in the ground, you will almost certainly reach this layer of clay and compounds. The different layers in soil are shown in Figure 12/12.

12.10 There are many types of soil.

Soils forming on granite, limestone, sandstone, or clay-rich glacial till will differ from each other. Be on the lookout for soil layers exposed at construction sites.
Concept: universality of change

The type of soil in an area depends largely on the climate of that area and the type of parent material there. The climate determines the kinds of weathering agents and how strong they will be. Where the weather is cold and damp, as in northern Canada and the Arctic, mechanical weathering due to freezing and thawing is very important. But the topsoil does not develop well because of the short growing season and the few plants. Instead, we find tundra soil, where the deeper material is frozen all the time. In the tropics, where the climate is hot and damp, chemical weathering is very important. The layer of topsoil is usually thick and red, and contains a lot of clay. The soils in the United States are more or less in between these two kinds.

There are many different kinds of soils in the United States as the map in Figure 12/13 shows. The map shows general

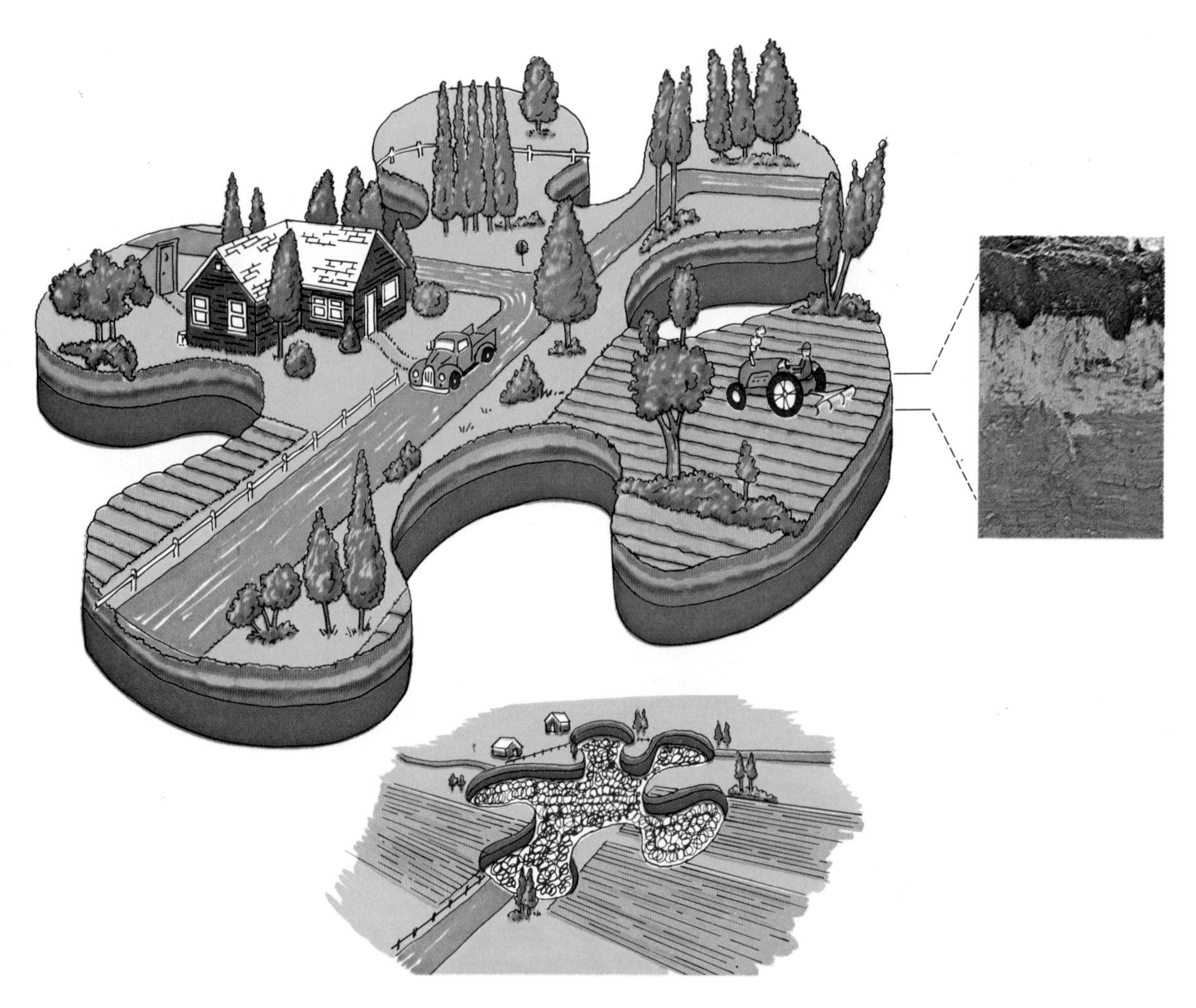

Figure 12/12 *Can you see the different layers of soil here? The topsoil layer is thick and dark.*

types. Desert soil is found in the areas of the western states. Water evaporates from the soil easily, often leaving a layer of calcium carbonate behind. As you would guess, where the climate is too dry for plants to grow in large numbers and too dry for much rotting anyway, there is not enough rotted plant matter to form a good topsoil. Prairie soil forms in the grasslands of the Great Plains and Middle West, where the climate is not so dry. A good, thick topsoil develops. In many places, it is rich and black and excellent for farming.

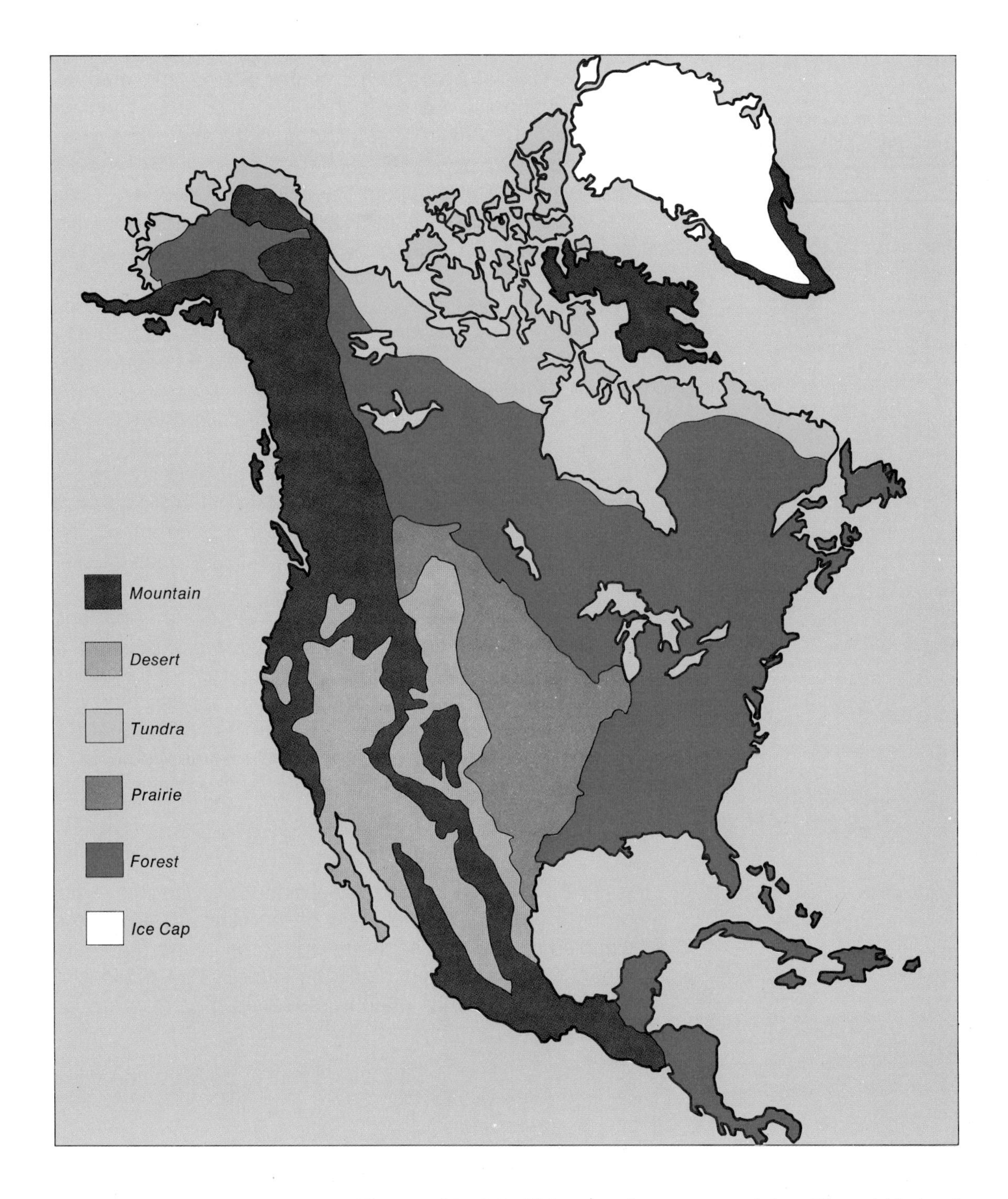

Figure 12/13 *This map shows some of the different kinds of soils that can be found in North America and Greenland.*

In the eastern part of the United States, the amount of rain and snow is greater than in the other areas mentioned so far. This greater precipitation helps the plant growth, and the soil type is called "forest soil." How would the greater precipitation affect the amount of leaching? Would the eastern soil have a well developed subsoil? Look again at Figure 12/12 and guess where the photograph was taken.

Mountain soils are usually rocky, and there are many different kinds, depending on the local conditions. Compare the soil map in Figure 12/13 with the climate map in Section 7.10.

Soil maps of local areas show that small areas can have a wide variety of soil types, too. Figure 12/14 shows a section of a soil map for the area around Syracuse, New York. The area covered by the section of map is about 256 square miles. Each color stands for a different soil type. Read the description of the types with the map. Look at all the different kinds.

Soil maps like the one in Figure 12/14 are important to many people. Not only scientists, but farmers, builders, and home buyers all need them. The farmer looks for dark soil containing large amounts of chemicals that can feed growing plants. Builders want soil that drains well so that foundations and cellars won't be damp. Some home buyers look for soil that will grow a garden and contains very little parent material.

Materials

soil sample, pill bottle

Activity Time: 20 minutes. If the students live in a city, they can collect the soil on a weekend trip, from local parks, from flower shops, athletic fields, or excavation sites. The layers that result from the shaking will vary in both color and thickness. Do not confuse this layering with the horizons. This is simply a crude analysis.

★ Bring in a soil sample and compare it with the samples brought in by other members of the class. Shake the soil samples in water to separate out some of the materials in them. Do the samples look the same? You could check to see if your results match a soil map of your area. ★

Did You Get the Point?

Weathering produces soil.

A soil usually develops slowly in several steps.

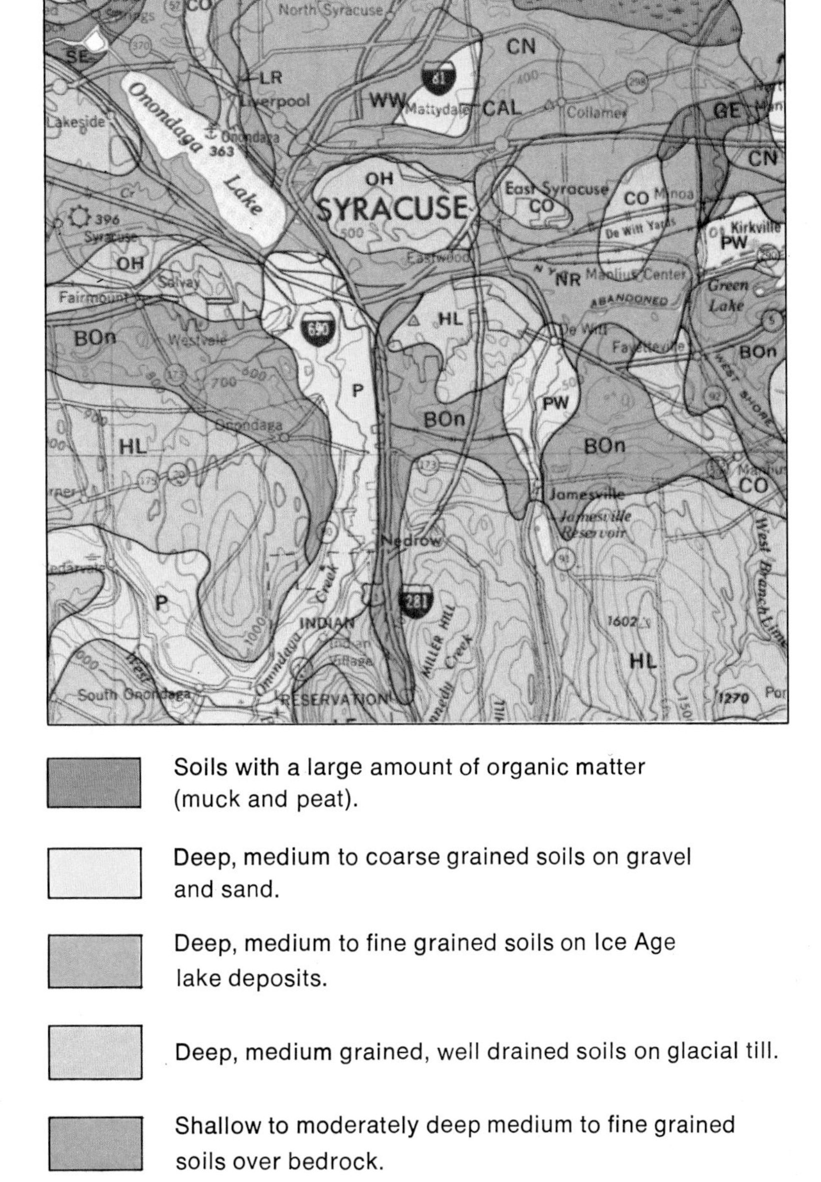

Soils with a large amount of organic matter (muck and peat).

Deep, medium to coarse grained soils on gravel and sand.

Deep, medium to fine grained soils on Ice Age lake deposits.

Deep, medium grained, well drained soils on glacial till.

Shallow to moderately deep medium to fine grained soils over bedrock.

Deep, medium grained soils on glacial till, less well drained than type 4.

Figure 12/14 *Each color on this map of the area around Syracuse, New York refers to a different kind of soil. Try to get a soil map of your area.*

The type of soil made by weathering depends on the type of parent material and on the climate of the area.

Soil maps show the different soil types in a given area.

Answers:

1. The sugar will be leached into the soil. You can do this as a demonstration, with a large graduated cylinder. Subsequent drying will show just where the sugar went.
2. a. Mountain areas for the first labeling; b. a low region in a known agricultural area for the second labeling; c. a wet region (e.g., Oregon) for the third labeling.
3. Any container that seals the item from the air is an attempt to keep chemicals from it. The container is also an attempt to keep entropy in check. Storing the item indoors and on a low shelf is an attempt at protecting it from mechanical weathering agents. Storage of the item can be improved upon by adding preservatives (a basic material if it is newsprint, oil if it oxidizes easily).

Check Yourself

1. Pour a cup of sugar or salt over some soil. Allow water to flow through the sugar or salt. What happens? Which step in the formation of a soil have you modeled?

2. Sketch an outline of the United States.
a. Label an area on it that might have soil with large amounts of parent material in it.
b. Label an area on it that might have a very thick topsoil.
c. Label an area on it where the soil would be leached more than the soil in most areas.

Can you explain how you decided which areas to label? Save the map. You may want to change it after studying Chapter 19.

3. Think of something you have put away to save for awhile.
a. How have you protected it from mechanical weathering agents?
b. What have you done to protect it from chemical weathering agents?
c. With your knowledge of mechanical and chemical weathering agents, what more could you do to protect it?

What's Next?

Concept: universality of change

The recent exploration of the moon has shown that the weathering process can be slowed down but never stopped. Although the moon has few weathering agents, the rock there still changes to dust over long periods of time. Even the tops and sides of the mountains on the moon are covered with dust.

On the earth, eroded material doesn't usually stay on the tops and sides of mountains because another process is at work. Erosion! It takes up where weathering leaves off. Instead of letting weathered material pile up and protect unweathered rock, the erosion process carries it away. In fact it is sometimes hard to tell the two processes apart. In many places the same agents that do the weathering do the eroding.

	15	16	17	18	19	20	21	22	23	24	25	26	27	28	29	30
1	E	A	R	T	O	P	S	O	I	L	E	C	C	S	C	Q
2	X	S	O	I	L	A	G	E	M	G	D	H	R	A	L	U
3	F	P	E	B	B	L	E	S	E	R	I	I	U	N	I	A
4	O	A	T	R	W	O	T	I	C	E	S	P	D	D	M	R
5	L	C	H	O	E	S	T	J	H	A	O	S	S	W	A	T
6	I	I	A	T	A	E	I	O	A	G	R	A	V	I	T	Y
7	A	D	W	A	T	E	R	E	N	T	D	A	C	N	E	D
8	T	J	O	Y	H	B	G	A	I	N	E	D	L	D	R	I
9	E	S	A	G	E	N	T	A	C	E	R	D	A	T	T	S
10	T	E	N	T	R	O	P	Y	A	M	A	P	Y	O	A	S
11	C	H	E	M	I	C	A	L	L	C	L	I	F	F	L	O
12	C	R	U	D	N	A	A	L	P	L	A	N	T	S	U	L
13	D	U	S	T	G	H	Y	D	R	A	T	I	O	N	S	V
14	L	E	A	C	H	E	S	A	L	T	S	I	M	E	D	E

Figure 12/15 *Use this mumbo-jumbo of letters to complete the Skullduggery.*

Skullduggery

Answers:
1. mechanical 2–23 down
2. disorder 2–25 down
3. ice 4–22 across
4. gravity 6–24 across
5. water 7–17 across
6. agent 9–17 across
7. chemical 11–15 across
8. wind 5–28 down
9. plants 12–23 across
10. talus 9–29 down
11. dissolve 7–30 down
12. salt 14–21 across
13. topsoil 1–18 across
14. hydration 13–20 across
15. acid 4–16 down
16. sand 1–28 down
17. exfoliate 1–15 down
18. clay 7–27 down
19. leach 14–15 across
20. pebbles 3–16 across
21. climate 1–29 down
22. lose 3–20 down
23. soil 2–16 across

Directions: Find one word to complete each of the following statements. Each statement is followed by a number. You will find the first letter of the word you are looking for in the row or column with that number in the mumbo-jumbo of letters in Figure 12/15. Watch out. The word can either run across or down.

Example: The process of making little pieces out of big ones is called _______ (4). Find the W in row 4. The word "weathering" is spelled out going down column 19. You try the others.

1. _______ weathering does not change the nature of the material. (2)

2. There is a natural trend toward _______ in the universe. (25)

3. Water can be a mechanical weathering agent when it changes to _______. (22)

4. _______ supplies a force needed to move pieces of rock from the weathering area. (6)

5. ______ can act both as a chemical and a mechanical weathering agent. (7)

6. Both types of weathering need a force and an ______ to carry them out. (9)

7. ______ weathering changes more than the size of the material being weathered. (11)

8. ______ is the weathering agent that needs to carry cutting tools with it. (5)

9. ______ can also act as both chemical and mechanical weathering agents. (12)

10. The pile of weathered rock at the base of a cliff is called ______. (9)

11. When water acts as the weathering agent, some of the weathered material may ______ in the water. (7)

12. Salt, sawdust, and sand were put on a section of garden. All were gone in a week. ______ was probably leached. (14)

13. The layer of soil containing decayed plant material and weathered rock is called ______. (1)

14. When water chemically combines with minerals in a rock, the process is called ______. (13)

15. ______ reacts with many minerals when it plays the role of a chemical weathering agent. (4)

16. ______ is usually made up of tiny pieces of the mineral quartz. (28)

17. If enough minerals in a rock hydrate with water, the rock surface may ______. (15)

18. When feldspar weathers by the action of carbonic acid, ______, sand, and a salt are formed. (7)

19. Water will ______ soluble salts and fine clay compounds. They help make up the subsoil. (14)

20. Rounded ______ are a sign that mechanical weathering has taken place. (3)

21. ______ mainly determines what a soil will be like. (1)

22. All things in the universe naturally tend to ______ energy. (3)

23. One end product of weathering is ______. (2)

For Further Reading

Wyckoff, Jerome. *The Story of Geology.* New York, Golden Press, 1962. A book about geology with good photographs and drawings. Covers rocks, water, and weathering.

Stone, A. Harris, and Dale Ingmanson. *Rocks and Rills: A Look at Geology.* New Jersey, Prentice-Hall, Inc., 1967. Things you can do to explore what happens in geology.

Chapter 13 Erosion

This is a good problem to have the class check. Weigh a shovelful of dirt and time a few students pantomiming pick and shovel work.
Concept: universality of change

Suppose a man wants to save money by digging the cellar of his new house himself. There is no reason why he and his trusty shovel can't do the job. The plan looks good on paper.

One shovel will hold about three quarts of damp soil. That's about nine pounds. Not very heavy at all. If the cellar is to be 50 feet long, 20 feet wide, and 10 feet deep, that means moving 10,000 cubic feet of soil. Since a cubic foot of soil weighs about 93 pounds, the total amount of dirt comes to 930,000 pounds, or 465 tons. A mere 103,000 shovelfuls will take care of the job.

Suppose our hero can work at a steady rate of one shovelful every four seconds even when the hole gets too deep to see over the sides. If he works day and night, with no breaks for food, water, or sleep, it should take only 114.4 hours. That's 4.76 straight days. All of Saturday, Sunday, Monday, Tuesday, and most of Wednesday.

But he has to eat and he has to sleep. Give him one and a half hours a day for meals and eight hours of sleep to recover from the day's digging. Then the job stretches to all of Wednesday, Thursday, Friday, and the next Saturday. If the cement for the cellar floor takes a week to deliver, he can order it the day he starts.

He could boast about his accomplishment for years. He would be an authority on the amount of energy it takes to move 465 tons of soil. The fact that the Mississippi River moves 20,000 times as much soil as he did in a week would really mean something to him. The fact that the Colorado River can move 1,650,000 cellarfuls of earth a year would be almost unbelievable. That's a cellarful every 19 seconds. WOW!

An activity might be made of the plotting of Colorado River data, month against amount of sediment. These data are available in USGS Water Supply Paper 998. Look for cyclical tendencies such as seasonal cycles.

Erosion by Ice, Wind, and Gravity

13.1 Erosion has the power to remove entire continents.

In case you haven't guessed it, this chapter is about earth moving but not just by the shovelful. For example, the whole North American continent is being carried to the oceans a little at a time. You can see this happening all around you. The entire process of carrying a continent to the sea is called **erosion** (eh-ROH-zhun). Running water in the form of rivers does most of the work. Other movers called **agents of erosion**, such as wind, ice, and gravity help too. Water, ice, gravity, and wind help move the land level to sea level.

If you figured out how fast North America is being eroded, you would find that about one foot of soil is carried away every 9000 years. Some mountains are even being worn down faster. Pike's Peak, a mountain in Colorado, has one foot of its surface removed every 1000 years. That's nine times as fast as the rest of the continent. At this rate, in five million years the mountain would be billiard-table smooth and barely above sea level. But that may not happen because the continent is rising at the same time as the mountain is wearing down.

13.2 Glaciers push, drag, and carry rocks and soil.

Parts of the world with temperatures below freezing most of the year have little rain. Most of their water comes down as snow. Since there is little chance for the water to melt and evaporate, the snow piles high. The pressure on the snow at the bottom of the pile eventually packs it hard like a snowball and turns it to ice. As the ice continues to build up, the pressure causes it to flow like water, but so slowly you can't see the movement. If the ice is in a valley, its flow is helped by the natural slope of the valley toward the sea. This moving ice mass is called a **glacier**.

A glacier may move only a few feet a year, but the glacial ice cuts much more than water. Like a rake dragged across a lawn or an unpaved driveway, a glacier tears rocks and soil from the ground.

A large sheet of ice called an **ice cap** may spread out in all directions over an area. Greenland and Antarctica are mostly covered by ice caps. However most glaciers flow down valleys that have already been cut by rivers. The ice flow scours the river valleys into a U-shape, like the one in Figure 13/1. Are there any U-shaped valleys in your area?

The valley in Figure 13/2 is nearly full of ice. Can you imagine the weight pressing down on the floor of the valley and against the valley walls? Rocks frozen in the ice cut and scratch the bedrock, and new material is added to the glacier. Any pieces loosened by ordinary weathering at the edge of the ice are frozen into the glacier. In this way the load carried by the glacier grows as it moves along.

Pieces of rock loosened by weathering from the walls above the glacier fall down along the glacier's edge. Instead of piling up and becoming talus, the rock is carried

Figure 13/1 *The U-shaped valley formed by a glacier moving through a river valley in Montana. Are there any U-shaped valleys like this in your area?*

Figure 13/2 *Where a tributary glacier joins the main glacier, two new bands of weathered rock are added to the flow.*

Longitudinal banding occurs when smaller glaciers join the main flow. Each band represents a single glacial tributary.

Demonstration Time: 20 minutes. If there is snow available, place a large pile on the high end of a tilted stream table. Mix the snow with sand or dirt outside before you put it on the stream table, or cover the front edge of the pile with sand or dirt, to simulate the dirty ice of a glacier. When the snow melts, it will leave lumpy terrain similar to that left by a melting glacier.

away by the moving ice. This material shows as dark bands on the surface of the glacier in Figure 13/2.

The front edge of a glacier can act both as a bulldozer and a wheelbarrow. Piles of rock and soil gather at the front of many glaciers or are frozen in the ice there. When the glacier reaches a warmer climate where the ice melts faster than it is being made, the ice drops its load. Rivers and streams then take over.

Much of the land in Canada and the north central and northeastern United States was covered by ice caps thousands of years ago. The piles of rock mixed with sand and clay which were dropped when these ancient glaciers melted are now hills, like those in Figure 13/3. A row of

Figure 13/3 *The dark ridges at the mouth of the valley are piles of rock dropped by the melting glacier. The one farthest out front marks the spot where the glacier stopped moving.*

Figure 13/4 *Scratched and polished rock surfaces like this one are evidence that a glacier passed through the area.*

Figure 13/5
Haystack boulders, large boulders located in unlikely places, are probably the work of glaciers.

such hills marks the exact line where the glaciers stopped moving south. Along the paths of the glaciers there may be scratch marks on rock surfaces as shown in Figure 13/4. Rock types not normally found in a particular area may be present there. Large boulders may be mysteriously located on smooth hilltops or flat plains as in Figure 13/5. All of these clues prove that something more powerful than running water has been working there.

13.3 Wind sorts out the smaller particles.

Glaciers are stronger eroding agents than water because ice is more rigid than water. For the same reason, wind does not erode as well as water. Wind is not as rigid as water. Wind can pick up small particles of weathered rock and carry some of them great distances. In desert areas this is an important means of eroding the land. Sand particles may collect in piles called **dunes** like those in Figure 13/6. A series of wind storms can eventually move the dunes many miles.

Some deserts have occasional rainfall that may cause flash floods. Such floods are also important in erosion.

In areas where soil has been left unprotected, the wind can carry away the best part—the topsoil. When it does, the sky may appear as in Figure 13/7. If the wind blows out over the sea, the soil is carried directly to the sea without the aid of the rivers.

Concept: interaction of matter

Figure 13/6 *The wind will move these sand dunes along and may change their shapes.*

Figure 13/7
Wind eroding the land can also cause severe duststorms.

13.4 The force of gravity may work as an eroding agent.

Mass wasting is erosion caused by gravity moving materials downhill without the help of ice, wind, or water. People in some parts of California are all too familiar with a slow kind of mass wasting called **hillside creep.** The soil on the hills where they have built their homes is slowly flowing toward the sea, just as glaciers and rivers do.

Can you see evidence in Figure 13/8 to show that the hillside is "flowing?" In California heavy rains may speed up this normally slow process. Once underground, the water lets the particles of soil slide past each other much more easily.

Sometimes the flow of large amounts of soil and rock even looks like a glacier as in Figure 13/9. Here, however, the rock glacier just moves on and on, pulled by gravity. It can't disappear by melting as ice glaciers do.

Figure 13/8 *As the soil in this California hillside creeps slowly downhill, the fence posts tip in the direction of the flow.*

Figure 13/9 *A rock glacier like this one may flow so slowly that trees have time to grow on it.*

Did You Get the Point?

The process of carrying the weathered rock material of a continent to the sea is called erosion. It depends on gravity.

Running water does most of the work of erosion, but glaciers are also good eroding agents.

Wind is another agent of erosion.

Mass wasting is erosion by gravity without the aid of the other agents.

Answers:
1. Ruts in roads or sidewalks which are not paved, puddle depressions, gullies, bare hillsides. You can see that water has washed the smaller particles away from the larger particles and deposited them elsewhere.
2. Dust storms leaving land barren, gullies on good farm land, cave-ins of banks along road cuts and rivers. Classic examples of cave-ins have occurred along shores of Lake Michigan and Lake Erie.
3. Rolling, hilly, lumpy, with hills not caused by bedrock.
4. Answers depend upon local conditions.

Check Yourself

1. Make a list of places near your home and school where you can see evidence that some weathered material has begun the journey to the sea. Explain why the material looks the way it does.

2. In what ways does erosion get in the way of man's plans for building on the property he owns?

3. Describe the landscape caused by glaciers.

4. Describe the eroding agents that cause an unpopulated area near your home to look the way it does.

Erosion by Water

13.5 More water falls on the land than runs off.

If you trace water erosion back far enough, you will probably say that the rain starts it all. A few claps of thunder, a few flashes of lightning, and then the downpour comes. As the water rolls off the land, a little more of that "one foot per 9000 years" is washed away.

Some students may ask what the units in Figure 13/10 are. The data are hypothetical. Units are deliberately avoided to keep students from being sidetracked into trying to figure out how a cm of rain in a stream can be measured.

Activity Time: 15 minutes.

★ But it is not all that simple. Figure 13/10 is a table of some data taken during a rainstorm. It shows the amount of rain that fell over a certain period of time. The table also shows the water flowing in the stream during the same period of time. This stream dries up after rainstorms. Plot the data as two separate lines on a graph. If you need help with graphing, see Appendix F. Make each line a different color. The result will be easier to understand if you put water on the horizontal axis of your graph and time on the vertical axis.

Why did the stream take so long to swell? Why didn't the stream carry away as much water as the storm brought? What limits the amount of water available for carrying away the land? ★

The answers to these questions are the subject of the next few sections.

Figure 13/10
Compare the rainfall and the stream flow.

Time	Amount of rainfall	Stream flow
12:00	0	0
1:00	26	2
2:00	47	6
3:00	48	11
4:00	20	20
5:00	2	30
6:00	0	20
7:00	0	10
8:00	0	5
9:00	0	5

13.6 Soil can soak up large amounts of water.

The most important thing that affects how much water will be around to carry away the land is climate. No rain would mean no streams. But, even in areas with plenty of rain, water cannot begin to carry away the land until there is **surface runoff,** water that runs across the surface of the land. Some soil is hard and tightly packed, and water runs off the surface easily. Most soils, however, are **permeable** (PER-mee-uh-bul), meaning that water will flow through them. If the water seeps into the soil faster than it falls on the surface, there can be no surface runoff.

Permeable soil can store large amounts of water between soil particles in spaces called **pores.** A measurement of the amount of water a soil can store in these pores is called the **porosity** (poh-ROS-uh-tee) of the soil. Water will continue to flow into the soil until all of the pores are full and there is no longer a place for the water to flow to. Then surface runoff will take place.

Of course you could argue that the stored-up water could flow through the permeable soil and then leak out through the side of a hill to join a stream as in Figure 13/11. But soils can still hold water even after much of the water has leaked out. It isn't like pulling a plug. You just can't get as much water out of the ground as was sent in, and the actual amount left behind depends on the type of soil.

Figure 13/11
Water flows down through the permeable soil and along the top of a non-permeable bedrock. Here, the water falls over the edge of the cliff to help form a stream.

Which soils, fine-grained ones or coarse-grained ones, have the most pore space? Which kind of soil holds moisture best even after permeable flow has stopped? Which type of soil helps prevent erosion by holding back the water that could become surface runoff? You can find out the answers by measuring these properties for yourself.

13.7 You can measure pore space and particle size.

Materials
screen sieves, test tubes, sand, grease pencil

Activity Time: 30 minutes.
Process: observing, measuring

★ Sand is usually a mixture of particles of different sizes. Carefully sift a few handfuls of sand through a set of screen sieves. Choose two different particle sizes of sand for your comparison of particle size and pore space.

Completely fill two large test tubes with the samples. Fill one test tube with the larger particles and the other test tube with the smaller particles. Lightly tap the two test tubes so that the particles will pack as tightly as possible. Add more sand as the grains pack. With a grease pencil, make an "L" on the tube of large particles and an "S" on the tube of small particles.

Place your thumb or finger over the mouth of one of the test tubes. Turn the test tube upside down and submerge it in a container of water. Do not move your thumb or finger away until the mouth of the test tube is beneath the surface. Allow the sand to fall from the test tube into the container of water. Before you take the test tube out of the water, mark how high the water rises in the test tube when the sand has left. Remember, the water is replacing the sand. It is not replacing the air that was trapped in the pores between the particles. The air stays in the tube.

Both water levels should be about the same, indicating that pore space does not change as particle size varies.

Finer particles used in the tubes will require more tapping to get them to fall out.

Repeat the experiment with the sand particles in the other test tube. What happened? In which test tube did the water rise higher? Which size sand particles trapped the most air? Did fine-grained or coarse-grained soil contain the most pore space? ★

13.8 Flow rate and water held back can be measured.

The pores are small in fine-grained soils. Lots and lots of small pores, however, can hold just as much water as a few large ones. But how does the size of the pores affect the rate at which water flows through them?

Materials

glass or plastic column with hose drain and clamp, beaker, 1000 ml graduate, stopwatch, can, crushed stone

Activity Time: 40 minutes. Sand will also work, but the flow rate of water through the sand will be extremely slow. There are prepared kits for this experiment, but it is best to use real earth material. Beads are too artificial, and are subject to being stolen for use in peashooters. A stop clock can be shared by two groups.

★ Get some crushed stone of the kind used in driveways and separate it into two different sizes. Place each size of stones in a different container. Label the containers "small" and "large." If the stones are not dry, heat them and let them cool again before you make your measurements. The crushed stone will be a model of soil particles.

Fit a drain to the bottom end of a long, waterproof tube. Clamp the drain shut. Fill the tube with stones of one size. Pack the stones as tightly as possible. Now pour a known amount of water, measured in milliliters, into the tube. Put a can under the drain. Figure 13/12 shows how to set up the materials.

Using a stop watch, you will measure how long it takes the water to flow from the tube when the drain is opened. Start the watch at the instant the water begins to flow and stop it the instant the water stops flowing in a steady stream. Save all the water that comes out. How long did it take the tube to empty? Did as much water come out of the tube as went in? How much water remained in between the stones?

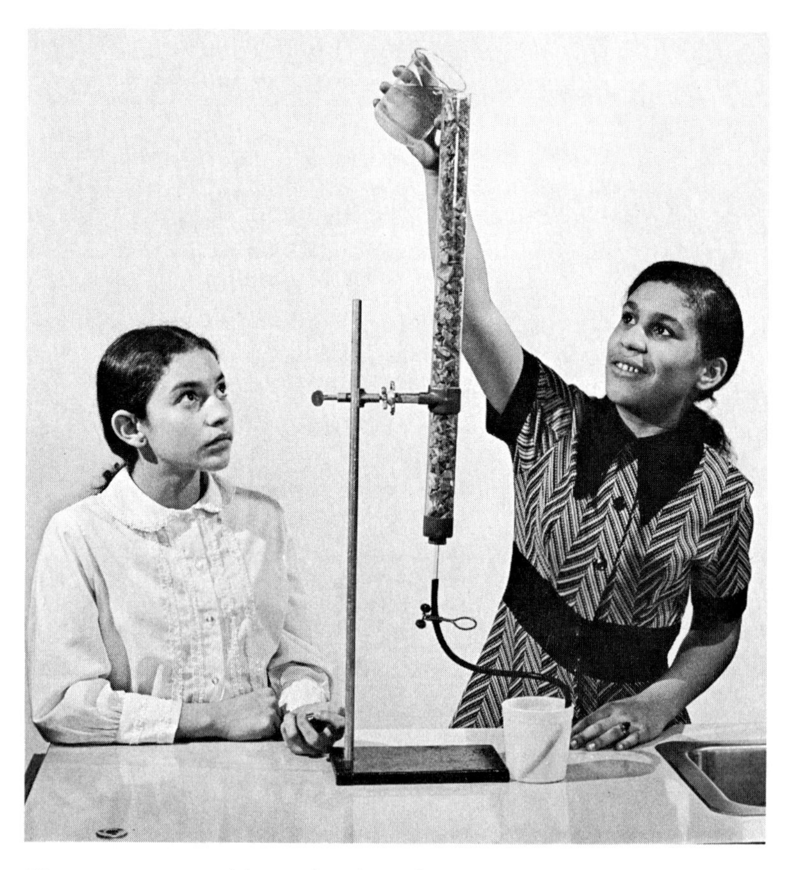

Figure 13/12 *Measuring how fast water can flow through model soil.*

You may have to demonstrate that surface area increases with the breaking of large pieces into small ones. Use a pile of boxes or blocks, separated block by block, to do this. Explain that it is this increased surface area of fine-grained soil particles that causes them to impede water flow, due to molecular attraction.

Repeat the experiment with the other stone size. Use the same amount of water as before. Compare your results with those from the other stone size. Does water flow fastest through fine-grained or through coarse-grained soils? Which type of soil holds back the greatest amount of water? Why? ★

Let's look back to Figure 13/10, which showed how the amount of water in the stream related to the amount of rainfall. Remember that the rainfall data were recorded during a rainstorm. One reason that some of the water was held back may be the type of soil in the area. You know from the experiment you just did that particle size does determine the rate at which water flows into and through the soil. So, maybe the soil is fine-grained in the region where the data of Figure 13/10 were recorded.

But that's only the beginning. Could plants have used some of the missing water? Like you, they need water to live. Could some of the water have evaporated before it had a chance to run off or soak into the soil? And what about the soil itself? Could something about the soil have changed during the storm? Could changes in the soil change the rate of runoff? Look a little closer at the soil itself.

The answer to each question is "yes." Promote discussion.

13.9 It is possible to make soil soak up water better.

Particles of undisturbed soil that contains a lot of clay have a tendency to clump together. The clumps are called soil **aggregates** (AG-ruh-getz). They make fine-grained soil behave like coarse-grained soil as long as the aggregates hold together. Soil that is protected by thick grass, or soil in a freshly-plowed field usually has these clumps.

This can be demonstrated in the classroom with real soil.

When raindrops hit the soil, the aggregates break up into small pieces again. These pieces fall into the pores between other aggregates and plug them up. The soil becomes almost waterproof, and runoff occurs. Look into a mud puddle sometime. Why does it hold water when the grass-covered soil around it lets the water soak in? On the bottom of the puddle, aggregates break up into very fine-grained sediment. This sediment plugs the pores. However, the soil covered with grass lets water flow into the ground because its aggregates are whole.

Some kinds of clay can absorb much water and swell greatly as they hydrate. This closes the pores soon after the rainfall begins, and reduces the absorption of water.

Farmers do not want their fields waterproofed. Runoff would carry away good soil. This is why farmers put straw around the roots of some plants, especially tomato and melon plants, that need large amounts of water. When raindrops hit the straw, they break into a fine spray. The spray is less likely to break up the soil aggregates. The plants can then use the water that drips through the straw and flows into the ground.

Did You Get the Point?

More water stays on the land than runs off because soils are porous.

Permeable soil absorbs water and keeps it from becoming part of the runoff.

The water-soaking properties of a soil can change during a rainstorm.

Some of the water that falls on the land evaporates.

Plants may use up some of the water that falls on the land.

Check Yourself

Answers:
1. The land would have to have non-permeable soil. There would be no plants. It might be cold, to slow evaporation. On the other hand, the soil and bedrock might be already saturated. The land slope would have to feed water directly to the stream.
2. Consider permeability, slope of land, vegetation, number of tributaries.

1. Suppose a stream was found that swelled at the same rate the rain fell, with no lag like the one in Figure 13/10. What would the land in the area have to be like to make this possible? Why?

2. Imagine a street in your town or city suddenly turning into a stream. All side streets that cross it will lead into it. How fast would it collect rainwater to erode the land? Why?

Different Kinds of Streams

Activity Time: 20 minutes. Work with stream tables can be messy, so choose your groups carefully for students who will work together rather than clown. You may want to split up various stream table activities among groups or have everyone do them. No special order of activities is necessary. All the activities in the text can be done at one time. Encourage inventiveness and open-ended experimenting.
Process: observing

Materials

stream table, sand, water jar, food coloring, blocks for props

13.10 Small streams are usually parts of larger streams.

At the beginning of this chapter you read about a large stream that does a good job of eroding the land, the Colorado River. But not all streams do such a good job. You can make a model of a stream to help you understand how streams cut away the land.

★ Set up a stream table. A wooden box with a drain at one end will do. Make the drain high up on the end edge of the box to leave space for a pool of water to collect. Fill the box with an even 3-inch layer of sand, but have about 6 inches bare near the drain. Be sure to use dry sand.

As a source of water, use a plastic jar with a hole the size of a pencil lead in the bottom. Set the jar so that the water from the hole will fall directly down into the sand at the head of the table. Prop the stream table on blocks at a slight angle, and let the water flow as in Figure 13/13.

An angle of approximately 4° works best.

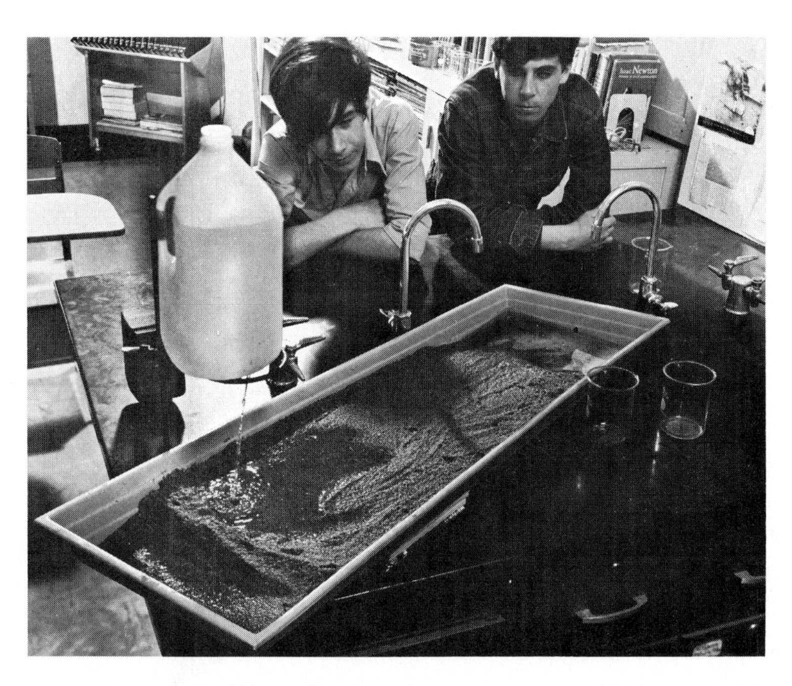

Figure 13/13 *Water flowing down a stream table is a model of a real stream flowing across the land.*

It soaks in. Runoff has stopped for a while. Runoff will start again when the pores are full.

The water from the jar is like surface runoff from an area farther upstream. What happens to that runoff when it hits the dry sand in the box? What will happen when all the pores between the particles have been filled with water?

Catch the overflow from the drain and use it again. That way you need not have running water in the room.

Refill the jar when the water level gets low. After the sand has been soaked with water, a pool will form. As water begins to flow from the drain spout, small streams will develop on the sand. Squirt a few drops of food coloring into the water just below the jar to make the streams easier to see. How many streams come from one big one? ★

13.11 A small stream can be either a tributary or a distributary.

You may want to use a drainage map of your own state.

Look at the map in Figure 13/14. Do the streams in your model look like any of the streams on the map? Some of them may look alike, but there is a difference between the streams on the map and your streams in the sand.

The streams in the map all flow *into* larger streams. Your streams in the sand all flow *from* larger streams. The

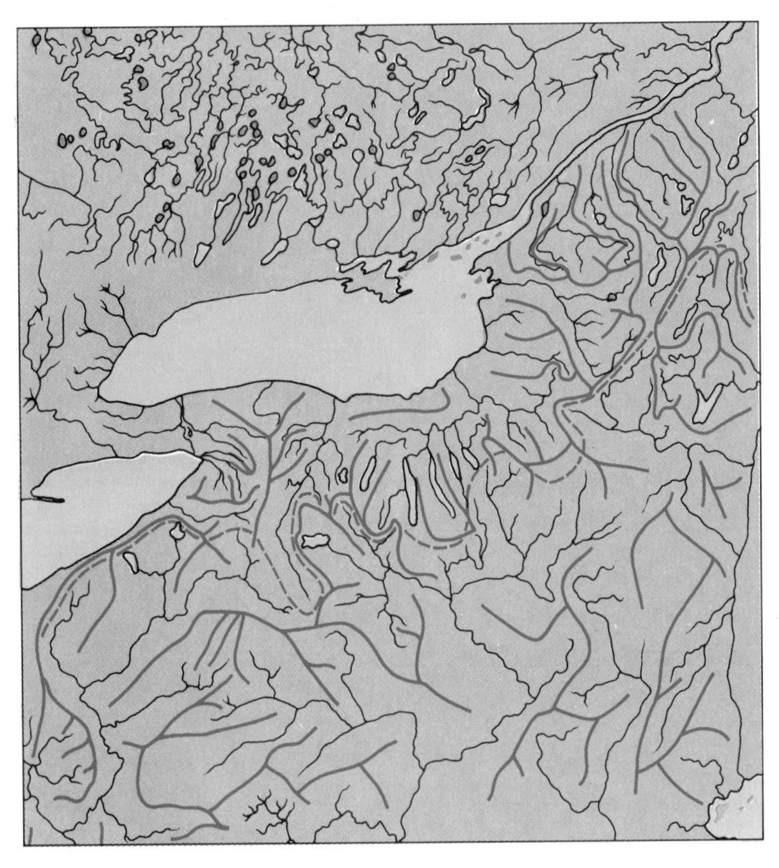

Figure 13/14 *Look at the stream patterns on this map to figure out where the higher land must be.*

Streams flow from high ground to low ground. On the map, the small streams originate at higher altitudes. In the stream table model, the reverse is true.

"ends" of the small streams shown in the map are really the beginnings. Does this tell you where on the map the areas of high ground have to be? Where are the areas of high ground on your model?

The little streams going into larger streams on the map are called **tributaries** (TRIB-you-tair-eez). They contribute or add water to a larger flow. Your little streams are called **distributaries.** They distribute or spread out water from a larger flow over a wider area. A river with all its tributaries and distributaries is called a **stream system.**

The high spots between the small streams also have special names. They are called **watersheds.** The hills shed water in the form of runoff to start streams. A major watershed that supplies water to stream systems on either side of it is called a **divide.**

Activity Time: 15 minutes.

Make a copy of the map in Figure 13/14. With a crayon, draw lines to show where ridges of high ground such as hills or mountains have to be to make the streams flow as they do. Notice how most of the hills and mountains you draw join together. They are the watersheds. Find the divide on the map and mark it with a dotted line.

13.12 Distributaries develop at the ends of some streams.

Next make a model of a single stream. You will be able to see some interesting things that did not show up before in the model of a stream system.

Materials

stream table, sand, water jar, food coloring

Activity Time: 40 minutes. Long running is not necessary if the volume of water delivered is large enough to carry the sand easily.

★ Set up your stream table again, but don't put the water jar in place yet. With your finger make a straight path in the sand from the place where the water will start to where the pool will form at the other end. This path is called a **channel.** Do not dig all the way down through the sand to the bottom of the box. Then fill up the jar and put it in position. Pour the water.

Figure 13/15 *Sediment deposited at the mouth of this model stream forms a delta.*

This is a good place to discuss the pros and cons of erosion controls. Consider the relative merits of large dams and reservoirs as opposed to small ponds on areas between tributaries.

Watch the long, straight stream carry more and more sand to the pool and drop it there. The pile of sediment at the mouth of the stream is a **delta.** Deltas can be found where streams flow into lakes and oceans. The slowing of the current at the stream's mouth causes sand to stop moving. The path of the stream will change as new sediment is added to the delta. As the stream runs, what is happening at the mouth, the end closest to the pool? Does the single path stay the same? Add some food coloring to see more clearly what happens. What kinds of streams are formed, tributaries or distributaries? Figure 13/15 may help you decide. ★

Did You Get the Point?

Small streams are usually part of a larger stream system.

A small stream is usually a tributary but could be a distributary, depending on its location in a stream system.

A watershed supplies a stream with its water.

Check Yourself

Answers:
1. In the vast majority of cases, the closest stream will be a tributary, so you can stress the relative numbers of the two kinds of streams.
2. Tributaries would be sewers and drainage ditches. They would be considered harmful if they polluted the stream system. Distributaries would be canals and irrigation ditches. They could rob downstream areas of needed water.

1. Find the stream or river closest to your home on a map. Is it a tributary or a distributary? How do you know? Describe the stream system it is part of.

2. Man sometimes makes tributaries and distributaries where there were none before. Describe some of them. Do any of them harm the stream system? Do any of them benefit the stream system?

The Eroding Power of Streams

13.13 The faster the stream, the more it erodes.

Most of the stream channel you dug with your finger will stay straight for some time. But some changes will begin to take place, too. First of all, the stream gets wider as the water continues to flow. Can you find the places along the stream that are widening the fastest?

Most of the cutting away of the stream channel occurs near the **source,** the high end of the stream. The farther away from the source, the less the erosion. In fact, there is no cutting away at the mouth of the stream. Sediment is piling up there instead. Some more food coloring will help you find out why.

Materials

stream table, sand, water jar, food coloring

Activity Time: 10 minutes. This can also be done as a demonstration.

★ This time when you put food coloring into the stream, watch the speed of the water, called the **velocity.** Notice where the coloring moves fastest. Does it seem to move fastest where the most or the least sand was carried away? What can you tell about how the velocity of a stream affects the amount of soil eroded? ★

13.14 Stream speed depends on the angle of the stream bed.

What determines the velocity of a stream? To find the answer, smooth out the sand in the stream table and cut a fresh, straight channel. You will measure the change in

Figure 13/16 *Measuring stream velocity as the angle of the stream bed is changed.*

stream velocity when you change the angle of the stream bed as the students are doing in Figure 13/16.

Materials

stream table, sand, water jar, food coloring, markers, protractors, stop clock

Activity Time: 40 minutes.
Concept: energy in motion

★ Place two markers next to the stream, one near the source and the other near the mouth. You can use pencils or sticks as markers. Next, measure the angle of the stream table. Record this measurement in a data table in your notebook. Three people can do the experiment together. One person can start pouring the water from the jar. Make sure that you start with the same amount of water in the jar each time. You will see why later.

When the water has flowed to the other end of the table, another person can squirt in some food coloring above the first marker. A third person can time how long it takes the food coloring to go between the two markers. As soon as you make the time measurement, stop pouring the water immediately. Otherwise the stream bed will not be the same for each measurement. Record the time measurement with the angle measurement in your notebook.

Measurements have to be quick, or the nature of the channel will change too much to produce usable data. A trough will not work as a channel. The channel must be rough like the one dug in the sand. A stop clock can be shared by two groups.

Sample data table:

Angle (degrees)	Time (seconds)
2	3
3	4
4	3
5	3
6	3
7	3
8	2.5
9	2
10	1.5

Do not move the markers. Raise the end of the table one or two degrees. Record the new stream table angle and do the experiment again. Record the time again. Repeat the lifting and pouring about nine more times.

Make a graph of your data. Did it take less or more time for the water to cover the distance between the markers each time you increased the angle? Did you find that your stream had to reach a certain angle before the velocity increased? It seems strange, but that's the way real streams behave also. Even the great Amazon River in South America obeys the rule you just discovered. Can you explain why the rule works this way? ★

13.15 Changing the angle of the stream bed changes stream flow.

Water molecules, like marbles rolling down a hill, would flow straight down the stream if they could. But just looking at a few zigzag streams tells you that it doesn't always happen that way.

When the angle of a stream bed is steep, the water molecules flow fast. But the molecules closest to the rough sides and bottom of the stream channel are held back. The ones that go the fastest are those near the top of the water at midstream. Instead of going in a more or less straight line, the water molecules follow irregular, twisting corkscrew paths. When the water flows faster, eddies are formed. They are circular flows of water in the stream like eddies in the air.

In your experiment with the stream bed angles, the speed of the water at different angles of the stream bed was affected by swirls. Your very first measurement was for water that went almost straight down the stream like the water in Figure 13/17. Increasing the angle caused the

Figure 13/17 *The water molecules on the right move faster than the water molecules on the left. However the swirls formed by the faster current make it appear to move more slowly.*

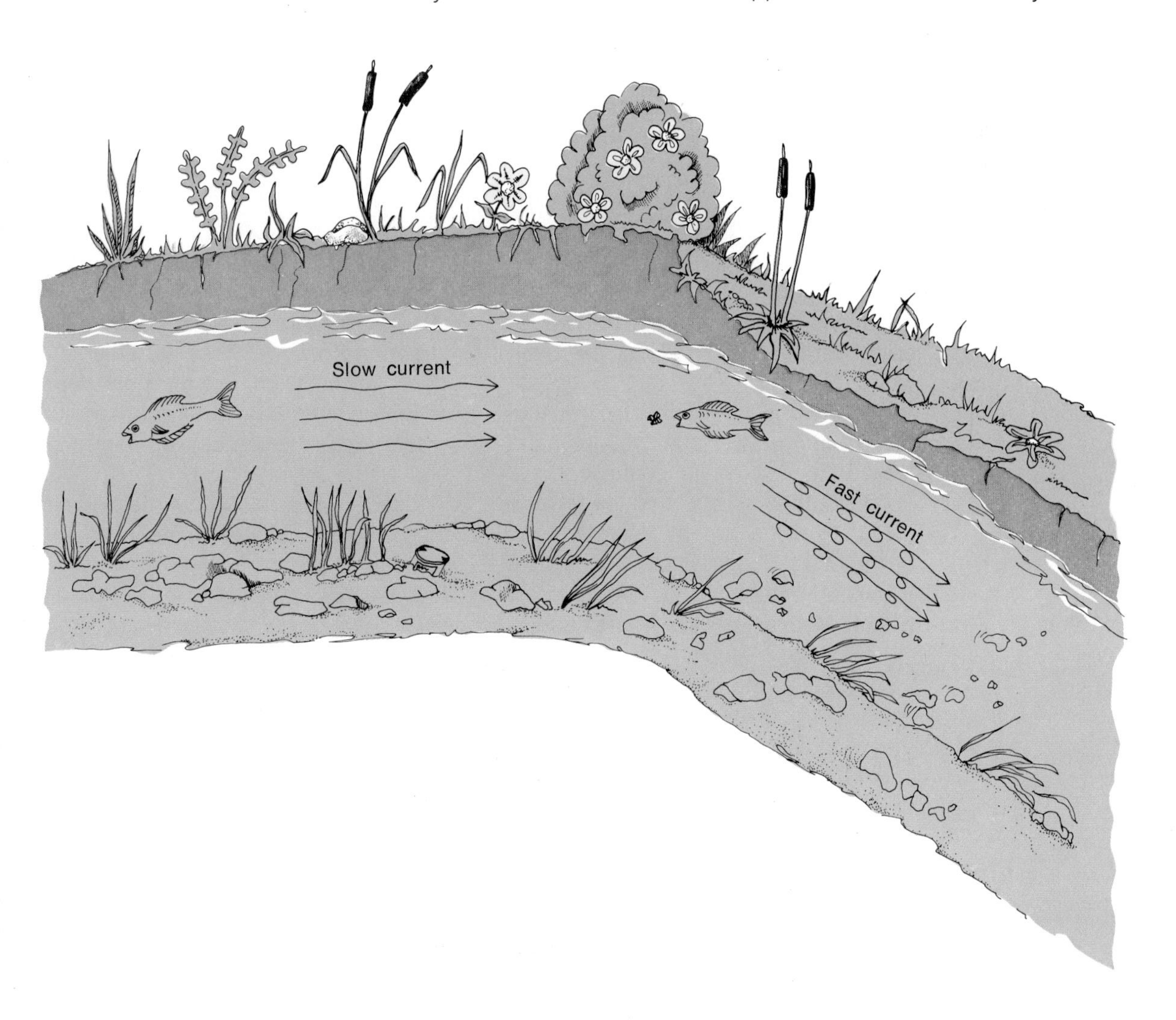

Investigating a Stream

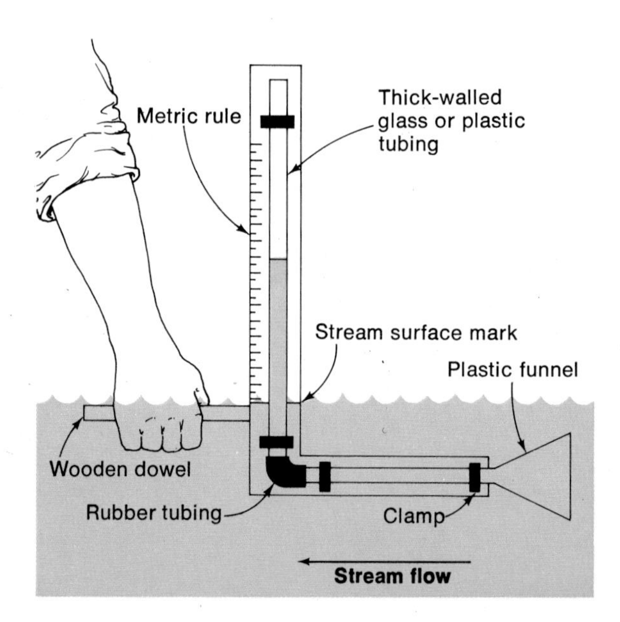

Facts about streams do not have to come from books. Whatever has been written about streams had to be observed and measured directly. You can make your own observations and measurements while enjoying the "great outdoors" at the same time.

Plan to visit a stream before the end of the school year. Add a few simple pieces of equipment to your picnic basket. You will need a homemade or store-bought velocity-measuring device like the one shown above, three plastic pill bottles with lids, a thermometer, and a plastic bag or two.

Find a bend in the stream. Lower the stream-velocity indicator into the water with the funnel pointing up stream until the rod in the front and the bottom of the metric rule are level with the surface. The height to which water is forced up the glass tube indicates how fast the stream is flowing. Measure the stream velocity at the inside of the bend, in the middle of the bend, and at the outside. Based on your results, where should the greatest amount of erosion be taking place?

Measure the velocity at the surface of the water and compare it to the velocity at some point below the surface. Where is the velocity the greatest? Where does the stream carry most of its load?

Take a sample of water at the surface of the stream, another halfway between the surface and the bottom, and a third just above the bottom. What differences do you detect?

Measure the temperature at different depths and at different locations along the stream. How might some of the differences be explained?

Collect samples of pebbles and plant

life in and around the stream. Record where each sample was found. What size pebble is usually found along the shore? What size pebble is usually found at the bottom of the stream? What type of sediment is found at the inside of a bend in the stream? What type is found at the outside of the bend? Why? Where is most of the plant life in a stream found?

Save one of your water samples. Date it and plan to get one sample each month for the next few months. You may end up with the only record of what is happening to the stream over the next year or two.

---

molecules to go faster. Larger and larger swirls formed. But suppose you could unwind the swirling path in Figure 13/17. Isn't it a bit like a race between the slowest and fastest runners in school? If the fast runner has to zigzag while the slow runner is allowed to run a straight line, who will probably win? How high did you have to raise your stream before the swirls no longer affected the speed of the water?

The swirling molecules help the erosion process. More water molecules come in contact with the stream bottom to pick up loose material. More water molecules come back to the bottom for a second pass. The greater the angle of the stream bed, the more erosion takes place.

13.16 Changing stream volume changes stream velocity.

Materials

stream table, sand, water, jar, markers, protractors, stop clock, nail

Activity Time: 40 minutes. This can be tried if interest and time warrant.
Process: observing, measuring
Concept: energy in motion

★ What happens to the flow of a stream when the volume of flowing water is increased? You can find out by using your stream table again. Each time you pour the water, make the hole in the bottom of the jar bigger. Keep the stream bed angle the same. Measure the speed of the water using food coloring as you did before. Record the times in your notebook. What can you tell from your results? ★

This is why flood seasons are times of increased erosion. When a stream floods, the erosion may even change the path of the stream. New channels sometimes form on land that was dry before, and bridges are left crossing over old deserted channels.

13.17 Stream erosion decreases with age.

Remember the man digging the cellar by hand at the beginning of this chapter. Did you wonder how old he was? It was assumed that he could move one shovelful of dirt every four seconds. This would have been a ridiculous rate if he were 85 years old. No matter how much a man wants to work, he will slow down with age. How does age affect a stream?

Materials

stream table, sand, water, jar, food coloring, protractor

Activity Time: 40 minutes. You may want to use more time. The stream tables can be left with the water running overnight if they are fed by faucets.

Stress the scaling down of the model.

★ You know that the streams you made were very young because you dug the channel only moments before you took any data. Now dig a fresh channel and watch this one grow old. You can make hundreds of thousands of years go by in an hour or so. Tilt the stream bed to an angle of 8 to 10 degrees by putting blocks under one end. Use lots of water to increase the stream velocity. Since your model represents an entire stream, 10 centimeters on the stream table could represent 2 miles. An average stream flows less than $2\frac{1}{2}$ miles per hour. The velocities you get could be more than 100 miles an hour in a real stream. This will certainly make things happen fast.

In your model, did you notice any change in the velocity of the stream? What about the amount of material moved by the stream? Where does water flow fastest around bends in the streams? What happened in the pool at the end of the stream? ★

A delta forms as the stream runs. Bends will tend to straighten and the stream bed will widen. Mass wasting will be seen along the stream course. A decrease in velocity will be indicated by a decrease in eroding ability. Stream distributaries on the delta will switch locations—move back and forth across it.

You may want to have students put an obstruction or rock halfway across the stream where it appears to be straightening. A new bend will develop.

You may want to show ground water movement by putting a drop of coloring in wet sand next to the stream. Watch movement of the spot.

Young streams are relatively straight and fast moving. They do the best job of cutting down through the land, leaving steep valley walls like those in Figure 13/18. The stream valley will continue to become deeper while the weathering process widens it.

Eventually, enough sediment has eroded from the source of the stream to make the level of the source lower. The build-up of sediment at the mouth of the stream raises that end slightly. These changes decrease the angle of the stream bed. The velocity of the water decreases. The valley becomes wider and the valley walls less steep. The stream, like the one in Figure 13/19, enters middle age. The middle-aged stream cuts the channel away more slowly.

Figure 13/18
Young streams like this one in Ausable Chasm, New York, do the best job of cutting downward through the land, and have the steepest valley walls.

Figure 13/19
What clues in this photograph of the Kicking Horse River in British Columbia, Canada, tell you that it is a middle-aged river?

Figure 13/20 *This old river in Arizona has weathered the valley down to a wide flat flood plain.*

Finally, when the stream is old like that in Figure 13/20, the stream bed has very little slope left. The weathered valley is very wide. The delta, the triangular pile of sediment at the mouth of the stream pictured in Figure 13/21, is covered with distributaries. The water moves so slowly that the stream wanders all over the floor of the widened valley. Fallen logs, rocks, and tree stumps can easily change the

Figure 13/21 *The delta formed where the Nile River flows into the Mediterranean Sea.*

path of the slow stream and form U-shaped bends called **meanders** (mee-AN-derz).

Sometimes a stream will cut a new path straight through the neck of the meander, like the stream in Figure 13/22. Then sediment piles up at the ends of the old meander and cuts it off from the main stream. This leftover bend is called an **oxbow lake.** Up and down the valley on both sides of an old stream, dried-up oxbow lakes can be seen in Figure 13/23 that look like old scars. Quite often, the stream will overflow its banks to smooth the valley floor even more. The part of the valley that is covered by water when the river floods is called a **flood plain.**

It is difficult to make meanders develop by themselves. It is best to cut one. If conditions are right, the meander will cut itself off into an oxbow lake. Sometimes introducing fine soil to the pool under the jar will speed up the cutting.

You may want to experiment with a single meander. Just cut one in the sand with your finger. A squirt or two of food coloring will show you where around a meander the water moves fastest, and how a meander can move across a flood plain. Watch for erosion in one part of the meander and build-up in another.

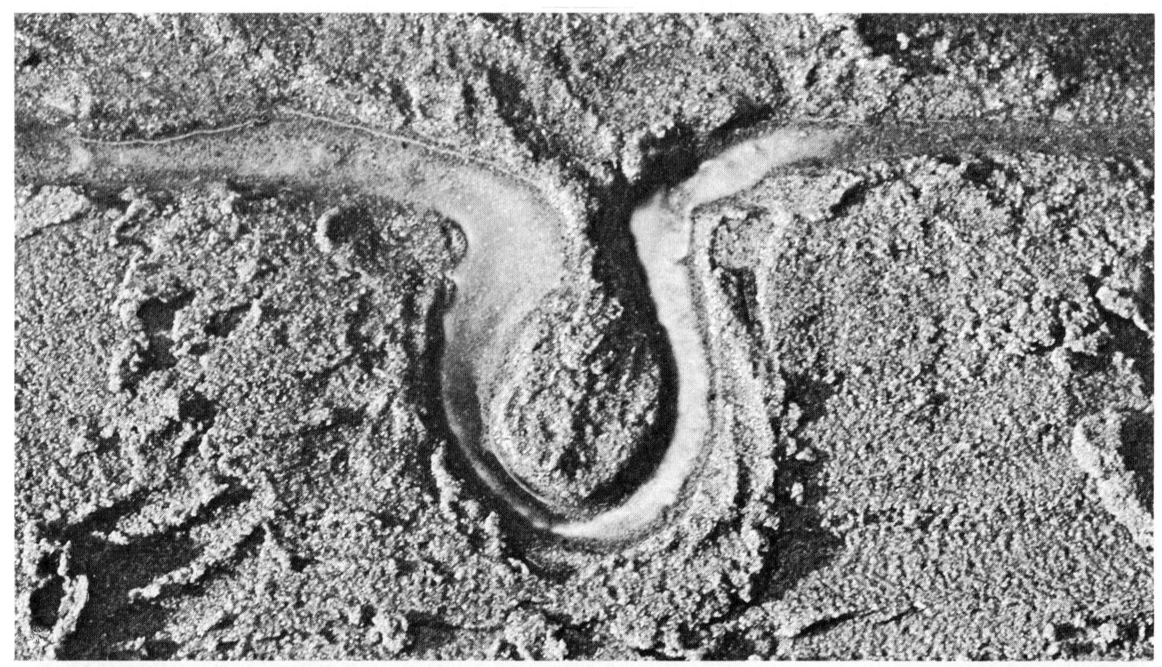

Figure 13/22 *Sometimes a stream will continue to erode across the neck of a meander. The cut off meander that results is called an oxbow lake.*

Figure 13/23 *The Mudjatik River in Canada is an old river with many U-shaped meanders. Can you find any oxbow lakes?*

Answers to questions are in the following paragraph. The whole Colorado Plateau has been uplifted since the Colorado River formed its meanders. If a steroscope and stero photographs are available, have students use them to study all the stages of stream development.

13.18 Old streams can become young again.

A very old stream with many meanders can suddenly begin eroding again. Look at Figure 13/24. Can you see the signs of old age, the meanders? What signs do you see that the old river has become young again? What do you think happened? Can you set up the problem in your stream table?

Figure 13/24 *How can you tell that the San Juan River, an old river in Utah, has become young again?*

Since the stream is now in a very deep and steep-walled valley, it must have begun to erode again. For this to have happened, the area where the stream has become young again was probably raised up. Grand Canyon is America's most famous example. Rivers like this one are a reminder that erosion and uplift are at work at the same time.

Answers:

1. Look for meanders, deep, steep-walled valley, etc.
2. For example, it might originate in mountains and travel the last 100 miles (161.5 Km) across plains and drop to the sea from a high, steep shoreline.

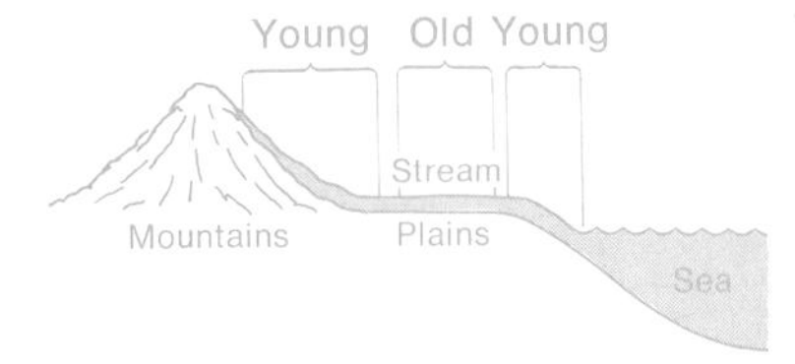

Did You Get the Point?

The ability of a stream to erode depends on the velocity and volume of the water, and on the age of the stream.

Changing the angle of the stream bed and the volume of water it carries changes the velocity of a stream. Both changes take place as the stream ages.

Check Yourself

1. On a map, find the stream closest to your home. Is the stream young or old? How do you know?

2. How can a river have parts that are called old and parts that are called young? Use diagrams to help explain.

What's Next?

Even the swiftest stream can carry eroded material only so far. Whatever a stream picks up on the way to the sea it must finally put down. Where does a stream dump sediment? Is the sediment piled in random heaps? Does the kind of sediment have anything to do with the way it is dropped? Who cares where and how the sediment is dumped?

Remember, the sediment started as rock. *Some* of it came from volcanoes. *Most* of it helped carry earthquake waves around the world. *All* of it plays a part in the rock cycle. Where and how it is dumped must have something to do with how soon it will return as rock again. Scientists call the next step in the rock cycle "deposition."

Skullduggery

Directions: Copy the mumbo-jumbo grid of letters in Figure 13/25. (Do not write in this book.) The letters contain a hidden reminder of the material covered in this chapter. You will find the SECRET MESSAGE when you answer the numbered questions.

Decide what word is missing in the following statements and write your answer on a separate piece of paper. Then cross out the letters that spell that word in the column or row that has the same number as the question. *Do ALL of the "Across" questions first.* If the letter you need does not appear, cross out a ? in that column or row instead.

Figure 13/25
Copy this grid of letters for the Skullduggery.

	12	13	14	15	16	17	18	19	20	21	22	23	24	25
1	Y	O	A	C	?	N	E	L	I	R	R	S	R	W
2	T	S	A	U	I	R	B	G	R	E	E	F	Y	T
3	A	M	F	R	O	P	B	O	T	S	E	E	L	E
4	S	A	E	?	P	R	K	H	O	O	U	F	I	N
5	O	P	?	N	A	I	D	W	E	V	S	R	O	R
6	S	N	P	Y	D	G	R	I	E	L	?	T	S	A
7	P	L	D	N	S	I	M	N	A	R	R	U	L	E
8	G	G	G	A	R	?	O	E	E	A	?	N	?	T
9	R	?	T	H	O	U	S	C	E	A	N	D	I	D
10	T	E	W	D	?	?	C	H	N	A	G	O	O	R
11	?	Y	?	?	D	O	E	A	R	I	S	L	N	I

Example:
1. (Across) _______ are examples of fine-grained soils.
C-L-A-Y-S
Cross out the letters
C-L-A-Y-S in row 1 across.

Answers:
2. tributary
3. permeable
4. porous
5. winds
6. delta
7. small
8. aggregate
9. ice
10. water
11. old
12. pores
13. sand
14. speed
15. eddy
16. sea
17. age
18. rock
19. high
20. in
21. rivers
22. larger
23. runoff
24. erosion
25. wind

Across

2. A small stream that feeds a bigger one is called a _______.

3. A soil that allows water to flow through it is said to be _______.

4. A soil that has large openings between soil particles is very _______.

5. _______ are the least effective eroding agents.

6. Distributaries can be found on the _______ that forms where a stream empties into a large body of water.

7. The particle size with the most surface area for water to stick to is _______.

8. A clump of fine soil particles that can form in soils is called an _______.

9. When water turns to _______, it is a strong eroding agent.

10. _______ is responsible for the most erosion, but only if it is allowed to run off the land.

11. The streams that do the least amount of cutting are the ones in _______ age.

Down

12. The spaces between soil particles are called ______.

13. ______ is an example of a relatively coarse-grained sediment.

14. "Velocity" means the ______ of a stream or streams.

15. A circular flow of water is called an ______. It helps to cause the velocity of a stream to decrease for a while as the bed angle increases.

16. Eventually, much of the material cut away from the top of a mountain can end up in the ______.

17. The ______ of the stream is a factor that determines how well a stream will cut away the land. (Hint: See Question 11.)

18. Eventually, the material carried to the sea by rivers will return in the form of ______.

19. Stream velocity is highest when the water level is ______.

20. Water flows slowest on the ______ side of a bend in a river (a meander).

21. Valleys are cut by ______.

22. Sand holds back water better than soils with ______ particle sizes.

23. There can be no erosion due to water until there is surface ______.

24. The process of cutting away and removing the land is called ______.

25. A dune is a pile of particles that has been moved by ______.

For Further Reading

Mather, Kirtley F. *The Earth Beneath Us.* New York, Random House, 1964. This book contains exceptional photos of glaciers and erosion.

Harland, W. B. *The Earth: Rocks, Minerals, and Fossils.* New York, Franklin Watts, Inc., 1960. From this book you can read how the earth was formed and how it has been changed by erosion and weathering.

Chapter 14 Deposition

Earth science class had just ended. The period had been spent doing a messy experiment to answer the question, "How is sediment sorted into separate layers of sand and clay after it reaches the sea?" Kim and Dixie were making a mad dash to their lockers to get rid of their books before lunch.

"You and your dumb questions," Dixie said as she piled her books in her locker. "You should know by now that Mr. Edleman isn't going to come right out and give you the answer. Look what your question did to my sweater."

"It's just water and a little sand," Kim assured her. "Give you water in a steel bottle and you'd find a way to spill it."

Dixie slammed her locker shut and stood there, brushing off her clothes with the back of her hand. "I don't see what dropping sand down a tube of water has to do with whatever that was you asked about. And that brick-throwing bit. If I could have used Mr. Edleman as a target, I could have thrown them farther."

"Better get a raincoat for earth science class, Dixie," was Kim's reply. "Come on. Let's go to lunch."

They chose a table. Roberta, another earth science classmate, joined them shortly with her usual double lunch all but hanging over the sides of the tray.

"Hi, Bobbi. Maybe you can help me get through to Dixie. She really doesn't get Edleman's brick bit." Roberta adjusted her chair and began to eat.

She stopped eating only long enough to comment. "Well, the wooden blocks and the bricks stood for two different materials in a load of sediment."

"I got that much," Dixie interrupted.

"Mr. Edleman just showed us something we already know." Roberta stopped for another mouthful, then continued. "You can't throw a brick as far as you can throw a block of wood. The brick always lands close to the thrower.''

''So?'' Dixie asked.

''A stream or river always drops the heaviest pieces of sediment first, even if all the pieces are the same size. The heavy pieces are like bricks being thrown by the river.'' Kim took a breath and waited to see if Dixie understood.

''What about the sand and water?'' Dixie looked at the nearly dry spot on her sweater.

''Mr. Edleman was just trying to show us another way particles of sediment get sorted.'' Kim stopped to see if Roberta could add anything.

''If you had made sure the clamp on the tube was tight, you wouldn't have gotten wet. Try it again and see. There! The end!'' And with this she stuck a forkful of her second dessert into her mouth.

Sediment Reaches the Sea

14.1 Sediment is sorted after it reaches the sea.

Process: observing

You know that sediment deposited in the sea may eventually become sedimentary rock. Look at the picture of shale in Figure 9/19. All of the particles are too small to see. Then look at the picture of sandstone in the same figure. You can see the particles that make up the rock. They are all about the same size. But were the particles of sand you examined in Section 13.7 all the same size? The stream that brought the material to make the sandstone and the shale probably carried both large and small particles, all mixed together. When they reached the sea, somehow the particles got unmixed. Before they dropped to the bottom of the sea, they were sorted, with large particles in one place to make sandstone and small ones in another place to make shale. Many sedimentary rocks are similar. They do not contain mixtures of particle sizes. This process of dropping off and building up of layers of sediment in a body of water is called **deposition** (dep-uh-ZISH-un).

14.2 Different particle sizes settle at different rates.

Try the experiment on sediment sorting for yourself. As you do the experiment, try to imagine a stream carrying its load of sediment across a delta and into the sea. Picture the sediment load being dropped when the stream meets the larger body of water and slows down. In your model you will watch the particles of sediment from close by as they fall toward the bottom.

Materials

screen sieve, stop clock, ringstand, clamp, column with drain tube, paper cups, sand, spoon, graph paper, lead shot

Activity Time: 30 minutes.
Concept: interaction of matter

★ Put a cup of clean quartz sand in the top section of a set of screen sieves. Shake the sieves to separate the sand into different-sized particles. These will be the sediment particles brought to the sea by a stream. Put the different particle sizes into different paper cups and number them in order of size. Make the largest particles number 1. Put the smallest particles away. You will use them later in the chapter.

Then get a long transparent column fitted with a drain tube. Clamp the drain shut. Fill the column with fresh water and clamp it to a ringstand in an upright position. The column will represent a large, calm body of water. Make sure the clamp on the drain tube is tight. Remember the trouble Dixie had.

Have one member of the team, the timer, operate a stop clock. The second person will watch the column and signal when to stop and start. The third will pour the sediment into the model sea. Figure 14/1 will help you with this experiment.

The larger (heavier) the particle, the faster it sinks. The dust may float because of trapped air and the surface tension of the water.

You may want to put a sample graph on the chalkboard if the students need help with the graphing.

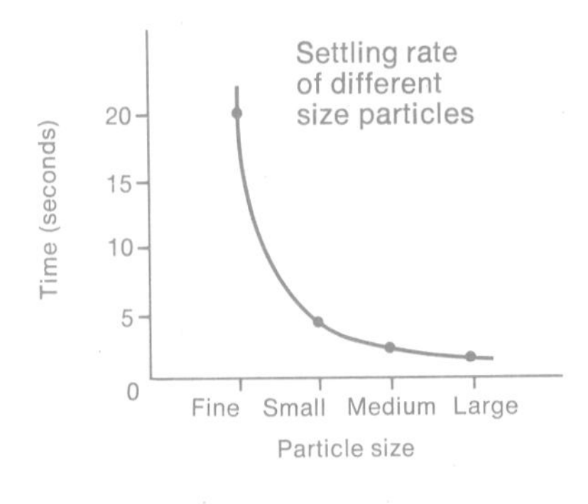

When the column watcher says "Go," pour a spoonful of the largest sediment into the column. At the same time the timer starts the clock. When almost all of the particles have reached the bottom of the tube, the column watcher says "Stop," and the timer turns off the clock. Record the settling time in your notebook on a data table. Label the first entry "1. Largest particles."

Repeat the experiment for each of the different particle sizes. Number the next size particles number 2, the next size number 3, and so on. Plot the finished data on a graph to give you a picture of just how the settling rate changes with particle size. See how to make a good graph in Appendix F. Use your number labels as the particle size

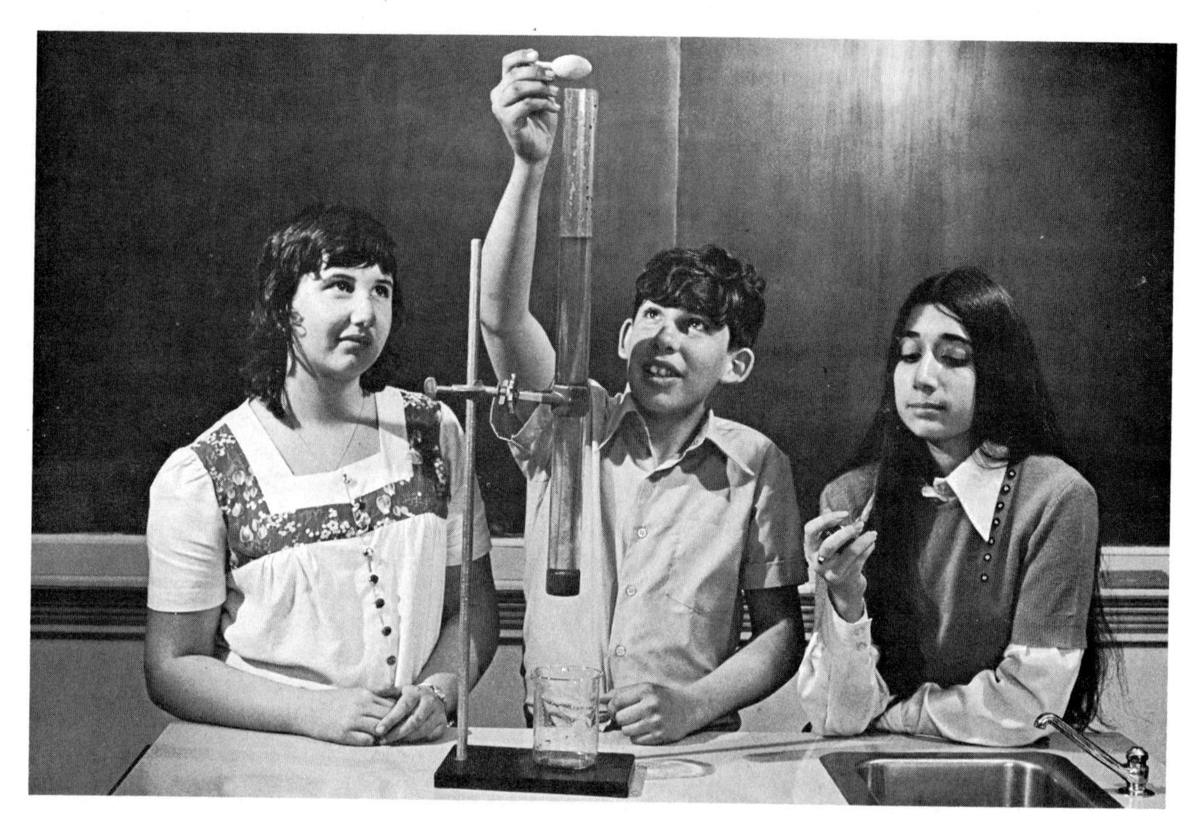

Figure 14/1 *Measuring the settling rates of different-sized soil particles in water.*

on the graph. As a final check of your results, drop a mixture of the different particle sizes into the column. Try to identify which particle size reaches the bottom first. ★

Water carrying a mixture of particle sizes doesn't stop when it reaches a large body of water. The stream water moves out into the sea or lake. You can see this movement when the two waters are different colors. Water can also move along the shoreline, or even diagonally across an ocean, like the Gulf Stream. People in boats can sometimes feel these movements. They are called **currents.**

The line in Figure 14/2 labeled A through E shows the path stream water might take on entering the sea from the left. Pieces of rock that are from the size of a pea to the size of a golf ball (4 millimeters to 64 millimeters) are

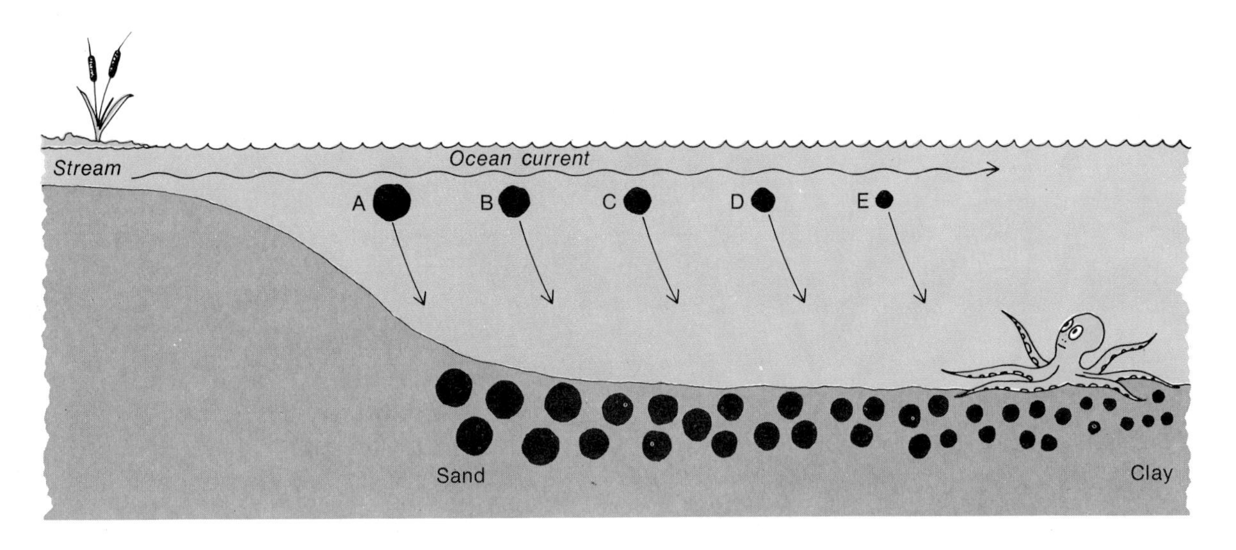

Figure 14/2 *Sediment being carried by a stream into the ocean. Particles of each size are dropped at a different distance from the shore.*

officially called **pebbles.** Pieces of rock that are pebble size and larger would have been left behind before the stream water got into the sea. The pebbles and larger rocks, when mixed with a little sand, form the material called **gravel.** Gravel can be found in river channels. When the stream water enters the sea, all sediment starts to fall toward the bottom and to the right at A because of the current. But only the fastest-settling particles hit the bottom directly under A. The other falling particles are swept along by the current.

The next particle size might not reach the bottom until the mixture of sediment particles is somewhere under point B on the line in Figure 14/2. Because the smallest particles take so long to settle, they are swept beyond point E before finally reaching bottom. The sediment has been sorted.

The circles under the ocean bottom in Figure 14/2 show how the particle size might change from shore out to deep sea because of the different settling rates. If you could walk along the bottom with bare feet, you would notice that it got less sandy as you went away from the beach. When you walked beyond point E, you would find something resembling a mixture of dirty flour and water. As it squished between your toes, you would know that the stream water had contained a mixture of larger particles (sand) and smaller particles (clay). The ocean currents had separated them.

There is a kind of sediment found in many parts of the world which has not been sorted according to particle size or type. Since it has not been sorted, you would assume it probably wasn't brought by a stream. Most of this sediment has been deposited by glaciers and is called **till.**

The pebbles in till are usually widely scattered in a steep bank or fresh outcrop, but sheet erosion may concentrate them on a horizontal surface or gentle grade.

Glaciers mix *all* the different particle sizes and types together and drop them all mixed up. You can recognize till because it has everything from the smallest to the largest size, and the pebbles are scattered throughout. That is, the pebbles in till are not all very close to each other or touching each other as the pebbles in gravel are.

14.3 Some particles are heavier for their size than others.

Sometimes, small particles are found where you would expect to find only large particles. See if you can discover why by doing another experiment with your column of water.

Materials

same as 14.2

Activity Time: 2 or 3 minutes. This can be part of the previous activity. The lead shot will settle first. You may have to put a sample graph on the chalkboard.

Process: measuring, observing

★ Select some sand particles that are about the same size as lead shot. Which is heavier—a sand particle or a piece of lead shot the same size? Drop both the sand and the shot into the column at the same time. Which kind of particle reaches the bottom first? Which kind of particles of the same size will settle faster in a body of water—the heavier ones or the lighter ones? ★

This, of course, is the concept of density. The word "density" is not used because some students cannot handle the idea of two variables (mass and volume) combined in one word. Avoid using the term unless you are sure your students understand the concept.

If the sediment load dumped into the sea in Figure 14/2 were made up of particles all the same size, would the heavier ones or the lighter ones be found closest to A on the line? Would the lighter or the heavier ones be found farther out to sea?

The heavier particles would be closest to A, and the lighter ones would be farther out to sea.

Mr. Edleman showed his class how this might work by having students throw bricks and blocks of wood cut the same size as the bricks. Both the bricks and the blocks of wood occupied the same amount of space, but the bricks were heavier. The same amount of effort was put into

Activity Time: 20 minutes. This can be tried if the class is interested. Use milk cartons—some filled with sand and some with sawdust.

Figure 14/3 *The dark material is made up of heavy iron minerals sorted from quartz grains along the shore of Lake Ontario.*

throwing the bricks and the blocks of wood. But it takes more energy to move a brick a little than it does to move a block a lot. The bricks landed closer to the thrower than the blocks of wood did. In terms of the line in Figure 14/2, the bricks would be found under the A and the blocks under the D or E.

Grains of iron minerals are heavier than most other minerals. They will tend to be deposited close to the shore (near A in Figure 14/2) even though their size is small. The dark band on the shoreline in Figure 14/3 is made up of very small grains of an iron mineral. Grains of sand larger than the iron ones are continually washed away, but the dark grains stay in place.

Prospectors used this concept when they panned for gold. Gold is much denser than most rock. Students may want to try panning for iron minerals on the next field trip. Put some dirt and water in a pie tin and move it in a small circle. Water will spill over the edge, carrying lighter particles with it.

Frederick Fuglister— Tracker of the Gulf Stream

Perhaps the most famous of all ocean currents, if you live in North America, is the Gulf Stream. The Gulf Stream flows northward along the east coast of the United States and then curves eastward toward Europe. Until recently scientists thought of the Gulf Stream as a kind of river in the sea, an unbroken flow of water that moved at a steady rate of between one and two miles an hour.

During the 50's and 60's, scientists began to realize that this was a false picture of the Gulf Stream. Not only did it flow faster than they thought but from its edges swirled eddies that made great loops of rotating currents up to 125 miles in diameter. Some of these eddies lasted about a year before they faded away.

Surprisingly enough, the man who first accurately described these Gulf Stream eddies had neither a college degree nor any formal training as a scientist. He is an artist and a musician named Frederick Fuglister (FYOOG-liss-ter), who took up oceanography by chance at the age of 30.

At the suggestion of a friend who was an oceanographer, Fuglister agreed to go on a cruise for the Woods Hole Oceanographic Institute on Cape

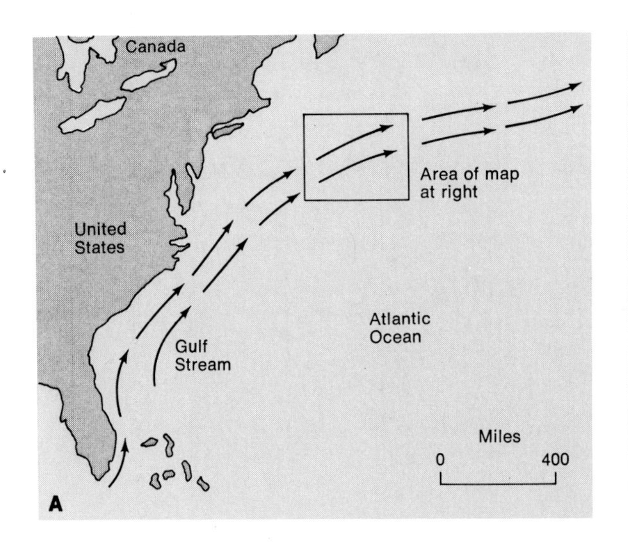

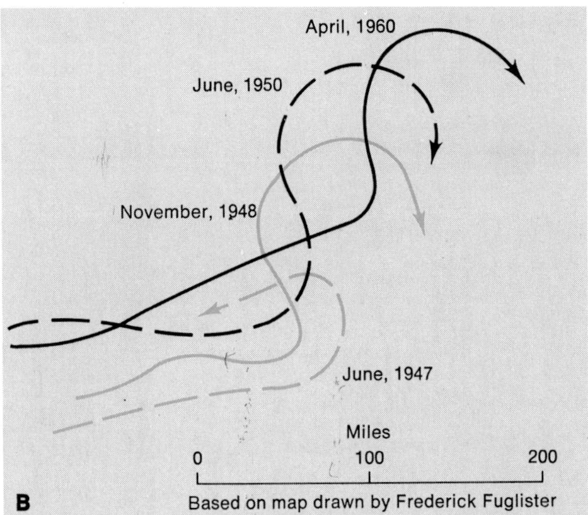

Cod in Massachusetts. Soon he became a regular on cruises from Woods Hole. Fuglister's main job on these cruises was to take temperature readings of the ocean at various depths and record them on charts of that part of the ocean.

After many cruises and many long hours studying data on ocean temperatures and currents, Fuglister saw that charts then in use had no reliable information on the exact course of the Gulf Stream. He realized too that the wanderings of the Gulf Stream were much greater than anyone had suspected. The chart in B indicates how the eddies in one part of the Gulf Stream change from year to year.

In 1969 the National Academy of Sciences awarded Fuglister its highest award for oceanography, the Alexander Agassiz Medal, for his nearly 30 years of work on the Gulf Stream. Probably his artist's eye and imagination helped Fuglister solve the riddle of the Gulf Stream. But perhaps most important were his long years of direct experience with the sea.

Did You Get the Point?

Sediment is sorted after it reaches a large body of water, such as a lake or sea.

Sediment can be sorted because large particles settle faster than small ones.

Sediment can be sorted because heavy particles settle faster than light ones.

Figure 14/4
How has the water level changed at this location? The different layers will give you the answers.

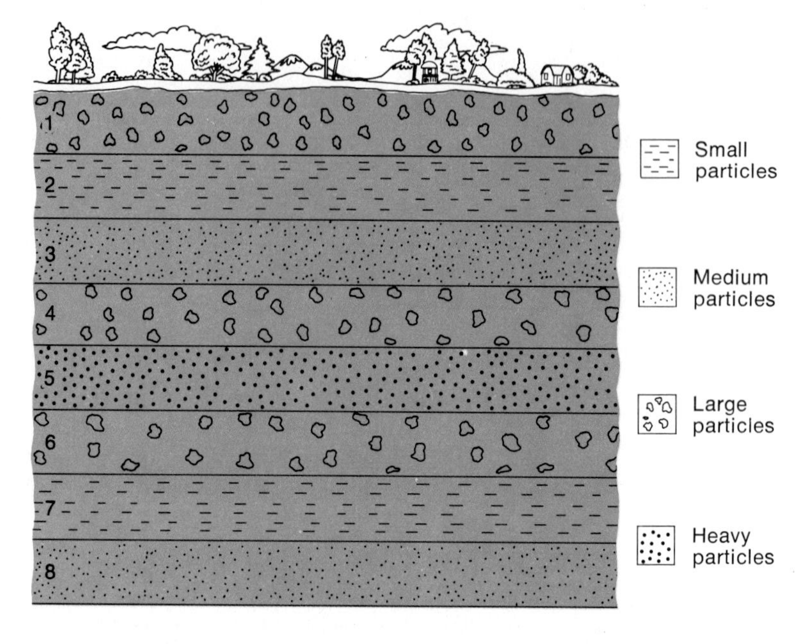

Heavy and large particles are deposited close to shore. Small and very light particles are deposited farther out to sea.

Glaciers leave sediment that is not sorted.

Check Yourself

1. Figure 14/4 is a cross section of an area that was once the bottom of a large body of water. You are looking at it from the side where some of it has been scraped away. As you can see, layers of sediment built up over the years before the body of water dried up. From time to time, as the sediment was being dumped in, the shoreline changed. Sometimes the plot of land in Figure 14/4 was at the shoreline, and at other times it was far out from the shore. Using the key, determine whether the shore must have been very close, a long way off, or in between, when each layer was deposited.

Water depth:
1. shallow, near or at shore
2. deeper, away from shore
3. medium depth
4. shallow, near or at shore
5. shallow, near or at shore
6. shallow, near or at shore
7. deeper, away from shore
8. medium depth

Features of the Sea Floor

14.4 The sea floor is a whole new world.

Suppose you had a special suit that would let you survive the tremendous pressure of a column of water seven miles high over every square inch of your body. That would be a pressure of 160,000 pounds (72,000 kilograms) per square inch instead of the 14.7 pounds (6.6 kilograms) per square inch you are used to on the surface of the earth. With a lamp attached, this suit would enable you to explore the dark, deepest part of the ocean floor, the Marianas (mar-ee-AN-ehs) Trench. It is just a little under seven miles deep. If the suit were flexible enough, you could get samples of the material on the sea floor from any place you wished.

With the millions of tons of sediment being dumped into the sea as the continents wear away, the sea floor itself must be changing. Your special suit with its built-in lamp would let you take a good look around the undersea world and investigate some of these changes. The map in Figure 14/5 shows some of the things you might see along the way.

The width of the continental shelf varies. Show this on a relief globe if one is available.

The deep sea floor does not begin until you are some way from shore. The shallow part before the drop-off is called the **continental shelf.** It is nothing more than the outside edge of the continent extending out to sea. It has very little real slope. If you were standing on a part of the continental shelf, it would look as flat as a table. All of the continents have edges like this that are flat and covered with water.

Try making a clay model as a visual aid.

Eventually, the continental shelf ends and the sea floor slope gets steeper. It is almost as if there had been a sharp edge there that had weathered away. The ending of the continental shelf is called the **continental slope.** Toward the bottom of the slope there is usually a bulge resembling the talus piles found at the feet of weathering cliffs. The bulge makes the slope more gentle and is called the **continental rise.** The end of the rise marks the beginning of the deep sea floor. The areas described can all be seen in Figure 14/6. The continental shelves and slopes and the sea floor are all cut by underwater valleys. The deepest ones are called **submarine canyons.**

The sea floor is covered with fine sediment that has been brought out from land. Here and there deposits made up of dead sea animals, plants, and minerals have been laid down. But even these get mixed with the continental mud. The mud is everywhere.

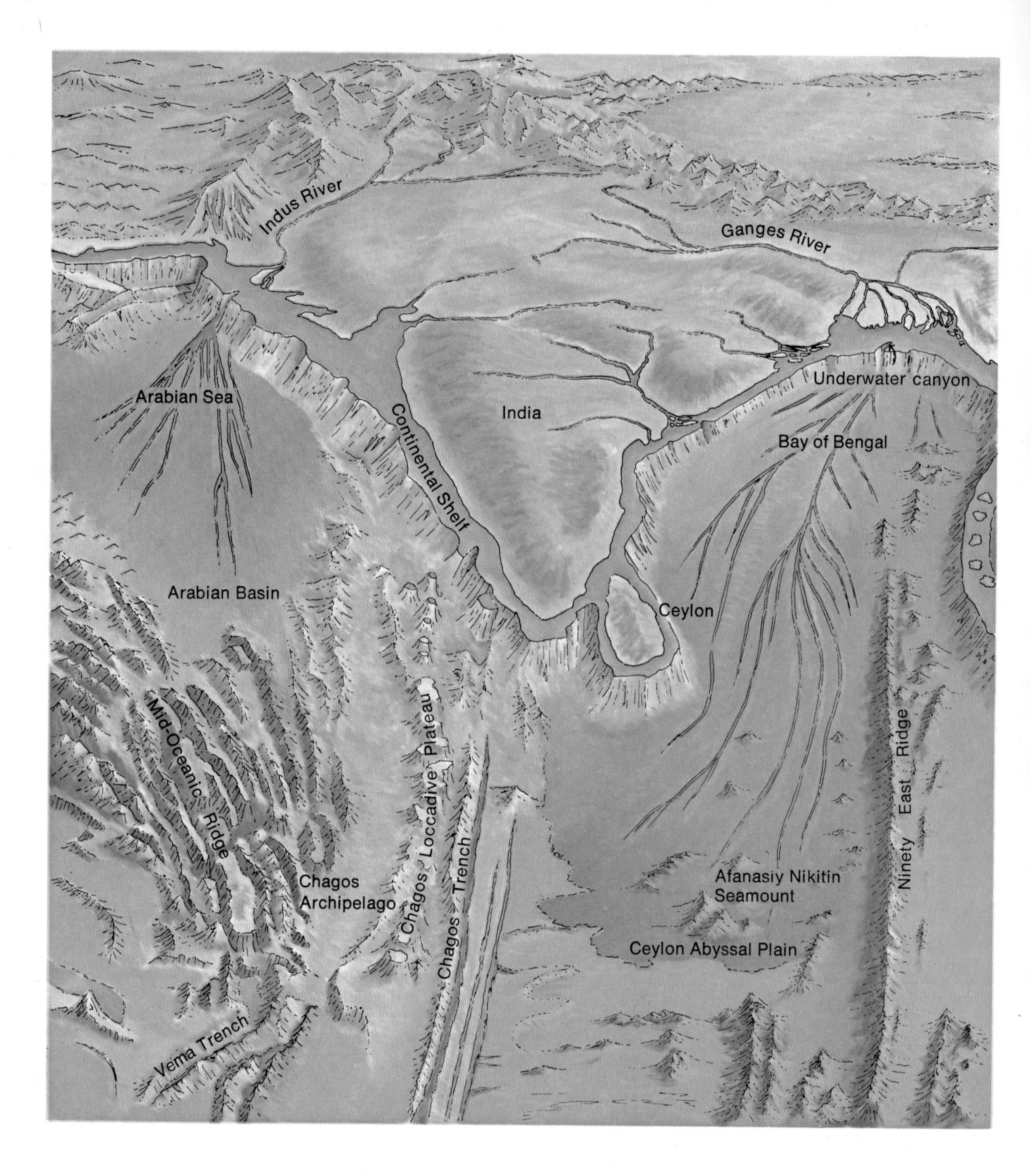

Figure 14/5 *A view of the ocean floor off the coast of India. Pick out as many different undersea features as you can. The Arabian basin is a trench, as much as 5203 meters (17,070 feet) deep between the two ranges of volcanoes.*

Figure 14/6 *The different parts of the underwater landscape from the edge of a continent down to the sea floor.*

The land masses of the earth have flat areas called **plains**, but the ones on the ocean floor are much flatter for their size. A real mystery begins when you try to figure out how the sea plains were made. There are wind and fast-moving streams to erode the land. But the water on the bottom of the ocean does not erode the way rivers do, and there can certainly be no wind. How are flat plains made, at the bottom of the sea?

14.5 Currents carry sediment far out to sea.

If you stopped to examine underwater canyons while on your undersea field trip, you might find a clue to the eroding agent at work on the bottom of the sea. Take a look at the underwater canyon labeled on the map in Figure 14/5.

The canyon is V-shaped like a river valley and seems to have been carved out of the continental shelf. Many canyons extend all the way back from the floor of the sea to the dry land. How many can you find on the map?

At places where some of the canyons on the map in Figure 14/5 begin, streams empty into the sea. But water from a stream can only cut the canyon if it is heavier than the water in the sea. The water would have to sink and flow under sea water to stay in contact with the continental shelf. Can fresh water be made heavier than sea water? The answer is in the smallest particles you sorted from the sand in the settling-time experiment.

Materials
dust, spoon, test tube

★ Add about three spoonfuls of the fine dust you saved from the sieves in Section 14.2 to a test tube three-fourths

Activity Time: 40 minutes.

full of water. Shake the tube until the fine particles mix with the water. ★

What you have in the test tube is a **suspension.** A suspension is a mixture of solid particles in a liquid, in which the solid material can usually still be seen. Given enough time, it will settle out. The particles of solid material are large compared to the molecules of the liquid. Do not confuse a suspension with a solution. A **solution** is a mixture in which the solid material has disappeared into the liquid. The solution may be colored, but the particles in the solution will be invisible because they are the size of molecules. A solution will not settle out no matter how long it sits.

Suspensions like the one in your test tube are sometimes called slurries. **Slurry** is a special name for the suspension of a solid in a liquid. Muddy rivers and streams carry some of their sediment in suspension. Rivers and streams could be said to dump slurries into the sea.

Materials

column, clamp, and ringstand; or earth box or aquarium, salt water, food coloring

The angle should be about 45°. Slurry will not mix. It will sink. Suspension is heavier than fresh water.
Process: observing

★ Fill a column with fresh water and clamp it at an angle, or fill an aquarium or earth box about a quarter full of water. Dump your slurry into it. Does the slurry mix with the water? Does it sink to the bottom? Is the water with the smallest particles suspended in it heavier than the fresh water?

Did the swirling motion of the suspended material as it moved down the tube look like Figure 14/7? This is called a **turbidity** (tur-BID-uh-tee) **current.** A turbidity current forms whenever a slurry is dumped into clear water.

How does salt affect the weight of water? Replace the water in the container with fresh water. Slowly dump 500 milliliters of colored salt water at room temperature into it. Where does the salt water go?

The salt water goes to the bottom.

Replace the water with real or fake sea water. Prepare another slurry, and pour it into the water. Is the slurry heavier than sea water? ★

The finer material in the slurry may remain suspended, but the rest will sink.

See Teacher's Manual for supplementary activity, involving a saturated salt solution.

Turbidity currents will continue to move as long as they are heavier than the water they are moving through or until friction stops them. Then suspended material can be moved by other ocean currents. In this way earth materials can be carried far out to sea.

Figure 14/7
Make a turbidity current by dumping a slurry into fresh water.

Demonstration Time: 15 minutes. If possible, several students should help you set up and carry out this demonstration.

Materials

aquarium or earth box, funnel, clamp, ringstand, hot water, food coloring, ice

14.6 Turbidity currents can be influenced by conditions in the sea.

Temperature seems to affect the movement of turbidity currents.

★ Fill an aquarium or earth box with water at room temperature as shown in Figure 14/8. Clamp a funnel close to the edge so that the bottom of the funnel stem is just below the surface of the water. Pour in 500 milliliters of hot water with food coloring in it. Where does most of the colored water go? If the water from a stream is warmer than the sea it empties into, is a turbidity current likely to develop?

Hot water floats.

No, a turbidity current will not develop.

What do you think would happen if the stream water were colder than the sea? Use your aquarium or earth box again. Do not change the water. Put 500 milliliters of ice-cold water with food coloring in it through the funnel. Where does the water go this time? If the water in a muddy stream is colder than the sea it empties into, is a turbidity current likely to develop? ★

Cold water sinks, so the cold stream would promote a turbidity current.

Figure 14/8
Will the colored salt water rise or sink when poured into the aquarium?

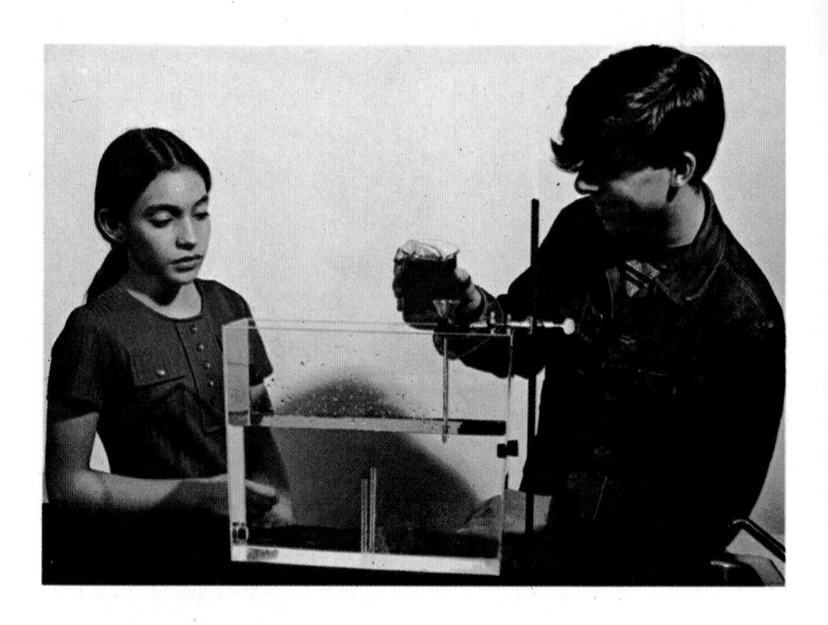

14.7 A strong current may spread sediment over the sea floor.

You can watch the process of sediment deposition on a small scale.

Materials

stream table, plaster of Paris, Portland cement, soil, test tubes, sand, spoon, trough

Activity Time: 60 minutes. The pans will have to be stored overnight in a place where they will not be disturbed. Rest the trough on the bottom of the pan at a 30° angle and clamp the trough to hold it steady. Some classes can do the gallon jar activity in Section 14.9 while one class finishes the stream table activity.

★ Mount a trough or column in a stream table that has been filled with water instead of sand. The trough will be your model of a stream emptying into a pool of still water like a sea or lake. To make the model more like the real thing, use a variety of slurries. Slurries made of plaster of Paris, cement, and a mixture of fine dark soil and sand will stand for different kinds of stream sediment.

Fill a large test tube about one-third full of powdered cement. Dump in a spoonful of unsorted quartz sand to give the sediment load a few large particles. Fill the test tube with water. Put your thumb over the top and shake it. Pour everything in the test tube down the trough at once. Carefully watch what happens.

When the turbidity current has carried the finest particles almost to the end of the stream table pan, mix another slurry. Use plaster of Paris with a spoonful of sand this time. Shake and pour it down the trough.

When the turbidity current has carried the finest particles nearly to the end of the pan, send down a slurry made from a half-and-half mixture of the sand and soil. Be careful not to jiggle the water in the pool.

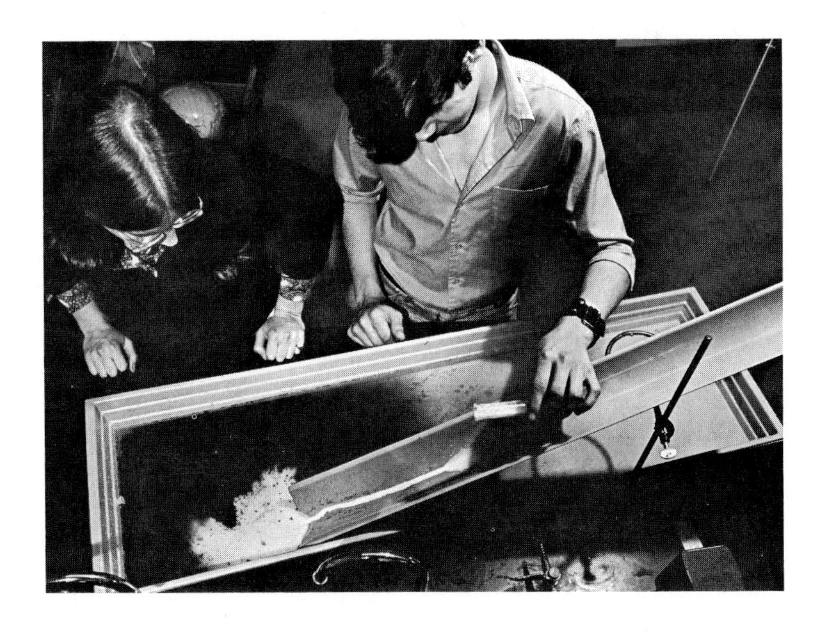

Figure 14/9
Building up layers of sediment on a model sea floor.

Repeat the series, changing the order if you wish, several more times. Try to get a build-up of sediment on the model sea floor as the students are doing in Figure 14/9.

When there is a deposit one-fourth inch thick on the bottom of the pan, siphon off the water or drain the pan, if it has a drain opening. Leave the last bit of water to evaporate overnight. Do *not* disturb the sediment until it has hardened.

The result will be a brittle slab showing distinct layers of light and dark material. The larger particles will be near the trough, the layering develops away from the trough.

When the plaster and the cement have dried, carefully lift the slab from the pan. Some of it will crumble, but you should be able to get out a few large pieces. Examine the edges of the pieces. What do you see? How would you describe the particle size making up the slab? Where are the larger particles? Is there any layering near the trough? Draw a cross section of the slab in your notebook. ★

14.8 The ocean floor builds up slowly.

As you can see, sediment may be deposited far beyond the end of the stream. The most rapid build-up occurs where the stream first meets the larger body of water. But sediment deposition happens in other areas, too.

By looking at the slab of sediment you deposited, you can see that only the smallest particles travel far out to sea. As you could have predicted from your experiment with settling rates, the larger particles dropped close to the mouth of the stream (the trough). Everywhere in your model sea, the sediment was laid down in layers. The layers alternated as you changed types of slurry. By looking at the layers you can tell the order in which you dumped in different slurries.

Perhaps you also noticed that the layers of sediment thinned out as they got farther from the trough. What could you say about the rate at which the deep ocean floor is built up, compared to the rate at which the area near the shore is built up?

The continents are worn down one foot per 9000 years, and the oceans have more than twice the surface area of the land. So, the oceans should theoretically gain about 29″ of sediment in 50,000 years. Along with the reason presented here, it is interesting to note that Chapter 15 will show that sections of the ocean floor may not remain exposed for the full 50,000 years. Some sections of the sea floor are sinking into the earth's crust and being replaced with new material.

The real ocean floor builds up so slowly that it takes an average of 50,000 years to lay down one and one-half inches of sediment in the deepest parts. One reason for the slow rate is that the amount of sediment is small compared to the very large area it spreads over. Not all particles of sediment produced by the weathering process are small enough to be carried to the deep ocean floor. Most of the particles are so large that they are deposited close to the continental shelf. Still another reason will be taken up in Chapter 15.

14.9 Sediment deposition can change the shape of the sea floor.

Your model had a sea floor that was perfectly smooth. What would have happened if the floor had been rough? Would flat, horizontal layers still build up? By changing your model a little, you can see for yourself.

Materials

jar, crushed stone, paper cups, dark soil, sand, spoon

Activity Time: 30 minutes.

★ Cover the bottom of a large glass jar with rocks. They should be about one-half inch to one inch in diameter. If this layer of rock stands for the sea floor, it would indeed have to be called rough. Fill the jar one-fourth full of water. Now you are ready to begin depositing sediment.

This time, make the slurries in a paper cup. Use half a cup of dark soil and half a cup of water for the first one. Stir the slurry and quickly pour it into the jar before the soil particles settle in the cup.

Allow a minute for the slurry to settle. Then add a light-colored sand slurry made the same way. Keep adding the dark and light slurries until the jug is filled all the way to the top.

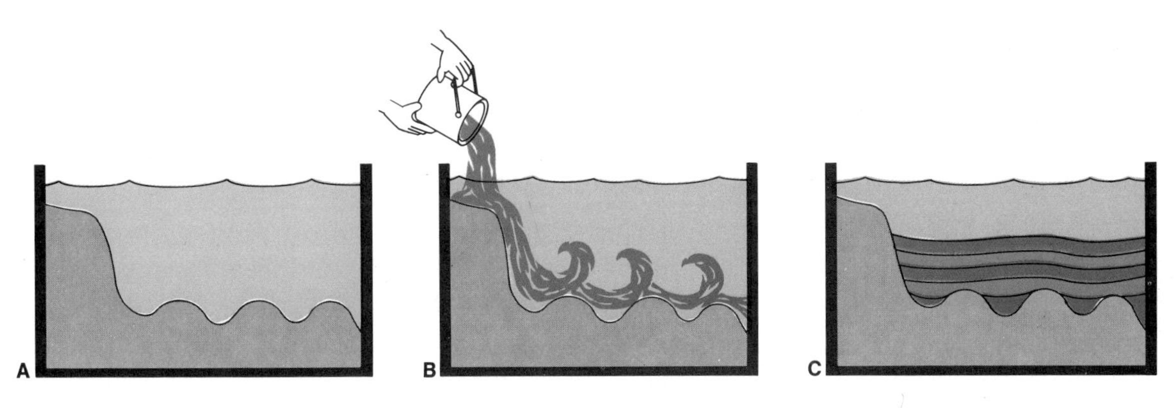

Figure 14/10 *Sediment poured into a tank with a rough bottom will smooth out the uneven places. The sea floor becomes smooth in the same way.*

The jar may have to sit overnight before the water clears. Settling time depends on the type of soil used. Bends in the layers will be gradual.

When the water has cleared enough to see inside, examine the layers that have developed. Is the surface of the sediment deposit smooth or rough? Look at the way the layers "flow" over the tops of the rocks that touch the sides of the jar. Are the bends in the layers sharp or gradual? Even the roughest sea floors become smooth after enough sediment is deposited on them. The layers bend a little where they meet the rocks, but the layering is mostly horizontal. Figure 14/10 shows another experiment that was conducted to study the deposition of sediment on an uneven surface. Notice the slight bending of the layers as they meet the sides of the low spots. Also notice how they straighten out when the low spots are full, to make the floor of the tank smooth. ★

Did You Get the Point?

Some type of eroding agent which does not occur on land appears to be working in the sea. Underwater canyons and the flatness of the sea floor are evidence of the agent.

Muddy water sinks through lighter water to form turbidity currents.

Turbidity currents can carry sediment great distances underwater.

Turbidity currents can be stopped or their courses changed by differences in the temperature of the water.

Sediments carried by the currents are laid down in nearly horizontal layers, regardless of the shape of the sea bottom.

Answers:

1. This is a thought question; no sure answer can be given. Perhaps they are extensions of streams which have disappeared. Perhaps they were formed by local currents washing sediment away. Maybe they represent underwater erosion along faults, and so on.
2. This depends upon the relative densities of the slurry and the salt water. Show the students how to make a hydrometer by sticking a thumbtack into the eraser of a pencil. The pencil will float, eraser end down, at a level that depends on the density of the liquid.
3. Two possibilities are to cool the water or add salt.
4. The layers will bend near obstacles but the upper layers will be flat.

Check Yourself

1. Some underwater canyons in the map in Figure 14/5 do not have streams emptying into the ocean where they begin. How do you think the canyons got there?

2. Suppose you have a slurry of sediment in water, and a tank filled with salt water. How can you predict ahead of time whether or not the slurry will cause a turbidity current if it is dumped into the tank?

3. Suppose you have a weight that can just barely float on pure, fresh water. You put it into a cylinder of hot water and it sinks. Describe two ways you can get the weight out without dumping out the water or in any way reaching into the cylinder.

4. Draw a back yard as it would be seen from the side. Include things like trash cans, clothesline poles, and swimming pools. Then, assume that the back yard suddenly becomes part of the ocean floor. Draw the layers of sediment that might be laid down until everything, including the house, is completely covered.

The Sinking Ocean Floor

14.10 The thickness of sediment layers on the ocean floor can be measured.

Since erosion is constantly wearing away the continents, you might expect the layers of sediment on the sea floor to be very thick. Machines have been built to measure the thickness of the sediment layers.

One machine sucks sediment into a pipe as it drives the pipe into the sea floor. It would be like hooking a vacuum cleaner to a straw and then driving the straw into the ground. Because the driving and sucking take place at the same rate, the operation does not disturb the layering of the sediment. When the pipe is split open, the thicknesses of the sediment layers can be measured. Such samples are being examined in Figure 14/11.

Activity Time: 10 minutes. The students may want to try to make core samples of their deposits from the activity in Section 14.9, using plastic straws. It is hard to do.

Another method of determining the thickness of the sediment layers uses echoes. From your own experience you

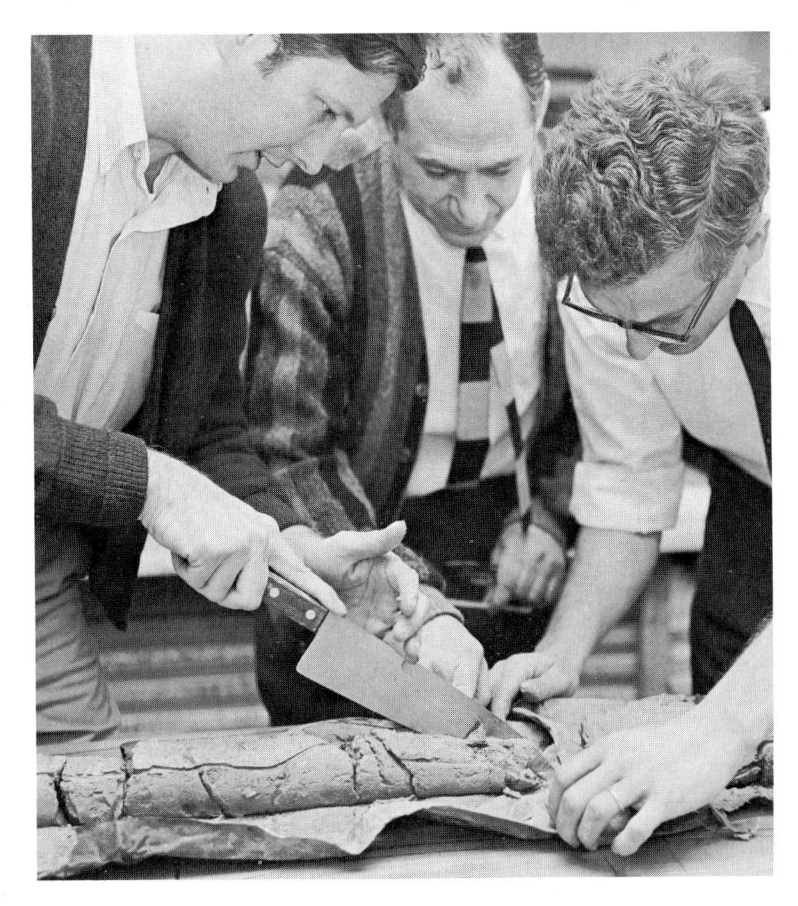

Figure 14/11 *Oceanographers cutting up a core of sediment layers from the ocean floor.*

know that some materials absorb sound better than others. Brick walls bounce sound back and cork walls make a room quiet. In the same way, sound should be absorbed more easily by soft sediment than by solid rock.

Large explosive charges are set off in the sea. A machine records when the charge went off, while microphones pick up echoes of the explosion as they come back from the bottom. The echoes arrive at different times depending on how deep the sound traveled before bouncing back.

Process: measuring

Figure 14/12 shows a section of the chart on which such an "echo machine" records this information. The top of the pile of sediment shows up quite clearly. Even differences in the types of sediments show up, as different degrees of blackness in the tracing. The sound not sent back

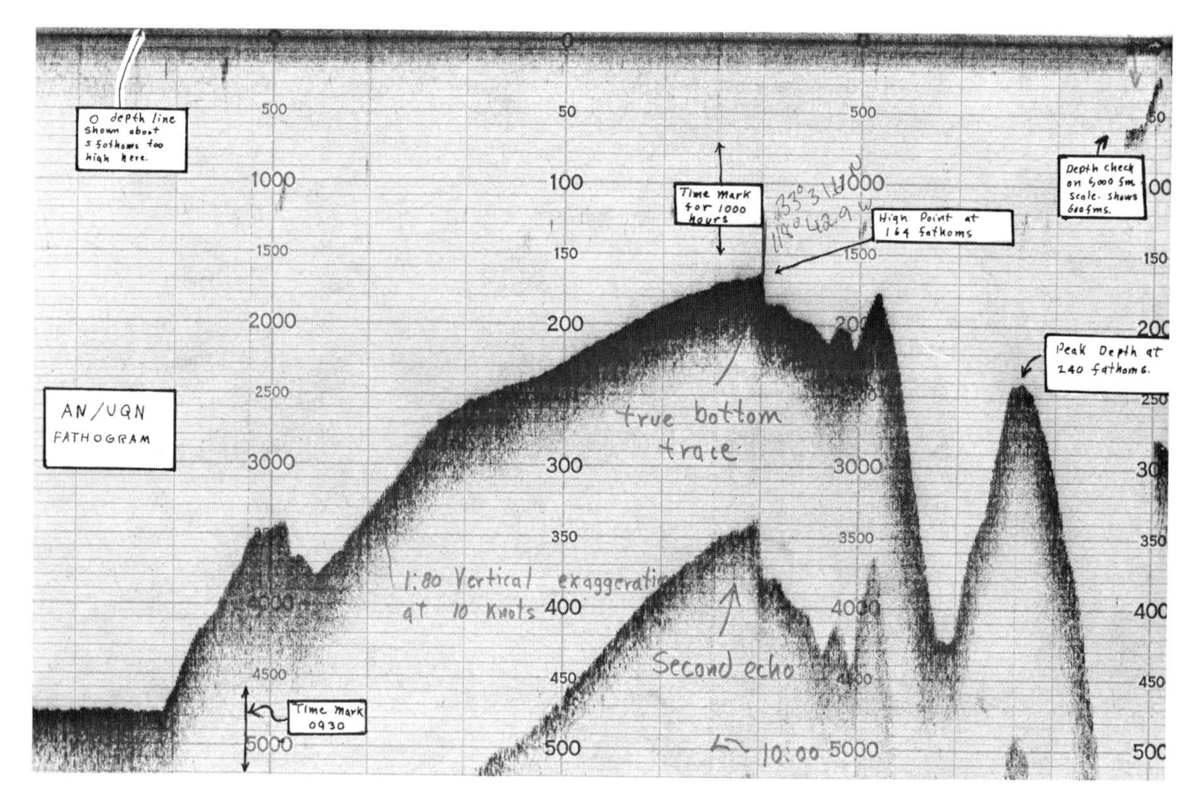

Figure 14/12 *An actual echo sounding chart like this one can provide information about the shape of the ocean floor.*

by the sediment eventually reaches the rock bottom of the sea floor. Echoes from the rock mark the bottom of the sediment pile.

Measurements like these show that the sediment layer on the floor of the ocean is not all the same thickness. In some places it is 1600 feet thick. In others it is only 900 feet thick. In a few places, sedimentary rock layers found high up on mountains go down nine miles. That *really* confuses the issue.

14.11 Parts of the ocean floor are sinking.

Measurements of the thickness of sediment deposits in the sea raise several questions. If the ocean floor today is seven miles deep at its deepest part, does finding sediment layers nine miles thick mean that the ocean had nine-mile deep trenches at one time? If we could know the rate sediment is laid down, could the age of the ocean floor be found? How much of the sediment came from the continents?

Scientists studying the sea have found a few clues to help answer some of the questions. "No," they say, "there was not a trench nine miles deep. The age of the ocean floor is difficult to determine, and not all of the sediment is from the continents."

Some of the evidence that parts of the sea floor are sinking was provided by a scientist named Charles Darwin. (You will meet him again in Chapter 18.) After studying coral reefs and **atolls** (A-tallz), circular reefs like the one in Figure 14/13, he proposed a theory. He knew that reef coral is a kind of animal that grows only in water less than 250 feet deep. He knew that the animal fixes itself to the bottom of the ocean and grows upward like a plant. When the first animals die, their skeletons remain to form a base for the later coral animals. As time goes by, a large mass

Figure 14/13
This Australian island is entirely surrounded by a coral reef or atoll that extends thousands of feet to the ocean floor. How did the large area of shallow water form within the reef?

of old skeletons is formed, with the living animals growing upward and outward from it. The ocean floor in the area of some reefs and atolls was deeper than 250 feet. The live coral at the surface must have grown up at the same rate the coral beneath it sank. The coral which sank below 250 feet died. Recent drillings have shown that many coral reefs and atolls have 250 feet of living coral growing on a column of fossilized coral extending almost a mile beneath them.

At about the same time Darwin made his discovery, odd-shaped peaks were found way below the surface of the sea, such as the one in Figure 14/14A. They seemed to be like volcanic islands except that they had flat tops. Scientists think that at one time the flat topped seamounts extended above the surface of the sea like volcanic islands. Figure 14/14B shows how the waves in the sea eroded the peaks away, leaving each with a low, flat shape. Then as the sea floor sank, the eroded mountains sank with it.

Figure 14/14
Mountains with flat tops like the one in A can be found far below the surface of the ocean. B shows how waves may have formed the flat tops by eroding volcanic islands.

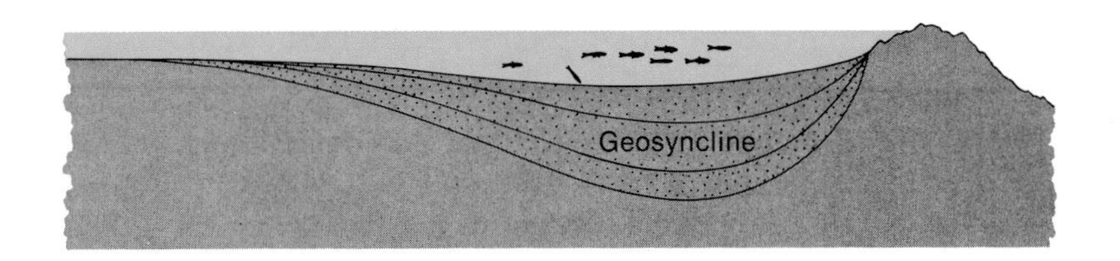

Figure 14/15
This cross-section of a geosyncline shows how the sediment piles up as the sea floor sinks.

14.12 The sinking floor raises new questions.

An area of sea floor that seems to be sinking as fast as sediment is piled into it is called a **geosyncline** (jee-oh-SIN-cline). A cross section of a typical geosyncline is pictured in Figure 14/15. The weight of the sediment alone is not enough to cause the sinking. Some other force or forces must be acting, too. Geologists have found evidence that there have been many geosynclines in the past. Some areas today may be modern geosynclines, such as the Gulf of Mexico and the Atlantic Ocean off the east coast of North America. Volcanic rocks are common in old geosynclines. Volcanoes are common in the western Pacific, as are earthquakes. Maybe they have something to do with a new geosyncline there. Whatever the force is that causes the floor to sink, it seems to be related to earthquakes and volcanoes.

Of course, you might argue that the mile-deep coral skeletons and the flat-topped seamounts are caused by water rising in the sea instead of the sea floor sinking. But as far as we know, the water level has not risen enough to explain nine-mile-thick sediment layers, mile-deep columns of coral, and flat-topped seamounts several hundred feet beneath the surface.

What is the force that causes the sea floor to sink? Will it continue to make the sea floor sink?

Did You Get the Point?

The thickness of sediment deposits which form on the floor of the ocean can be measured.

Measurements of the sediments and mountains on the ocean floor indicate that parts of the floor are sinking.

The age of the ocean floor cannot be measured by knowing the rate of sediment deposition.

Check Yourself

This letter will be a review of the chapter. Letter should note that: (a) turbidity currents carry some sediment far out to sea; (b) sea floor will sink at same rate sediment settles (geosyncline).

1. Imagine that a pen pal of yours lives near a place where a very large river empties into the sea. He has written to you about floods that make the waters muddy and carry great amounts of soil to the sea. He is worried about the area off shore that his family depends on for fishing. It might be filled in by the time he has to take over the job of supporting a family! Write him a letter convincing him that he has nothing to worry about. Tell him about the things that affect the rate of deposition.

What's Next?

The sediment on the sea floor came from the weathering and erosion of all kinds of rocks on the continents. Some of it started as magma that reached the surface of the earth to become lava. The lava cooled into rock. The new rock weathered along with the old rock and eroding agents carried the sediment to the sea. Finally, the sediment was sorted and piled into geosynclines and spread over other parts of the ocean floor. But is that the end of the story? Remember the rock cycle in Chapter 9!

There is evidence that the building up and breaking down by volcanoes, weathering, erosion, and deposition repeats itself over and over again. The sediment at the bottom of the pile may eventually change back to rock by cementing or melting under extreme pressure. It may then be forced to the surface for another "go round."

Many places on the earth are building up and others are breaking down. As mountains are rebuilt, other parts of the earth are changed. There are interesting new theories to explain how the changes are related to each other.

Skullduggery

Directions: The missing words to the statements below are given in code. When you have the first few answers, the ones you KNOW are correct, you will have enough of the code to help you figure out the answers to the other questions. The code is a simple one. The letters of the alphabet have just been rearranged. Q is used in the code instead of A, and so on. Any letter of the alphabet could be any other letter in the code. Make yourself a chart as you find out. Write your answers on a separate piece of paper.

Answers:
Code: (1) A = Q; (2) B = W; (3) C = E; (4) D = R; (5) E = T; (6) F = Y; (7) G = U; (8) H = I; (9) I = O; (10) J not used; (11) K = A; (12) L = S; (13) M = D; (14) N = F; (15) O = G; (16) P = H; (17) Q not used; (18) R = K; (19) S = L; (20) T = Z; (21) U = X; (22) V = C; (23) W = V; (24) X not used; (25) Y = N; (26) Z not used.

1. large or heavy
2. sandstone
3. turbidity currents
4. cold water
5. colder
6. more
7. flat
8. where the continental rise ends
9. smooth sea plain
10. the surface becomes smooth
11. wrong
12. the ocean floor is sinking
13. a) flat topped seamounts
    b) nine-mile thick layers of sediment
    c) deep coral
    d) fossils in deep sediment
14. cycle
15. rock
16. The next chapter will bring the sediment to the surface again.

1. The particles of sediment that settle out in water fastest can be described as SQKUT GK ITQCN.

2. The sedimentary rock LQFRLZGFT would be found close to shore because of the way sediment is sorted.

3. The eroding agent at work on the bottom of the sea is ZXKWOROZN EXKKTFZL.

4. EGSR VQZTK is heavier than warm water.

5. If a stream carried no sediment in suspension, its water would have to be EGSRTK than the sea it emptied into in order to set up a turbidity current.

6. If fresh water weighs one gram per cubic centimeter, salt water weighs DGKT than one gram per cubic centimeter.

7. The continental shelf would be described as being YSQZ.

8. The sea floor begins VITKT ZIT EGFZOFTFZQS KOLT TFRL.

9. About 66 percent of the sea floor is LDGGZI LTQ HSQOF.

10. When sediment is deposited on an uneven surface, ZIT LXKYQET WTEGDTL LDGGZI.

11. The statement that "turbidity currents lay down equal thicknesses of sediment all over the sea floor" is VKGFU.

12. It is possible to have piles of sediment nine miles thick because ZIT GETQF YSGGK OL LOFAOFU.

13. Some of the bits of evidence that the floor of the sea is sinking are (a) YSQZ ZGHHTR LTQDGXFZL (b) FOFT DOST ZIOEA SQNTKL GY LTRODTFZ (c) RTTH EGKQS (d) YGLLOSL OF RTTH LTRODTFZ.

14. The deposition of sediment seems to be part of a ENEST.

15. It is believed that the sediment will again appear on the surface as KGEA.

16. Try your code on the message below.
ZIT FTBZ EIQHZTK VOSS WKOFU ZIT LTRODTFZ ZG ZIT LXKYQET QUQOF.

For Further Reading

Mather, Kirtley F. *The Earth Beneath Us.* New York, Random House, 1964. Beautiful photographs of many different land forms including the effects of deposition on the face of the earth.

Dugan, James, Robert C. Cowen, Bill Bavada, Luis Marden, and Richard M. Crum. *World Beneath the Sea.* Washington, D.C., National Geographic Society, 1967. An exciting voyage undersea with a group of oceanographers.

When the teacher told us that the continents floated all I could think of was . . .
HALP!
STOP
NEWYORK 30 mi.
LONDON 30 mi.

Chapter 15 Mountain Building

Dear Cindy,

Here are the earth science notes you asked me to send. I hope you can read them. H
Hope your enjoying your vacation.Too bad it has to be in a hospital though--and so far away.A

Alot of these notes are missing. Mr. Preising has got me all confused. I don't know what to take down half the time. But you a said you wanted them, so here they are. I'll send better ones q next time.

mMaureen and I are still going together.

Oh. Almost forgot. Tommy Chin drew a few crazy sketches about that continental drift/stuff. Mr. Preising didn't like him dumping on his pet theory like that at first but he finally gave b in and let us hang them on the side z board. I'm sending them for your to see.

Take care of yourselff. I'm keeping my eye on your boyfriend for you. k.

See ya, Ted

What Makes Sediment Sink?

15.1 The continents are in balance with the ocean floor.

After reading Chapter 14, you should be almost convinced that parts of the sediment-covered ocean floor are sinking. That there are layers of sediment nine miles thick, as mentioned in Section 14.10, is another story. Gravity alone cannot cause layers of sediment to sink to such a great depth, according to geologists. So what *does* cause geosynclines?

As Ted hinted in his letter to Cindy, there is no easy answer to the question. But you can make a model to help explain how nine miles of sediment can settle on the sea floor.

Materials

aquarium or earth box, blocks, grease pencil

Demonstration Time: 20 minutes. Get as many students involved as possible. If you have more than one container, split the class into groups and do this as an activity.

★ Fill an aquarium or an earth box half full of water. Float blocks of wood on the surface. Add more to the first layer of blocks, but make the pile highest on one side of the tank. Draw a "sea level" line on the front of the aquarium or earth box, level with the tops of the low blocks. Some of the blocks in the high stack will extend above the line. The rest of the blocks will be below the line as in Figure 15/1.

The stack of blocks in Figure 15/1 represents a mountain. The single layer of blocks to the right stands for the sea floor. You can now act out the erosion process. First, draw a line on the glass along the bottom edges of the floating blocks. This bottom line is your record of the blocks before the change you are about to make.

As the mountain weathers and the pieces are carried to the sea, the top of the mountain will become lower. Take one block from the mountain pile. Deposit it on the model sea floor. Keep taking blocks from the pile and placing them on the sea floor. Put them wherever you think streams and turbidity currents would place sediment. Watch what happens to the bottom edges of the floating blocks as you make the transfer.

As you remove blocks from the mountain, it seems to rise. As you pile more blocks on the sea floor, the sea floor

Figure 15/1
Wooden blocks floating on water show how land may rise or sink as erosion takes place.

sinks. Draw a new line along the bottom edges of the blocks to show the change in level that took place. ★

In nature the rising and sinking are believed to take place at the same time. This movement is called **isostasy** (eye-SAHS-tuh-*see*). Isostasy may be one clue to an explanation for geosynclines. You can see that the weight of sediment carried into the ocean would push down on the rock below on the ocean bottom. Now is it possible to find out whether or not the unseen parts of mountains, called **mountain roots**, look and behave like the bottom of the original block pile in the aquarium?

15.2 Mountain roots rise as the sea floor sinks.

Believe it or not, we can look at the roots of mountains without having to dig them up. It is all done with data on gravity. Measurements can be made from the surface of the earth.

Figure 15/2
This man is using a gravity meter to identify different types of rocks under the surface of the earth.

The force of gravity is the strange, unexplained attraction between any two objects. The heavier the objects, the greater is the attraction between them. Remember Newton and gravity from Section 4.3? You weigh what you do because of this attractive force between you and the earth. Astronauts weigh less on the moon than they do on the earth because the moon is lighter than the earth. There is less material in the moon to pull on the astronauts.

The attraction of gravity can be changed in another way. Increasing the distance between two objects makes the pull between them weaker. Bringing the two objects closer together makes the pull stronger.

Process: measuring

The instrument in Figure 15/2 is called a gravity meter. A **gravity meter** can identify different types of rock by using the fact that the force of gravity changes with mass and distance. Figures 15/3 and 15/4 show how it works. Keep in mind that the stronger the attraction of gravity, the higher the readings on the meter.

In case you haven't guessed, mountains generally are made up of lighter rocks, or mostly lighter rocks. A high gravity reading would mean a lot of heavy rock below sea level. A lower reading would mean less heavy rock and more light rock below sea level. Where there is more light rock, there must be a downward extension of the mountain below sea level, or a mountain root.

A gravity meter reading taken at point A in Figure 15/3 will be the same as the meter reading taken at point B. The same amount of heavy material and light material are under both points. Also, both points are the same distance from the center of the earth.

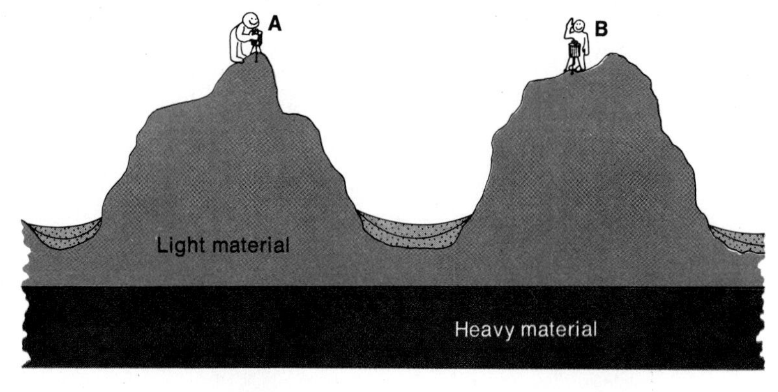

Figure 15/3
Gravity meter readings taken at the top of mountain A and mountain B would be the same. Why?

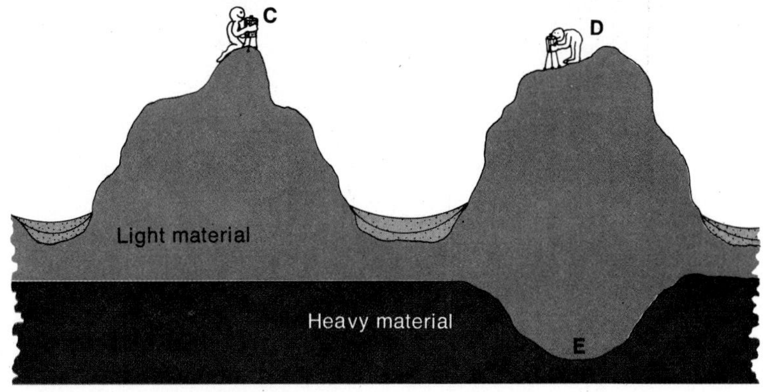

Figure 15/4
Gravity meter readings taken at the tops of mountains C and D would not be the same.

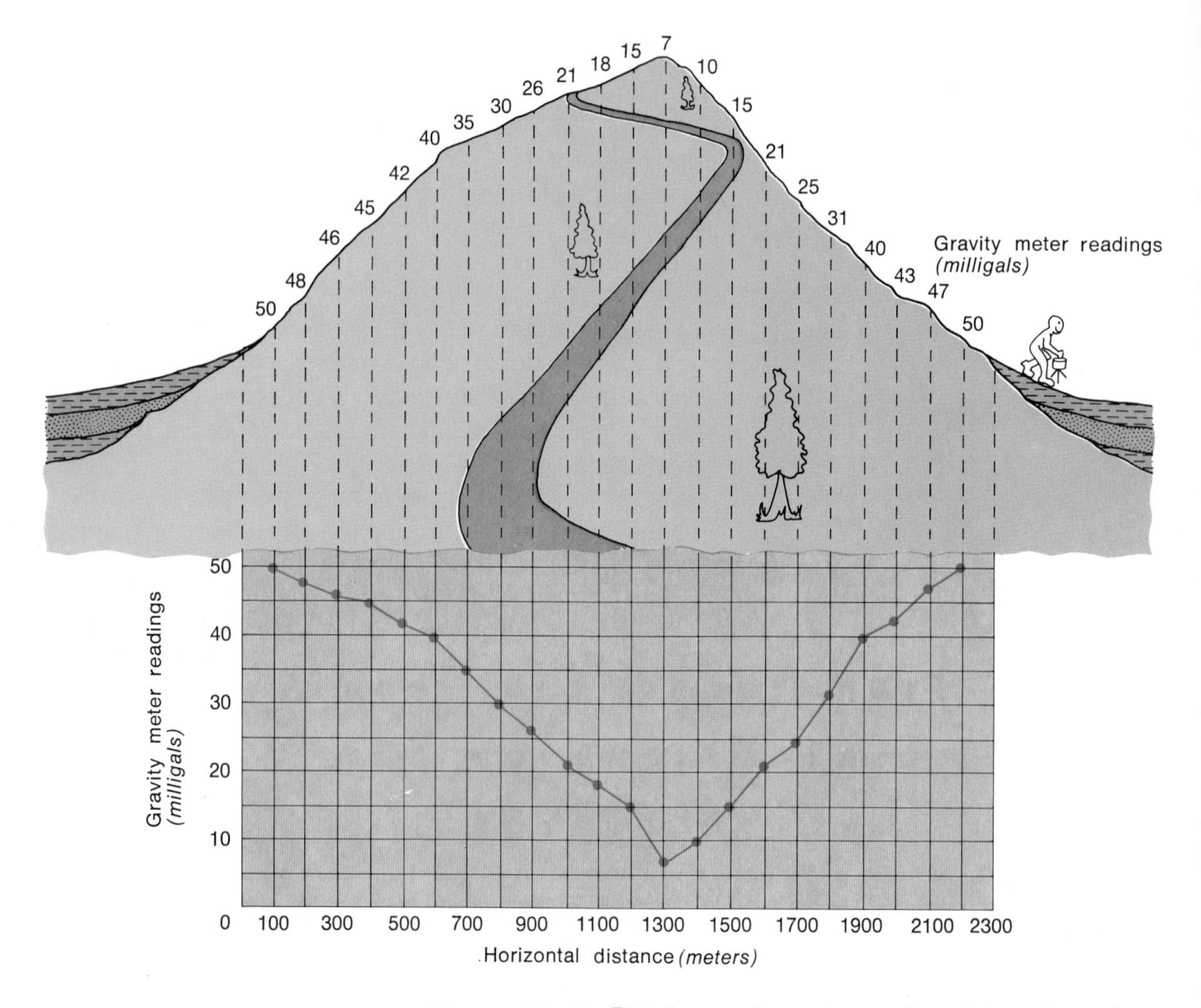

Figure 15/5 *Plot the gravity meter readings taken along the mountain on a copy of the grid. Do you see a mountain root?*

The reading at C will be higher, because there is more heavy rock there.

The rocks under point C in Figure 15/4 are the same as the rocks under point A in Figure 15/3. But the amount of light and heavy rock under point D is different from that under point C. The bulge of lighter rock labeled E in Figure 15/4 is a mountain root. If gravity meters are taken to the tops of mountains C and D, will the readings be different? Which reading do you think would be higher, the one at C or the one at D?

Materials

copy of Figure 15/5

The numbers on the figure are not actual readings from a real mountain. They are simplified to avoid student confusion. Actual readings are in milligals. A gal (named for Galileo) is a unit of acceleration equal to one cm per second per second, and a milligal is a thousandth of a gal. The standard, at Washington, D.C., is 9.8.

Activity Time: 10 minutes. This can be assigned as homework.

The floating cork can be demonstrated. Make a hole in the bottom of a plastic cup. Seal nylon line in the hole with airplane glue. Tie the line to a cork. Fill the cup with water to the desired height.

★ With readings from the gravity meter, scientists can tell the amounts of heavy and light rocks beneath mountains. Look at the cross section of a mountain in Figure 15/5. Below it is a framework for a special kind of graph. The numbers along the bottom mark off the horizontal distance in meters. A scientist wanted to know if there was a downward bulge of light rock under the mountain or not. As he walked over the mountain, he took a gravity meter reading every 100 meters of horizontal distance. He wrote in his notebook the readings you see on the drawing of the mountain.

Make a graph of the gravity readings. You may use graph paper if the squares are the same size as those in Figure 15/5. If you do not have graph paper, copy or trace the grid in the book, and label the lines.

Plot the gravity data on the squares. Connect the points with a line. Place your graph directly below the mountain, matching up the meter marks. Do you see the outline of a mountain root? Does it look like the line of block bottoms in the aquarium? ★

Measurements like these have shown that most big mountains have big roots and most small mountains have small roots. Isostasy seems to explain these findings. Areas of the sea floor next to continents sink as sediment builds up. And gravity measurements show that the root of a mountain should rise as the peak of the mountain is eroded away.

Figure 15/6
Can you make a cork float in water the same way as this one does?

15.3 Forces inside the earth help the crust sink.

Figure 15/6 shows a cork floating on water. Do you notice anything unusual about what you see? Don't corks usually float higher in the water than the cork in Figure 15/6? What you can't see in the figure is the fishline pulling down on the cork.

Sediment piled in a geosyncline looks as strange to a geologist as the cork riding low in the water did to you. The deeper-than-usual slump beneath the sea floor (the nine-mile-thick sediment deposit) can't be caused by the

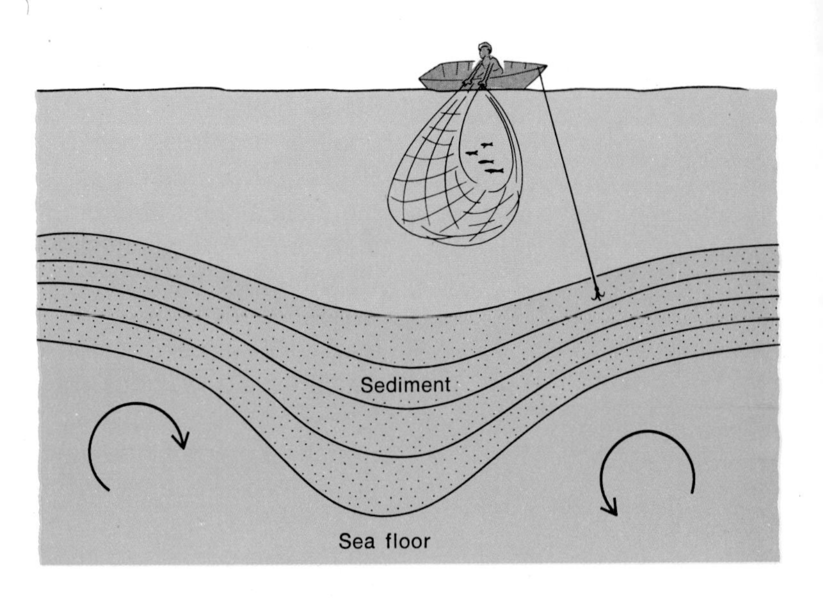

Figure 15/7 *Some scientists believe that convection currents cause sediment layers on the sea floor to sink.*

weight of sediment alone. Something more than isostasy is at work. Something has to be pulling sediment down in the same way the fishline pulls down on the cork.

Concept: energy in motion
Process: formulating hypotheses

At first, scientists thought that convection currents might be doing the pulling. As you learned in Section 7.2, the word "convection" refers to a process in which molecules rise as they heat up and sink as they cool down. The moving molecules form a current traveling through a gas or a liquid. According to the convection theory, convection currents can even move in the material below the crust of the earth.

The theory says that a pair of convection currents may be flowing toward each other in the area of every geosyncline. This movement is pictured in Figure 15/7. The movement of the currents toward each other plus the weight of the sediment can allow layers of sediment nine miles thick to build up on the sea floor.

The sediment at the bottom of a geosyncline would be pinched between masses of the earth's crust carried by the convection currents. As miles of rock above the geosyncline's bottom layer pushed down and the two convection

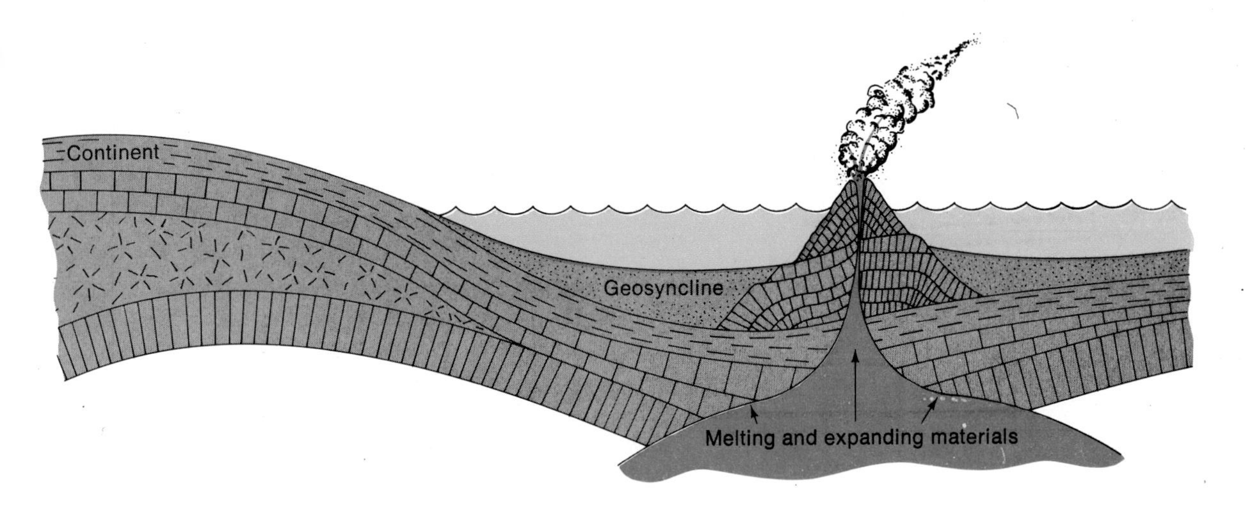

Figure 15/8 *The thick layers of sediment piled in a geosyncline may melt at the bottom. As the melting rock expands, it may then push up on the land above or break through at the surface to form a volcano.*

currents pushed in from the sides, the lowest part of the geosyncline could enter the zone of great heat below the crust. The sediment would melt to form magma at the bottom of the pile. Expanding minerals could exert a pressure, as shown by the arrows in Figure 15/8. The pressure could force some of the magma upward along cracks in the earth's crust to form volcanoes. It might make bulges on the surface.

The bulges on the surface might be big enough to be called mountains. If the melted magma stayed below the surface and cooled, it would form a rock such as granite. The granite might be pushed upward by new magma, or by other forces which you will study later in this chapter. If the rock on top of the hardened magma eroded away, the result might be mountains of granite, like many of the ones in New England. Figure 15/9 illustrates the making of a range of granite mountains.

Some geologists even believe that the convection currents under the surface of the earth form a pattern. The places where currents flow toward each other could make a pattern of geosynclines around the earth.

Figure 15/9
When minerals melt and expand they can push up to form bulges. When the rock that covers the granite is eroded away, a granite mountain range will be left.

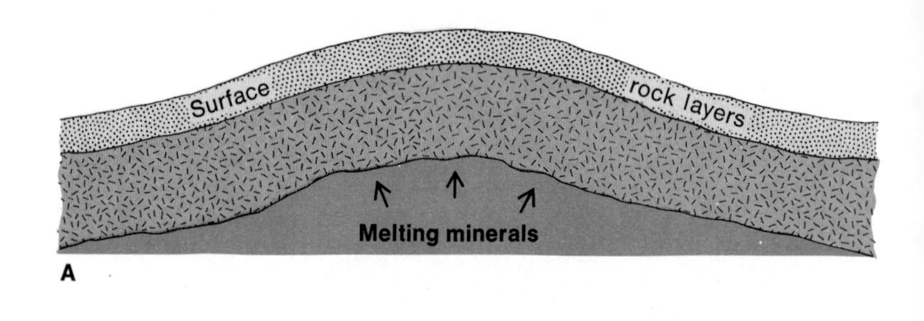

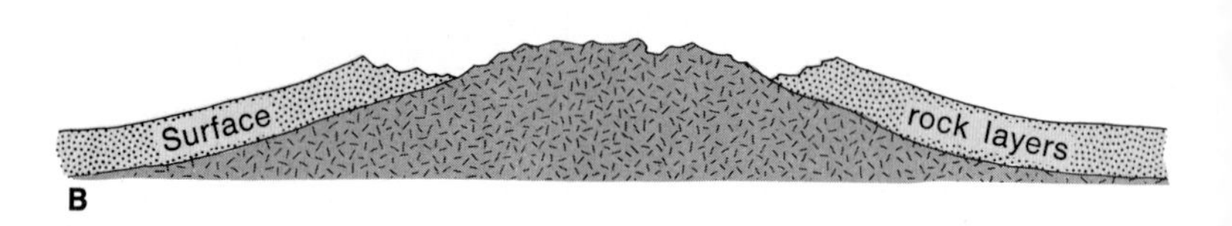

But there are several big "ifs" to the isostasy–convection current theories. To have convection currents, there *must* be a source of heat somewhere deep within the earth. The material in the current *must* be able to flow. And for isostasy to work with the currents, the layers of sediment and rock in the earth's crust *must* be able to bend without breaking. The sea floor and mountain chains would be changed with each little bend. Some scientists even think the continents would have to be floating around on denser rock. Sure looks like a lot of "ifs," doesn't it?

Did You Get the Point?

When sediment is carried from one place to another, the rising and sinking of different parts of the earth at the same time is called isostasy.

Because of isostasy, big mountains have big roots and small mountains have small roots.

As a big mountain gets eroded away, its root rises and becomes smaller.

The isostasy theory and the convection theory help to explain unusual amounts of sinking in geosynclines.

New mountains can form from old geosynclines, because of isostasy and convection.

Figure 15/10
What is wrong in this picture?

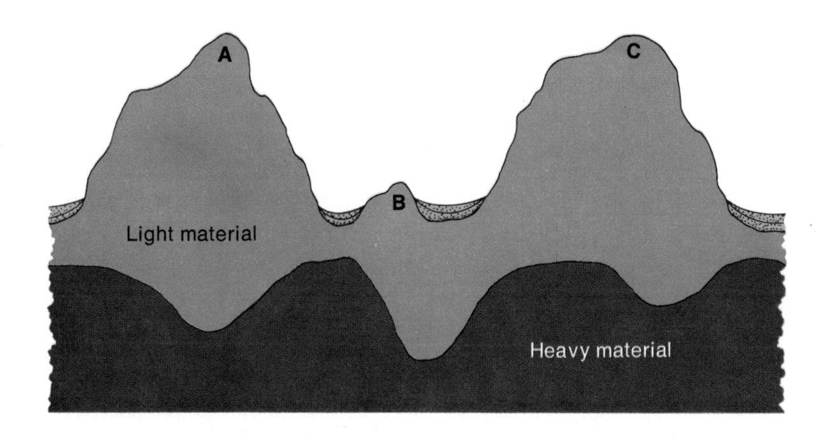

Answers:
1a. The first mountain has the right root size. The second one has too large a root (erosion ahead of isostasy). The third one has too small a root.
b. This question is intended to promote discussion. Maybe something is squeezing the mountains and displacing the roots. Unknown forces could be pulling down the second mountain and pushing up the third. Erosion could be faster than normal on the second mountain and slower than normal on the third.
c. As balance is restored, land will shift. Some land will sink and some will rise. The result will be faulting and earthquakes.

Check Yourself

1. Figure 15/10 is a cross section of three mountains and their roots.
   a. What is wrong with the drawing? Where is isostasy lagging behind erosion?
   b. There are places where strange situations like the one in Figure 15/10 actually exist. Make a list of things that might have caused them.
   c. What might happen as the isostasy catches up with the erosion? Make a list and discuss the possibilities with the rest of the class.

Are Continents Moving Around?

15.4 A theory of drifting continents was proposed many years ago.

In 1912 a student of the earth named Alfred Wegener (VAY-guh-ner) proposed a theory that the continents float. As you will see in Section 15.10, the crust is lighter than the mantle. You know that something like a cork will float on top of something heavy. According to Wegener's theory, at one time all of the continents were part of a large continent. He called the "super continent" Pangaea (pan-GAY-uh). About 200 million years ago, Pangaea began to split. First, it split into two big pieces that Wegener called

Laurasia (lor-AY-shuh) and Gondwanaland (gond-WAH-nuh-land), and later into several smaller continents. The pieces floated away from each other until they reached their present locations. The theory also says that the floating continents are still moving apart.

You can see convection currents at work in a simple model. They will move very quickly. You won't have to wait the millions of years it takes convection currents to move the earth's crust around.

Materials

turpentine, gold or silver paint, small bottles

Activity Time: 20 minutes. See note in Teacher's Manual. This is a model of a convection cell. The paint sinks at the top of the bottle and rises from the bottom, where it is warmed by the fingers. The cooled paint is denser than the warm paint. The flakes move with the current, as anything floating on the surface would.

★ Put a drop or two of gold or silver paint into a small glass bottle full of turpentine. Seal the bottle and shake the mixture. Hold the bottle upright at a slight angle by pinching the bottom end with the thumb and fingers of one hand. Watch what happens within the mixture after about 30 seconds' worth of heat from your body.

Is there any movement? Where is the sinking taking place? Why? What would happen to an object floating on the surface? ★

15.5 Early evidence supported the theory of continental drift.

The gap between Australia and Asia would not be as large on a globe. Its extent in Fig. 15/11 is partly due to the projection.

The first evidence to support the theory of continental drift was based mostly on observations, not on actual measurements. For example, if you cut the continents from a globe, they fit together like the pieces of a puzzle as shown in Figure 15/11.

Another kind of evidence came from rocks and fossils on either side of the Atlantic Ocean. In some places, rocks and fossils on separate continents across from each other are surprisingly similar. For instance, Brazil and the part of Africa across from it are very much alike, geologically. They look as if they might have been together at some time in the past.

Even scratches left on rock during an ancient ice age support the theory that all the continents were once part of the same continent. The scratches run in all directions if you look at one continent at a time. But when you plot the direction of the ice flow on each southern continent on the map of Pangaea, as in Figure 15/12, the scratches line up toward one central point. That suggests parts of South

Figure 15/11
Some scientists believe that the modern continents were once joined as a giant continent that they call Pangaea.

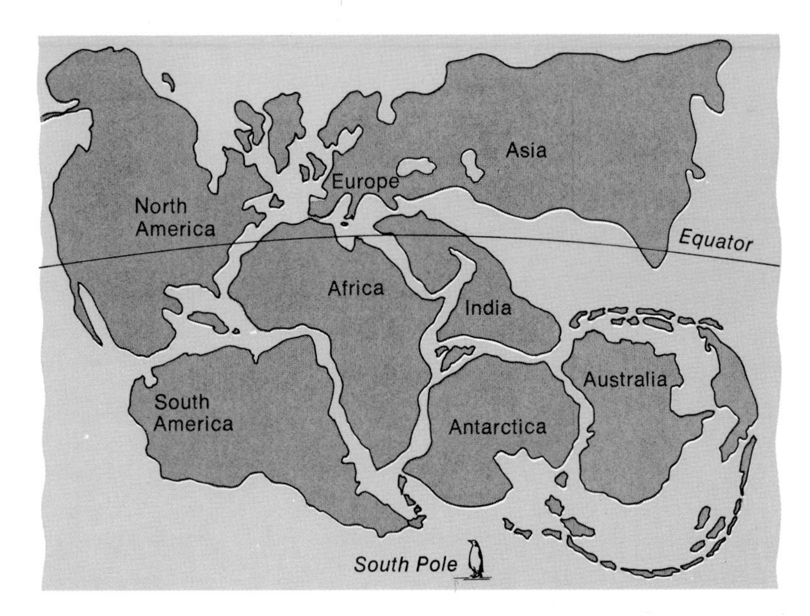

Figure 15/12
Scratches in rock made by ancient glaciers are evidence that all the continents may once have been joined. The scratches line up as if they were all made by the same ice sheet that flowed outwards from a central point.

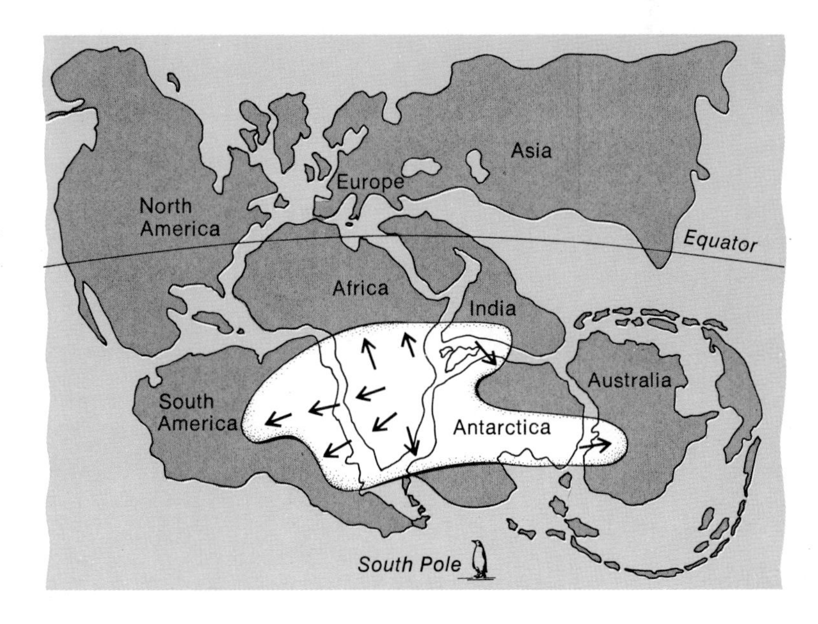

Process: forming hypotheses, observing

America, Africa, India, Antarctica, and Australia could have been together under a single ice cap.

The early evidence for continental drift did not convince many geologists. Most of them called the data "pure chance." What do you think? Could huge, solid continents move around?

Glaciers in the Desert?

The hammer marks a large groove carved in sandstone by a glacier. Glacial grooves like this one are evidence that the Sahara Desert may have once been at the South Pole. Can you find any smaller grooves in the picture?

Don't look now, but the land that used to be at the South Pole 450 million years ago has moved to the Sahara Desert! It has traveled 5500 miles to its present location. Further evidence for the continental drift theory!

Magnetic minerals in the rocks of Australia, Africa, and elsewhere suggested that the Sahara Desert was at the South Pole during the Ordovician Period of geologic time. Geologists searching for oil deposits in the desert found scratches left by glaciers on flat rock surfaces. To gather more evidence, Dr. Rhodes Fairbridge of Columbia Univeristy and an international team of scientists spent nearly a month in the Sahara. They studied the rocks from the air, from cars and trucks, and finally on foot.

The scientists found some rocks that had formed from glacial till. They found others that were originally deposited by rivers that flowed from melting glacier ice. Fossils whose ages are known were found in rocks below and above the glacial ones. The fossils showed how long ago the ice formed.

The evidence for glaciers extends from Morocco to Chad, indicating a great ice cap like the one now covering Antarctica. Geologists assume that the climates of the tropics and the poles have not changed. The only other way to explain glacial features in the Sahara is that the land must have moved.

15.6 Magnetic minerals support the theory. For 50 years the continental drift theory was not accepted by many scientists. When the theory was first proposed, they had little valuable evidence from the sea floor. Few measurements had been made of the rocks that were supposed to be moving. But within the last few years, new instruments and techniques have added more information to support the theory of drifting continents. The new information has won the interest of people all over the world.

One of the new methods uses the magnetic records in rocks and sediment that contain **magnetite,** a mineral that is naturally magnetic. Most of the rocks containing the records are hardened lava flows or other igneous types. Most of the sediment containing the records is from the ocean floor. The records begin when the magnetite crystals were free to move about.

Before the lava hardened and while the sediment was drifting to the ocean floor, the magnetite crystals could twist and turn. Like little compass needles, they lined up and pointed in the direction of the earth's magnetic poles. When the igneous rock hardened and when the magnetite became part of the sea floor, the twisting and turning stopped. The crystals were locked in position, their south ends still pointing north. Geologists can date the rocks and sediment containing the miniature compass needles. The dating tells them where the earth's magnetic poles seem to have been at different stages of the earth's history.

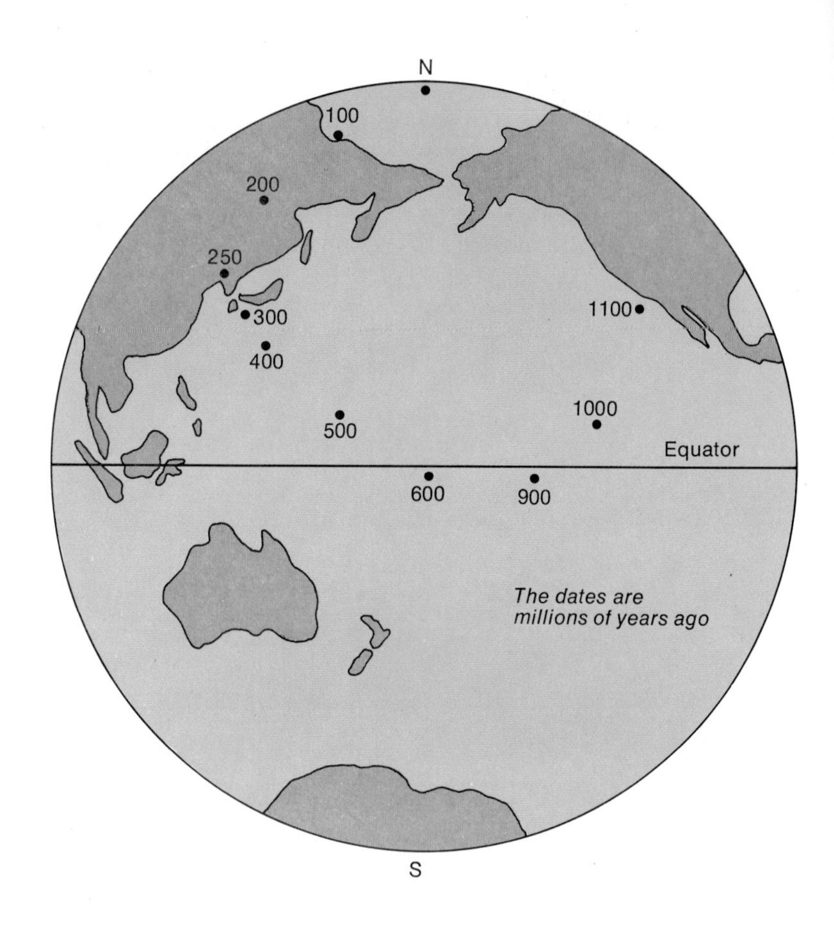

Figure 15/13 *If you do not believe the continents have moved, the earth's north magnetic pole must have followed this path during the past billion years.*

If you make a map showing where the magnetic minerals point, like the one in Figure 15/13, it looks as if the north magnetic pole and the nearby geographic North Pole were in the middle of the Pacific Ocean 500 million years ago. By the Dinosaur Age, 100 million years ago or so, they seem to have moved into eastern Siberia. And somehow they finally got to the positions they now occupy.

But scientists don't think the North Pole was ever near the place where Hawaii is now. Other discoveries show that there is reason to believe the earth's poles have not changed that much. Is there any other way to explain the evidence of the magnetic minerals? Maybe the poles have stayed the same but the continents have moved around!

15.7 New evidence from the sea floor supports the theory.

Now we are getting more and more information from the ocean bottom. Scientists have drilled into it and brought up samples. They can tell the ages of the sediments from tiny fossils of marine organisms.

There is a ridge on the ocean bottom called the East Pacific Ridge, which extends southward from California. People who have been drilling into the ocean bottom have found that as you go west from the East Pacific Ridge, the rocks of the ocean floor get older. They also found that as you go eastward from the same ridge, the floor gets older in that direction, too. It seems that the ocean bottom material has been pushing away from the ridge in both directions.

This fits the theory of convection currents. Look at the three currents in Figure 15/14. There could be a geosyncline at A. But what happens at B? Are there places on the earth where convection currents push up the crust? And what about the crust *between* A and B? If the crust is pulled down at A and pushed up at B, wouldn't the material between A and B be moving from right to left? A continent on the crust between A and B would be carried along with the flow. What about a continent that was exactly at B before the push up? It might be split in two. The two halves might even move away from each other.

In areas where two currents swing up together, new material would be brought to the surface. This happened at point B in Figure 15/14. As new material is brought up, the old material is pushed away to either side.

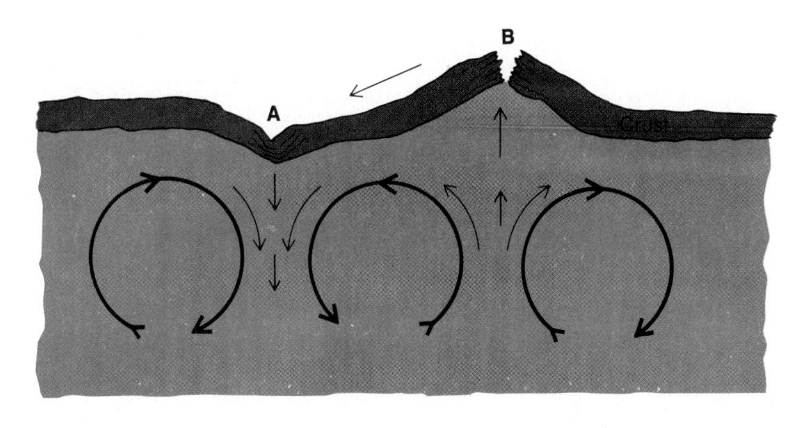

Figure 15/14
Convection currents can work together to push some parts of the crust up and pull other parts down.

Activity Time: 5 minutes. The youngest part of the sea floor will be where the two pages meet.

You can picture how this finding from the East Pacific Ridge illustrates the drift theory. Have a classmate hold two sheets of paper vertically between the palms of his hands. Pull the top edges of the pages outward in opposite directions from between his hands.

Let the surfaces of the pages pulled out represent the sea floor. Pretend that the pulling of the pages takes millions of years. Where will the youngest part of the sea floor be when you stop pulling? Remember, your classmate's hands model the area under an ocean ridge, an area where two convection currents meet and push up together.

Demonstration Time: 5 minutes. Oceanic ridges can be shown on some relief globes.

The East Pacific Ridge is only one of many mountain ranges running through the seas and oceans of the world which could have been originally formed by upward currents like the one in Figure 15/14. Continents could have moved away from them. Such underwater mountain ranges are called **oceanic ridges.** Notice in Figure 15/15 how the Mid-Atlantic Ridge seems to follow the general shape of the coasts of Africa and South America.

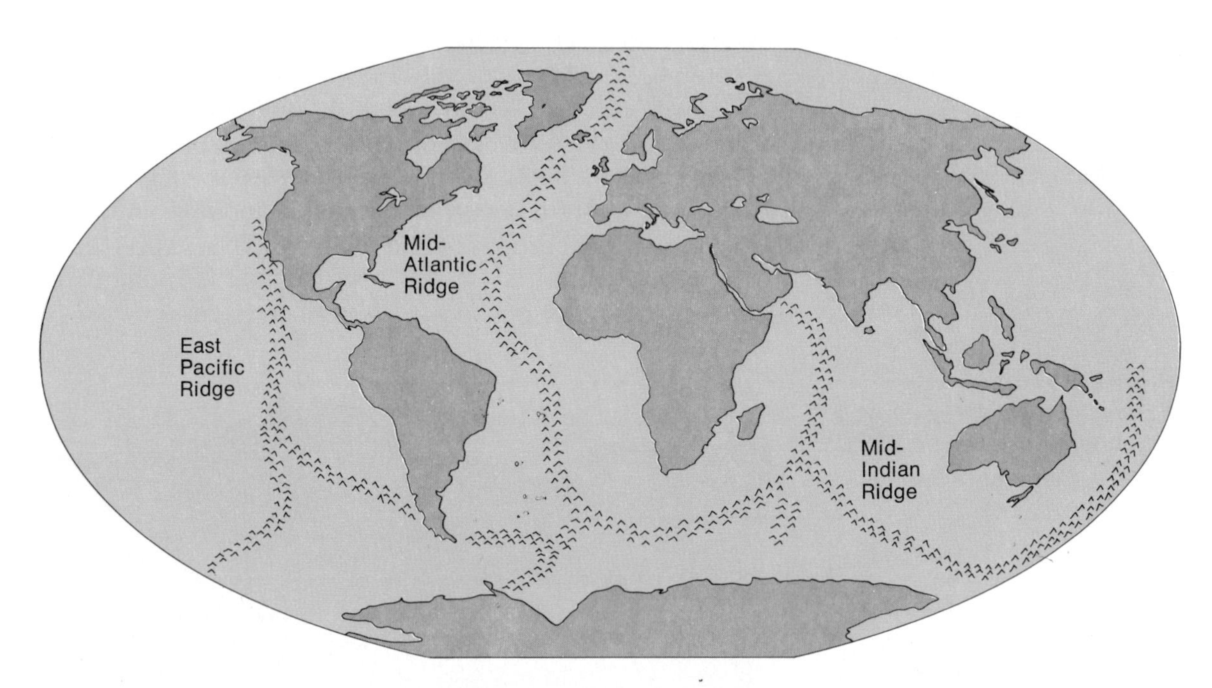

Figure 15/15 *The major underwater mountain ranges are called oceanic ridges.*

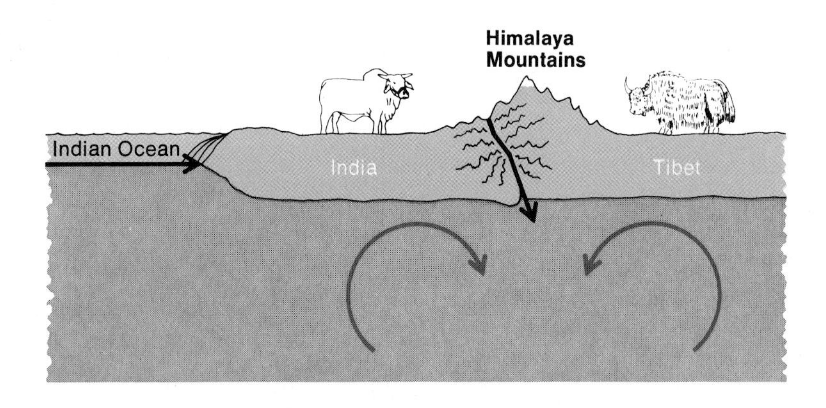

Figure 15/16 *The Himalaya Mountains may have been formed by the collision of India and the original Asia.*

Other recent evidence supporting the drift theory appeared when a computer showed how the continents could be fitted together to make Pangaea. The computer used information about the outside edges of the continental shelves. It showed that the continents fit together even better along their continental shelves than they do along their shorelines.

Coal and fossils of tropical animals found in Antarctica could only mean that the continent once had a warm climate. Antarctica must have wandered a great distance to reach the South Pole.

India could have collided with Asia. See "The Breakup of Pangaea," by Robert S. Dietz and John C. Holden, *Scientific American,* October 1970, pp. 30–40.

And when two continents bump into each other, would there be signs of collision? Figure 15/16 shows what would happen if two land masses collided. Mountains! Look at India on a relief globe and then on a map of Pangaea. What do you see? Could India have collided with Asia?

Can you see why more and more geologists believe in the theory of continental drift?

Did You Get the Point?

There is much evidence that the continents have moved across the earth and are drifting away from one another.

The continental drift theory seems to support the convection current idea. The two may be part of the same theory.

The Himalayas may have been formed because of continental drift.

Check Yourself

Answers:

1. Perhaps the continents behave like a glob of tar on a wheel. Spinning would tend to make the tar spread out. Could the spinning of the earth be doing the same thing to the continents? While this can't replace the convection theory, it can add to it.

2. This question is meant strictly as a thought provoker. There is no real answer. It is meant to show students another property of a floating object and the material it is floating in. Feel free to mention surface tension. (In short, the cork is trying to float as high as it possibly can. The shape of the surface can make the high-water level move about.) You may want to supply students with corks or cork chips to take home.

1. Perhaps you can come up with your own theory of how and why the continents are drifting apart. First, examine a globe. What is on the exact opposite side of the globe from each of the continents? How can this observation be worked into the continental drift theory?

2. Fill a paper cup half full of water. Try to make a small cork or wooden ball float in the center. What happens to the cork? Now pour water into the cup with the cork still in it until the water just begins to flow over the sides. What happens to the cork? Try the experiment with several floating corks equally spaced around the sides of a half-filled cup of water. What happens as the cup is filled again? Can this observation be worked into the continental drift theory?

A New Theory

15.8 The earth's features form a pattern.

Concept: universality of change

The earth is a constantly changing ball of rock. Rocks are rising and layers of sediment are sinking. Land masses may be floating around like king-size ships on a very thick sea. The sea floor is being rolled out in some places, and reeled in, in others. Mountains rise where there were never mountains before. And there is a good chance that rock beneath the crust is flowing in convection currents. Is there a pattern to all these motions?

You have seen that volcanoes and earthquakes seem to go together, and they occur in similar patterns. If you plot on a world map the ocean trenches, the oceanic ridges, and the major modern mountain ranges that used to be geosynclines, you can see that they are all more or less connected to each other or next to each other. They do form a pattern. The areas between the zones of volcanoes, earthquakes, mountain ranges, trenches and so forth are called "**plates**" by geologists. You can see in Figure 15/17 that North America is on one plate, and Eurasia is on another. Africa is on a third.

According to the modern version of the continental drift theory, the plates are large pieces of the earth's crust which are being carried around, very slowly, by convection currents underneath them. Where two plates separate, a ridge

Figure 15/17
The plate theory says that the earth's crust is made of large sections that slide around. The map shows how the plates fit together.

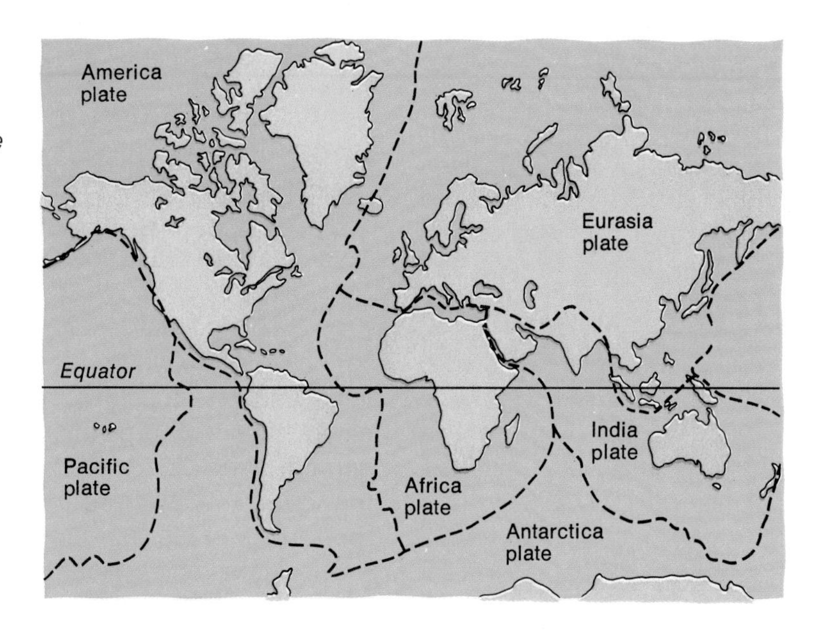

of volcanoes is formed. Where two plates grind against each other, lava is erupted, earthquakes shake the crust, and mountains are formed.

15.9 The movement of plates can explain the origins of mountains.

As you know, many mountains are made up of sedimentary rock that must have come from geosynclines in the sea. How did this rock get thousands of feet above sea level? The collision of two floating land masses as shown in Figure 15/16 explains how the Himalayas might have been built. But what about other mountains? How were the Rockies and the Appalachians formed?

The Rocky Mountains are mostly made up of sedimentary rock, along with some granite. The old sediments have been greatly folded and faulted. According to what you read in Section 15.8, it is easy to guess that the famous San Andreas fault system of California is at the edge of a plate which includes much of the Pacific Ocean. Could another plate, with North America on it, be moving westward, its edge pushing on top of the edge of the Pacific Plate? Could this cause the earth's crust to buckle and break, and form mountains from the Pacific Coast to the Rockies?

Figure 15/18
Fault-block mountains like these form when large sections of rock are cracked and tilted.

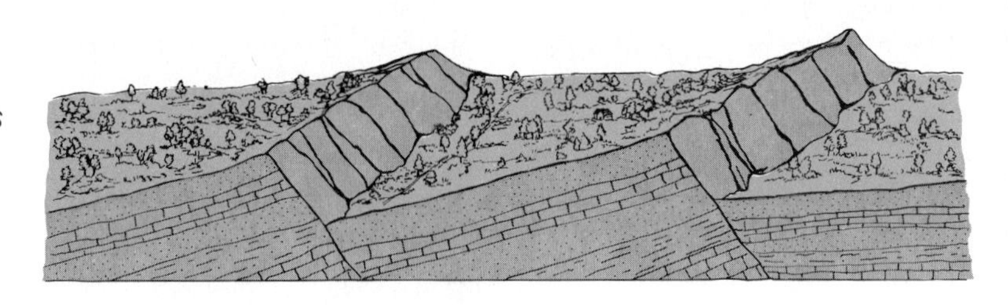

Books can be used to represent tilted fault blocks.

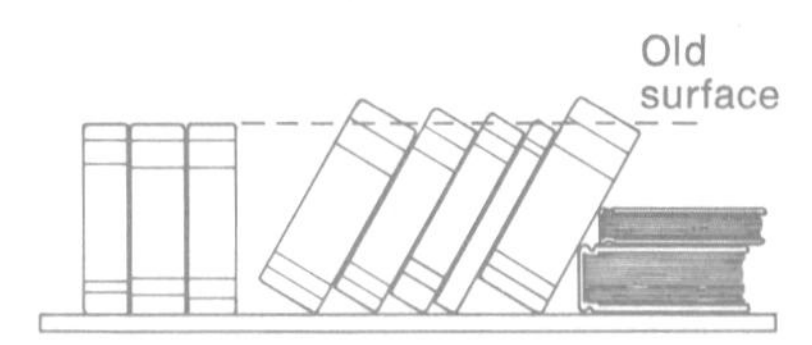

Fold a sheet of paper once down the middle and crease it; then push it from side to side into an Appalachian type fold. Tip the fold, and cut horizontally. The result is a model of what is pictured in Fig. 15/19A. In the photograph, the curved ridges at the lower right are presumably from a tilted down fold; the curved ridges at the upper left are from a tilted up fold. Both folds are parts of the same series of folded rocks.

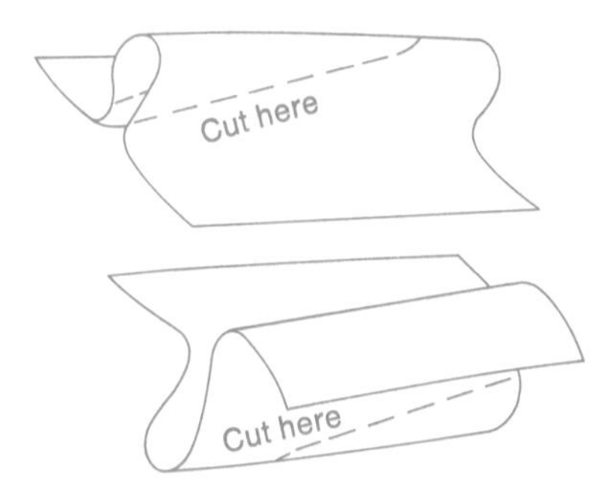

Process: inferring

The pressure where the plates meet could explain mountains which have been faulted but not folded. The pressure could break off large blocks of rock from their neighbors and tip them like the ones in Figure 15/18. Mountains formed in this way are called **fault-block mountains,** and the ranges in Nevada are possible examples.

The Appalachian Mountains appear to have been formed when a great pressure squeezed the rock layers into wrinkles or folds. Notice how the Pennsylvania mountains in Figure 15/19A seem to line up in rows. The ridges are made up of hard rock layers and the valleys have been eroded in soft rock layers, but all the layers are part of a single fold. The curved ridges show that the fold is tilted downward toward the right. Can you demonstrate this with a piece of paper? Figure 15/19B shows a section of folded rocks in Pennsylvania. You can see how much has been eroded away.

The Appalachians are older mountains than the Rockies. The ages of the rocks show that Appalachian rock layers were folded before the beginning of the plate movement that formed the Rockies. Geologists believe that the mountains in eastern United States may have been caused by an earlier movement of the American Plate.

Figure 15/20 is the map of Pangaea, with modern mountains shown as they are on the modern continents. You will notice how many of them are along edges of continents. When Pangaea started to break up, the present ranges were not all in existence, so some must have been formed after the break-up began. This suggests that mountain ranges are formed where land masses split apart and where continents slide on top of ocean basins.

For example, the diagram in Figure 15/21 is a cross section of the earth's crust, showing what might have happened as a part of Pangaea broke up. What is now North

Figure 15/19
Only rounded ridges are left from the folding of the Appalachian Mountains that began 280 million years ago.

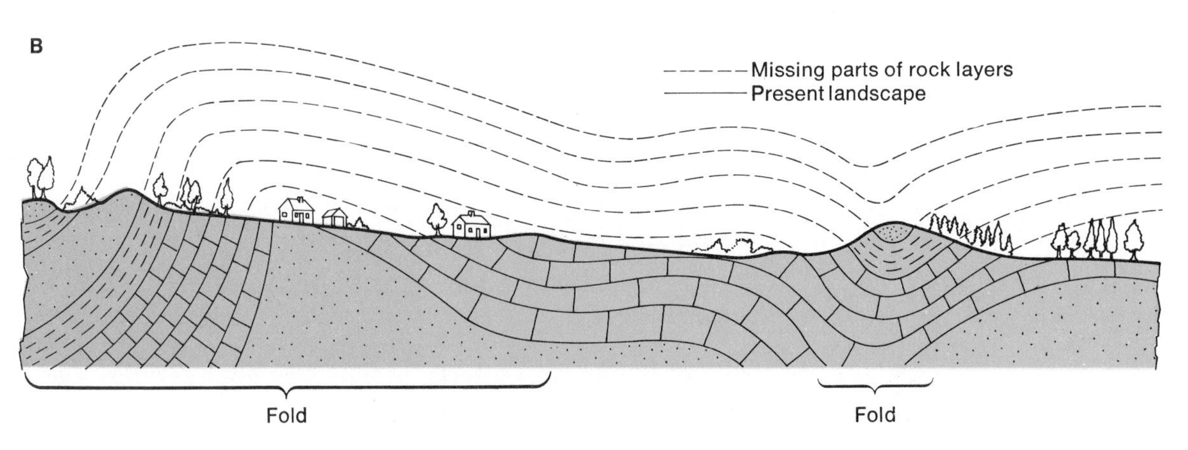

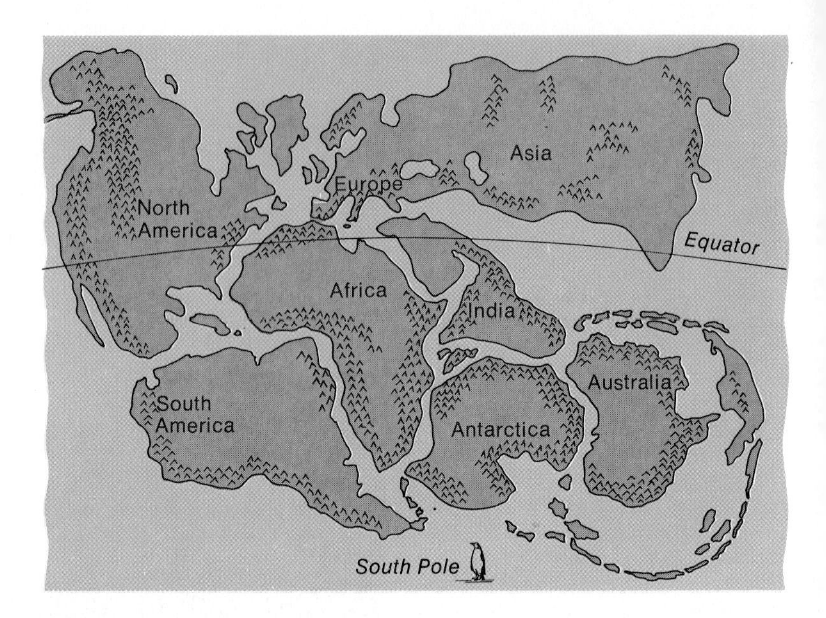

Figure 15/20 *Some mountain ranges of modern continents seem to be along cracks formed as Pangaea split up. Others are at edges that have pushed against ocean basins.*

America moved westward and pushed up over the edge of the Pacific Plate. Where the two met, according to the theory, the crush has caused faults, earthquakes, trenches, volcanoes, and folded mountains. Can you see why there are volcanoes and earthquakes in Japan?

In the middle, lava is pouring out from below onto the sea floor, in between North America and Eurasia. This fills in the crack left as the continents move apart, and the volcanoes make up the Mid-Atlantic Ridge.

The theory of plates is one you can watch during your lifetime. Maybe the theory will be proved completely wrong and need to be replaced. You missed a lot of the debate over the original continental drift theory, but you can become a part of this one.

Did You Get the Point?

There is a pattern to the features of the earth, but it is not a simple one.

The earth's crust can be divided into six sections, called plates.

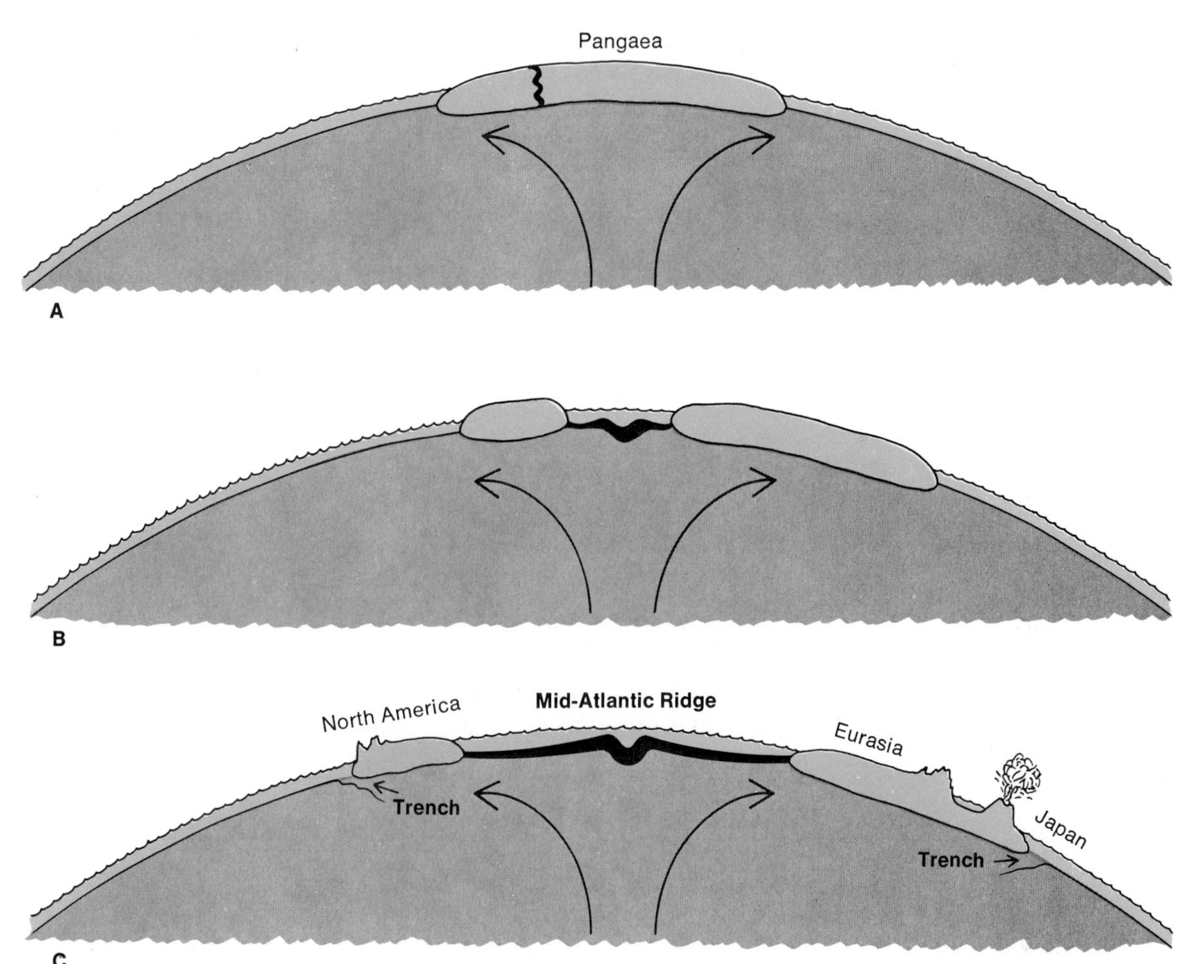

Figure 15/21 *Geologists believe that the break-up of Pangaea into North America and Eurasia created mountains and volcanoes. The Mid-Atlantic Ridge is the site of the break.*

Mountains and earthquakes on the earth's surface may be explained by overlapping plates.

Answers:

1. The zones outlining plates could be places where convection currents meet and work together.
2. Yes, many mountain ranges are found where there could have been geosynclines. Others are in areas that were covered by inland seas, or older geosynclines that once existed where there is now land.

Check Yourself

1. Compare the plate theory to the convection theory. How can the two be fitted together?

2. Examine a relief map of the world. Are most of the mountain ranges on the continents located at places where they could have been caused by geosynclines? How can you account for some of the inland mountain ranges?

Inside the Earth

15.10 Speeds of earthquake waves change in the earth.

Sections 15.10 to 15.13 are optional. You can describe the inner earth without going into the details upon which the assumptions are based. However, many states may wish to use this material. The sections constitute an excellent example of how we can gather information about things we can't see and test.

Plan to lecture this material, using the chalkboard. It is not difficult but it requires careful ordering of observations.

The speeding up and slowing down of S waves can be shown with a 50 ft (15.2 m) length of clothesline stretched tightly by two students. Slack in the rope represents soft rock, etc. Pluck rope as in Chapter 12.

At the end of Section 15.3, there were some "ifs." The modern theories might work if there is a lot of heat inside the earth. The theories might work if material below the surface can flow, and carry continents. Volcanoes tell us a lot about heat below the surface, and earthquake waves tell us even more.

You know from Section 11.1 that earthquake waves start inside the earth. On their way to the surface they pass through materials deep within the earth that we cannot study directly. The speed of the waves and even their paths can change before they reach the surface.

Section 11.11 explained the differences between the two kinds of earthquake waves, P waves and S waves. The data in Figure 15/22 show how fast both types of waves travel in different materials. Compare the speeds of each type of wave in soft clay and in hard granite.

The different behaviors of the two types of waves tell us something about how the earth is put together. They give us some clues about what materials make up the different parts of the earth. The waves may end up proving that convection is possible. Figure 15/23 and other ones to follow will show you how the clues help solve the mysteries of the earth's makeup.

Earth materials	P wave speed (km/sec)	S wave speed (km/sec)
Gravel, clay	.3–2.0	.17–1.1
Sandstone	1.0–4.3	.8–2.5
Limestone, dolomite	1.7–6.0	1.0–3.7
Granite	4.0–5.7	2.1–3.3
Ice	3.5	1.7
Water	1.5	No waves
Air	0.33 (speed of sound)	No waves

Figure 15/22 *S and P wave speeds in different materials.*

Figure 15/23
Seismograph stations A through J are standing by to record earthquake activity. (The shortest distances from an earthquake at X to the first seven stations are: XA = 1600 km, XB = 3000 km, XC = 4700 km, XD = 6800 km, XE = 8800 km, XF = 11,600 km, XG = 12,400 km.)

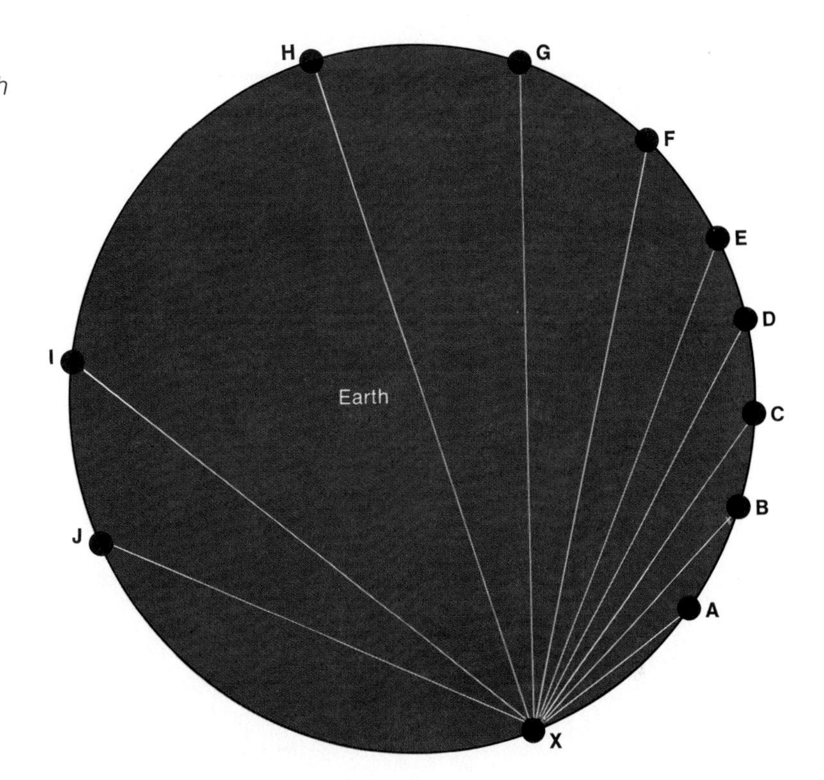

Activity Time: 20 minutes

Students should get:

Path	P wave travel time (seconds)	S wave travel time (seconds)
XA	320	533
XB	600	1000
XC	940	1567
XD	1360	2267
XE	1760	2933
XF	2320	3870
XG	2480	4133

For both types of waves, all actual travel times are shorter than calculated times.

We would expect abnormalities along paths XF and XG.

The material beneath the surface must be harder (denser) than granite.

The circle in Figure 15/23 is a cross section of the earth. The letters around the edge of the circle mark the locations of various seismograph stations. All of the stations are directly north and south of one another. The X on the diagram is the epicenter of an earthquake that will be recorded and studied by each of the stations. The lines show the straight-line paths, the shortest routes, from the epicenter.

Pretend that the earth is made entirely of granite. Find how long it would take S waves and P waves to reach stations A to G. Assume that the S wave travels at a rate of three kilometers per second in granite. Assume that the P wave travels at a rate of five kilometers per second.

Figure 15/24 gives the actual times it takes P and S waves to reach stations A to G. At which stations did the S wave arrive too soon for the whole earth to be made of granite? At which stations did the S wave arrive too late for the whole earth to be made of granite? Which of the S waves might travel through a material other than granite? Is the material beneath the surface harder or softer than granite?

Figure 15/24
Actual travel times of S and P waves.

	Actual travel time (seconds)	
Path	**P wave**	**S wave**
X A	220	420
X B	370	670
X C	530	960
X D	690	1270
X E	830	1500
X F	1180 (weak)	No wave
X G	1380	No wave

15.11 Earthquake information locates zones in the earth.

Geologists at first tried to explain why the S waves arrived early by using information about the speed of waves in different materials. Later, with the help of computers, they were able to figure out how much heavy and light material the different waves traveled through to get to each station. The results showed that the farther the wave traveled, the more "heavier-than-granite" material it must have passed through. Figure 15/25 illustrates the increase in heavy material.

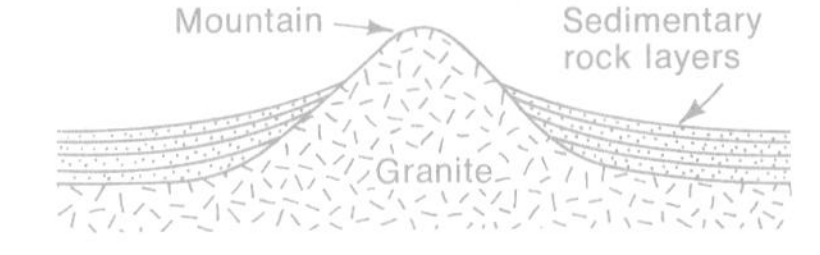

Eventually, geologists had enough information to draw a diagram of the earth's outer portion as in Figure 15/26. The waves told them that the continents are made up mostly of sedimentary rock and loosely packed sediment, resting on bases of granite. Underneath continents and oceans is a layer of basalt. All these layers together are called the earth's crust. The crust is twenty-five miles thick under the continents and three to five miles thick under the ocean. It is thicker under the continents because of the mountains and their roots.

The earth's crust appears to float on the even heavier material underneath it. This zone of heavy material is called the **mantle** (MAN-tul). The early wave arrivals are due to the speeding up of waves by the heavier material of the mantle.

Computer-predicted travel times take into account the heavy material in the mantle. Compare them with the actual travel times in Figure 15/27. Notice that the P wave is 240

Figure 15/25
The S waves must have traveled through different amounts of heavy material to have arrived at the seismograph stations at such different times.

Wave path	Material passed through
X A	loose-granite-loose
X B	loose-granite-heavy-granite-loose
X C	loose-granite-heavy-granite-loose
X D	loose-granite-h e a v y-granite-loose
X E	loose-granite-h e a v y-granite-loose
X F	loose-granite-h e a v y-granite-loose
X G	?????????????????????
X H	loose-granite-heavy-???????-heavy-granite-loose

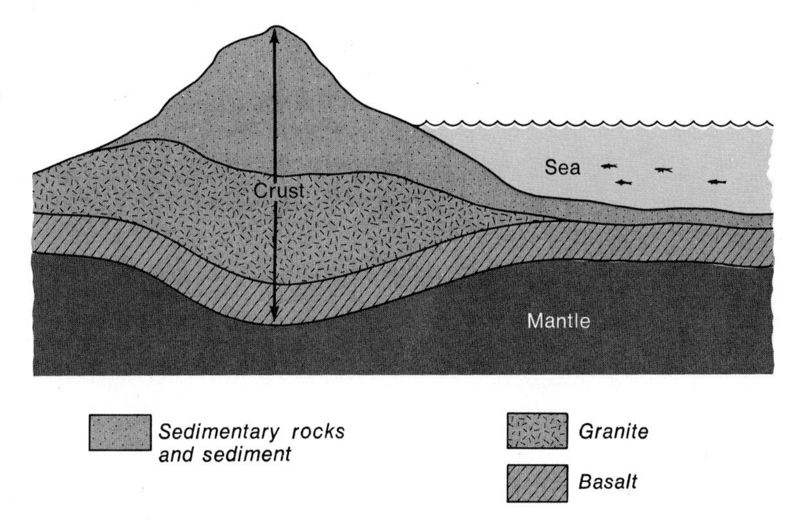

Figure 15/26
The earth's crust is made up of several kinds of rock but is very thin compared to the mantle below.

Figure 15/27
The actual and calculated P wave travel times.

	P wave	
Path	Actual travel time (seconds)	Calculated travel time (seconds)
X A	220	220
X B	370	370
X C	530	530
X D	690	690
X E	830	830
X F	1180 (weak)	1180 (weak)
X G	1380	1140

seconds later in getting to station G than the computer predicted. Why is it late? Why are the P waves weak at F and not at G? Why can't S waves get through to F and G? There must be another zone that needs to be studied.

These questions will be answered in the next section.

15.12 The earth has an inner core and an outer core.

Materials
boards, cloth strips, thumb tacks, pencils or dowel rods

Activity Time: 5 minutes.

★ Changes in the speed of a wave also cause the wave to change direction. Roll a round pencil down a slanted board. Its path is straight. Now cover one half of the board with cloth so that the pencil will be half on cloth and half on the wood as it rolls. Roll the pencil. What happens? Try rolling the pencil at different angles. What happens now? Similar things happen to an earthquake wave as it passes from one kind of rock to another, or from loosely packed rock to tightly packed rock. ★

The farther material is beneath the earth's crust, the tighter it is packed. This is due partly to the weight of what is above it and partly to the kind of material it is. The farther down you go, the heavier and tighter the material gets. So earthquake waves should travel faster the deeper they go. This change in speed bends the paths of the earthquake waves, as the change in speed from board to cloth changes the path of the pencil. Instead of the straight-line paths in Figure 15/23, the waves would follow curved paths like the ones that are drawn in Figure 15/28.

You remember that the P wave appeared to be late in getting to station G, and the S waves did not get to F and G at all. Why not? There is a hint in the cross section of the earth in Figure 15/28. Perhaps waves going toward stations F, G, H, and I have to go through another material, which slows them down. Perhaps the S waves that hit the new material did not get through it. Data from Figure 15/22 indicate that the new material must be some kind of a fluid. S waves won't travel through a fluid such as water or air. Liquids are just not rigid enough.

Any P wave on the way to F would hit the boundary of the new material, slow down, bend, and come out at G as shown in Figure 15/29. Using computers, scientists found that the extra 240 seconds of travel time noticed at G just fits the slow-down in speed through the fluid zone.

So you would not expect any waves to get to station F. But in fact, weak P waves were picked up at F, as Figures 15/24 and 15/27 show. This can be explained if we guess there is a solid zone at the center of the earth which could speed up the waves slowed down by the fluid zone.

Figure 15/28
S waves from the earthquake at X did not get to stations F through I. Since S waves can't travel through liquids, the material inside the inner circle in the diagram may be a fluid.

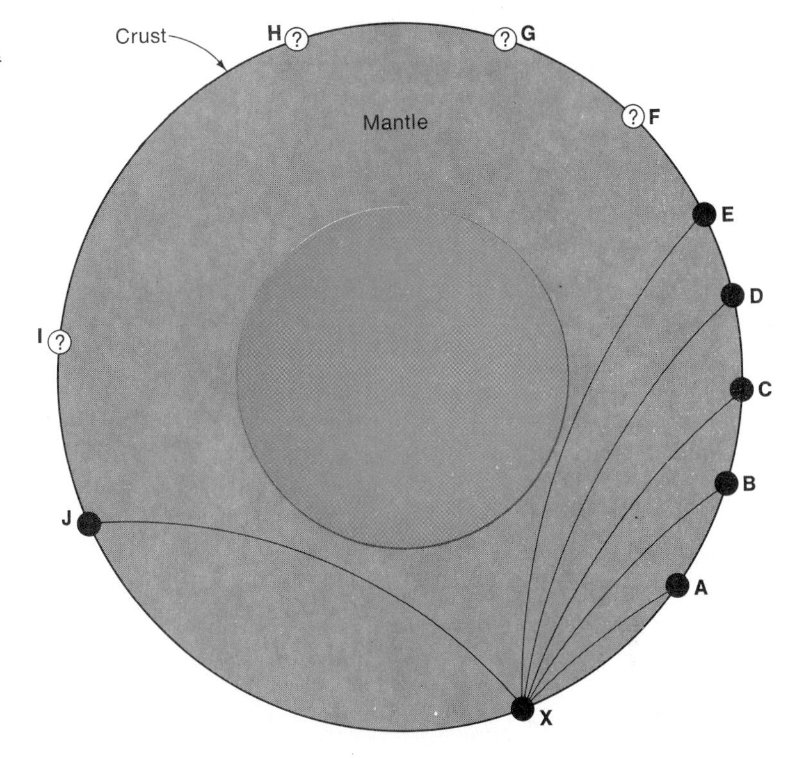

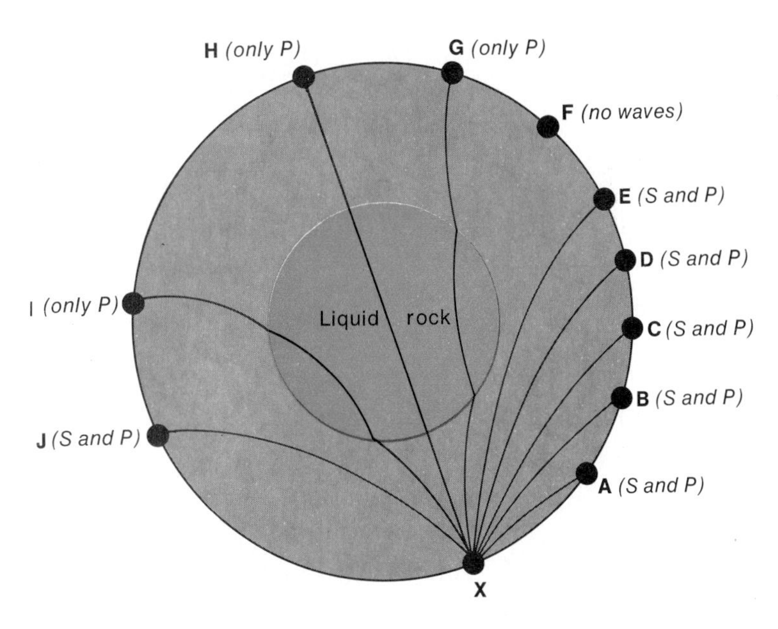

Figure 15/29 *P waves headed through the liquid zone at an angle are both bent and slowed down while the S waves don't go through the liquid zone at all.*

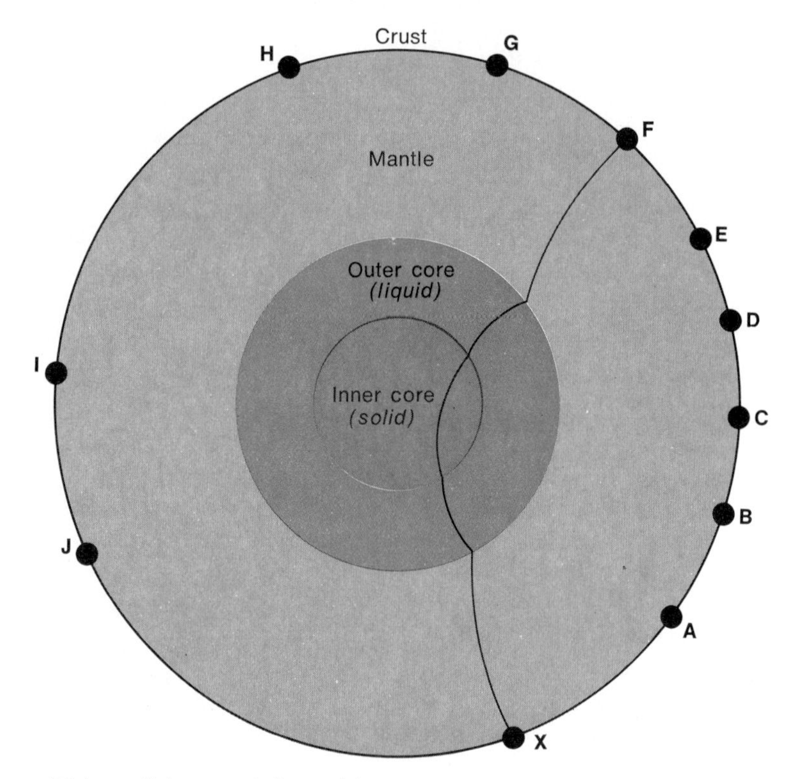

Figure 15/30
Weak P waves received at station F hint that a very hard, solid zone lies within the liquid zone of the earth.

This solid material could

1. catch a wave headed for G in Figure 15/30,
2. bend it back,
3. speed it up to make up for the 240 seconds lost traveling through the fluid outer core,
4. and send it on to F.

The fluid zone, most likely liquid metal, is called the **outer core.** The solid central zone is the **inner core.** Studies at stations on opposite sides of the earth from earthquakes (such as H, I, J, and K in Figure 15/30) have backed up this theory.

Finding the outer core supports the convection theory. Since the outer core is fluid, it must be very hot. It could be a heat source for the convection currents.

15.13 Convection is possible.

The geologists gathered more and more information for the computers. With results from the computers they could

Appendix S contains a chart of information about these zones, including densities.

guess at some of the material making up the zones within the earth. The way earthquake waves behaved going through the inner core told them it was like a mixture of iron and nickel, which are heavy minerals. The waves told them that the outer core seemed like a melted mixture of the same materials. The mantle appeared to be more like rock made of slightly less heavy minerals such as augite and olivine (page 264). The crust under the continents seemed like granite, which has the lighter quartz and orthoclase.

If the outer core is hot enough to be fluid, it seems reasonable that the mantle is hot enough to be "plastic." The plates bearing the "light" continents can be moving on top of the heavier, hotter mantle material.

Recent studies seem to tell us that the mantle is not the same from top to bottom. The top part seems to flow more than the bottom part. Convection currents might mix only the top part. If this is true, and if rock can be bent like the cigarette in Section 11.1, then convection currents may indeed be helping to cause geosynclines. The hot fluid outer core could be the heat source. High pressure on all sides could keep the rocks of the upper mantle and lower crust from breaking. (See Appendix S.) After all, to form a geosyncline, rock only has to flow about two inches a year. And this just happens to be the amount the sea floor is spreading!

Answers:

1. Surface to mantle = 25 mi, 40 km
   Mantle to outer core = 1800 mi, 2900 km
   Outer core to inner core = 1300 mi, 2100 km
   Inner core to center of earth = 900 mi, 1450 km
2. About 111 km/degree.
   X to A = 3660 km
   X to B = 5770 km
   X to C = 7440 km
   X to D = 9215 km
   X to E = 10,770 km
   X to F = 12,990 km
   a. More than 11,000 km
   b. More than 11,000 km but less than 13,000 km
   c. 12,990 km

Did You Get the Point?

Varying speeds of S and P waves give us a picture of the inside of the earth. The earth has a thin crust, a mantle, a fluid outer core, and a solid inner core.

This model of the inner earth shows how convection currents could be possible. Convection currents help explain how geosynclines form and plates move.

Check Yourself

1. Look up the radius of the earth. Using the scale drawing in Figure 15/30, figure out the thickness of the outer and inner cores and the mantle. You can check your answers in the Appendix, but explain to your teacher how you got them.

2. Look up, or figure out the circumference of the earth. Figure out the number of kilometers per degree around the circle in

Possible questions:
1. Draw a pair of convection currents that might be pulling down on the earth's crust.
2. Draw a pair of convection currents that might be responsible for creating an ocean ridge.
3. Draw a cross section of a geosyncline.
4. Draw two mountains of different sizes and show how their root sizes might differ.

Figure 15/30. Use this number of kilometers and a protractor to find out how far from the earthquake epicenter stations A through F are. Use actual distances to answer the following questions:

a. How far from the epicenter must a seismograph be for it *not* to pick up S waves?
b. How far from the epicenter must a seismograph be to receive only weak P waves?
c. How far from the epicenter is the station that receives strong waves coming through the inner core?

What's Next?

5. Draw a cross section of a land area showing three places with different gravity readings.
6. Draw a cross section of the earth showing the four major zones discovered so far.
7. Draw the paths of several earthquake waves from an epicenter to four different stations, assuming the earth is made entirely of granite.
8. Draw a diagram showing the last wave to travel through the mantle without hitting the outer core. Be sure to show the area that will receive only weak P waves.

With this, the raising of mountains and the rearranging of the continents, you're back to Chapter 9. The process begins all over again. New rocks are formed, volcanoes bring them to the surface, and earthquakes accompany the change. The new rocks will weather. Whole mountains will be eroded away. The rocks just made will again be deposited as sediment, and new geosynclines will form again as the process continues to go on and on. The earth's history tells us that.

How does the earth's past tell us that the rock cycle will go on and on? That's a whole unit in itself. It begins with a study of how events that happened millions of years ago can be dated.

Skullduggery

9. Draw a diagram showing a wave passing through the inner core of the earth.
10. Draw a map showing the placement of today's continents in the continent of Pangaea.
11. Draw a map showing some evidence that South America and Africa were once joined together.
12. Draw a diagram showing how the ages of the rocks on either side of an oceanic ridge might change.
13. Draw a diagram showing how convection currents might create an ocean trench and a mountain range.
14. Draw a map showing the plates in the plate theory of continental drift.
15. Draw a sketch showing how mountains can be built in areas bordering geosynclines.

Directions: Each of the drawings in Figure 15/31 is an answer to a question. Write a question to go with each answer on a separate piece of paper. The questions should be ones that would make a good exam.

For Further Reading

Reinfeld, Fred. *Treasures of the Earth.* New York, Sterling Publishing Co., 1962. A chapter on how mountains are formed and another on how they are worn down. The rest of the book tells about volcanoes, earthquakes, and glaciers, and ends with a chapter for collectors of minerals.

Strahler, Arthur N. *The Story of Our Earth.* New York, Home Library Press, 1963. Very clear drawings and descriptions help to explain convection currents, mountain building, and continental drift.

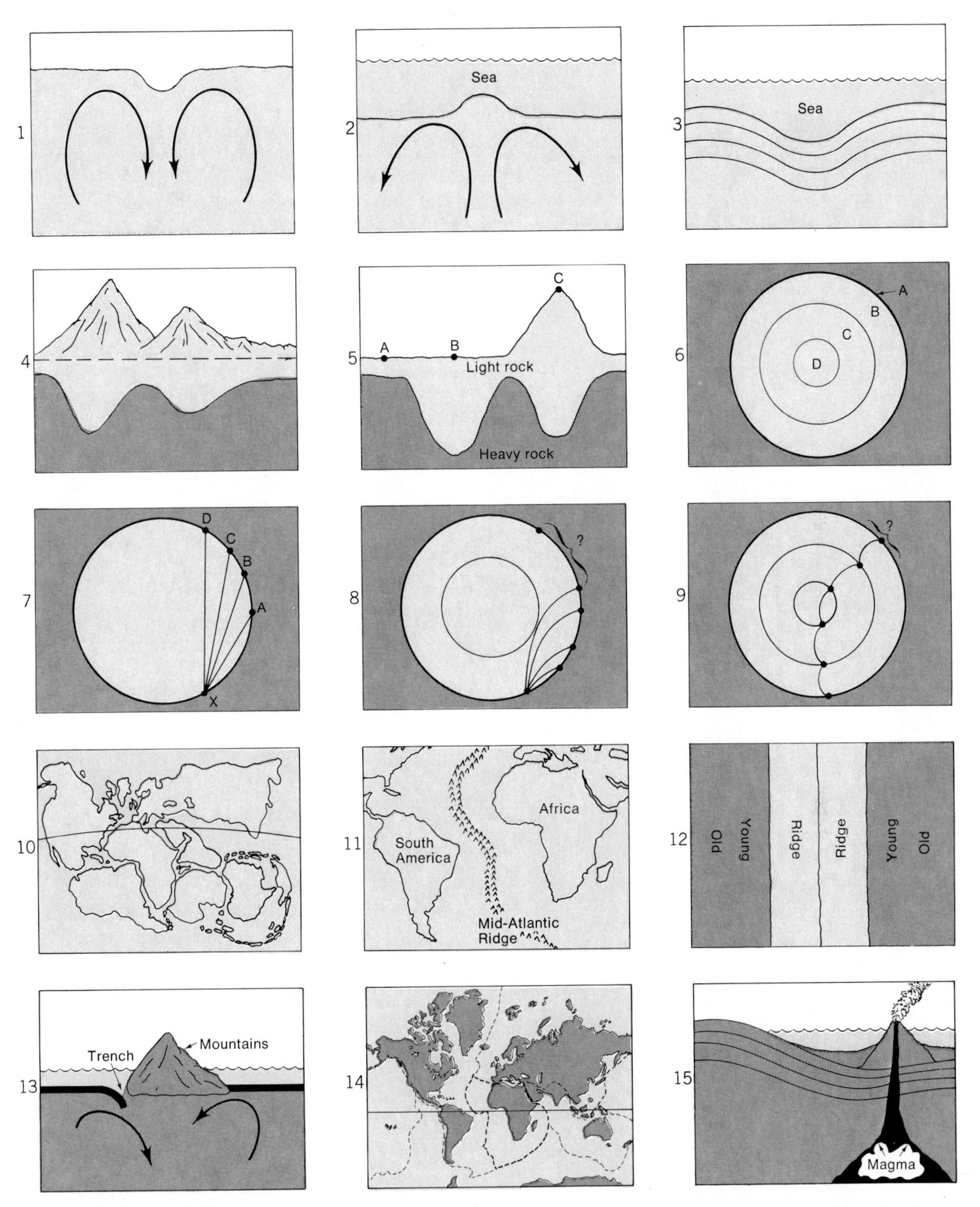

Figure 15/31 *Write questions to go with these answers.*

Unit Five
The Earth's History

Chapter 16 Geologic Time

An extra activity suggestion is to count sand grains and see how big a pile 1,000,000 grains would be. Do this by sampling rather than actual counting. Count the grains in one ml and extrapolate.

Have you ever seen a million of anything? A million all at once? You might be tempted to say that you have seen a million grains of sand, but just how much of the beach did the 1,000,000 grains take up?

Could you carry 1,000,000 pennies all at one time? Just how many lifetimes are there in 1,000,000 years?

Materials

ditto master, typewriter, paper, tape, marker

Activity Time: 20 minutes to tape the pages up. Prepare master and copies beforehand. Have students estimate the number of asterisks there will be on a page and calculate the number of copies they think will be required. Use the same pages for all your classes. Take them down at the end of each period, so that each class can put up the whole series. Doing the activity helps the students get a "feeling" for a million. The series will fill one wall. Copies can be made on scrap paper.

About 12.7 days nonstop would be required to count all the asterisks.

★ Before you can study the events that have taken place in the history of the earth, you should have some idea of how many a million of something really is. At this point, you will need to have 1,000,000 asterisks. How many copies of a master filled with asterisks would you need to have 1,000,000 asterisks?

After the copies are made, tape the pages to a wall of the classroom. Do not let the pages overlap. With a marker, circle the number of asterisks equaling the total years you expect to live. If someone had forgotten to type the circled asterisks in the first place, would these years be missed?

How long do you think it would take to count all the asterisks? If you can count one number per second, how many days would it take? Figure it out. ★

Students might determine how many walls would be needed to display a billion asterisks. Are there enough walls in the school for this?

One million of anything is a lot, especially if you have to count or carry a million of something. And one million years is many times greater than the number of years in the average lifetime. It is hard to imagine that much time. So whenever you read the number 1,000,000 in this chapter, remember the asterisks on the wall of the room.

Measuring Time Without Clocks

16.1 Indirect methods are used to measure time.

Prehistoric people had a number of ways to record events that happened to them. But they were not able to record *when* the events happened. They had invented methods of recording, but their ways of measuring time were different from ours. As a result, the history they recorded is hard for modern man to understand.

The Chinese were among the first to record how many years ago an event happened. But their records only go back 2500 years. Events that happened more than 2500 years ago may have been recorded but were not dated.

Older records may only be speculation. Such writings are often transcriptions of verbal accounts, which may not be accurate.
Process: measuring

Process: observing

Then how do scientists tell the ages of, or date, things as old as mountains? How can they tell the age of the sea floor and the fossils pressed in rock? Mountains, fossils, and the sea floor may be much more than 2500 years old, so scientists had to find a way to date such old things.

What scientists do is measure time by measuring something else. They measure how long a certain process takes in years, using clocks and calendars. Then they use the process in a kind of "time yardstick." For example, before children learn to tell time with a clock, they know it's time for supper when the sun sets. They even know when it's time for bed by which television program is on. Some people have found that a movie theater intermission is about as long as a candy bar followed by a soft drink. They don't often miss any of the movie. How could a burglar tell that the owner of the house in Figure 16/1 has been away for almost a week? Do you see how such indirect ways of telling time can be just as useful as reading a clock?

16.2 Some sediment layers can be used to measure time.

The sediment in some bodies of water tells of changes in seasons just as tree rings do. During the spring thaw in colder climates, the velocity of streams is high due to the

Figure 16/1 *Do you think that anyone is at home?*

large amount of water from melting snows or glaciers. At that time, larger or heavier particles are deposited in lakes downstream. The smaller or lighter particles are kept in suspension by waves and currents.

These particles don't settle out until winter, when the lakes or ponds freeze over. As a result, the year's deposit of sediment is actually a double layer. The coarse particles laid down in the spring are on the bottom and the finer particles laid down in the winter are on the top. Since the fine particles are laid down in winter, when plant life is dying, dark rotted material often colors the winter part of the layer as seen in Figure 16/2. One set of light and dark bands, called a **varve,** marks the passing of one year. Can you tell how many years are shown in the pile of sediment in Figure 16/3? Different sets of varves overlap going back in time. So far, it is possible in North America to date things as old as 20,000 years by counting varves.

Figure 16/2 shows three complete varves, with a winter half at the bottom and a summer half at the top. Figure 16/3 shows one of a series of samples in a box taken from an Ice Age lake deposit. The bottom is a winter layer. There are 25 complete varves, some of which show indistinct secondary varves, and the top of the sample consists of a summer layer. The entire sample is one foot long.

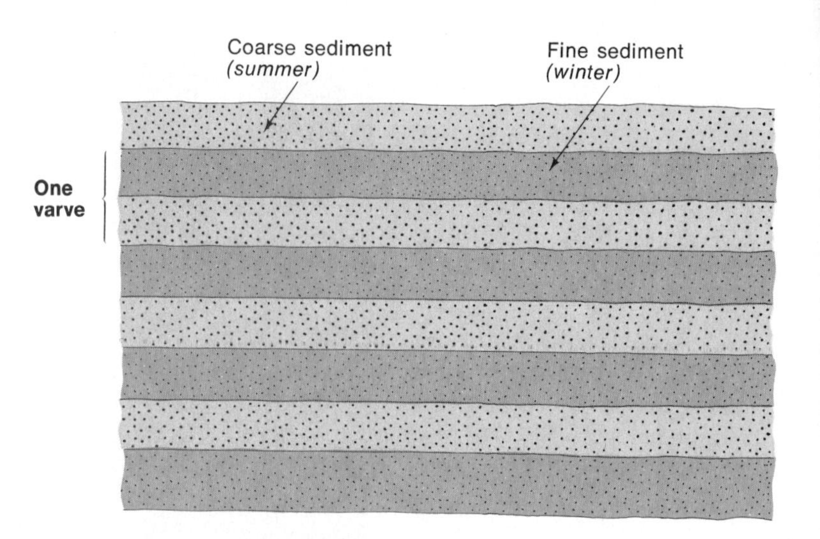

Figure 16/2 *Varves are sediment layers deposited at the bottoms of lakes or ponds. A coarse layer from spring and summer floods and a fine layer that settled in the winter marks one year.*

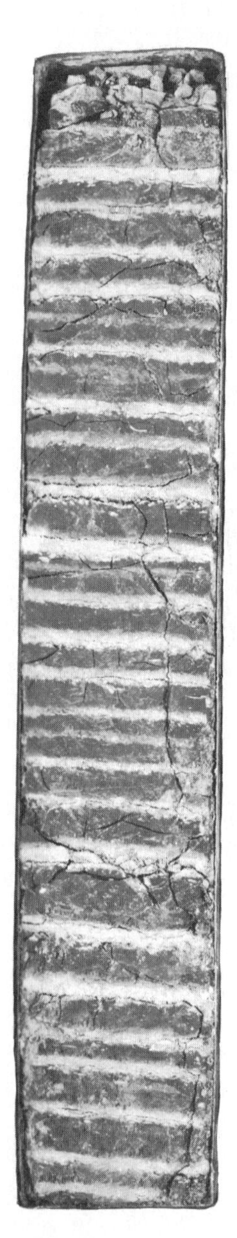

Figure 16/3
The amount of sediment deposited in lakes and ponds can vary greatly from season to season.

16.3 Trees place events in time.

A tree grows by adding a ring of new wood each year around the wood already there. The age of a tree can be found by cutting it down and counting the rings as in Figure 16/4, or by using a hollow drill to pull a small core of wood from the trunk as in Figure 16/5. How old were the trees in Figures 16/4 and 16/5 when the rings were exposed?

The thickness of each ring tells something about the climate during the time the tree was adding the ring. A wide ring means warm and wet weather during the growing season. A thin ring shows a dry or cold growing season. Can you find good and bad growing seasons in Figures 16/4 and 16/5? If the tree in Figure 16/5 was cut down in December, 1970, what year was the worst growing season?

If enough cross sections of trees from the same area were available, it would be possible to trace certain events back hundreds of years. An example will show you how the tree-ring method works.

Suppose there is evidence that a well-preserved log like log VII in Figure 16/6 was used as a marker on a pioneer grave. If the log can be dated, so can the grave on which it was found. Tree rings from other logs in the area can be used to find the age of log VII if at least one ring can

Figure 16/4
A single tree ring has a dark part and a light part. How many rings can you count in the close-up of a cut log?

The tree from which the magnified section was taken must have been at least 15 years old. The shallow arcs of the growth rings indicate the tree was much older. The large log in the woodpile was about 40 years old when it was cut. The tree in 16/5 was about 55 years old.
Process: measuring

In 16/4, the third and fourth seasons were good ones. The eighth and ninth were poor ones. The worst growing season in 16/5 would be 1953.
Process: observing

Figure 16/5 *This core sample of wood was pulled from a tree trunk. How old was the tree when the sample was taken?*

be found in one of the logs for each year of time that passed from the burial to the present. Counting backwards from a recently made ring should place the event in history. The following exercise will illustrate how this works.

★ The cross section of the tree labeled I was cut in 1901. It has three interesting rings labeled A, B, and C that show odd changes of climate. Cross section II is a little older than cross section I. It has three rings that show the same odd changes of climate. If rings A, B, and C on tree I were made at the same time as rings A, B, and C on cross section II, when could ring D have been made? Simply count backwards from 1901.

Rings D, E, and F on cross section II match rings D, E, and F of cross section III. Rings G, H, and I on cross section III match rings G, H, and I on cross section IV, and so on until you get to rings P, Q, and R on the log you are trying to date. Trace the sequence of tree rings from tree to tree until you can tell the year tree VII began to grow. If the tree was cut as part of a burial, how old is the grave site? ★

Activity Time: 15 minutes. Discuss the method. Ring D was made in 1887.

Tree VII began to grow in 1835, and was cut down in 1850.

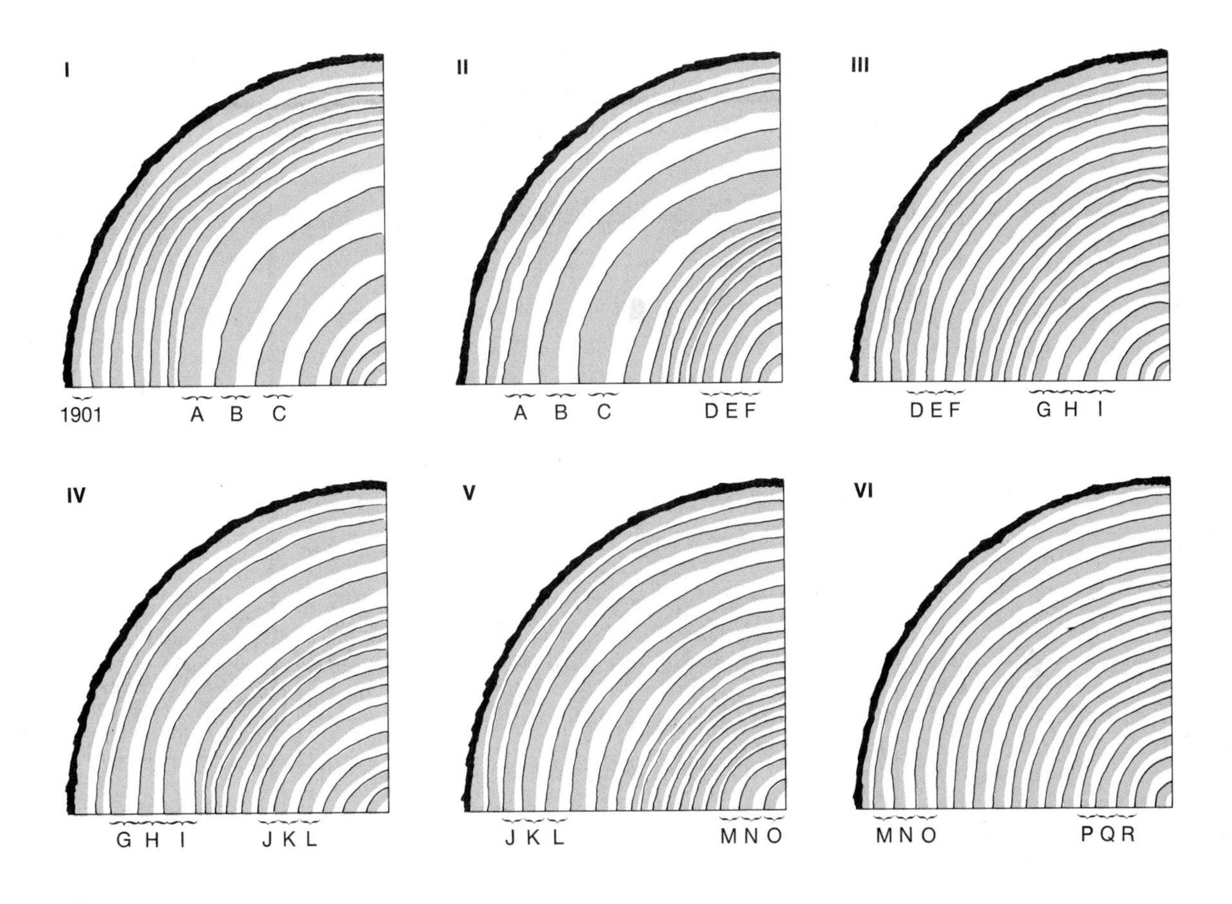

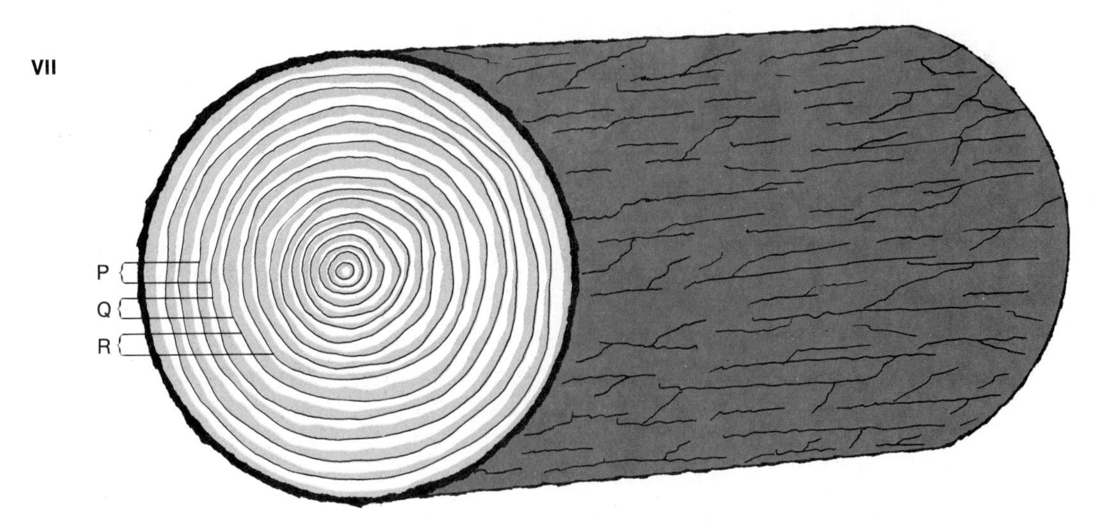

Figure 16/6 *By matching similar tree rings and counting backwards from 1901, you can find the age of log VII.*

Answers:

1. Is the bed messed up? Is it warm?
2. You might feel temperature changes in the walls—warm during the day, cold at night. You might tell by sounds—traffic noises or birds. Estimate how long it takes for exposed food to spoil. Time some process with your pulse and use the process as a clock. For example, put an empty glass on the floor and a glass full of water on a pile of books next to it. Arrange a strip or cloth with one end in each glass. The combination of gravity and capillary action will slowly move the water into the lower glass.

Did You Get the Point?

Indirect methods are used to date events in the history of the earth.

Varves date sediment layers.

Tree rings can date logs several hundred years old.

Check Yourself

1. What are some of the things Flo might tell Andy to check to answer his own question in Figure 16/7? Assume that he doesn't have a watch and there is no clock in the room.

2. If you were locked in a room with no windows (but plenty of air, food and water), how could you measure the passage of time so that you would know the day of the week when you would finally be let out?

Figure 16/7 *How might Andy answer his own question?*

Atoms Date the Past

16.4 Trees contain two kinds of carbon atoms.

The trouble with tree-ring dating is that very few trees still around are old enough to have rings grown before the invention of writing. Once a tree dies, the wood usually rots quickly. But if the conditions are right, the wood may remain unchanged for thousands of years.

Scientists do sometimes find very old wood. It could be an ancient Egyptian mummy case, or a piece of a roof beam from the remains of an early Arizona Indian building, or even charcoal from a campfire which burned thousands of years ago. Fortunately, there is a way to date wood without studying tree-rings.

Use the word "isotope" only if you are sure the class can handle it. Too much new information distracts from the point being made.
Concept: particulate nature of matter

Even though wood dies out and changes in some ways as years go by, the carbon atoms it contains stay pretty much as they were. Two kinds of carbon can be found in wood. The main type is called **carbon-12,** and the other kind is called **carbon-14.** The numbers mean that an atom of carbon-12 weighs 12 atomic mass units and an atom of carbon-14 weighs 14 atomic mass units.

While a tree is living, it takes in a little of each kind of carbon and gives off a little of each kind. The amount of carbon-14 in the atmosphere is always very small, so the amount in the tree will also be very small. And as long as the tree is living, the amount of one kind of carbon compared to the amount of the other kind will stay about the same.

Demonstrate a Geiger counter if one is available.

Aside from the atomic weights, the only noticeable difference between the two kinds of carbon is that carbon-14 is radioactive and carbon-12 is not. An instrument known as a **Geiger** (GUY-gur) **counter** can be used to detect the radioactive carbon.

Concept: levels of organization

Like all radioactive atoms, carbon-14 **decays.** When it decays, the nucleus gives off a particle, and the atom changes into a different material. This process changes the radioactive carbon-14 to nitrogen gas, and the gas is not radioactive. As time goes by and more and more carbon-14 decays, Geiger counter readings of the amount of radioactivity in a pile of pure carbon-14 will become less.

When a tree is living, the Geiger counter reading will always be the same because the carbon-14 atoms are replaced as they decay. After the tree has died, the carbon-14 atoms are no longer replaced, but the decay process goes on. The amount of radioactive carbon gets smaller, giving the lower readings as shown in Figure 16/8. Geiger counter readings of the dead tree will be lower than the readings taken when the tree was alive. Scientists know just how high the Geiger counter reading should be from a piece of living wood, so they can figure out how long the

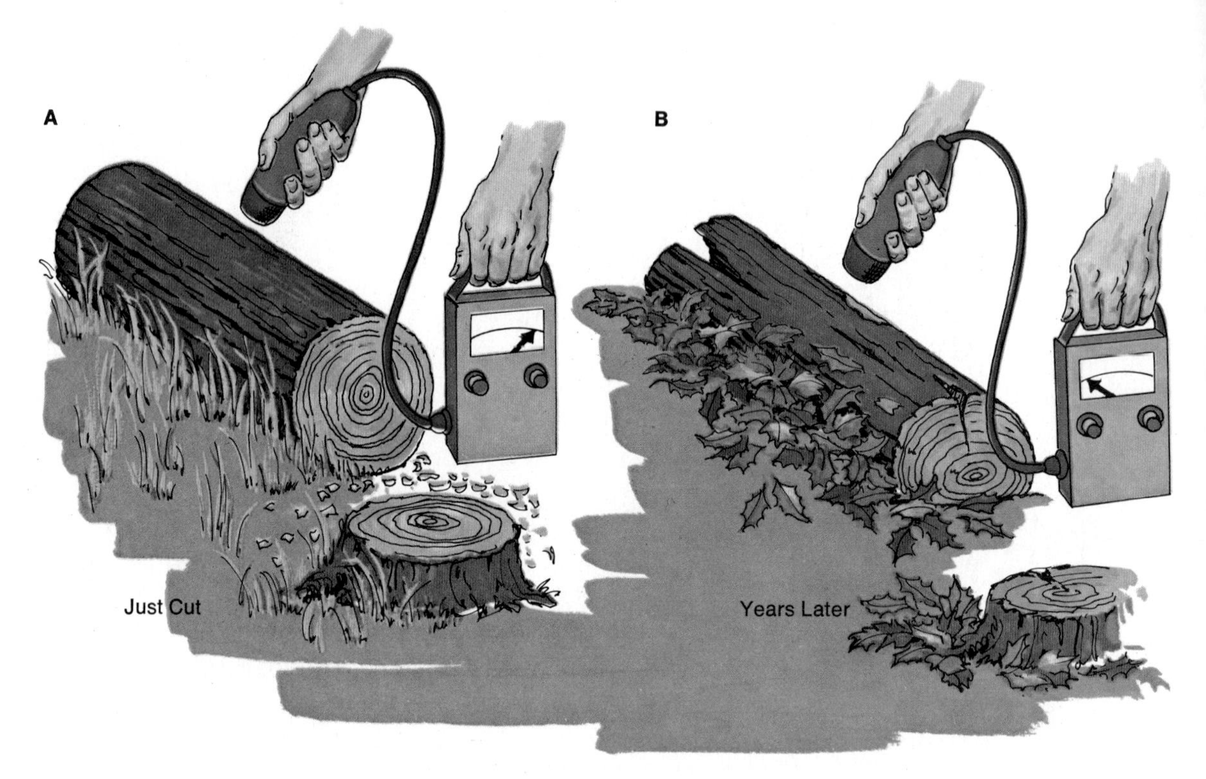

Figure 16/8 *When a tree dies, the radioactive carbon-14 continues to decay, but is no longer replaced. So the longer the tree is dead, the lower the Geiger counter reading will be.*

wood they are testing has been dead. A game with pennies will show you how this fact can be used to tell time.

Materials
cardboard box, pennies, BB's

Activity Time: 30 minutes.

16.5 Carbon atoms place events in time.

★ Put 100 pennies and several hundred BB's in a cardboard box. The BB's will represent the carbon-12 (nonradioactive) in a tree, and the pennies will be the carbon-14 (radioactive). Put a top on the box and shake it several times. Lots of noise will indicate a good shaking.

The only rule of the game is that all pennies that come up heads must be removed from the box after each series of shakes. Open the box and remove the pennies that come up heads after the first shaking. They stand for carbon-14 atoms that decayed to nitrogen, lowering the Geiger counter reading. Count them and record the number of pennies that remain in the box.

Shake the box again. Then open it and take out the second batch of heads. How many pennies are left in the box? Record the information in your notebook. What would be happening to the Geiger counter reading if real carbon-14 were being used? What would be happening to the amount of carbon-12?

Continue to shake, remove, and count two more times. In your notebook draw a graph of the information. Put the shake number along the horizontal line at the bottom and the number of pennies left in the box on the vertical line up the side. Connect the points with a line.

You can get more accurate results by starting with more pennies. Try averaging the data of the entire class—the result will probably be very close to losing half the pennies each time.

Notice that about half of the pennies were removed after the first shaking. Half of the half that was left were taken out after the second shaking. What did that leave? The third shaking resulted in the half of the half of the half being taken out. What did that leave? Should you lose exactly half of the pennies each time? How could you do the experiment to get results closer to half the pennies? ★

If you were to use real carbon-14 instead of the pennies, you would have to wait about 5730 years for half of the carbon-14 to decay into nitrogen. The number 5730 is called the **half-life** of carbon-14. The half-life is the amount of time it takes for half of the radioactive material present at any one time to decay. Now relabel your graph by putting the half-life numbers 5730, 11,460, 17,190, 22,920, and 28,650 in place of the shake numbers 1, 2, 3, 4, and 5. You now have a graph of the rate of carbon-14 decay in years.

Suppose a scientist figures that a certain log should contain one gram of carbon-14 if it were living. Instead, he finds that it contains only half a gram. The tree must have been dead one half-life—5730 years. If the log contained one-fourth of a gram (half of a half), the tree must have been dead two half-lives, or 11,460 years. Figure 16/9 shows two things that have been dated using this method.

Dates approaching 50,000 years are often subject to large errors. You might wish to refer to the comparison of radiocarbon dating and bristlecone pine dating. See "Carbon 14 and the Prehistory of Europe," by Colin Renfrew, *Scientific American,* October 1971, pp. 63–72.

16.6 Uranium can be used for dating.

Carbon-14 dating also has some drawbacks. First it can only be used to date things that were once alive. And

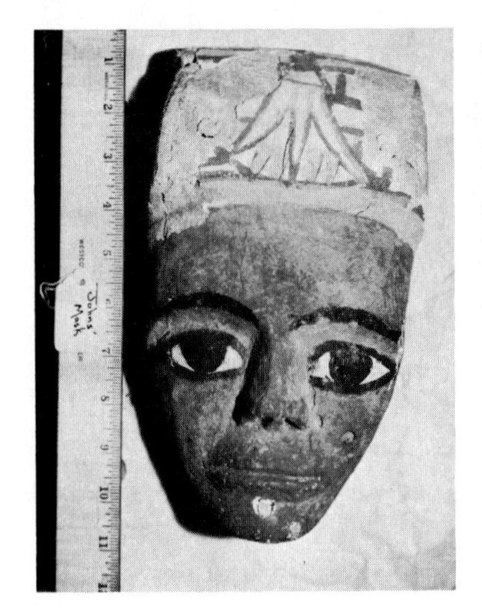

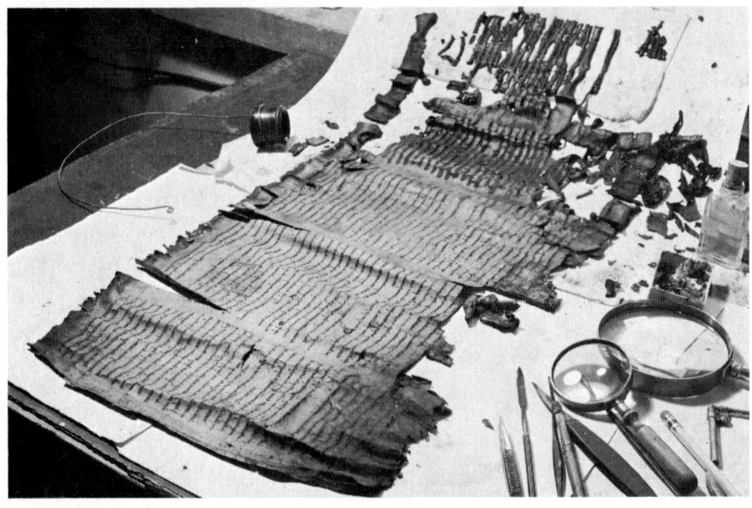

Figure 16/9 *The carbon-14 method was used to date both of these objects. The scrolls found near the Dead Sea are over 2000 years old, and the Egyptian mask is about 2800 years old.*

second, living things have very little carbon-14 to begin with. After about 50,000 years, too little is left to make accurate measurements. So carbon-14 dating of objects more than 50,000 years old is mainly good guessing.

For more information on the decay of uranium, see Cotton and Lynch, *Chemistry, An Investigative Approach,* Houghton Mifflin, 1970, Chapter 25, or the supplement to the 1973 revision, *Chemistry, An Investigative Approach, Supplementary Readings and Investigations,* by Cotton, Darlington, and Lynch.

Many rocks, however, contain radioactive minerals such as uranium, which decays in steps. Uranium changes from one radioactive element to another until it finally becomes a kind of nonradioactive lead. The half-life of uranium as it changes to lead is 4.5 billion years. Unlike carbon-14, radioactive uranium is around long enough to measure millions of years and hundreds of millions of years, not just thousands.

Scientists can tell the difference between lead that came from uranium and lead that was there all the time, because the weights are different. The original lead weighs 204 atomic mass units and lead that formed from the decay of uranium weighs 206, 207, or 208 atomic mass units. Therefore, if you find a rock which contains lead weighing 206, 207, or 208 along with uranium, that lead must have come from the decay of uranium.

A Geiger counter can measure the amount of uranium in a rock, which can be compared to the amount of lead 206, 207, or 208 in the same rock. Using a half-life graph or simple arithmetic, scientists can tell how many years ago the rock cooled from magma or lava. The uranium dating method provided identification for the items in Figure 16/10.

Not all rocks contain uranium, but many contain other radioactive elements. Potassium, for example, may be radioactive. Radioactive potassium decays into nonradioactive

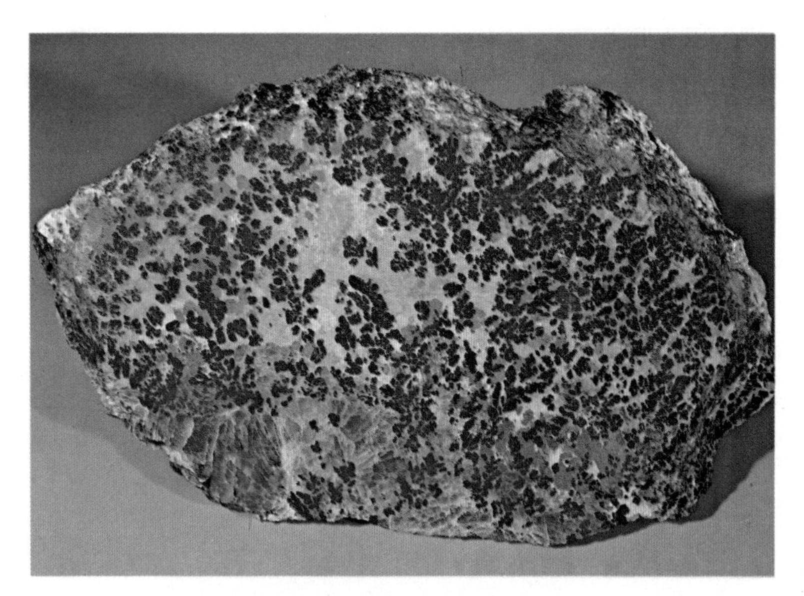

Figure 16/10 *The rock at the top, from the Ruggles Mine in New Hampshire, was found to be 304 million years old by using the uranium dating method. The rock at the bottom is a meteorite found in Australia that was dated to 4.6 billion years by the same method.*

argon. The half-life of radioactive potassium is 1.3 billion years. Comparing the amount of radioactive potassium to the amount of argon in the rock can give its age just as comparing the uranium and lead does in other rocks.

Did You Get the Point?

Radioactive carbon is used to date events that happened up to 50,000 years ago.

Uranium and other radioactive elements are used to date events that occurred more than 50,000 years ago.

Check Yourself

1. Figure 16/11 shows a series of events that have happened in the past. Which dating technique would you use to date each event? Explain why you would use each method.

A Diary of the Earth

16.7 Important events help organize the earth's history.

People all over the world have been finding fossils for hundreds of years. Even cavemen might have wondered, from time to time, just when those shells and leaves in the rocks lived. Mountains, too, have long been a subject of interest. Before the discovery of the Geiger counter and radioactive dating, scientists could not be sure about how long ago mountains were formed.

Materials

adding machine tape, meterstick, colored pencils

Activity Time: 30 minutes. Students may go out in the hall to do this. Tape can be fastened to the wall.

★ Now both fossils and mountain building can be put on a kind of geologic calendar. To make a calendar of your own, you will need a meterstick and a strip of adding machine tape five meters long. If you have colored pencils, use them for marking.

The age of a fossil can be assumed by dating the rock in which it was found. Large dinosaurs, like the ones in Figure 16/12, seem to have been common in many places

Figure 16/11
What methods would you use to date these objects?

Answers:

A. Carbon-14, written records.
B. Since the fossil is in a rock, shale, it might be dated by potassium-argon. If it were in unconsolidated clay, it might be dated by carbon-14 or varves. The solid rock implies an age beyond the range of carbon-14. Clay would imply a more recent lake deposit.
C. Carbon-14; likely to be within the range.
D. Potassium-argon. The area has been under ice for more than 50,000 years.
E. Tree rings.
F. The age of the point would probably be determined by carbon-14 from the bones or charcoal. If enough wood is around, tree ring dating might work. If students suggest dating the point itself, note that the rock from which the point was made could be a billion years old.
G. Carbon-14 from wooden stock, written records.
H. Uranium-lead or potassium-argon, since evidence indicates that the moon is much older than 50,000 years.

Wooden mummy case from Egypt

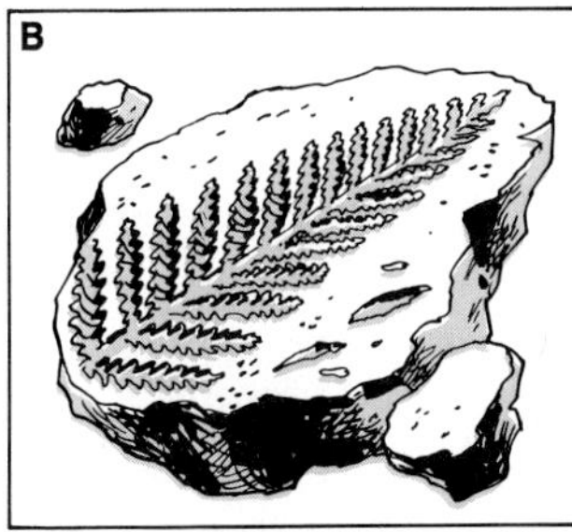

Fossil fern found in shale

Mammoth found frozen in glacier

Coal found in Antarctica

A very old sequoia tree

Spear point from a prehistoric campfire

"Fire-arrow," oldest known Chinese gun

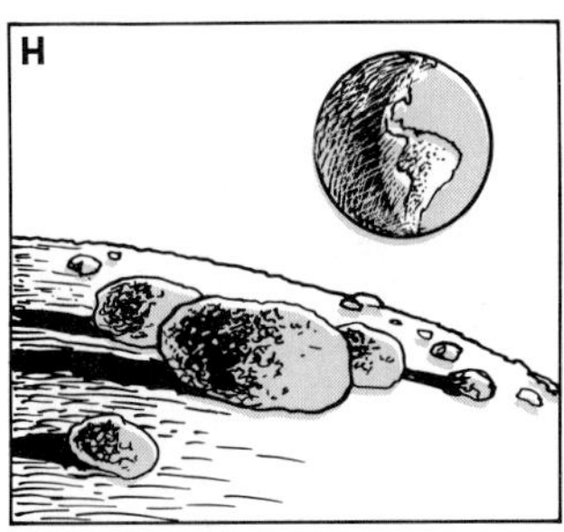

Moon rock

Figure 16/12
Large and powerful dinosaurs like these roamed the earth millions of years ago.

around the world, for millions of years. Yet they all disappeared, that is, they became **extinct** (EX-STINCT). When did this happen? Since the most recent rock containing a dinosaur fossil is about 70 million years old, the last dinosaurs must have died that long ago.

You might dwell on "milli-" and "centi-" here, and compare millimeter with million.

Draw a line across one end of the piece of adding machine tape and label it with today's date. Let one millimeter of paper stand for one million years. Then you can place the extinction of the dinosaurs on your calendar by drawing another line across the paper 70 millimeters from the "Now" line. Be sure to label the new line with the event it is marking.

You can tell how old a mountain range is by dating pieces of its rock. The *youngest* rock which has been folded or faulted or otherwise changed by the mountain-making process would date from a time just before the mountain rose.

The Appalachian Mountains are made up of sedimentary rock layers which were folded by a horizontal pressure into very large wrinkles. Review Figure 15/19 which shows the Appalachians. The youngest rocks which were folded are about 270 million years old, so the folding must have started about that long ago. Mark that event on your calendar.

Fossil algae have been found in rocks 1.8 billion years old, and fossil bacteria in rocks 3 billion years old. See "The Oldest Fossils," by Elso S. Barghoorn, *Scientific American,* May 1971, pp. 30–42.

Another important date in the history of the earth marks the time when the first abundant fossils of animals appeared in the rocks. This date has been put at some 600 million years ago. Add another line to your adding machine tape calendar and label it with this event.

The time when the earth was first formed is very hard to place exactly. It is impossible to dig up a rock that has been around since then. The original surface rocks were weathered, and the products of the weathering may have been carried to the sea, deposited in layers, melted, and pushed back up to the tops of mountains more than once. Each time the rocks were melted, the amount of lead, uranium, potassium, and argon in them changed. The oldest known rocks are thought to be 3.5 billion years old.

But we should have some kind of date for the time when the earth was first formed. Using the information from old rocks, scientists have been able to put the age of the earth at about—ABOUT—4.5 billion years. (This is 4500 walls full of asterisks.) Scientists will be making new estimates during your lifetime as they gather more and more information. Since the number "4.5 billion" might change, put it on your calendar as a dotted line.

Process: measuring

Geologists have used the rock record to divide the history of the earth into four major divisions of time. The time from the beginning of the earth until the time of the first abundant animal fossils is called **Precambrian** (pree-KAM-bree-*un*). It is the longest of all the divisions in years. Scientists know much less about this time division than about the other three. Fossils are few and the rocks are often greatly changed from their original forms. Label the section of your tape that covers the Precambrian.

The next three divisions are called **eras** (EAR-ahz). The one which followed the Precambrian continued to the time just after the Appalachian Mountains had been formed.

Events, plants, and animals	Number of years ago
Blue-green algae lived	3000 million
Trilobites lived	600 million
Water covered almost all of North America	430 million
First land plants appeared	500 million
First fish appeared	500 million
Early amphibians developed	375 million
First birds appeared	150 million
Early horse developed	70 million
First seed plants appeared	400 million
First reptiles appeared	325 million
Sierra Nevadas formed	25 million
Coal-forming plants lived	300 million
Last ice age ended	10 thousand
Mount Vesuvius erupted	1893
Early man appeared	2 million
Continental drift began	150 million

Figure 16/13 *Plot these events on your paper calendar.*

This era is called the **Paleozoic** (*pay*-lee-uh-zo-ik) Era. Mark this time span on your tape calendar, too.

The **Mesozoic** (*mes*-uh-zo-ik) Era came after the Paleozoic. The era ended with the extinction of the dinosaurs. Again identify this part of your calendar by name.

The final era is also the shortest one. It is the era in which we are living, the **Cenozoic** (*sea*-nuh-zo-ik) Era. Put this last name on your calendar in capital letters.

The dates for the plants and animals in Figure 16/13 came from dating the rocks in which fossils of these organisms were found. Students will comment on the difficulty of placing recent events. Stress that what we think is old may be comparatively young.

All of the other things that have happened on the earth since its beginning can be placed in one or more of the four time divisions. Figure 16/13 shows some of these things. Mark them on your calendar using lines and labels. When you come to the appearance of man, make the line and label stand out. ★

16.8 The calendar can become a ruler for measuring time.

If you found an old newspaper nailed between the walls of a house, could you tell when the house was built? If the paper had no dates on it, could you figure out the age of the house? How old would the house be if the newspaper told about a Civil War battle?

Now that your calendar is complete, you can use it to date rocks and events in the earth's past in the same way newspapers can date an old house. Suppose you found a rock layer with the fossil of a dinosaur in it. During which era was the rock layer probably laid down? What would be the youngest possible age for the rock layer?

Mesozoic. 70 million years.

Figure 16/14 shows some of the information you plotted alongside the names of time divisions within the eras. Geologists use these names the way historians use phrases like "the Dark Ages" and "the Renaissance" when they talk about history. The time divisions are called **periods.** What kinds of fossils would a sedimentary rock have to contain before a geologist can say that it was laid down in Pennsylvanian time?

The rock would have to contain fossils of organisms unique to that period, or typical of the period if they are not unique. A good combination would be tree ferns, which are typical, and first reptiles, which are unique. More on this in Chapter 17.
Process: inferring

Be sure to save your calendar. In later chapters you will use it instead of the Geiger-counter method to date many other things. It will be your ruler for measuring time.

Did You Get the Point?

The date of the origin of the earth is difficult to know exactly.

Rocks from the early history of the earth are hard to find.

Events in the earth's past enable scientists to divide earth history into eras and periods.

Fossils can be used to date events in the earth's past.

Check Yourself

1. Write an article for the school newspaper using the names of all 12 of the different time periods from Figure 16/14.

2. Using your calendar and Figure 16/14, determine what is wrong with the pictures in Figure 16/15.

3. If you could spend any amount of money or use any amount and kind of equipment, how could you better determine the age of the earth?

Answers:
1. Use the time chart, Fig. 16/14, to check the articles. An imaginative alternative, such as a satire or a nonsense poem, will fulfill the same purpose—getting the students to use these new terms.
2. Man came after the dinosaurs, and flying reptiles came after trilobites.
3. You might drill deep into the earth for unchanged rock, or mantle or core material that can be dated. You might get dates from the moon and other planets. You might assume the rocks on Mercury have not changed as much as those on Earth.

Figure 16/14
A condensed history of the earth.

Era	Period		Millions of Years Ago
Cenozoic	Quaternary		
			2
	Tertiary		
			70
Mesozoic	Cretaceous		
			135
	Jurassic		
			180
	Triassic		
			225
Paleozoic	Permian		
			270
	Carbonif-erous	Pennsylvanian	
			325
		Mississippian	
			350
	Devonian		
			400
	Silurian		
			440
	Ordovician		
			500
	Cambrian		
			600
Precambrian time			
			4500?

Plants and Animals	Geologic Conditions in North America
man appeared; mammoths became extinct	four ice ages; then climate warmer; Sierra Nevada Mountains formed; active volcanoes in Cascade Range
mammals widespread; modern forms of most living things developed	climate first warm, then cooler; many mountains in west; volcanic activity in northwest; North America approaching its present shape
first abundance of flowering plants; dinosaurs became extinct at end of period	inland seas and swamps; chalk, shale deposited; mountains in far west growing; Rockies forming at end of period
flying reptiles; larger dinosaurs; simple mammals appeared; first birds	shallow seas over much of western U.S.; mountains forming in far west; continent drifting
first dinosaurs; early amphibians became extinct; turtles, crocodile-like reptiles widespread	much of continent above water; eastern climate warm and humid, western climate warm and drier
many ancient animals such as trilobites died out; amphibians and mammal-like reptiles abundant	much of continent rose; Appalachians formed
first reptiles; spread of amphibians and insects; tree ferns abundant	eastern and western mountains continuing to grow; great coal swamps; mountains forming in west; seas becoming smaller
many sharks; seed plants widespread	climate warm and humid; large seas remain; eastern mountains becoming more extensive
first amphibians; first seed plants; first forests; many kinds of fishes	large seas remain; mountains forming in east and west at end of period
first land plants	mountains reduced by erosion; large seas over the continent
seaweeds, first fishes	warm climates; seas becoming larger; mountains forming in east at end of period
algae, brachiopods, trilobites	land low; seas large; climate mild
blue-green algae, fungi, bacteria; and then worms and mollusks	great volcanic activity, deposition, erosion, glaciation

The Drifting Continents

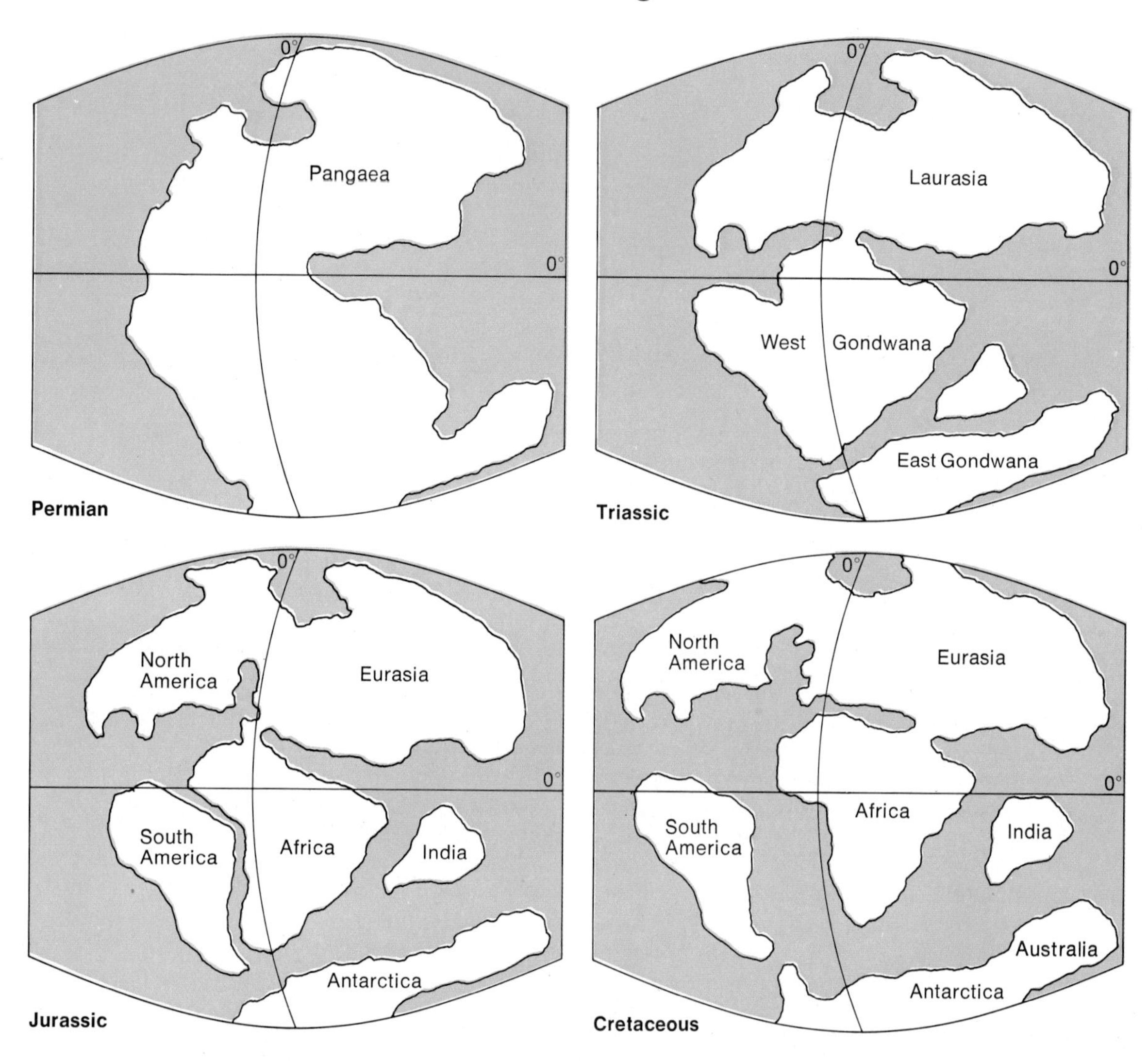

Robert S. Dietz and John C. Holden of the National Oceanic and Atmospheric Administration recently used clues from rock layers, fossils, and radioactive minerals to chart the course of the drifting continents. The maps show how the continents have moved throughout geologic time. Notice how Africa has moved in relation to where the equator and the Prime Meridian cross. Note how long Australia hung on to Antarctica, and how long eastern Canada and northern Europe stayed together. And how about India's travels? Would you believe 5000 miles!

Figure 16/15 *Use your calendar to figure out what is wrong with this picture.*

What's Next?

The animal in the flying reptile's mouth is a trilobite; trilobites are mentioned in Figure 16/14 and two are drawn in Figure 17/4, as typical fossils of the Cambrian and Devonian Periods.

If you tape your calendar to the wall so that you can see all of it at once, you will notice that even if the very first man to walk the earth could read, write, and measure time, he could tell us very little of the earth's history. In terms of the total age of the earth, man is a recent event. He has been around for only 0.044 percent of the earth's lifetime. And that is why much of your calendar is blank.

The number of events that still have to be discovered and understood is very large indeed. Even new dating methods will fill in the blank spots very slowly. All of the asterisks

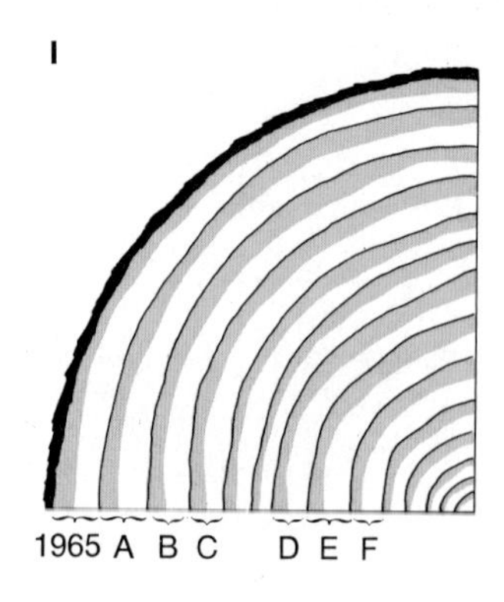

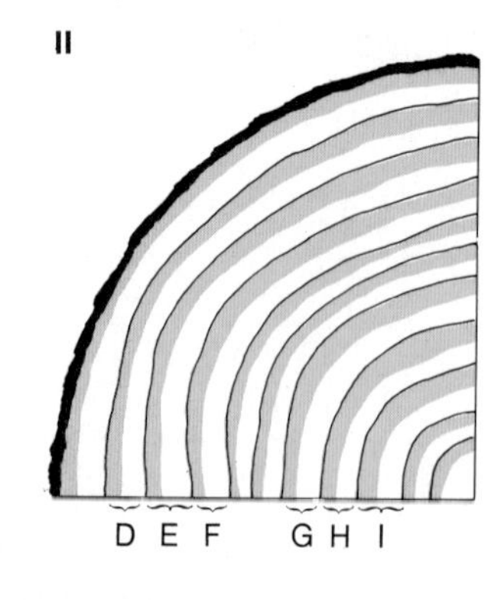

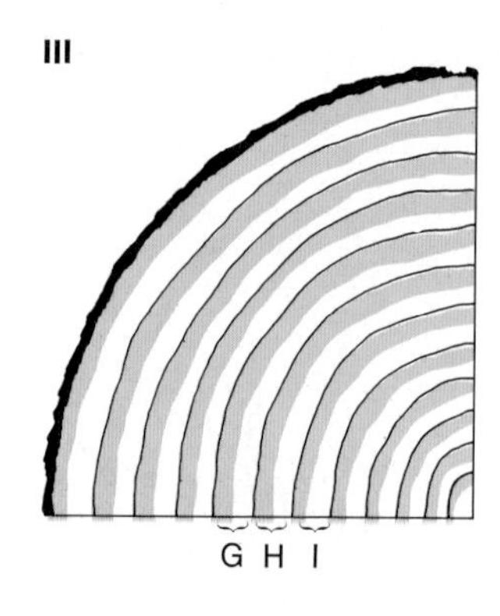

Figure 16/16 *In what year was tree III cut down?*

you taped to the wall stand for all the years in one millimeter of your calendar. How many events can take place in that one millimeter? How do you read the records of past events? Where do you look for them?

Skullduggery

Answers:
1. Tree III was cut down in 1958.
2. The bone chip dropped on top of the 1800 winter layer, and was covered by the 1801 summer layer; it was presumably dropped in the spring of 1801.
3. The log was 17,190 years old; or three half-life periods.

Directions: Answer the following questions.

1. In Figure 16/16 growth rings marked with the same letter were made in the same year. Determine the year tree III was cut down.

2. A fragment of bone was found in the varves in Figure 16/17. Using the two dates given in the figure, find in what year the bone fragment was deposited.

3. A flood washed an old log out of a river bank. It should contain 0.28 grams of carbon-14 if the log were still alive. Geiger counter readings showed that the log contained 0.035 grams of carbon-14. How old was the log?

Students can use graph paper. The animal which appears is a fossil trilobite.

The following questions are multiple choice. Only one choice is correct in each case. The teacher will give you a page of squares covered with numbers similar to the page in Figure 16/18. For each question, blacken all of the squares on the page that have the same number as the correct answer. The blackened squares will trace out a fossil animal.

Answers: (A) 3; (B) 6; (C) 10; (D) 13; (E) 20; (F) 23; (G) 25; (H) 30; (I) 34; (J) 40; (K) 43; (L) 45; (M) 49; (N) 55; (O) 60.
Note on N—the crust, including mountain roots, has been changed many times, and the core is mainly iron and nickel, which are not useful for dating. The mantle might contain the oldest material which can be dated.

Example: The answer to question A is choice 3. Fill in all of the squares with the number 3 in them.

A. Which of the following is *not* used to measure time indirectly?
1 varves **2** tree rings **3** a clock **4** a Geiger counter

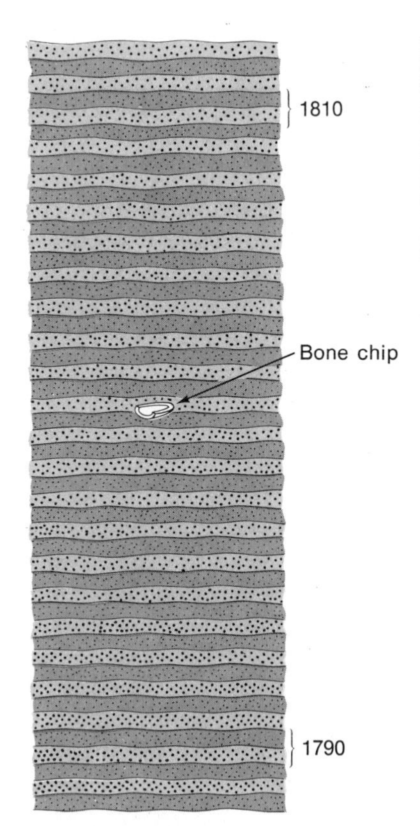

Figure 16/17 *How old is the bone chip buried in these varves?*

1	1	1	1	1	1	1	1	2	2	2	2	2	2	1	1	1	1	1	1	1	1
2	2	2	2	2	2	2	2	1	3	3	3	3	1	2	2	2	2	2	2	2	2
4	4	4	4	4	4	4	3	3	4	4	4	4	3	3	4	4	4	4	4	4	4
5	5	5	5	5	5	6	5	5	5	5	5	5	5	5	6	5	5	5	5	5	5
7	7	7	7	6	6	8	8	8	9	9	9	8	8	8	8	6	6	7	7	7	7
8	8	8	10	9	9	9	6	9	9	9	9	9	9	10	9	9	9	13	8	8	8
11	11	10	12	12	12	6	6	11	11	11	11	11	11	10	10	12	12	12	13	11	11
7	10	7	7	7	7	6	8	8	8	9	9	8	8	8	10	7	7	7	7	13	7
10	7	11	11	11	11	9	9	9	9	8	8	9	9	9	9	11	11	11	11	7	13
10	7	14	14	14	14	15	15	13	13	13	13	13	13	15	15	14	14	14	14	7	13
20	20	20	20	20	20	20	20	15	15	15	15	15	15	20	20	20	20	20	20	20	20
20	20	16	16	17	17	17	18	18	18	18	18	18	18	18	17	17	17	16	16	20	20
13	19	19	19	19	19	19	19	13	13	13	13	13	13	19	19	19	19	19	19	19	13
13	21	21	23	23	23	23	23	21	21	21	21	21	21	23	23	23	23	23	21	21	13
13	13	13	22	22	22	22	22	21	21	21	21	21	21	22	22	22	22	22	13	13	13
24	25	24	24	24	24	24	24	25	25	25	25	25	25	24	24	24	24	24	24	25	24
26	25	26	30	30	30	30	30	27	27	27	28	28	28	30	30	30	30	30	26	25	26
29	25	25	29	29	29	29	29	31	31	31	31	31	31	32	32	32	32	32	25	25	32
33	34	35	35	35	35	35	35	34	34	34	34	34	34	36	36	36	36	36	36	34	37
38	34	38	40	40	40	40	40	39	39	39	39	39	39	40	40	40	40	40	38	34	38
41	41	40	41	41	41	42	42	42	43	43	43	43	42	42	42	41	41	41	40	41	41
41	41	40	41	43	43	43	43	43	41	41	41	41	43	43	43	43	43	41	40	41	41
44	44	43	43	44	44	44	44	44	45	45	45	45	46	46	46	46	46	43	43	47	47
48	48	49	49	49	49	49	49	49	48	48	48	48	49	49	49	49	49	49	49	48	48
50	50	50	49	50	50	50	50	50	45	45	45	45	51	51	51	51	51	49	51	51	51
52	52	52	49	49	49	49	49	49	52	52	52	52	49	49	49	49	49	49	52	52	52
53	53	53	53	49	54	54	54	54	54	60	60	53	53	53	53	53	49	54	54	54	54
56	56	56	56	55	55	55	55	55	55	56	56	55	55	55	55	55	55	56	56	56	56
57	57	57	57	57	55	55	58	58	58	58	58	58	58	58	55	55	59	59	59	59	59
1	1	1	1	1	1	60	60	2	2	60	60	2	2	60	60	1	1	1	1	1	1
4	4	4	4	4	4	4	4	60	60	5	5	60	60	4	4	4	4	4	4	4	4
8	8	8	8	8	8	8	8	8	8	60	60	8	8	8	8	8	8	8	8	8	8

Figure 16/18 *Copy this grid for the Skullduggery.*

B. Which dating method covers about the same amount of time as written history?
5 varves **6** tree rings **7** carbon dating **8** uranium dating

C. Which dating method would you use to date a log cabin used by settlers in California?
9 varves **10** carbon **11** uranium **12** radioactive potassium

D. Which dating method would be used for dating rocks from the Cambrian period?
13 uranium **14** carbon **15** varves **16** tree rings

E. Why is carbon dating limited?
17 Carbon is rare. **18** There are two kinds. **19** The object being dated by carbon has to be alive. **20** Radioactive carbon has a short half-life.

F. Which of the following would be dated using varves?
21 geosynclines **22** mountains **23** glacial lakes **24** dinosaurs

G. If you start with a pile of carbon-12 weighing eight grams, how much would you have left after 11,460 years? Assume that it is protected from weathering and thieves.
25 8 grams **26** 4 grams **27** 2 grams **28** 1 gram

H. How much lead would result if 16 grams of uranium decayed for 4.5 billion years?
29 16 grams **30** 8 grams **31** 4 grams **32** 2 grams

I. Why can't uranium dating be used for all types of rock?
33 Not all uranium is radioactive. **34** Not all rocks contain uranium. **35** Not all rocks contain lead. **36** The method takes too long.

J. Which of the following is not an era?
37 Cenozoic **38** Paleozoic **39** Mesozoic **40** Cambrian

K. Evidence for dating the creation of the earth comes from all of the following except
41 old rocks. **42** meteors. **43** fossils. **44** the moon.

L. The amount of earth history witnessed by man compared to the earth's age is like
45 2 millimeters compared to 4500 millimeters. **46** the life of a dog compared to the life of a man. **47** one inch compared to one yard. **48** one inch compared to one light-year.

M. According to your calendar, which came first?
49 trilobites **50** dinosaurs **51** birds **52** man

N. The true date for the origin of the earth could be determined by direct study of
53 the earth's crust. **54** mountain roots. **55** the earth's mantle. **56** the earth's core.

O. If Precambrian rocks of a certain area do not have fossils in them, the possibility of some kind of life there cannot be ruled out because
57 animals without shells might not leave fossils. **58** the fossils could have been destroyed by heat. **59** Precambrian rocks that contain fossils may not have been uncovered yet. **60** all of the above.

For Further Reading

Poole, Lynn and Gray. *Carbon-14 and Other Science Methods That Date the Past.* New York, McGraw-Hill Book Company, Inc., 1961. A look at archaeology and how we find clues to the past. You learn how Dr. Libby discovered carbon dating and what other methods of dating have been found since.

Wyckoff, Jerome, *Marvels of the Earth.* New York, Golden Press, 1964. An introduction to drifting continents and earth history. Also treats volcanoes, earthquakes, glaciers, and other events that shape the land.

JEWEL THIEF
AT LARGE

Chapter 17 Stories in Stone

Sherlock Holmes, the famous detective, and his partner, Dr. Watson, sat in Holmes' cozy little apartment in London. The good doctor was wrestling to get free from a Chinese puzzle that fitted over the thumbs. Holmes was hidden behind his newspaper. Smoke from his heavy, crooked pipe drifted around the room.

"Watson," Holmes said finally, "I can't understand the police. Like you with that puzzle, they make the same mistakes over and over. The jewel thief! He is right under their noses and they don't know it."

"I'm sorry, Holmes, old boy. What did you say? I'm trying to get this thing off." Doctor Watson continued to tug at the gadget between his thumbs.

Holmes put down the paper and turned to Watson. "You must have tried to free your thumbs from that thing a dozen times. Hasn't it occurred to you yet that pulling and tugging isn't the way to get free?"

"Hmm. I suppose you're right, Holmes." Watson stopped wrestling with the puzzle. He put his hands in his lap and looked at the device.

"Let me give you and the police some lessons in thinking clearly," said Holmes. "Listen to the latest account of the crime."

Police today disclosed what they believe to be the robbery plan. After taking the jewels from the safe, the thieves drove to a wooded

Solving the mystery makes a good homework assignment. For the answer, as explained by Holmes himself, see the Teacher's Manual. Read it to your class after the students have had time to think about it. You may want to act it out with real sand and clay.

area and tossed the bag of jewels from the moving car. A partner was to pick the jewels up and bury them in a nearby clearing.

"Now listen to this, Watson. See if you can spot the key to the crime."

An honest citizen spotted the bag minutes before the partner arrived. He inspected the contents and carried the bag into the woods. There, he found an animal burrow of some sort and dropped the bag into the hole. Using a stick for a shovel and his coat for a basket, he filled the hole with soil dug from behind a bush. The clay that underlies the layer of sandy topsoil in the area was still on his coat when police questioned him later.

"Now Watson, here's where the police made the same mistake you're making with that puzzle of yours."

When the police arrived on the scene, they found a pile of dirt next to the burrow. The loot had been stolen again. There was the clay right on top of the pile, the same kind of clay observed on the honest citizen's coat. The partner must have been watching as the jewels were being buried and dug the bag up again. No clues were left at the scene of the crime, and the police are at a loss . . .

"Well, Watson, did you figure it out?" Holmes asked.

"No, Holmes. I'm afraid I'm going to have to ask you to take it off for me."

"I meant the jewel theft, not your Chinese thumb puzzle. Did you catch the clue the police missed?" Somewhat discouraged, Holmes got up and went to where the doctor was sitting. "Think, Watson, think." Then, as if he had designed the puzzle himself, Holmes freed his friend with two quick movements of one hand.

"Thanks, Holmes. My thumbs were going to sleep. Show me that again, would you?" Watson looked amazed.

"Haven't time, Watson. I must go see the inspector. The thief must be picked up before he gets too far away." Holmes was already halfway out the door. "Coming, Watson?"

Dr. Watson fidgeted in his chair. He knew he had missed something. "Ah, er, no!" he said as he cleared his throat. "I think I'll stay here and read the evening paper." The door slammed shut. Watson put the puzzle in his pocket and began to study the account of the jewel theft.

Can you solve the mystery?

Clues to Earth History

17.1 Rocks contain clues to geologic history.

The geologic history of the earth is a big mystery. But rocks contain clues to erosion, deposition, and mountain building that took place in the past. Road cuts blasted through solid rock like the one in Figure 17/1 have exposed clues. Oil wells drilled down through rock layers bring up samples of deeply buried rock with more clues. Tunnels dug for subways or mines as in Figure 17/2 also expose buried rock where the clues can be observed close up.

Process: inferring

The trick is to be able to read enough clues to put together the geologic history of an entire area. You already know how to read some of them. To read the others, you need some practice first. Can you write about the history in the exposed rock layers in Figure 17/3? Take the clues one at a time.

Figure 17/1 *Rock layers exposed in a road cut.*

Figure 17/2 *This fossil clue was found deep within the earth in a Pennsylvania coal mine.*

The mystery clues should be used as the focus of the chapter. Have the students interpret the clues in ciass. It is not necessary to follow the clues in the text of the chapter in the order in which they are given. The clues can be put into the discussion as they come up.
Process: observing

17.2 Clue number one is the fossils.

Some types of plants and animals lived and died out within single periods of geologic time. This means that if you find a rock layer containing fossils of one of these plants or animals, you will know that the rock must have been deposited during that time period. Figure 17/4 gives some examples of fossil clues like this and the periods they come from.

Fossil types that can be found in layers from two or three different time periods can be useful for dating rocks, too. But fossil types that span many time periods aren't much use.

Fossils can give you another kind of information, too. They can tell you some things about how and where sediments were formed in the past. For instance, some of the

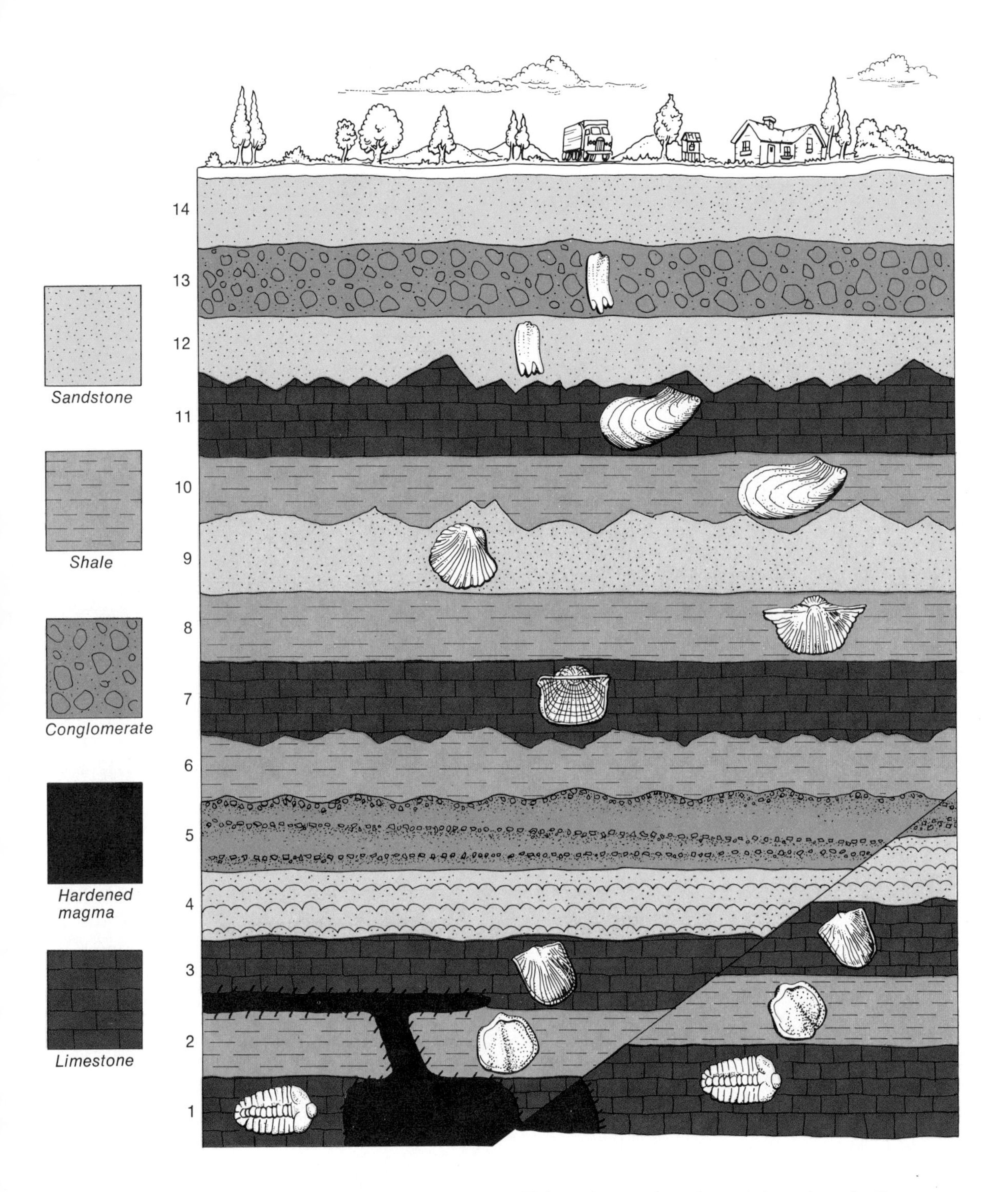

Figure 17/3 *This column of rock contains clues to the geologic history of the area. How many clues can you read?*

Layer 6 has no fossils, so you can't tell how old it is by the fossil method. Layer 7 is Pennsylvanian, 8 is Permian, 9 is Triassic, 10 and 11 are Cretaceous, and 12 and 13 are Quaternary. The other clues are the subjects of Sections 17.3 and 17.4. You can discuss at this point the kinds of animals represented by the fossils, and where they presumably lived.
Process: observing, inferring

Name of period	Millions of years ago period started	Names of typical fossils	How the typical fossils look
Quaternary	2	Equus	
Tertiary	70	Turritella	
Cretaceous	135	Inoceramus	
Jurassic	180	Eoderoceras	
Triassic	225	Monotis	
Permian	270	Neospirifer	
Pennsylvanian	325	Dictyoclostus	
Mississippian	350	Muensteroceras	
Devonian	400	Phacops	
Silurian	440	Pentamerus	
Ordovician	500	Rafinesquina	
Cambrian	600	Paradoxides	

Figure 17/4 *This geologic time chart includes names and drawings of fossil remains of animals that can be used to date the rocks they are found in.*

The *Equus* fossil is a horse tooth, *Turritella* is a marine snail, *Inoceramus* and *Monotis* are marine clams, *Eoderoceras* and *Muensteroceras* are cephalopods, *Phacops* and *Paradoxides* are trilobites, and the other four are brachiopods. The cephalopods, brachiopods, and trilobites were all marine animals. Point out that some types of fossil clams and snails were originally fresh water or terrestrial animals.

fossil animals in Figure 17/4 were strictly ocean animals. So you know the rocks in which they are found must have come from ocean bottom sediments. Some were land animals. The rocks that contain their fossils would be river, swamp, and lake deposits.

Can you see that the rock layers in Figure 17/3 contain the fossils shown in Figure 17/4? Working up from the bottom of the pile, the rocks are from the Devonian (dih-VOH-nee-un), Silurian (sy-LOOR-ee-un), and Ordovician (or-duh-VISH-un) Periods. How long ago were these layers put down? Can you determine during what time periods layers 6 through 13 were put down? Do these clues give you some other clues to the history of the rock layers in Figure 17/3? Trace the layers in your notebook and record the time period in which each layer was laid down. Decide in each case whether it was deposited in a lake or a sea. You will want to use your tracing as you read on.

17.3 Clue number two is missing layers.

The fossils in Figure 17/3 also tell you that rock layers from three periods are missing. What happened to the rock layers for the Mississippian, the Jurassic (joo-RASS-ik) and the Tertiary (TER-shee-air-ee)? There are two parts to the solution of this mystery.

First, notice that the surface of rock layer 9 in Figure 17/3 is very uneven. Since you have already found that sediment deposited by streams or turbidity currents normally builds up flat, horizontal layers, erosion must have taken place. In other words, weathering and stream action must have cut into the original flat top of the layer. The period of erosion after layer 9 was formed could have removed the Jurassic layer that you would expect to find on top of it. You can see in Figure 17/3 that the same thing happened again after layers 6 and 11 were laid down, just where the missing Mississippian and Tertiary layers should be.

Another possibility is that the uplift of layer 9 could have occurred just before the Jurassic; perhaps there never was a Jurassic layer. The assumption here is that a high land would be eroding, and not receiving sediment during the Jurassic Period. Such buried erosion surfaces are called "unconformities." There may be one or more evident in your area.

Secondly, in order for the deposition to stop and for erosion to begin, the uneven layers would have to be high and dry. Since it is unlikely that the seas and streams dried up, rock layers that had formed on the ocean bottom had to be raised to become exposed to weathering agents. After the uplift, deposition would stop and erosion would begin. Perhaps a pressure in the earth's crust caused a large area to rise high above sea level without being folded or tilted. That is what happened in the Grand Canyon region of Arizona.

17.4 Clue number three is rocks out of order.

Layers 1, 2, and 3 must have been overturned.

If you look at Figure 17/4, you will see that layers 2 and 3 in Figure 17/3 are on top of a layer laid down after them in time. The order of the layers from oldest to youngest should be layer 3 on the bottom, layer 2 next, and then layer 1 on top. What must have happened?

Figure 17/5 will help explain. As you look at the different parts of Figure 17/5, keep in mind that the mountain building forces you studied in Chapter 15 can exert the pressure necessary to fold and turn over the rock layers. Remember from Section 11.1 that even something as brittle as a cigarette can be bent like a thick ribbon if the pressure surrounding it is kept high. Watch how the order of the layers changes. Can you see what the fossil clues in Figure 17/3 tell about layers 1, 2, and 3? Were any other layers in Figure

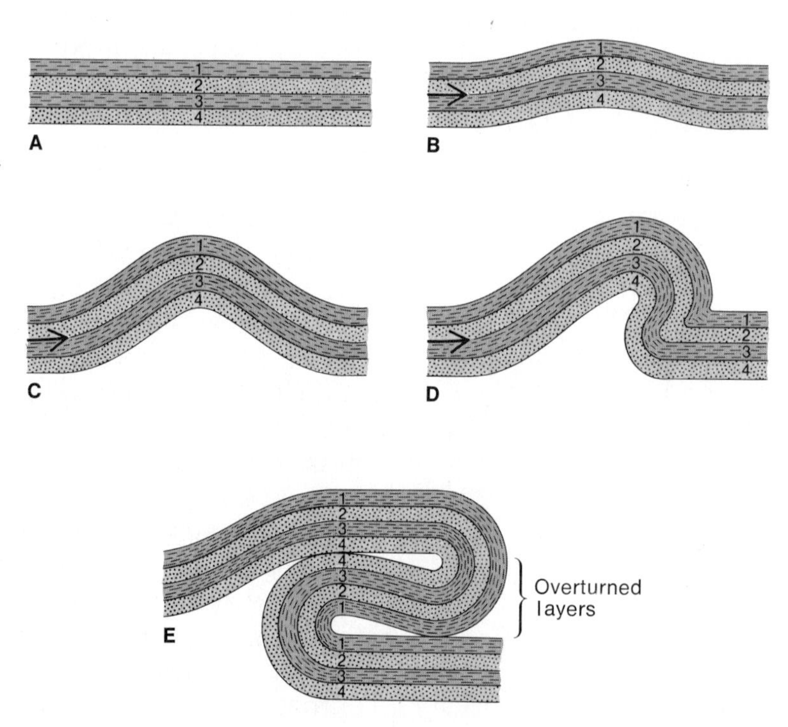

Figure 17/5
In some places older rock layers are found on top of younger rock layers. To see how this can happen, keep your eyes on the order of the rock layers as they are folded.

In an actual case, there could well have been layers above layer 1, that have eroded away. The rocks above and below layers 1 to 4 would have kept the pressure high. According to the fossils, no other layers were overturned, but there is other evidence to be considered.

17/3 involved in the folding process? Layers 4, 5, and 6 contain no fossils to date them. What other clues might help you answer this question?

17.5 Clue number four is the way certain layers are laid down.

Look closely at layers 4 and 5 in Figure 17/3. Layer 4 has ripple marks, which are illustrated in Figure 9/12. Layer 5 has smaller layers with larger particles of sediment on top of smaller particles.

Now look at the wavy lines in Figure 17/6. These represent ripple marks on the ocean floor. If you turn the book upside down, how will the drawing look? Do the ripple marks in layer 4 of Figure 17/3 look more like the right-side-up or the bottom-side-up position of the drawing? Is rock layer 4 overturned?

The bottom-side-up position of the drawing matches. Layer 4 is overturned.
Process: inferring

When a handful of sand is dumped into a container of water, which particles get to the bottom first, the big ones or the small ones? Look at layer 5 in Figure 17/3. Where are the big particles? Where should they be if the rock layer is right side up? Was rock layer 5 overturned?

If layer 5 were right side up, the large particles in each of the sublayers would be below the small ones. Layer 5 must have been overturned, too. Discuss the possibilities for layer 6. There is no evidence that it was overturned, and layer 7 is Pennsylvanian in age, so it may be Devonian or Mississippian, formed after the overturning. Consider how answers could be found (trace the layer to another area where there are fossils or other kinds of evidence).

According to the clues, then, two other layers were involved in the folding that turned over layers 1, 2, and 3.

Figure 17/6 *Compare this drawing of ripple marks in layers as they might be under a beach with the ripple marks in rock in Figure 9/12. Are the ones in the drawing overturned?*

The History of the Grand Canyon

The Grand Canyon of Arizona is one place where you can see a geologic cross section directly. Over a mile of the earth's crust is exposed for study. Can you date the rocks? What ages of rocks are missing? Find two surfaces that have been eroded. Find evidence of folding. What has happened since the Kaibab Limestone was formed? Use the geologic time chart in Figure 16/13 and the fossil chart in Figure 17/4 to help you.

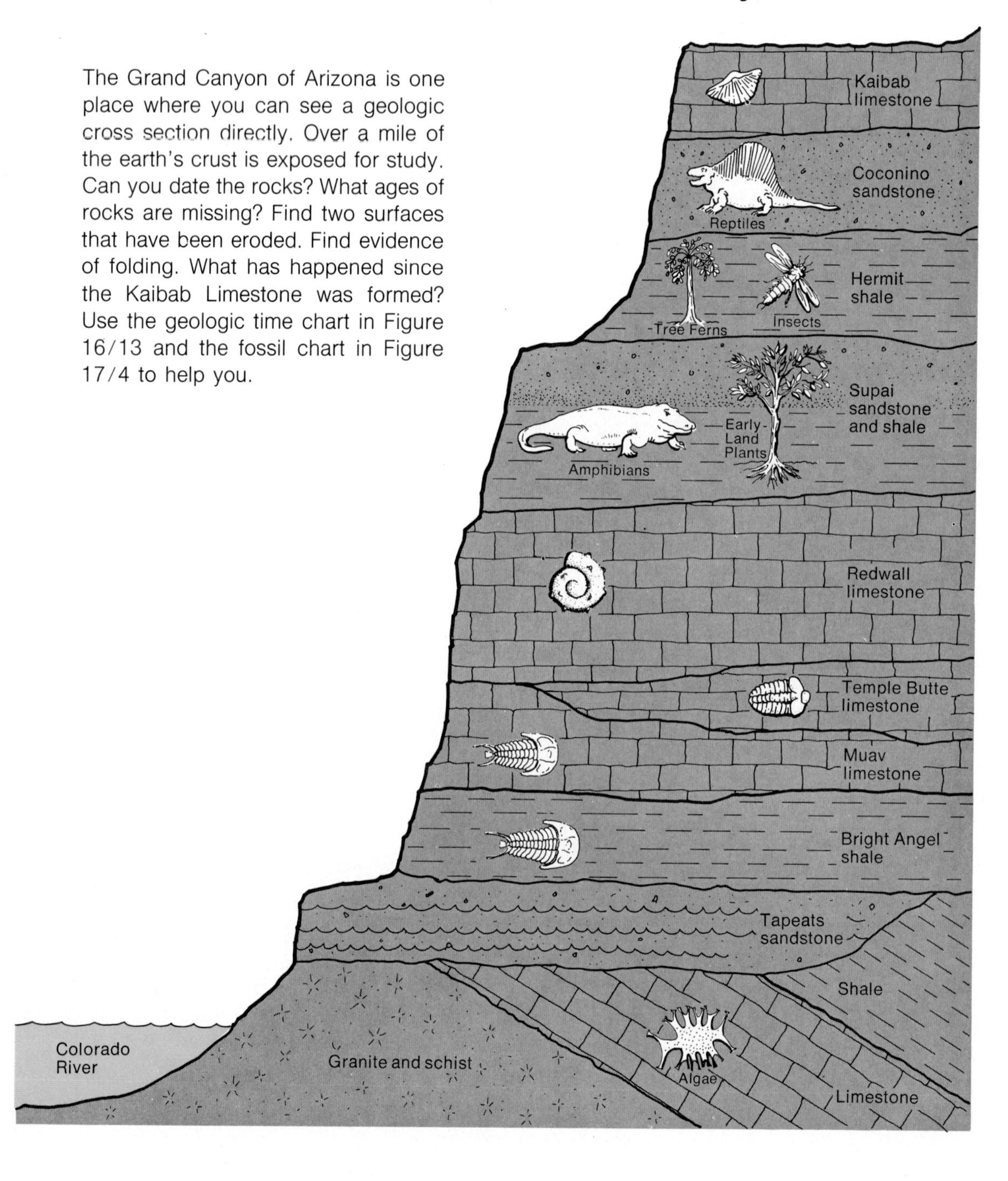

17.6 Clue number five is the type of rock in each layer.

You have learned how sediment is sorted by a stream before it is deposited. You learned that the largest particles of sediment are dropped first. Sand-sized particles are the next to settle out. Away from the shore, turbidity currents deposit the fine particles of clay that become shale. If the shoreline changes location, the places where the different sized particles fall will change, too.

For instance, imagine an ocean bottom where limestone is forming. Suppose a new mountain range begins to be pushed up on the nearest continent. As the new land weathers and erodes, clay might be carried out to settle on top of the limestone. As erosion continues, new sand being dumped into the ocean would build the shoreline out into the water. After many thousands of years, the shoreline could move very close to the area you originally imagined. The sand could even be washed out on top of the clay. When this deposition site becomes dry land, a road cut would show a layer of limestone at the bottom, a layer of shale, then a layer of sandstone.

Process: inferring

Can you see now how a pile of rock layers at one place, like the one in Figure 17/3, can have many different kinds of rock in it? Something must have changed between the deposition of different layers. When the sand for sandstone (layers 4, 5, 9, 12, and 14 in Figure 17/3) was laid down, there must have been a current of water strong enough to carry sand. The current slowed down and dropped the sand. So when each of the layers was formed, the area was probably a beach, or shallow water near the shore, or a river bed. (Remember what settles first from Figure 14/2?)

When the current slows down enough to drop sand, it can still carry clay, which is much lighter. As you know from Section 14.2, and as you can tell by looking at lake and ocean bottoms, the clay is carried farther away from the shore into the deeper water. So each shale layer (layers 2, 6, 8, and 10 in Figure 17/3) must have been originally a muddy ocean bottom.

Limestone is mostly formed in ocean water, and many limestones contain fossils of ocean animals. Therefore, limestone layers (1, 3, 7, and 11 in Figure 17/3) show that

Figure 17/7
The dotted line shows where the shoreline around North America and Greenland would be if all the glaciers were to melt.

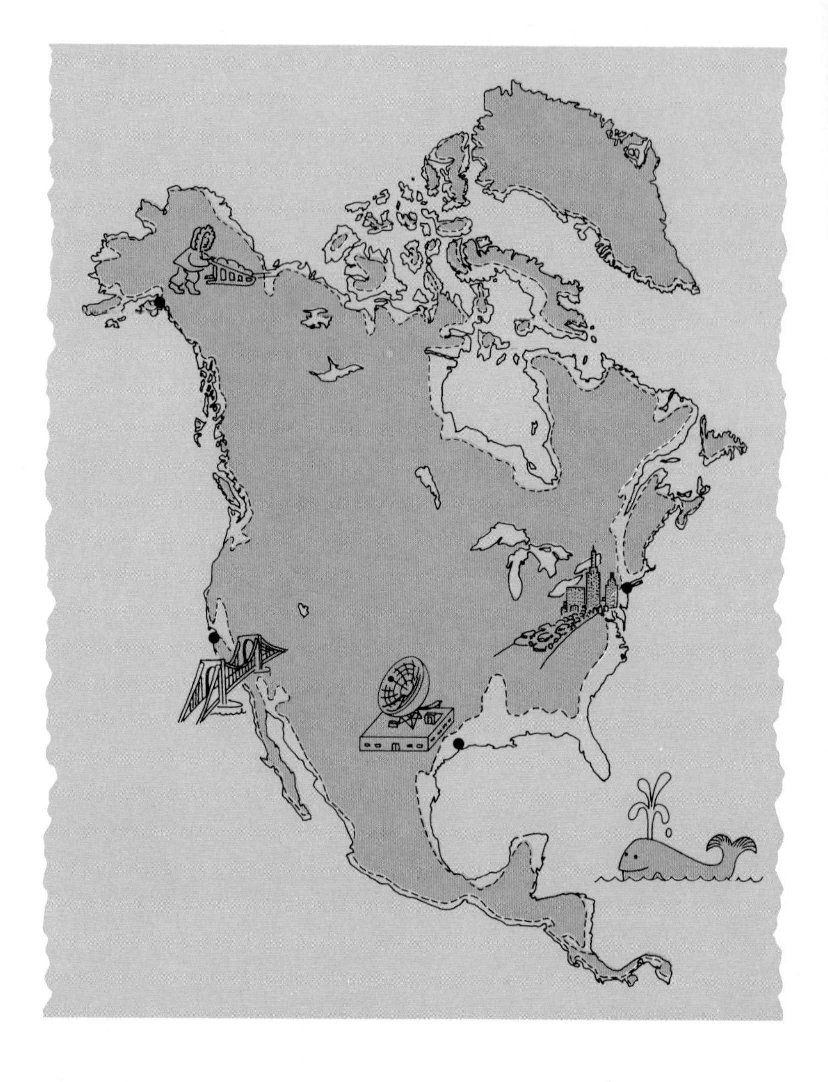

when they were formed the deposition area was the floor of the sea.

1, 3, 7, and 11 represent clean deeper ocean bottom. 2, 6, 8, and 10 represent muddy deeper ocean bottom. 4, 5, and 9 were formed in shallow ocean near or at the shore.

Look at Figure 17/3 again. Next to the sketch you made earlier in your notebook, describe the kind of area in which each ocean layer was put down. Mention the depth of the water at the time of deposition.

Even if shorelines aren't moving back and forth, the depth of the ocean can change. One way for sea level to rise is if glaciers melt. Look at Figure 17/7. If all the glaciers in the world today were to melt, the level of the sea would rise about 200 feet. Coastal areas that are now out of the water would become areas of deposition. Suppose that the

Sea level would go down. What was formerly ocean bottom could become a beach, and sand could be deposited on top of mud.

number of glaciers suddenly increased, or existing glaciers became larger. What would happen to the level of the sea? Could sediment layers near the coast change?

17.7 Clue number six is putting the events in the order in which they happened.

For the history of Fig. 17/3, see the Teacher's Manual.
Process: inferring

Sometime during the deposition of the sediment that made the rock layers in Figure 17/3, hot magma was squeezed into layers 1, 2, and 3. This is called an **intrusion** (in-TROO-zhun). A fault also cut layers 1, 2, 3, 4, and 5. With a few hints, could you write a history of the events recorded in Figure 17/3 rocks? Try it.

Hints: Metamorphic rocks completely surround the intrusion. The rocks became metamorphic because of the heat from the magma and the pressure it exerted when it squeezed into the rock layers. The diagram shows that the magma came from below the rock layers, so the hardened intrusion must not have been overturned. The fault cuts through the intrusion as well as the sedimentary rock layers.

Compare your history with your teacher's. How much of the mystery did you work out?

Did You Get the Point?

Clues in the rocks can point out:

when and where the rocks were formed

rock layers that have been overturned

rock layers that are missing from a series of layers

moving shorelines

changes in sea level

the order in which certain geologic events happened

Check Yourself

Answers:
See Teacher's Manual.

1. Figure 17/8 shows several columns of rock. Refer to Figure 17/4 to give a brief history of what has happened in the area where each column was found.

Figure 17/8 *Write a history of the area around each column of rock.*

2. Figure 17/9 shows two cross sections of rock and rock layers. Write the history shown by each section. The slashed boundaries between igneous rock and the sedimentary layers indicate that heat has affected rock layers in that area. You will have to use the things you have learned from many chapters to do a good job.

Putting Sets of Clues Together

17.8 Hidden rocks can be used as clues.

Places where the solid rock of the earth's crust is uncovered are called **outcrops.** Sometimes no outcrops can be found in an area. There may be no oil wells, no mines, and no

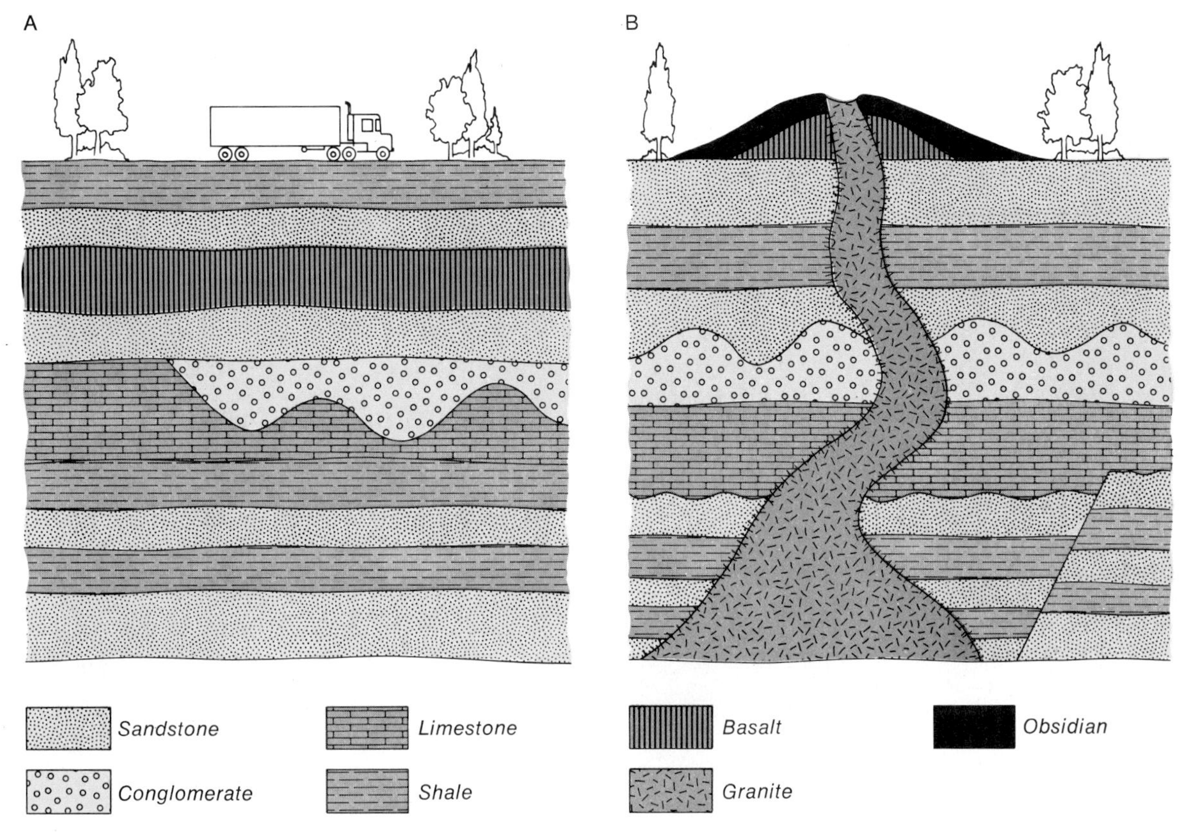

Figure 17/9 *Write the histories of these two rock columns by carefully reading clues in the layers.*

road cuts. The hidden geologic history of such an area must be read from rock clues that can be seen in nearby areas.

Materials

tracing paper, transparent tape, scissors

Activity Time: 40 minutes. Try this yourself first. See the key in the Teacher's Manual.

★ In the map in Figure 17/10 outcrops can be found near A, B, C, and D. The question marks show places where there are no outcrops. But geologists want to know about the rocks underneath so that they can figure out the geologic history of the area. Figure 17/11 shows small sections of the outcrops found at A, B, C, and D. Trace each of the sections, or **columns,** on tracing paper. Cut them out with a pair of scissors. Color them like the ones in the book if you want to.

Put the four columns on a clean page of your notebook so that they are an inch or two apart. Column A should

Figure 17/10
Outcrops of rock can be seen at points A, B, C, and D on the map. Geologists can use clues in the outcrops to figure out what kinds of rock are hidden at the question marks.

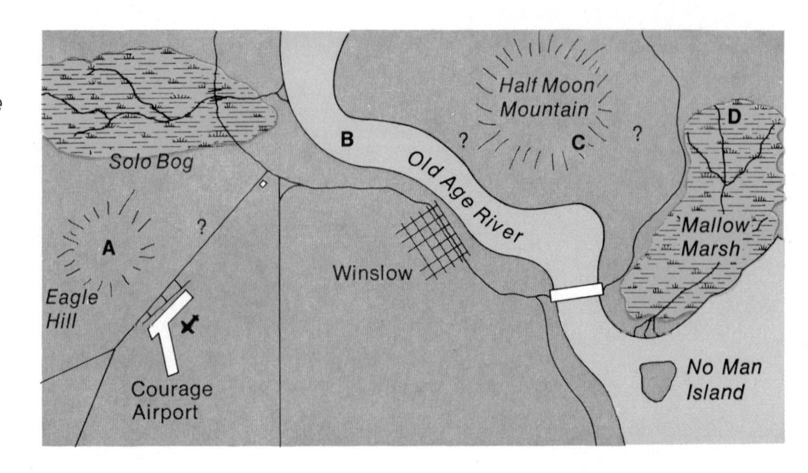

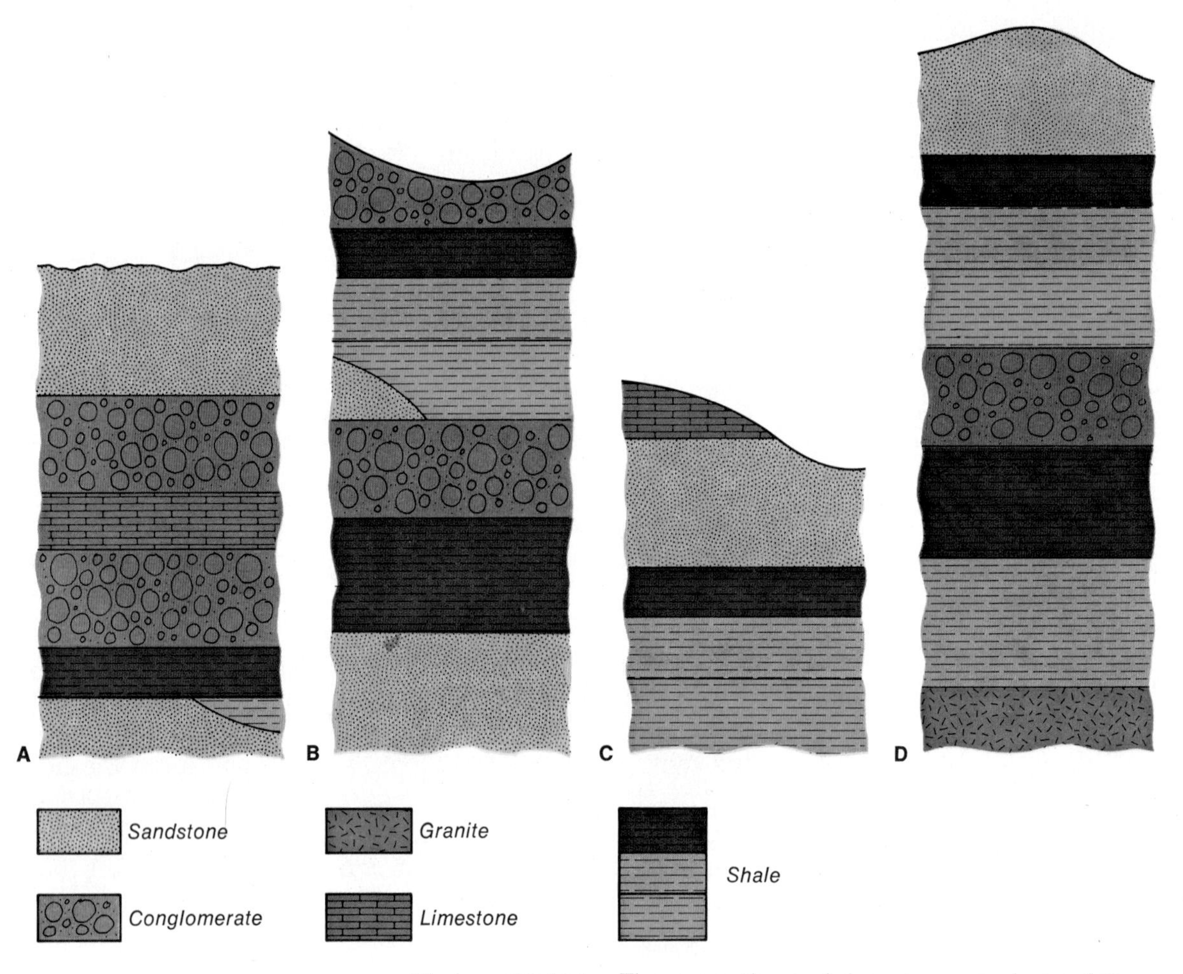

Figure 17/11 *These are the rock layers exposed at points A, B, C, and D in Figure 17/10.*

be on the left. Line up the columns so that identical layers are across from each other. (The top layer of shale runs all the way across the page.) Some of the columns will be higher than others.

Now tape the columns in place. With a pencil, draw lines to connect identical layers in neighboring outcrops. Your lines will run under question-mark areas shown on the map. Sometimes you will see that one rock type must change to another between outcrops. If you wish, your teacher will help you find where these changes take place.

When you have connected the outcrops, you can read the geologic history of the area from the clues. Place your completed series of outcrops in front of you. The story begins with the bottom layer. Number the layers as they are between A and B starting with the layer just above the granite. Try to write a brief history of the area shown in the map in Figure 17/10. Compare your history with your teacher's. ★

Make sure each student's correlation looks like the one in the Teacher's Manual before you go further. A history of the area is in the Manual.
Process: observing, inferring

17.9 Rock patterns can be used as clues.

In some places, rock layers or other masses of rock can be seen at the surface. In other places, rocks below the topsoil can be found by drilling. Maps showing the locations of rock masses and giving the time periods in which the rocks were formed are called **geologic maps.** Figure 17/12 shows a part of a geologic map. Each color stands for a rock layer from a different time period. Notice how the rock layers seem to form a pattern. You will need a model to solve this mystery.

Materials

clay, knife

Activity Time: 20 minutes. If only two colors of clay are available, alternate them. The oldest layer in the model is in the center. The oldest layer on the map is on the outside; the rock layers are shaped in a basin-like structure, but the surface is more or less flat, as you can show in a clay model. See the Teacher's Manual for the model of New York State.

★ Make a "layer cake" out of modeling clay. Use a different color for each of the rock layers shown in the map. (The colors don't need to match those on the map.) Push up the center of the layer cake. Now cut off the rounded dome with a knife. Your model should look like Figure 17/13. Notice the pattern the exposed layers make after the dome has been cut off. Which is the "oldest" layer in your model? Which is the oldest layer on the geologic map? Can you tell what has happened to the area shown by the map?

Figure 17/14 shows a geologic map of New York State. Using your "layer cake" and the knife, see if you can fold,

Figure 17/12
Rock layers from different geologic time periods can form a circular pattern with the help of erosion. This is a geologic map of Lower Michigan. Make a clay model to see how the rock layers have to be bent to make a pattern like this.

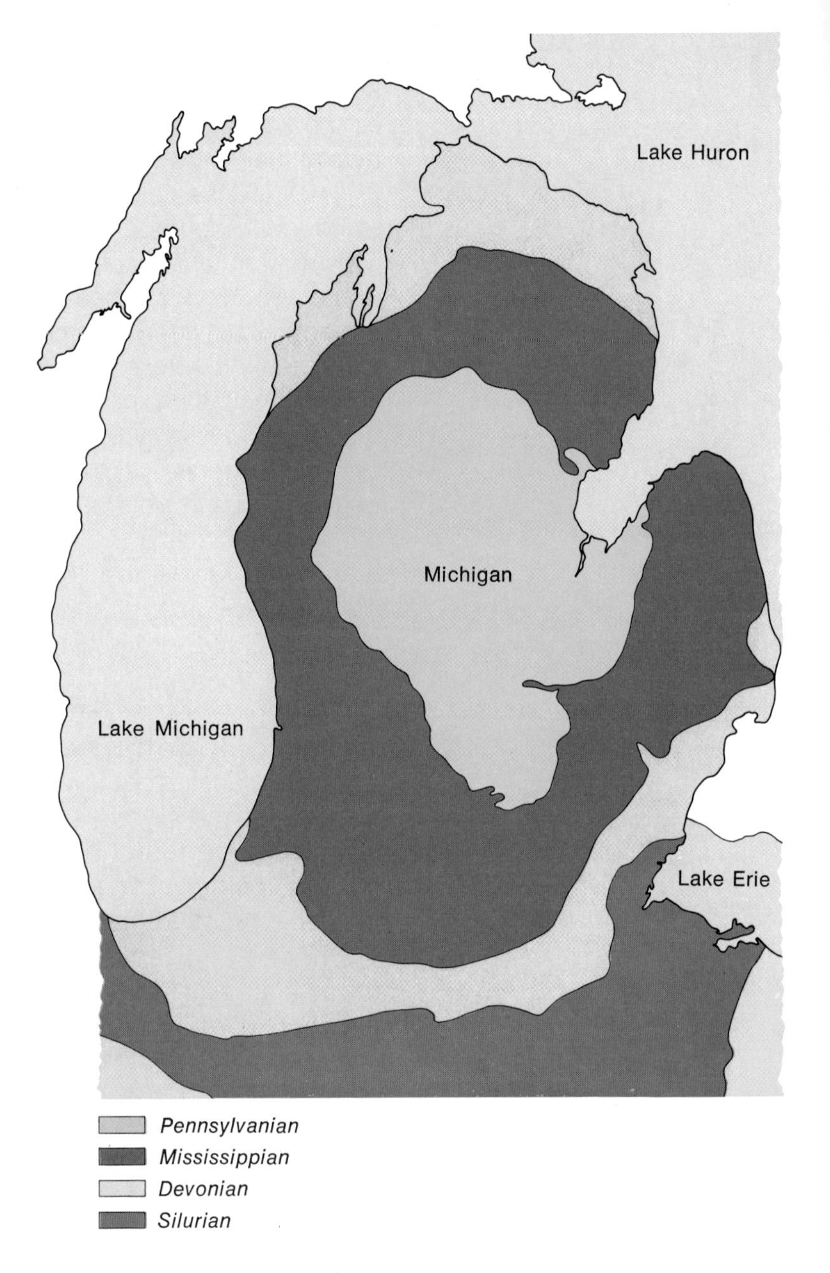

bend, and cut the clay to show how each of the patterns in the map was formed. ★

You can see that rocks contain clues to past movements in the crust of the earth. They can even help you trace mountains that are now gone. You can use these clues to draw a picture of the geologic history of almost any area.

Figure 17/13 *Cutting a clay model of pushed up rock layers.*

17.10 Sets of clues can help write a history of the North American continent.

You may wish to use the Houghton Mifflin Earth Science Overhead Visual 20, "Growth of a Continent," in the discussion of this section. It covers the presumed growth of North America before the Cambrian Period.
Process: observing, inferring

Thousands of wells and mines have been dug across the United States. Streams and men have cut through layers of rock, exposing them for geologists to look at. By reading the clues and reasoning just as you have done with Figures 17/3 and 17/11, geologists have put together a series of maps showing how the area of North America might have

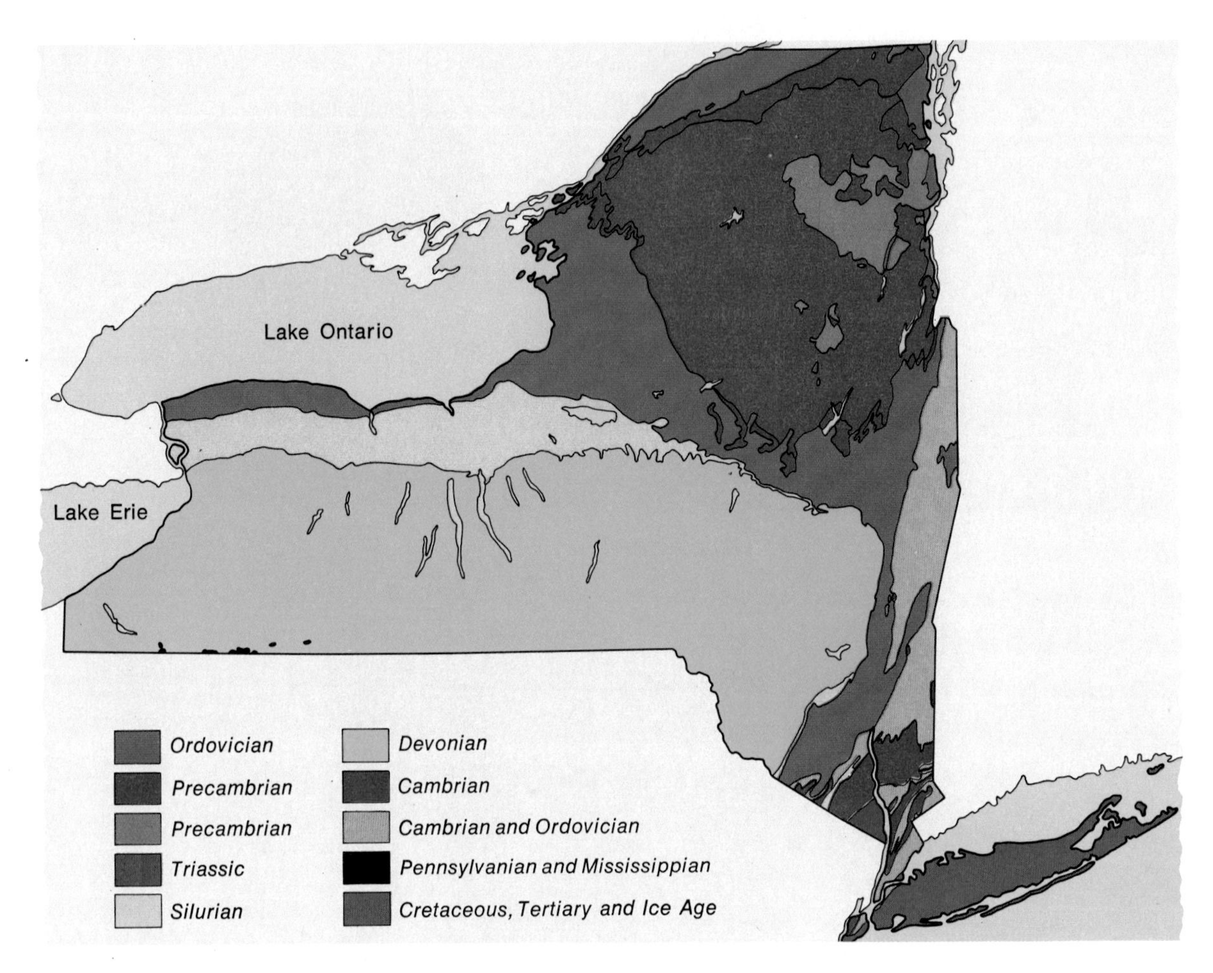

Figure 17 / 14 *Make a clay model to show how the rock layer patterns in New York State have been formed.*

looked at various points in geologic time. Figure 17/15 shows the area during the geologic periods since the Cambrian.

You can tell something about the history of your area, too. In your notebook, list the time periods shown by the eleven maps. Next find your home on each map. What was your area like during each period? Do you think your area contains sedimentary rocks for each period? If your local rocks are sedimentary, were they formed in fresh water or salt water? If your area does not have sedimentary rocks now, could it have had ocean bottom sediments in the past? If so, what could have happened to them? Write your answers in your notebook.

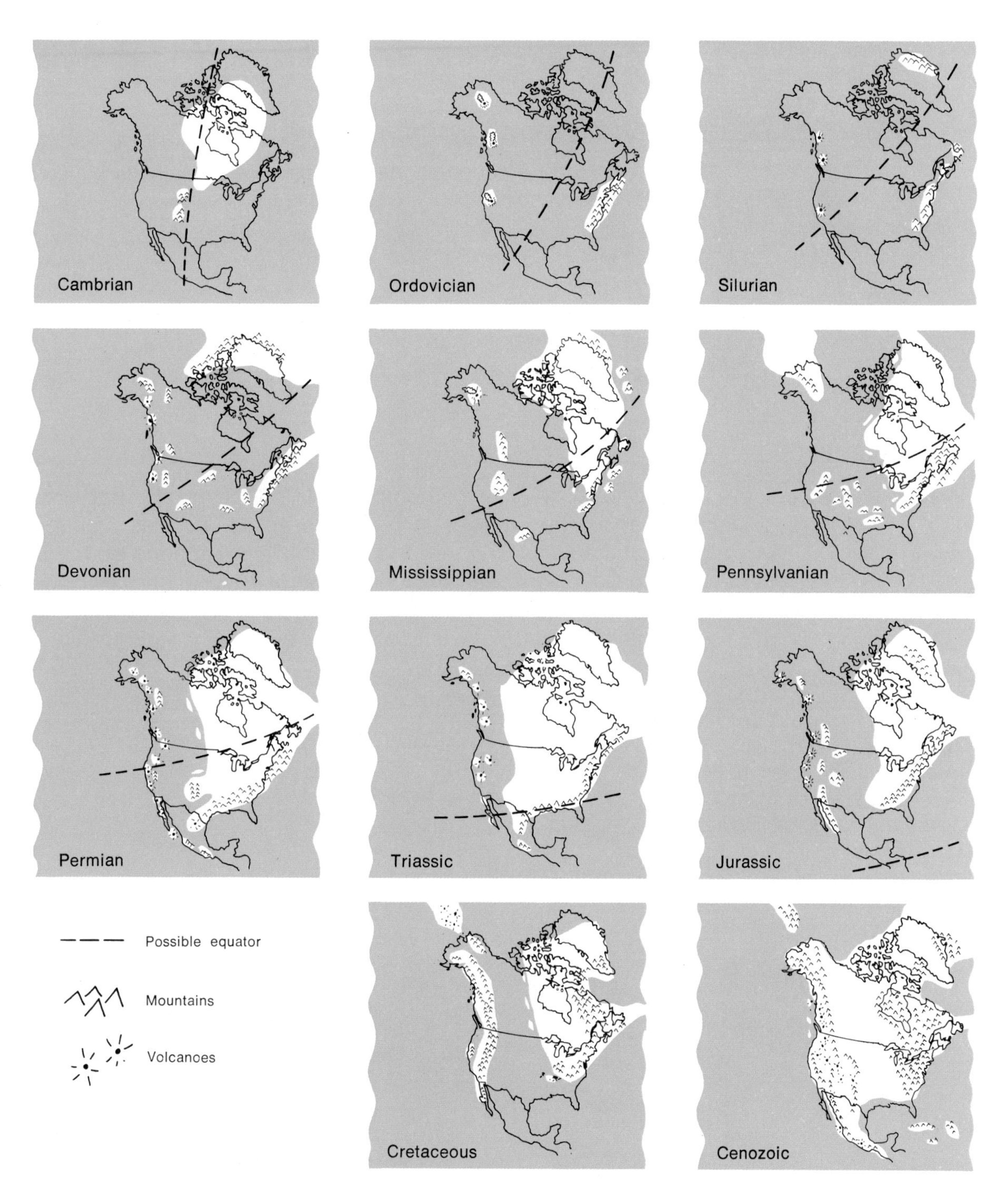

Figure 17/15 *The shape and face of North America and its position relative to the equator have changed many times.*

Answers:

1. Marine fossils and limestone would indicate the Permian ocean. In some areas, terrestrial rocks can be identified by rock characteristics and fossils. In some places, Permian and older rocks are folded in ways which indicate land was being uplifted. The large areas with no Permian rocks of any kind are presumed to have been above sea level.
2. Fossils show what organisms were living when the sediment was deposited. The organisms reflect the living conditions, including the climate. The rock itself can record the climate. A warm, moist climate would promote an abundance of green plants, which would form a rich, dark soil. This could ultimately become a dark, organic shale. Mud cracks in rock might indicate alternate wet and dry periods.

Did You Get the Point?

Clues found in rocks make it possible to look back at geologic history.

Geologic maps help interpret rock clues.

The North American continent has varied greatly during 600 million years.

Check Yourself

1. What types of clues might have helped scientists make the map of the Permian Period?

2. How can records of the type of climate be preserved in a rock?

What's Next?

The North American continent is still changing. But what about the changes in plant and animal life that occurred at the same time the land was changing? Do the life forms on our continent change as fast or faster than the shape of our continent? What happened to all the plants and animals that have left us their fossils? When did man appear? Can man become extinct? Read on!

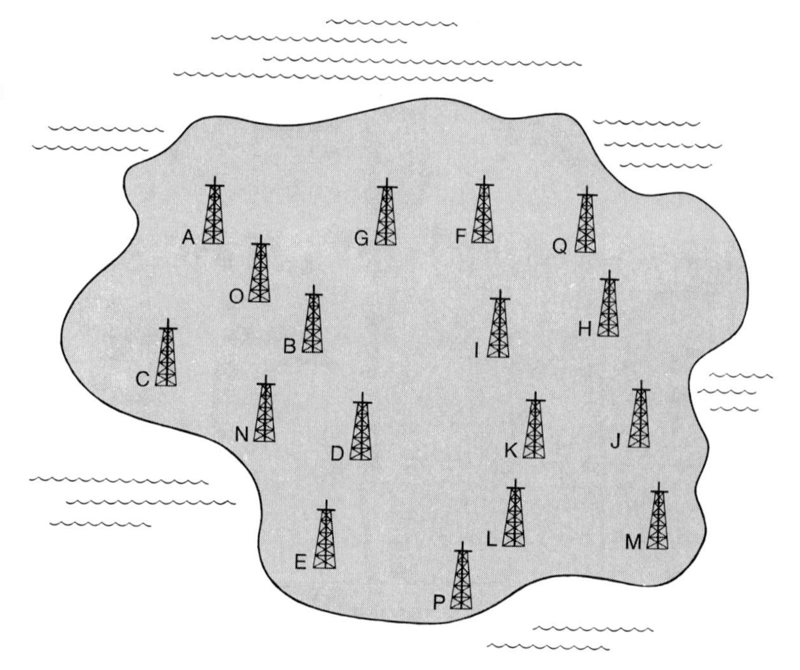

Figure 17/16
Using the information about each drilling site given in Figure 17/17, draw a map of the shoreline around this imaginary continent during each of the three time periods.

Skullduggery

1. Draw a cross section of rock layers of your own and include many mysteries to be solved. Exchange cross sections with a classmate. Solve each other's mysteries.

2. Figure 17/16 is a map of a make-believe continent as seen today. The letters mark the sites of several oil wells. Figure 17/17 provides information about sedimentary rocks found for three time periods at each site. Use these data to draw a map of the continent for each of the three time periods.

Answers:
2. Student maps should resemble these:

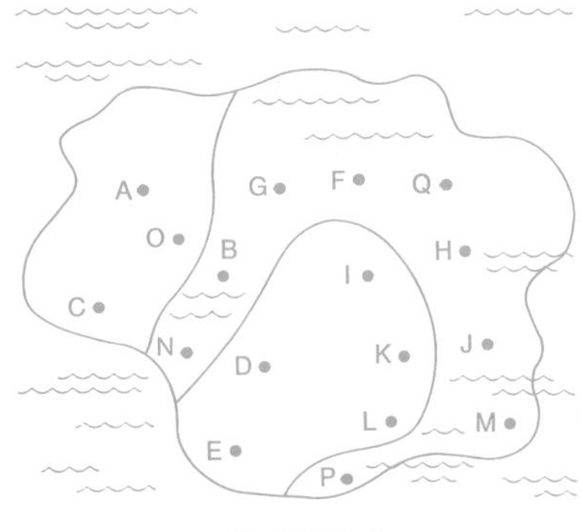

Mississippian

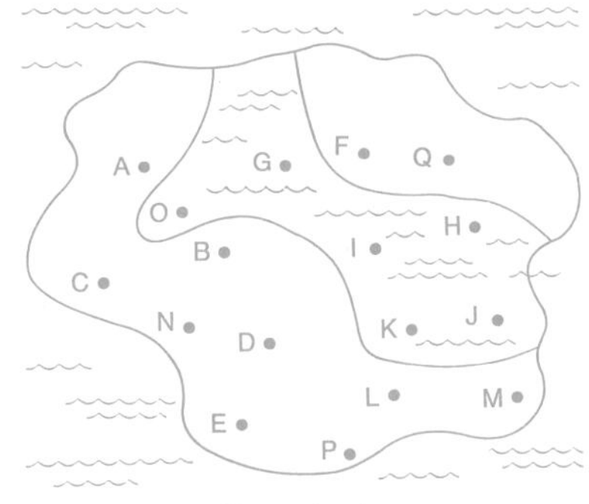

Pennsylvanian

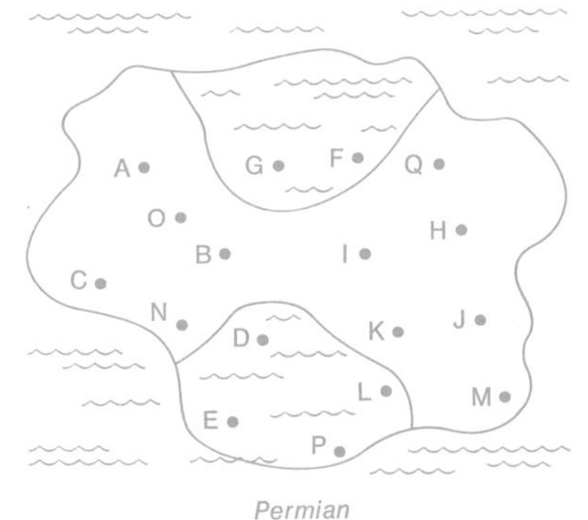

Permian

Oil well	Ocean rocks found		
	Mississippian	Pennsylvanian	Permian
A	no	no	no
B	yes	no	no
C	no	no	no
D	no	no	yes
E	no	no	yes
F	yes	no	yes
G	yes	yes	yes
H	yes	yes	no
I	no	yes	no
J	yes	yes	no
K	no	yes	no
L	no	no	yes
M	yes	no	no
N	yes	no	no
O	no	yes	no
P	yes	no	yes
Q	yes	no	no

Figure 17/17 *Use this information to map a history of the make-believe continent in Figure 17/16. A "no" means that no ocean rock for that time period was found.*

For Further Reading

Rhodes, Frank H. T., Herbert S. Zim, and Paul R. Shaffer. *Fossils—A Guide to Prehistoric Life.* Golden Science Guide, Racine, Wisconsin, Western Publishing Co., 1962. An excellent guide to fossils that are commonly found and used to date rocks.

Fenton, Carroll Lane. *Tales Told by Fossils.* Garden City, N.Y., Doubleday & Co., 1966. Do you doubt that sea bottoms rise and fall? You'll see lots of evidence in these pages.

Chapter 18 Development of Life

Birds: left, earliest known bird, *Archaeopteryx;* right, imaginary Chinese phoenix. Other animals: Above stream—left, winged dragon; right, Ice Age giant armadillolike *Glyptodon.* In water—left, Mesozoic swimming reptile, *Elasmosaurus;* right, imaginary Etruscan sea horse. Below stream—left, Permian reptile, *Ophiacodon;* right, imaginary gorgon.

Here's what man's imagination has created to explain the past, with scenes from the fossil record mixed in. Can you tell which creatures really lived on the earth and which are just bad dreams? Careful! Some of the real animals are just as fantastic as the imaginary ones.

Man invented the imaginary animals before the fossil record was understood. He needed demons to answer questions about life on earth. Now that we know about the fossil record, there are still questions to be answered. Several have probably occurred to you. Haven't you ever wondered why life millions of years ago was so different from what it is now?

A Quick Look at Earth Life

18.1 Fossils provide a history of earth life.

When people talk about earth life in general they are talking about the earth's biosphere. The **biosphere** (BY-uh-sphere) is a zone around the earth, where animals and plants live. Not only the earth's surface, but the topsoil, the lakes, streams, and oceans, and the bottom of the atmosphere are all parts of it. Compared to the whole earth and the entire atmosphere, the biosphere is a very thin zone.

The earth has had a biosphere for most of its history, but according to the rock records there were not many kinds of living things until relatively recently. Let the month of September represent the entire age of the earth. The organisms which became the oldest known fossils would be living by September 10. They were bacteria and the simplest algae. By September 17, there would be other kinds of algae. But the biosphere would not have a real variety of animals and plants until after the 27th!

The fossil record from before the Cambrian is poor, and the ancestors of Cambrian animals are not known, in most cases.

Since geologists have been studying fossils for over a hundred years, you would think that they would know all there is to know about earth life. But this is not the case. The fossil records have brought up many questions. Some have been answered, some have not been answered.

For example, it seems that all types of earth life didn't "make the scene" at the same time. Why? Some kinds of plants and animals seem to have become extinct long before there was a man around to see them. Why? Fossils show that some animals that geologists believe lived 500 million years ago looked just like sea animals of today. Other animals that lived 500 million years ago seem to have been mysteriously transformed into more complicated types. Why?

Concept: the universality of change

Take a look at the record for yourself. Figure 18/1 is a make-believe pile of sedimentary rocks. Assume that fossils were found in all the rocks. The chart shows some kinds of fossils that were found in the various layers. In that way, the chart shows some of the animal types that lived and died in the different geological periods. Scientists who have studied the fossils believe that certain animals are related to each other as the lines show. Notice that some kinds of animals disappeared. If the bottom layer of the rock pile was formed on September 27th, when did modern horses finally appear? What questions can you think of to ask about the fossil record?

Modern horses would appear about 11:45 P.M. on Sept. 30.

Questions might be:

1. Are there any other reptile types not on the chart that came from the first reptiles? *Answer:* dinosaurs, crocodiles
2. What animals did jawless fishes come from? *Answer:* lower chordates
3. Why would a fish have lungs? *Answer:* temporary drying up of ponds and streams—to live in mud or crawl to another body of water
4. Why did mammal-like reptiles disappear? *Answer:* mammals were better adapted to the areas in which they lived
5. Why haven't turtles changed in so many years? *Answer:* no serious enemies, not affected by changes in environments (they could always find adequate food and someplace to live)
6. How do they know the first horse was a horse? *Answer:* by the teeth, by the continuous fossil record between the first and the latest

18.2 Today's earth has many forms of life.

As Figure 18/1 suggests, there weren't nearly as many life forms around 600 million years ago as there are today. Today we have so many different kinds of life that no zoo

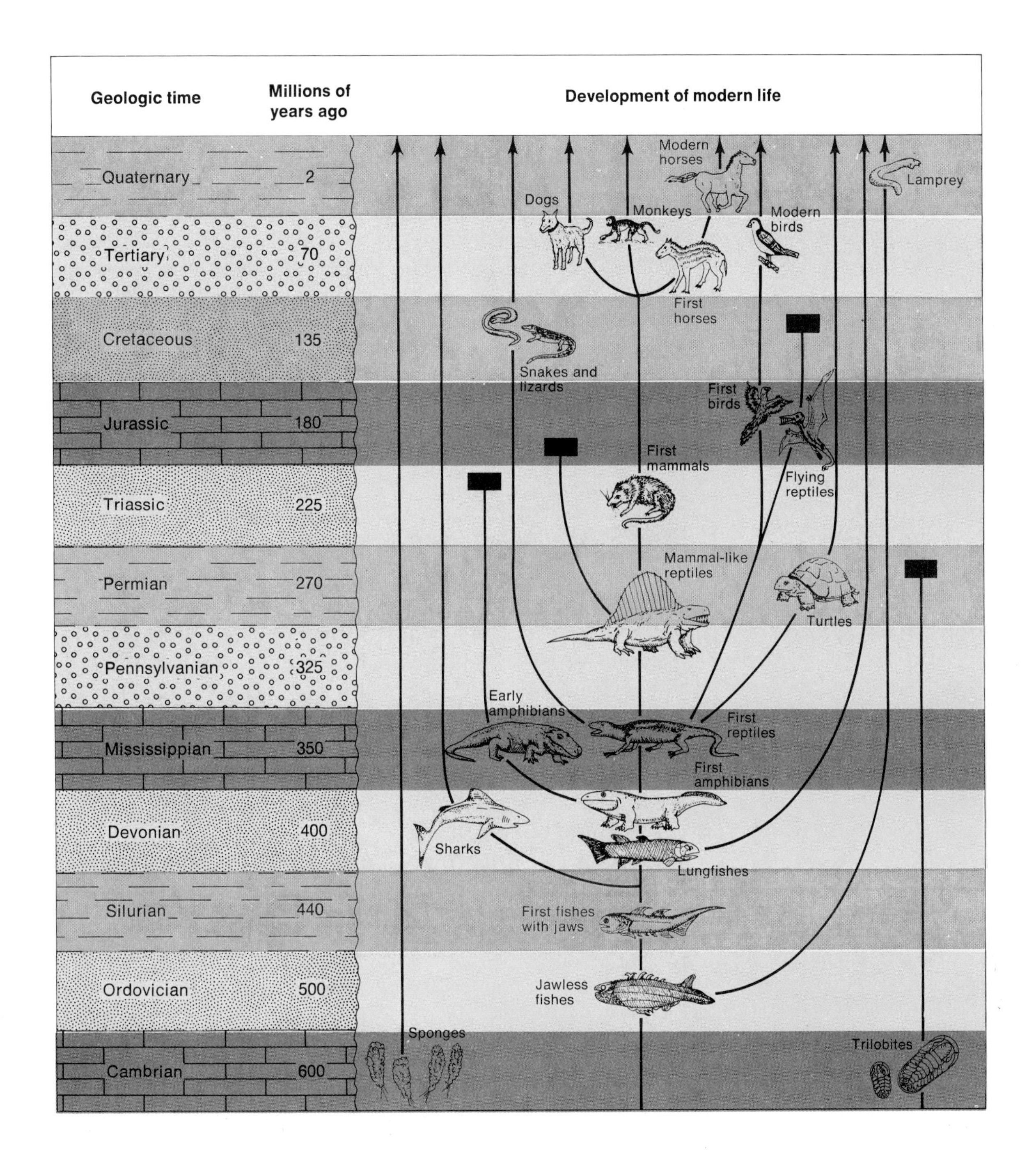

Figure 18/1 *Fossils found in sedimentary rock layers are evidence of the animals that lived before modern times.*

Figure 18/2
Some unusual earth animals: (A) mandrill, (B) lion fish, and (C) Hercules beetle.

The numbers of animals are vague because estimates vary greatly from source to source.

Concept: the universality of change

could possibly have one of everything. There are over 30 thousand different kinds of fish, almost 9 thousand kinds of birds, over 5 thousand kinds of reptiles, and about 3.5 thousand types of mammals. And don't forget the insects. So far, scientists have counted more than 800 thousand types, and they are still counting. Figure 18/2 shows some animals you may not have seen before.

Figure 18/3
A worm trail photographed on the ocean floor, two and a half miles down.

Plants? If you don't live in a desert or plains area, you could pass more than 100 different kinds of plants on a Sunday ride through the country. If you were able to identify the plants, you could possibly find 20 kinds of trees in the woods and fields, along with 60 kinds of wild flowers and weeds. You should have no trouble finding five types of grasses, and at least that many types of bushes and shrubs. Add a few ferns and ground pines. And don't forget the crops on the farms.

The earth has life almost anywhere you look. Wherever there is a place to live, something will call it home. Small, simple plants, the lichens, live on high mountain tops. The fresh worm trail in Figure 18/3 was photographed on the ocean bottom, two and a half miles down. Some starfish live almost five miles below the surface. The earth has forms of life that can live without oxygen. Some forms of life use sulfur and salts as food. We have animals that can survive freezing temperatures. Microscopic life forms have even been found in boiled water. How different our planet is from our satellite, the moon, where no life has been found.

Review the need for oxygen, and the number of organisms that need it, if it seems appropriate here.

18.3 Each continent has its own kinds of life.

The whys, wherefores, and how comes of the earth's biosphere are not restricted to old records in the rocks. Why aren't palm trees found growing naturally in Boston? Why do cactus plants grow only in dry regions? A short activity will raise a few more questions.

See suggested answers in the Teacher's Manual.

★ Figure 18/4 is a map of the world with numbered zones drawn on it. Each of the zones has its own special group of animals. Figure 18/5 is a list of animals that can be found in the zones, but they are not in any particular order. Try to place each animal in the zone where it is normally found. In your notebook explain why each animal lives where it does. Try to find something about each animal that would prevent it from living in or moving to other zones. (*Hints:* What does it eat? Does it live in forests? Can it stand cold weather? Can it stand hot weather? Can it swim? Does it need protection?) Have your teacher check your lists.

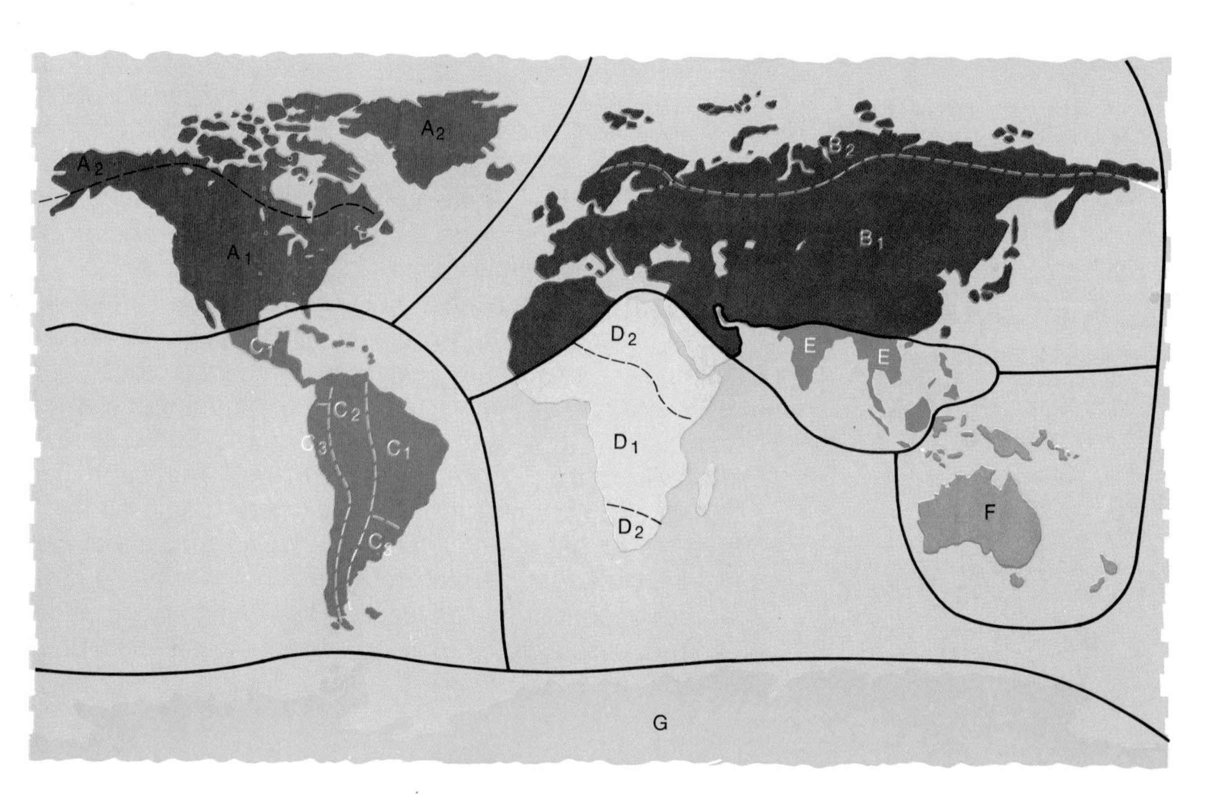

Figure 18/4 *Match up the animals listed in Figure 18/5 with the zones on the map.*

Figure 18/5 *Where do these animals live and why do they live where they do?*

As you can see, it is easy to place some animals in the zones where they live. Certain things about each animal tell you what it has to have to survive. Only certain places can provide what it needs.

Why aren't penguins found near the North Pole? It's just as cold there as it is in Antarctica.

Why aren't giraffes found in Pakistan? The climate is similar to that of East Africa and the trees are as tall.

Why aren't the sea snakes found in the Atlantic Ocean? When they first appeared both oceans were available.

North America had native wild horses before the end of the Ice Age. But they became extinct in the western hemisphere. Since wild horses can live here perfectly well today, what could have happened to them?

Reindeer and caribou are similar animals living in similar zones. Why two kinds of nearly the same animal?

How do eucalyptus trees and earthworms sort themselves on the globe if they can't observe, reason, and plan? ★

Answers can be found in the patterns which earth life seems to follow and the clues within these patterns. Different people, however, read the clues to say different things. As a result, there is no one answer yet. For the time being we will have to be happy simply knowing that weather is not the only thing that determines what lives where. Perhaps if you know what the patterns are, you can come up with some answers of your own.

Did You Get the Point?

Fossils provide a partial history of earth life.

Many questions about early earth life are still unanswered.

The earth has a variety of life forms.

Life can be found almost anywhere on the earth.

Different regions of the earth have their own types of plant and animal life.

Check Yourself

Answers should be suggested by the consideration of Figures 18/4 and 18/5.

1. Prepare a list of some plant and animal types near your home and name some places on the earth that do not have them.

2. What conditions near your home determine what can live there? Name some pets that you could never keep because of these conditions.

Patterns and Clues

18.4 Food places animals in certain areas.

All living things need energy to go on living. The energy comes from their food. And the energy in the food comes from the sun. Green plants take their energy directly from the sun, but animals have to take their energy second or third hand. Most animals eat some kind of plants. Those animals that don't eat plants eat animals that do. In this way, energy can be passed along from the sun to all kinds of earth life.

Man gets his energy second hand, sometimes third or fourth hand. Look at Figure 18/6. Notice the long route

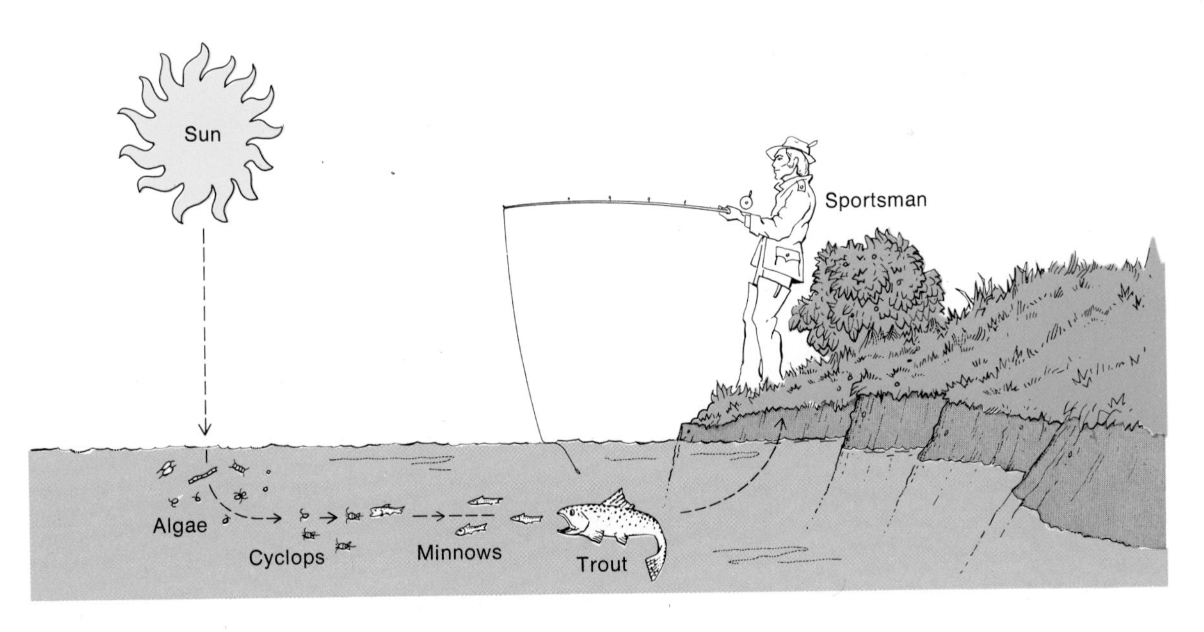

Figure 18/6 *Trace the sun's energy to man in this simple food chain. Where else does the energy go?*

the sun's energy takes before it gets to the man. Energy-filled plants in the water are eaten by tiny water animals, which are eaten by minnows. The minnows are eaten by bigger fish, trout. Finally, the man catches and eats the trout.

The route taken by the energy is called a **food chain.** Each kind of plant or animal is a link in a chain. One kind of organism can be a link in several food chains at the same time. Man himself is the end link in many food chains.

Because man can be part of so many food chains, he is more fortunate than other animals. For instance, Figure 18/7 shows that it takes the energy in 25,000 pounds of alfalfa to make a 150-pound person. If something happened to all the cows that pass energy on to man, man would get energy from other sources. Just look at the selection in any food market.

The loss of minnows could cause starvation of trout, if the trout could not find other fish. The loss of minnows might permit the numbers of *Cyclops* to increase, which would cause the algae to decrease until a new equilibrium was reached. The man would be less affected. He could look elsewhere for food.

Some animals live on only a few kinds of food. Some live on just one kind. The permanent loss of a single link in a food chain would cause these animals to starve. How would the loss of minnows from Figure 18/6 affect the other members of the food chain? Could the loss of a link in the food chain be what happened to some of the animals shown in Figure 18/1 that died out?

Figure 18/7 *It would take all the energy in this much alfalfa (turned into beef) to build the person standing on the pile.*

18.5 Only fittest organisms survive.

Concept: the nature of equilibrium

The entire biosphere is a crisscross of many food chains, that is, most food chains are connected to others. A more complicated diagram, and a more realistic one than Figure 18/6, is shown in Figure 18/8. At every stage of the original chain, several organisms are competing for energy from the same source. Trout, bass, and herons are all eating minnows. However, there is more than just competition between different kinds of animals. There is also competition between similar kinds of animals. Trout compete with other trout for minnows and herons compete with other herons for minnows. If one creature is not very good at getting food, it will have a harder time surviving. The survival of the most able is called **natural selection.**

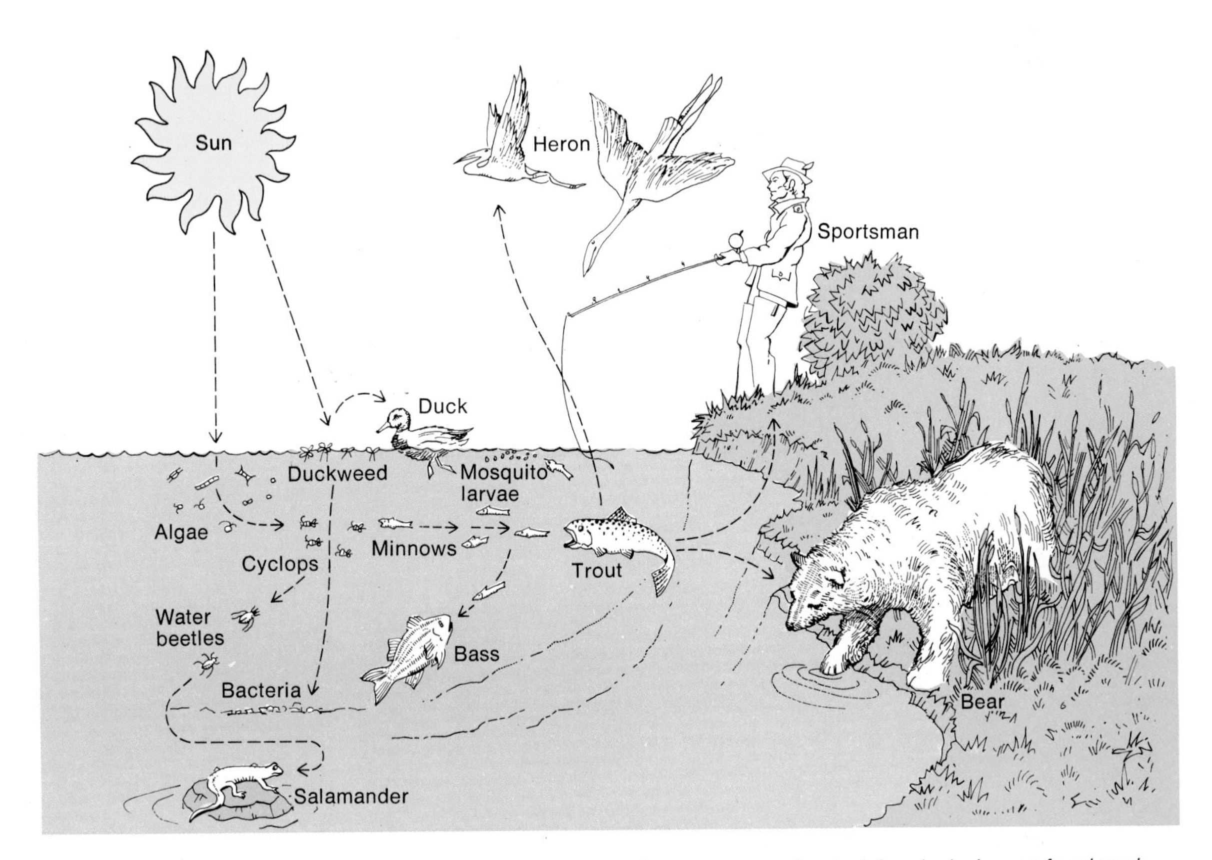

Figure 18/8 *A more complicated food chain, or food web. Could you add any more arrows?*

Materials

tracing paper

Activity Time: 15 minutes. The connections between beaks and habitats are outlined in the Teacher's Manual.

★ A puzzle will help you see how it works. Figure 18/9 is a map of a make-believe continent, Birdland. You are going to place a colony of 10 birds of different kinds at each of the locations marked on the map. For the time being, each kind of bird will be identified by its beak type only. (Study the bird beaks in Figure 18/10.) One kind of bird fits best at each location. As you can imagine, each bird will have a survival problem. The food is different at each location. The only bird in each location that will survive is

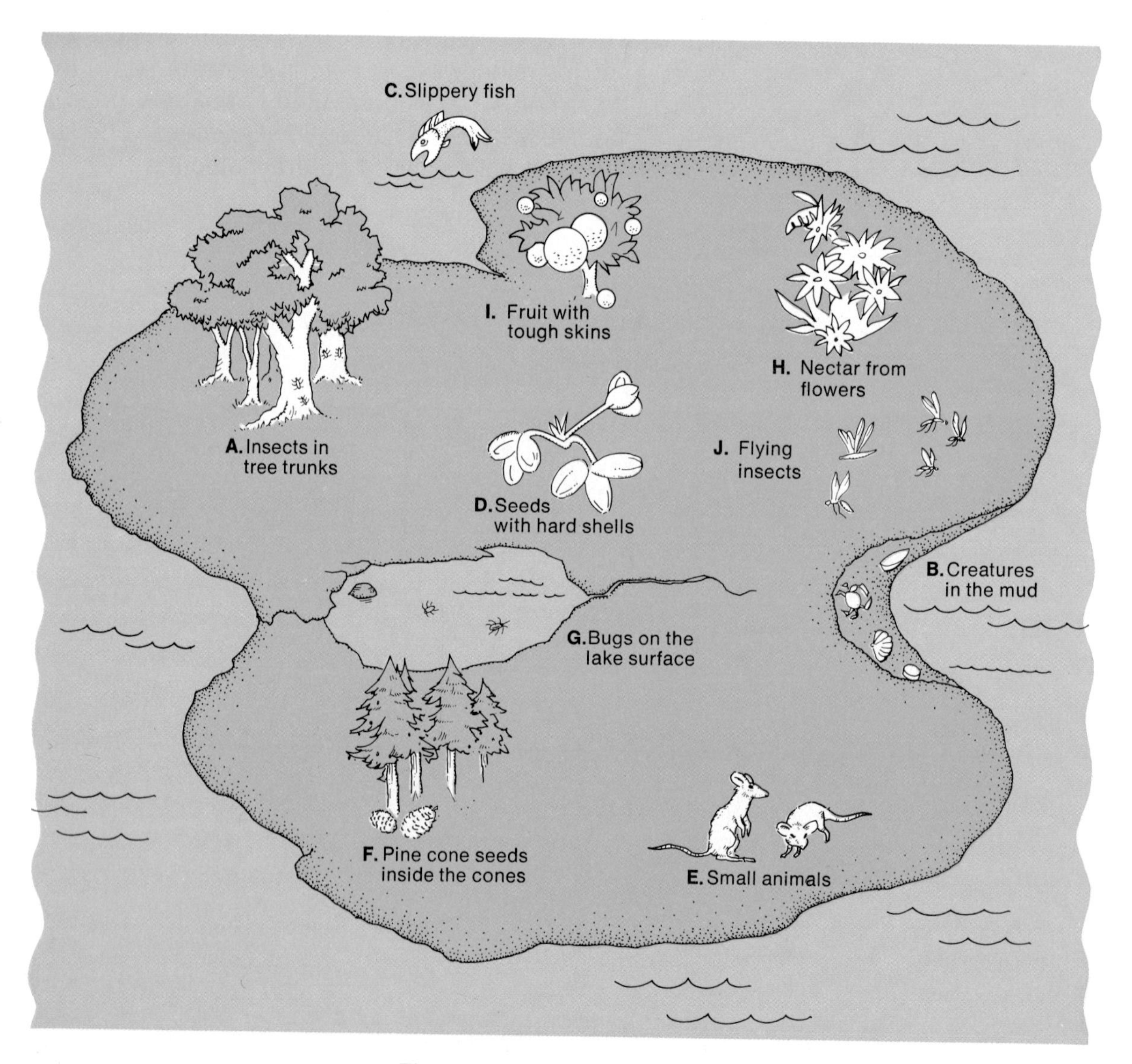

Figure 18/9 *The kinds of food available on Birdland.*

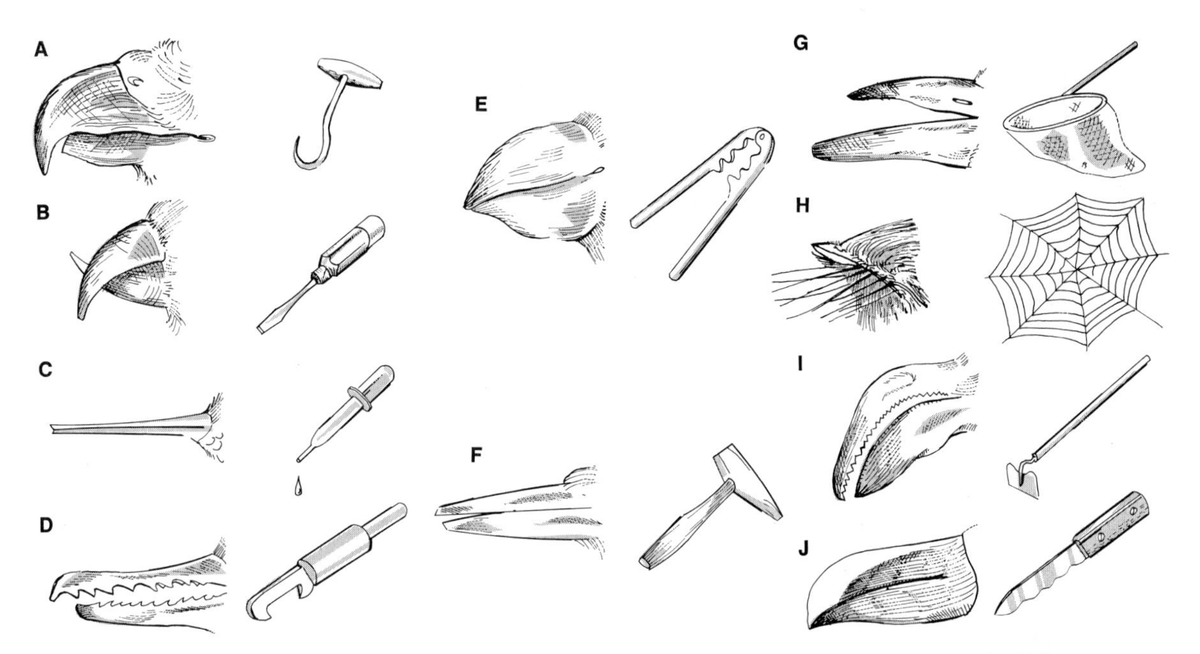

Figure 18/10 *Birds use their beaks like tools. Where on Birdland would you find the birds that have these beaks?*

the one whose beak is best suited for getting and eating the food at that place. The other birds will have to move or starve. Looking at the beak types, decide which bird in each colony will survive. Think of the beaks as tools. You may want to trace the beaks on paper. You could cut out your tracings and place them directly on the map.

Can you see how natural selection helps sort out which animals can survive best in which places? Figure 18/11 shows you the birds that the various beaks belong to. ★

There can be many reasons why some organisms survive and others don't. If a kind of organism cannot keep from being eaten, if it cannot reproduce as fast as other types, or if it cannot adapt to changing conditions, it may become extinct. Climate seems to do most of the controlling. A change to warmer climate might permit a warm-climate, meat-eating animal to move in from another area and eat the native plant eaters. A change to a cooler climate might slow down or stop reproduction in some animals. It might get too hot for a furry animal. It might become too dry for a plant that needs a lot of water. Could changes of climate

Figure 18/11 *The birds living on Birdland that match the beaks in Figure 18/10.*

be what happened to some of the animals that died out in Figure 18/1? Could this have something to do with deciding what animals are to live in the numbered zones in Figure 18/4?

18.6 Plants and animals themselves produce changes.

Concept: particulate nature of matter, levels of organization, interaction of matter

Every living thing contains special code molecules in its cells. These molecules hold the plans for making more of the same kind of living thing. Sometimes code molecules change. A change in such a molecule may cause an organism to be different in some way from its parents. The change in the code molecule is called a **mutation** (meu-TAY shun). Figure 18/12 shows organisms that have had mutations.

Some mutations make it harder for cells or organisms to survive under normal conditions. For example, a mutation in a corn plant could result, after reproduction, in new plants without chlorophyll. Such plants could not make their own food, so they would die. A disease that keeps human blood from clotting resulted from a mutation. A person with this disease may bleed to death from a small cut if he cannot get to a doctor in time.

The disease is hemophilia.

Sometimes, however, a mutation can help an organism to survive more easily than others. A series of mutations could result in a new kind of zebra that could detect and outrun lions more of the time.

A mutation of this sort formed a kind of wheat plant that is shorter than other wheat plants. The short plants are less likely to be broken by rain or hail.

Some mutations do not become useful unless living conditions change. For example, mutations have caused flies that are not hurt by DDT. Other mutations have caused kinds of germs that cannot be killed by medicines like penicillin. The first flies and germs with the mutations lived as well as the ones without mutations because the mutations were not harmful. Then along came DDT and penicillin. Only the mutant flies and germs survived and were able to reproduce.

Figure 18/12 *Some mutations: (A) A normally red geranium produced white flowers with smaller petals when exposed to X-rays; and (B) The two sheep on the right, called ancon sheep, have shorter legs that keep them from jumping over fences.*

Have a dentist irradiate some radish seeds for you. Plant them and see what happens.

Scientists have learned how to cause mutations in laboratories. They can change code molecules with certain chemicals, X-rays, and radioactive elements. But they can't always predict the results. A laboratory mutation could help an organism survive. But it also might kill the organism before it has a chance to reproduce.

A long period of time may be necessary for a mutation to have an effect on a whole population of organisms. If an organism with a mutation survives and reproduces, the mutation will be passed on to its offspring. Then the offspring must survive and reproduce to spread the mutation further. The plants or animals with the mutation must continue to survive and multiply for many generations in order to become a major part of a population.

These questions are answered in Section 18.8.

Can all this explain what you see in the record shown in Figure 18/1? Can you see a connection between time, mutations, and natural selection?

Answers:

1. Possibilities: heron, duck, and bass to man; man to mosquito; minnow to ducks; all organisms to bacteria. Hawks eat ducks and whippoorwills eat mosquitoes. To include the deer, arrows would go from the sun to the bush and from the bush to the deer and from the deer to the man.
2. 1-D (Ostrich)—useful only on ground, tough for dry rocky country
   2-E (Chimney Swift)—relatively flat, with claws for holding in crevices
   3-A (Parrot)—useful on tree branch, two toes on either side of branch
   4-C (Jacana)—toes elongated to distribute weight over greater area, for walking on lily pads
   5-B (Mallard)—webbed for swimming
   6-F (Pheasant)—average flat foot for grassland
3. Helpful mutations would be: ability to see infrared light for night vision, ability to make vitamins in the body instead of having to eat them, lungs less sensitive to polluted air, stronger knee connections for athletes.
4. Important human characteristics are the brain, stereoscopic vision, opposable thumb, and our type of legs. These give us the ability to think abstractly and reason, to examine things, and to move around easily.

Did You Get the Point?

Every animal or plant is part of one or more food chains.

Changes in food chains cause changes in the biosphere.

Some plants and animals are better able to survive than others. They may be better able to get a particular food, reproduce, or escape enemies. Or they may be better able to adjust to changes in climate.

Mutations can cause new forms of life to develop from old forms.

Some animals and plants with mutations live to reproduce and others don't.

Check Yourself

1. Figure 18/8 is not complete. How many arrows can you add between the organisms on it? Where would you add hawks, whippoorwills, and deer?

2. Figure 18/13 shows six different types of bird feet and a map of another Birdland with six different types of living conditions labeled. Which feet are best suited for living at each location?

3. Can you think of a mutation of the human body that would be helpful? How about a harmful one?

4. What body parts and special abilities make it possible for humans to survive in the world? How do these body parts and special abilities serve you?

5. Draw as many food chains with yourself at the end as you can. Start each chain at the sun.

Figure 18/13 *Match up the bird feet with the numbered homesites that would best suit them.*

Theories About Earth Life

18.7 There are theories to explain the beginning of life.

You know there are many unanswered questions in the rock record of life. One of the biggest questions is "How did life begin on earth?" The lack of evidence hasn't stopped people from making educated guesses at answers.

For centuries the most widely accepted idea was the one outlined in the Book of Genesis in the Bible. It states that a divine being, God, made the earth, oceans, the continents, plants, sea animals, birds, land animals, and man—in that order. Interestingly, the order in Genesis is similar to that in the rock record.

Process: inferring

Other theories have the earth forming when the sun did. Some scientists think that it formed along with the newly forming sun four and a half billion years ago and that it slowly cooled in orbit as the third planet. In the cooling earth drawn in Figure 18/14, convection currents are helping to sort the lighter minerals from the heavier minerals. Masses of lighter minerals could have cooled to form the beginnings of the continents. This theory goes on to say that heavy rains toward the end of the cooling period caused the ocean basins to fill and continents to stand out as dry land. Eventually the seas filled with the proper chemicals and the atmosphere became just right for life.

Since the rock record begins with bacteria and tiny sea plants, some scientists believe that molecules in the sea accidentally combining with each other resulted in the earth's first living things. According to this theory, it took one and a half billion years just to "set the stage" for life.

18.8 There is a theory to explain the further development of life.

Since scientists believe the early seas could produce only simple life forms, other theories are needed to explain how more complicated forms of life came into being. More observations have to be made.

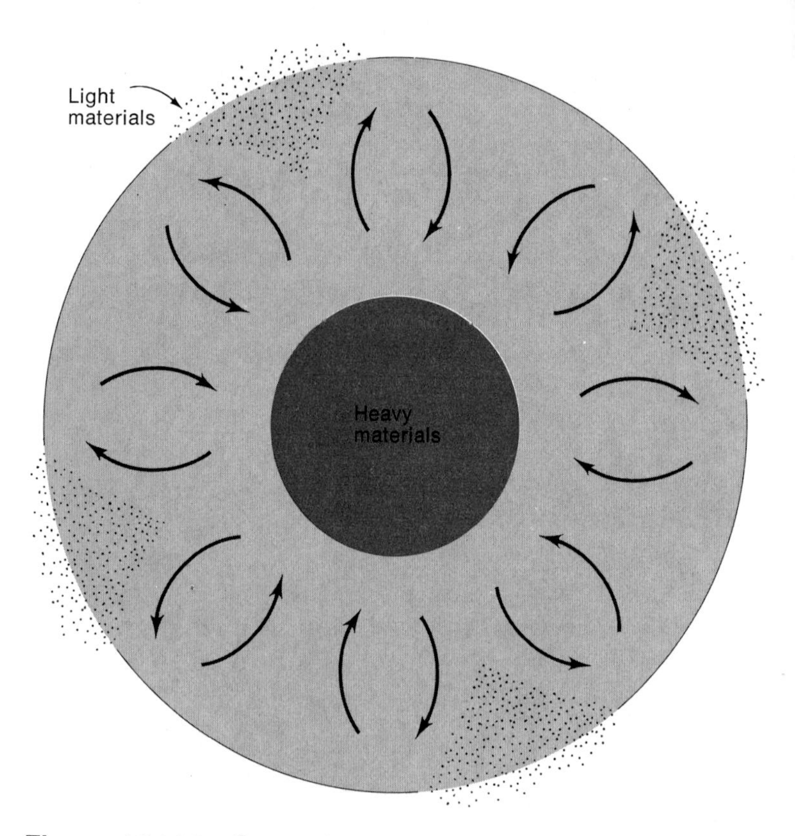

Figure 18/14 *Convection currents in a slowly cooling earth could have carried lighter minerals to the surface to form the continents.*

A. Cars: (E) 1915 Buick; (G) 1927 Jordan; (B) 1936 Terraplane; (H) 1946 DeSoto; (F) 1953 Mercury; (A) 1958 DeSoto; (D) 1963 Dodge; (C) 1970 Ford Torino

B. Men's fashions: (H) prehistoric; (E) 1300; (D) 1530; (A) 1590; (J) 1730; (B) 1800; (F) 1850; (C) 1900; (I) 1961; (G) 1972

★ Try to put the little pictures in Figure 18/15 in some kind of order. The row of cars from old to new can be referred to as the evolution of the automobile. The clothing pictures show the evolution of men's fashions. ★

Don't the two groups of pictures in 18/15 seem to follow patterns similar to those in Figure 18/1?

A scientist, Charles Darwin, noticed just such a pattern. In 1859 he published his theory of **evolution** by natural selection. Quite simply, the modern version of his theory suggests that mutations, or changes in plant and animal life, gradually led to different or more complicated forms of life.

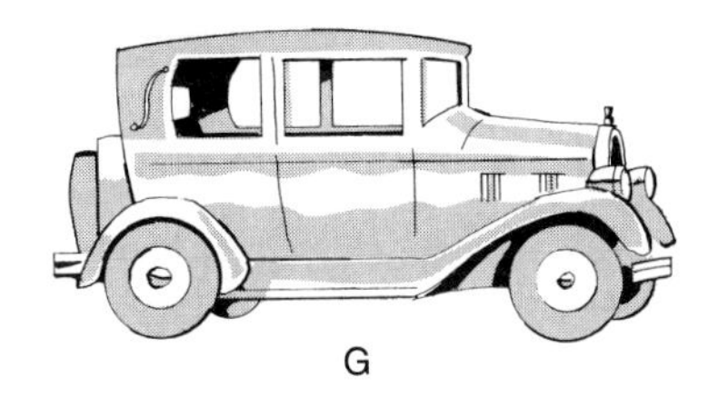

G

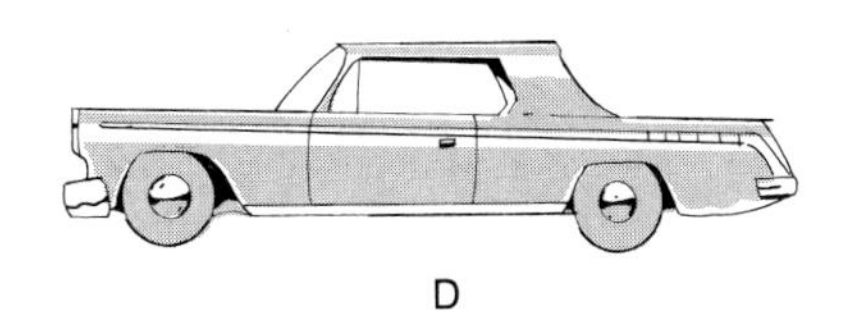

D

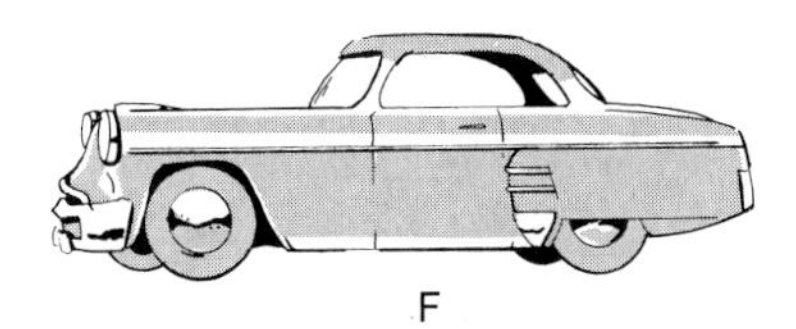

F

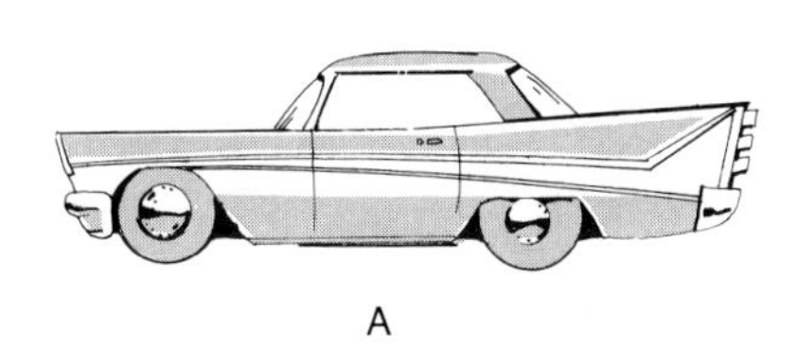

A

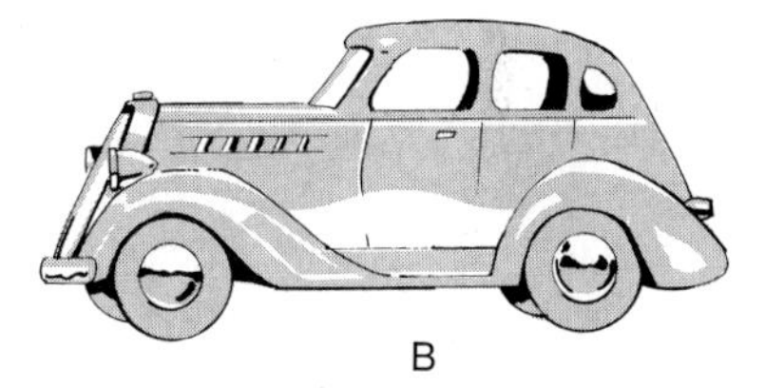

B

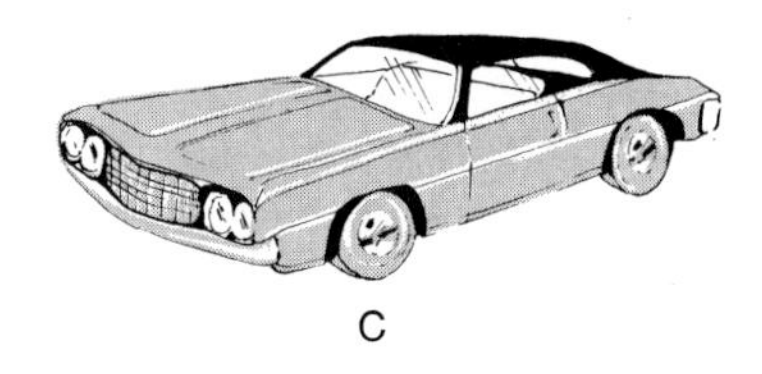

C

H

E

Figure 18/15A *Put the cars in the order in which they developed.*

Figure 18/15B *Put the costumes in the order in which they developed.*

Figure 18/16 *A 500,000 year old fossil skull of a man-like creature called "Java man." The dark top of the skull was part of the first discovery. The face and lower jaws have been restored in plaster of Paris from other fossil fragments of the Java man. The head on the right shows what this individual might have looked like.*

Natural selection kept those organisms that could survive in geographic areas and food chains that suited them.

The theory of evolution by natural selection suggests answers to many of the questions you can ask about Figure 18/1. But it does not explain or prove everything. For instance, fossils of humans have been found in sediments from the Ice Age. The one illustrated in Figure 18/16 is a famous example. However, fossils do not show a direct evolutionary line leading to modern man that all the experts can agree to. A great deal is still not known about patterns of life in the past.

Is There Life on Mars?

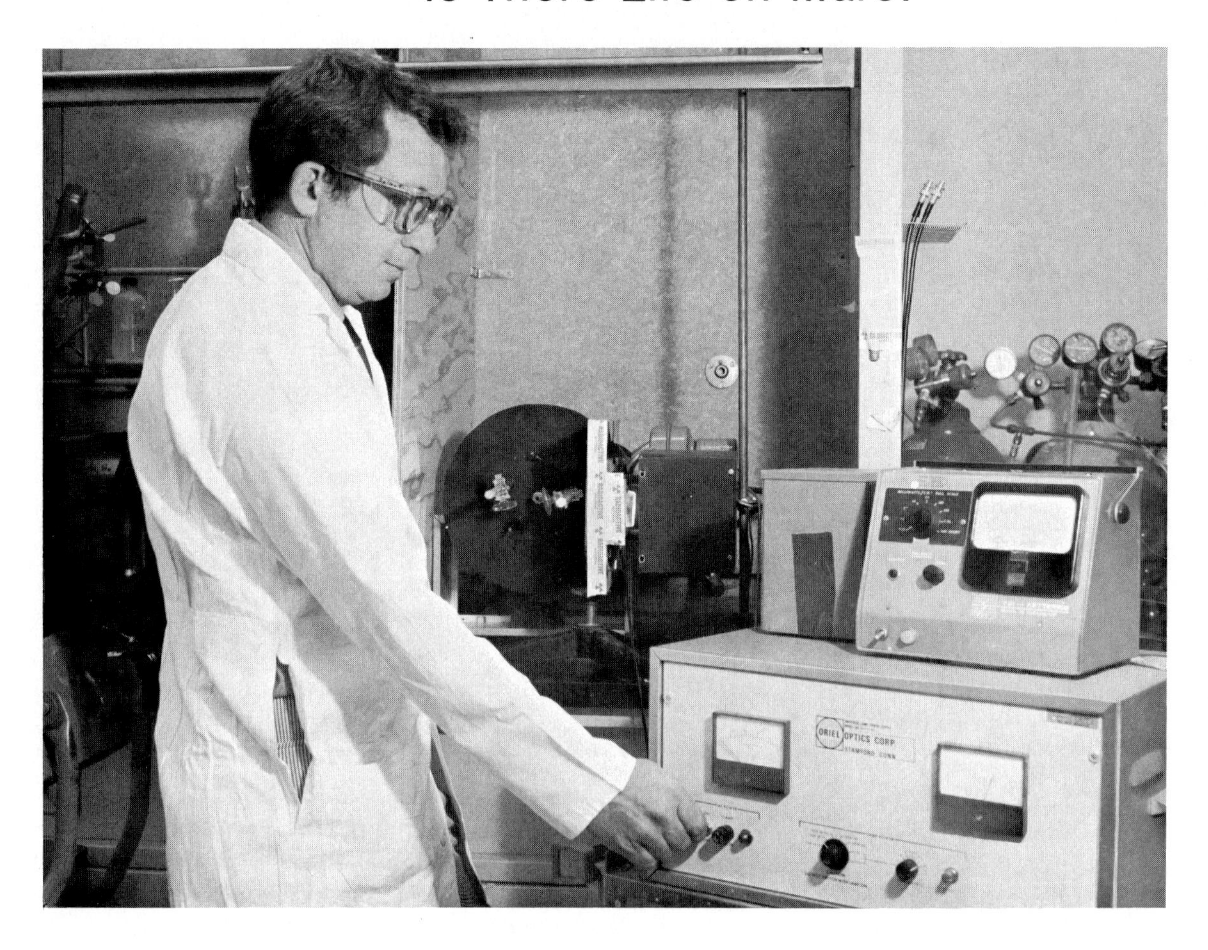

Is there life on Mars? This question can't be answered yet but experimenters at the Jet Propulsion Laboratory in Pasadena, California, have shown that simple organic compounds could exist on Mars. Organic compounds are compounds that contain carbon. They are found in all living things.

The experimenters mixed carbon dioxide gas with small amounts of carbon monoxide gas and water vapor, and added some fine, sterilized soil and ground glass. From Mariner fly-by missions in 1969 they had learned that the Martian atmosphere contained water vapor and carbon gases. They bombarded the mixture with the same kind and amount of ultraviolet radiation as probably reaches Mars from the sun.

After more than 200 tests, the research team reported that compounds more complex than the original gases were found on and just beneath the surface of the test soil. The organic compounds they found are like compounds that scientists believe were ancestors of biological molecules on earth.

But there are problems. Ultraviolet radiation can destroy life and certain organic compounds as well as create them. Thus on Mars the more complex molecules may be broken down as fast as they are made. Or perhaps they are protected as they sift deeper into the soil. Mars' atmosphere contains so little gas of any kind that organic compounds would be produced very slowly.

There is another problem in the speculation about life on Mars. Life as we know it on earth needs nitrogen. When the research was done, it was not known whether nitrogen could be found on Mars. Whether or not there is any organic matter on Mars will not be known until spaceships land on the red planet.

18.9 Some life disappears.

One of the many things Darwin's theory of evolution doesn't explain is why some plants and animals suddenly died out. Can you see some examples of this mystery in Figure 18/1? A popular example is the mystery of dinosaur extinction.

Animals on the chart that have become extinct are—trilobites, first jawless fishes, first jawed fishes, first amphibians, early amphibians, first reptiles, flying reptiles, mammal-like reptiles, first mammals, first birds, first horses.

The specific reason or reasons for the extinction of dinosaurs is not known. They must have been unable to adapt to some kinds of changing conditions.

At first, climate change looks like a good explanation, but it doesn't seem reasonable that changes in climate alone could have caused the extinction of all the dinosaurs. The same conditions would have hurt other animals at the same time, and some of them did not become extinct. Turtles, crocodiles, lizards, and snakes are still around.

Maybe animals that ate dinosaur eggs appeared. Some dinosaur eggs must have been even larger than the one in Figure 18/17, and they might be hard to hide. If all those eggs were eaten, there could be no large dinosaurs. However, it is difficult to explain why the smaller dinosaurs couldn't have survived an invasion of egg eaters, along with the turtles and crocodiles.

The size of some dinosaurs might have had something to do with their extinction, too. Maybe there were natural disasters like floods, droughts, or disease. Perhaps some dinosaurs, like the ones in Figure 18/18, were too big and heavy to find food or escape dangers. This sounds like a good theory, but the rock record shows that the largest dinosaurs were around for 60 million years. Since the dinosaurs were successful for so long, what could have happened that could kill them all in a relatively short time? No one knows.

Figure 18/17
The size of these fossil dinosaur eggs suggests that they may have been easy prey for egg-eating animals.

Figure 18/18
Dinosaurs, like this brontosaurus, may have been too large and heavy to survive in a changing environment.

Did You Get the Point?

Some scientists believe that it took over a billion years for the conditions of the earth to become just right for life.

Darwin's theory of evolution by natural selection explains the development of life on earth in terms of natural selection.

Studies of mutations show how changes can take place in organisms.

Any kind of living thing may become extinct.

Check Yourself

1. Many automobile manufacturers have gone out of business. Try to explain why, in terms of the evolution of the automobile.

2. How do you think modern animals got to be listed in "danger of extinction?" What might have happened to cut their numbers down to a dangerous level?

3. Write a short essay on how scientists think the earth developed into a planet suitable for life. Find references in the library.

What's Next?

Whatever the cause for the extinction of dinosaurs, you can be sure that it could happen to any kind of plant or animal. Within the last few years, many animals now living on our planet have been listed "in danger of extinction." Figure 18/19 shows only a small sample. In Chapter 19 you will see how man could get on this list.

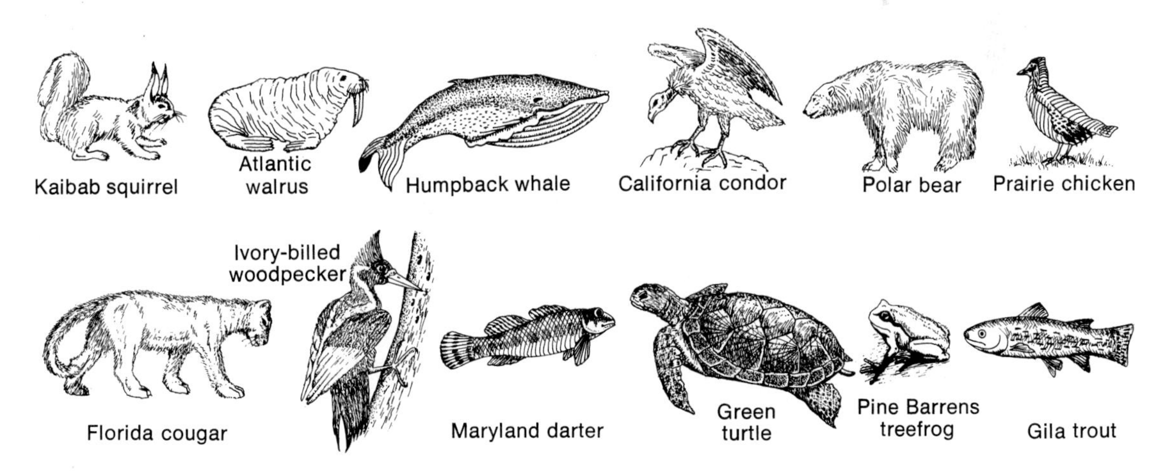

Figure 18/19 *A few of the hundreds of animals that are threatened with extinction in the near future.*

Skullduggery

Directions: Your teacher will give you a copy of Figure 18/20. Each of the following is a true or false statement. To finish the picture, you have to draw a line to the number for each true statement. Begin at the dot labeled "Start" and draw a line to the number of the first true statement. Then draw a line from there to the number of the next true statement, and so on. Whenever a statement is false, make no additions to the line. You will know if you were right about all of the statements by the picture your line traces out.

Answers:
(1)T, (2)F, (3)T, (4)F, (5)T, (6)F, (7)F, (8)T, (9)F, (10)F, (11)T, (12)F, (13)T, (14)T, (15)F, (16)T, (17)T, (18)F, (19)T, (20)T

1. There are more different kinds of insects than there are birds.

2. Life does not exist on ocean bottoms because of the great pressure of the water there.

3. The airplane probably has not finished evolving.

4. There are fossils of many kinds of animals and plants in rocks over a billion years old.

5. According to the record in the rocks, some animal types appeared rather suddenly.

6. The special code molecules in living things cannot be changed.

7. An animal cannot be part of more than one food chain at the same time.

8. Man can cause mutations in life forms.

9. There is little danger of extinction for modern animals.

10. Most animals get their energy directly from the sun.

11. Some animals can only live in certain places on a continent.

12. Horses have been part of the rock record longer than birds have.

13. Mutations could cause an animal type to become extinct.

14. Special abilities can determine where an animal will live on the earth.

15. The theory of evolution says that man evolved from the monkeys.

16. Survival of the fittest is the way nature selects the animals that will live in an area.

17. When an animal becomes extinct, there may be a number of different reasons for it.

18. Most animals do not appear to have developed from less complicated animals.

Figure 18/20
Copy this puzzle for Skullduggery.

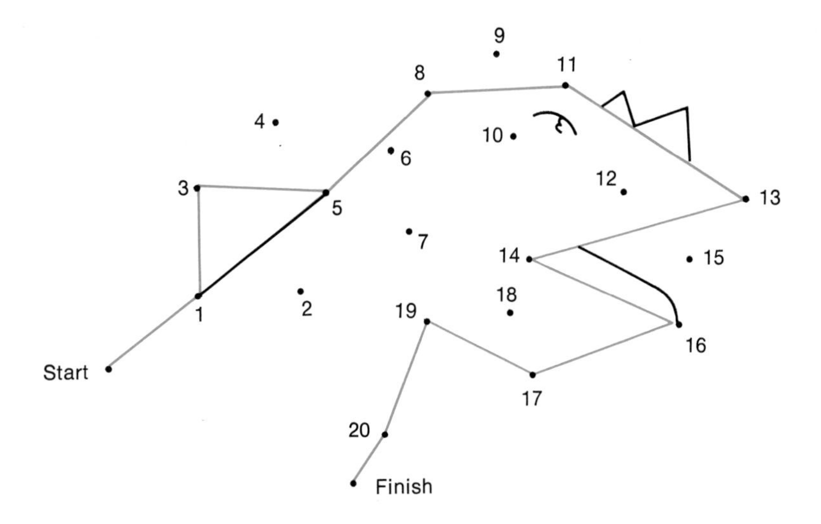

19. Ability to reproduce, keep from being eaten, and the ability to adapt to changing conditions are ways nature selects animals that will live in an area.

20. Man also comes under the control of natural selection. It is possible for him to become extinct.

For Further Reading

Moore, Ruth, and editors of Time-Life Books. *Evolution.* New York, Time-Life Books, 1968. The story of Darwin's voyage on the *Beagle* and the theory he formulated on survival of the fittest. Many pictures of special features that help animals survive. Also a chapter on the discovery of the special code molecules.

Ravielli, Anthony. *From Fins to Hands.* New York, The Viking Press, 1968. If you are interested in dinosaurs, you will like reading how they began. The first step was taken by a branch of the amphibian family that laid hard-shelled eggs that could hatch on land. Reptiles, including dinosaurs, developed from this branch.

BOB'S GROCERY

Chapter 19 Man in His Environment

The captain is speaking to an advanced civilization so the talk is designed to sound futuristic. Discuss some of the ideas raised with the class. For example, you may wish to explain that time capsules are storehouses of historical information put in protected places such as old mines. Future explorers may find them.

Captain's Log. Star Date: 1071.86: An exciting day! Since our arrival here, thick clouds had covered the sky above our excavation site. Today they moved apart. For the first time we could see light from the star around which this planet orbits. Eight other planets orbit around the same star.

After the clouds opened, the temperature dropped from 400 to 150 degrees in only two or three hours. Records show that the temperature drops like this only once a year. For about a million years the clouds didn't open up at all! What a relief to be able to cut down the flow of coolant through our suits for a while.

Until this trip, we had only traced the history of this planet from its birth to just before it no longer supported intelligent life. That was a period of five billion years. Much of the information had come from fragments in "time capsules." They told of stone ages, bronze ages, iron ages, ice ages, dark ages, an age of reason, and a space age. The written record ended at the space age. Our record will begin there.

We started the search by disintegrating a large stony slab that might have been part of a building. Beneath it we found a scientist's dream—thousands of different objects. Even a few samples of the material called "paper" were saved. The words BOB'S GROCERY were lettered on one piece. However, most of the samples brought back to the ship are containers made of a material called "plastic." We dated these items to what we have named the plastic age, the time just before the intelligent beings who made the containers died out.

Other items, also made of plastic, have survived the weathering process. We are the first to discover small plastic statues, complete with moving limbs and blinking eyes.

Plastic disks found near the statues may have been part of some religious ceremony, with a different disk for each prayer. One disk even has part of a paper label with a word written on it that we have not been able to translate—"Beatles."

We also found hundreds of fossils of small animals with long thin tails and long, sharp, curved front teeth. When we tested their bones, we found surprisingly large amounts of lead, mercury and other poisons.

Now, we are studying what appears to have been the bed of a small stream that ran through the site. If our conclusions are correct, we may be able to explain what happened here. It doesn't appear to have been a war. It was something more terrible. I wonder if it could have been avoided.

Equilibrium

19.1 A classroom battle can illustrate what happened.

Figure 19/1
The small team during a plastic foam ball fight.

Activity Time: 15 minutes. This activity can be a wild one. Try it with your class if you wish. You will have to referee.
Concept: the nature of equilibrium

Imagine you were to divide your earth science class into two teams and have a plastic foam ball fight. The purpose of the fight would be to help explain the starship captain's report of the dead planet.

Here is the way such a game goes. The class is divided into two teams, one-third of the students on one side of the room and two-thirds on the other. A strip of tape across the center of the room marks off the playing territories. Each team must stay on its own side of the room at all times.

When both teams have heard the rules, the teacher dumps 800 plastic foam balls on the floor of the bigger team's half of the room. At the signal "Go," the big team tries to throw as many balls as possible to the other side of the room in two seconds. The small team tries to send them back as fast as they come over the tape. Figure 19/1 shows a small team in action.

At the end of two seconds, the small team counts the plastic foam balls in its half of the room. All 800 of the balls are then piled in the big team's half of the room for another round.

Figure 19/2
Data from the plastic foam ball fight.

Throwing time	Balls on small team's side
0 seconds	0
2	40
5	92
10	204
15	321
20	460
25	602
30	598
2 minutes	604

A graph of the data will look something like this:

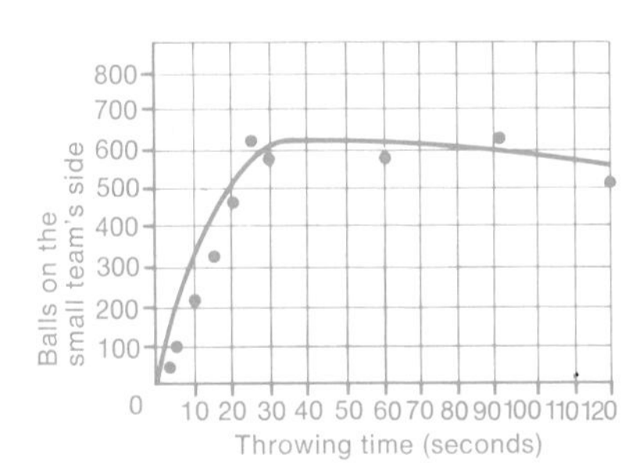

After a certain amount of time, there is no increase in the number of balls in the small team's side.

The second round lasts for five seconds. The balls that end up in the small team's half of the room are counted again. Even though the small team works as hard and fast as it can, the number of unreturned plastic foam balls seems to increase as the throwing time gets longer.

The balls are gathered in the big team's half of the room for another round, this one lasting 10 seconds. Eight more rounds are fought, each one longer than the one before it. The actual results of one class battle appear in Figure 19/2.

Materials

graph paper

★ Plot the data in Figure 19/2 on a graph. Measure time along the horizontal axis. On the vertical axis, plot the number of plastic foam balls left on the small team's side. What does the graph show? Can you explain why the curve levels off? What would have been the score at the end of a battle lasting 30 minutes? ★

The small team can always return *some* of the balls, so the large team will never get all 800 to the small team's side. A 30-minute battle would still have a score of about 600.

19.2 Similar "battles" take place in nature.

A box containing 800 plastic foam balls could be said to be in **equilibrium** (ee-kwih-LIB-ree-um) as long as the whole boxful of balls is not changed in any way. If you put 200 of the balls on one side of the room and 600 on the other side and just left them there, they would be in a new equilibrium, as long as nothing else changed.

The simple kind of equilibrium is static equilibrium: for example, a 1-kg weight balancing another 1-kg weight on an equal-arm balance.

The ball fight is a model of another kind of equilibrium, a **dynamic** (dy-NAM-ik) one. The balls kept changing sides of the room; the fact that they were moving makes the equilibrium dynamic. But remember, before the end of the fight, the *number* of balls in each half of the room *stopped changing.* Since the number on one side was always about 200 and the number on the other side was always about 600, the entire supply of plastic foam balls in the room could be said to be in equilibrium. Some people call dynamic equilibrium ''a change that isn't really a change.''

Dynamic equilibrium occurs in our surroundings, or **environment** (en-VY-run-munt), here on earth. For example, the oxygen in the air today is not all the same oxygen that was there yesterday, but the total amount is about the same. Animals breathed some in, and plants let some out. What about the amount of nitrogen in the air?

In each of these examples, there is a change in position or condition of individuals but no change in quantity.

Yes, the nitrogen in the air is in dynamic equilibrium. Nitrogen-fixing bacteria convert nitrogen gas to soil nitrates. The nitrogen may be incorporated into proteins. Eventually it is released by decomposition. Students should not be expected to know this.

Rabbits were born today and rabbits died. The total number of rabbits is about the same, but some of the individual animals making up the population are different.

The total amount of energy the earth received from the sun today must be close to the amount the earth lost yesterday. If this were not true, the earth as a whole would be getting warmer or colder. But, the energy the earth lost yesterday is not the same energy it got back today.

It rained in many places on the earth yesterday. But the total amount of moisture in the air has not changed much. Today water evaporated into the air. The water that evaporated today was not necessarily the same water that fell as rain yesterday.

How many other examples of dynamic equilibrium can you name? Can you think of anything that was in equilibrium in the past but is not now?

Other examples of equilibrium:
(1) plant life in a forest
(2) lake level (inlets vs. outlets)
(3) number of stars in the sky
(4) stocked shelves in a supermarket
(5) number of iron atoms (in ore, cars, rust, etc.)

The earth's human population is no longer in even an approximate equilibrium. The populations of animals that are becoming extinct are out of equilibrium in the other direction.

19.3 Dynamic equilibrium can be upset.

Suppose you ran the plastic foam ball experiment again but used 1000 balls instead of 800. How would it turn out? What would the graph look like?

The graph would flatten out at a larger number of balls: nearly 1000.

Now suppose you repeated the experiment with 800 plastic foam balls but put the same number of people on each team. How would it turn out? What would the graph look like?

The graph would flatten out at a smaller number of balls (about half the number used).

The possibilities are endless. The teams or the number of plastic foam balls may change and the old equilibrium will be upset. But as long as some balls and some people

Figure 19/3
No matter how the rules for the plastic foam ball fight are changed, the equilibrium curve always levels off.

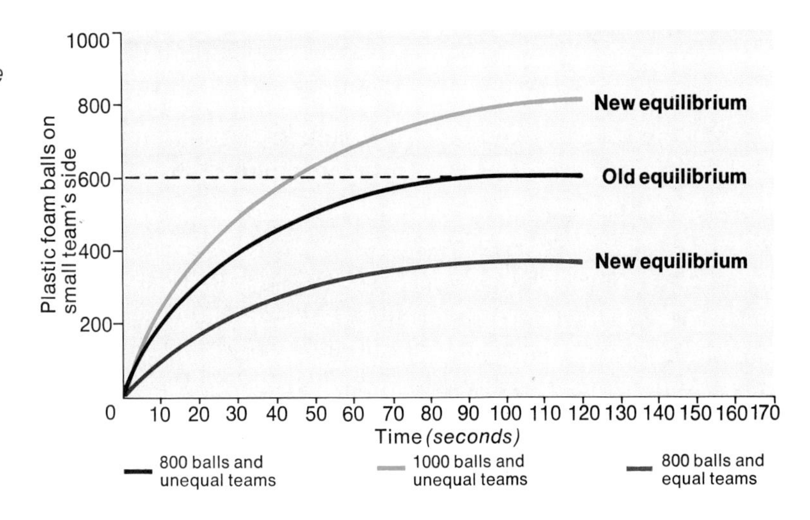

remain in the fight, a new equilibrium will be set up. If you graph the new data after each change, the curve will still level off, but the point at which the curve levels off will be different. Figure 19/3 shows how the equilibrium curve may change.

Man upsets the earth's equilibriums when he tampers with the air, the water, and the earth's crust.

When he upsets the equilibriums without replacing them with new ones that are just as good, the result is called **pollution.** And pollution is not good for living things, including people. A person's body is used to air made up of 21% oxygen. A person's body needs water that does not contain large amounts of lead, mercury, and other poisons. People depend both on plants growing in the fields and fish living in the oceans for food.

The human body cannot adjust to some of the equilibriums that may be caused by pollution. If man does not somehow save or bring back the old equilibriums—well, the starship commander's report could become real.

Did You Get the Point?

Many of the systems that man depends on to stay alive are dynamic equilibriums.

If you change some part of a dynamic equilibrium, the equilibrium will be upset.

A new equilibrium is set up whenever a dynamic equilibrium is upset.

Pollution refers to man's upsetting of old equilibriums in his environment.

Check Yourself

Answers:
1. a. Poisons in the air ⇌ pure air.
   b. Number of people ⇌ living space.
   c. Sound ⇌ the ears' ability to take it without physical damage to hearing.
   d. Number of people ⇌ money, energy, time, and care centers for the handicapped.
   e. glass, paper, wood ⇌ supply of raw materials, dumping space.
2. We, too, have a fixed amount of life-supporting materials. The amounts in a spacecraft are small, easier to control, and adjustments can be made more quickly. It is doubtful that any equilibrium can be upset without endangering an astronaut's life.

1. The following is a list of laws proposed by some antipollution groups. What equilibriums are the laws trying to bring back? Be careful! Not all of the equilibriums were mentioned in the text.
a. Driving a car is punishable by five years in jail and a $20,000 fine. (*Hint:* Cars produce gases that are not usually found in the air.)
b. A married couple will not be permitted to have any more than two children.
c. Making any sound over 60 decibels is punishable by a fine of up to $500. (*Hint:* Electricity makes it possible for the average rock band to produce 125 decibels of sound, but people have the same hearing equipment they had before they started to use amplifiers.)
d. Married couples may not have children unless tests done on the father and mother predict that the children will be healthy.
e. Everything we manufacture must be recycled.

2. Since the first space flights many people have compared living in a spaceship to our living on the earth. In a spaceship, machines maintain all of the equilibriums for the astronauts.

How is the earth like a spaceship? Do astronauts in a spacecraft have a better chance of controlling their environment than we have? How many equilibriums can be upset in a spacecraft without endangering the life of an astronaut?

Effects of Pollution

19.4 Population increase affects the environment.

Before the 20th century, most babies did not live long enough to grow to adulthood. Many of those who did live long enough to become adults died before they reached the age of 50. So the number of people in the world did not increase very fast. The environment could keep up with all the kinds of waste that man produced. Some of the waste changed into other materials. Some of it became food for plants or animals. A kind of equilibrium existed as people produced waste and the environment turned the waste into things the people might use again.

Now more people in many countries live to be 70 or older. About 95 percent of all babies reach adulthood. If you look at the graph in Figure 19/4, you will see that as the years went by, the number of people increased faster and faster. In A.D. 1 there were about a quarter of a billion people. It took about 1600 years to double the population. Then, in only 200 years, the population doubled again. The third time, it took only 130 years. That brings you to about the year 1930. Has the population doubled since then?

★ You may be able to illustrate this in your own family. Draw a "family tree" from your grandparents on one side to the people in your own generation. Let circles with names in them represent people. Make the chart like the one in Figure 19/5. ★

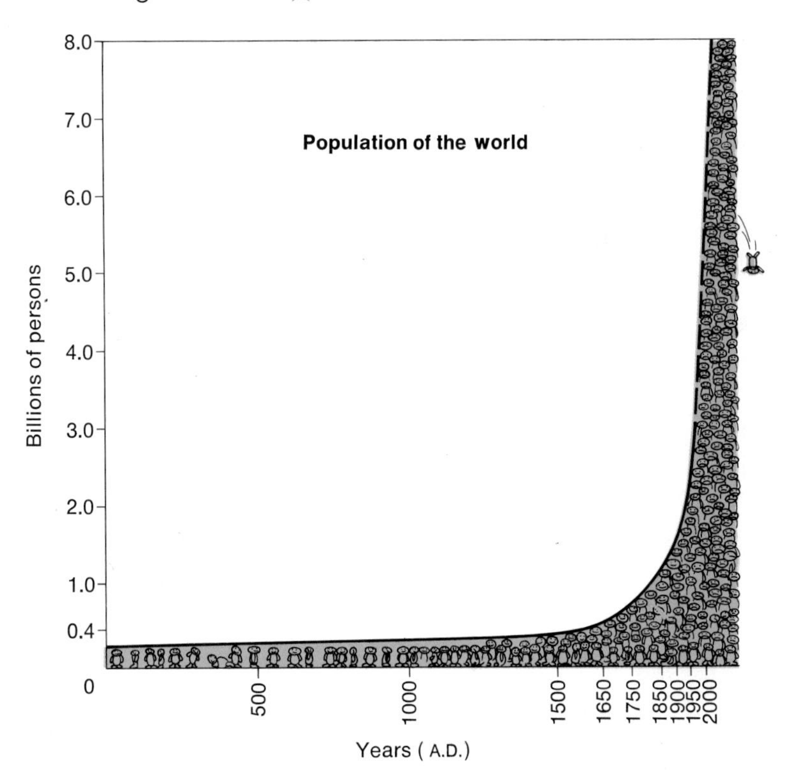

Figure 19/4
Look at what has happened to the world population since the year 1 A.D.

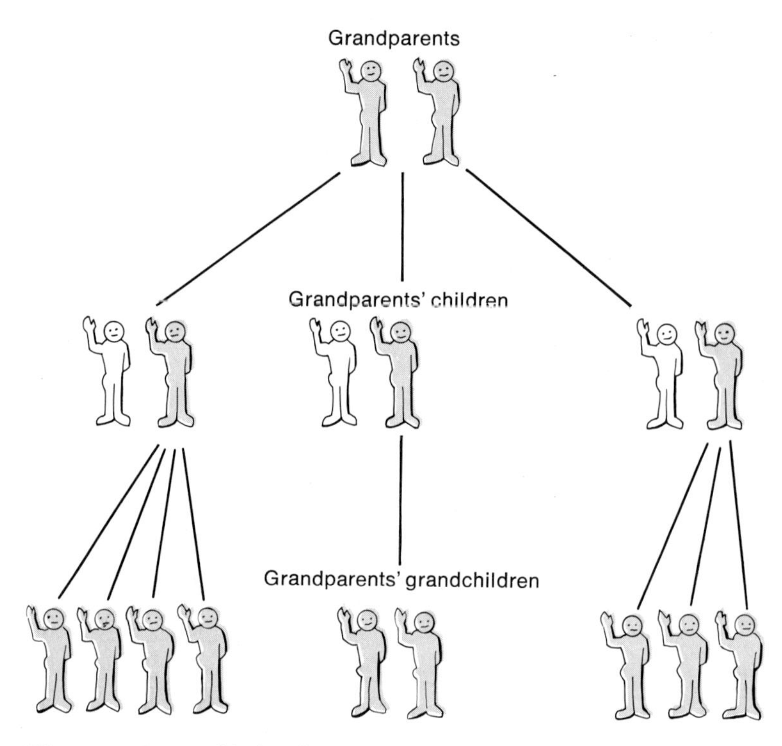

Figure 19/5 *Notice how fast a single family can grow. How many more people might there be when the grandchildren have children?*

The environment can't keep up with us any more. It's like doubling the number of people on one team in the plastic foam ball fight. In this country, the trash thrown out every day amounts to six pounds for each person. That comes to over 30 billion glass containers, 8 billion pounds of plastic, and who knows what else every year. You can see in Figure 19/6 what happens when the garbage in a big city is not collected for a few days. About 150 cars are abandoned on the streets every day in New York City alone. Junkyards like the one in Figure 19/7 fill up fast. To get an idea of how fast the waste is piling up around us, try making a model of pollution.

Materials

popcorn, frying pan, hot plate

★ Clear a space in the center of the room and put an ungreased pan on a hot plate. Have one person do the following: Pour in a cup of unpopped popcorn, cover the

Figure 19/6 *When the garbage collectors went on strike in New York City, it didn't take long for the garbage to pile up.*

Figure 19/7
Junked cars add to the pollution. The iron in them will rust but they pile up a lot faster than they rust.

Activity Time: 20 minutes. One student performs as the class watches. Be sure to use fresh popcorn seeds. If the container has been left open, the seeds won't jump from the pan. Remember not to use any oil.

pan, and turn up the heat. Shake the covered pan every 15 seconds. After the first kernels pop, remove the cover. As the popping continues, try to pick up all the corn and throw it in the wastebasket. When the popping stops, you stop.

The popped kernels represent the waste produced by man. The popped kernels you throw away stand for wastes that are removed from the air, land, and water by the environment. The popcorn left on the floor stands for waste that is piling up because the environment can't keep up with us.

Equilibrium will not be reached.

What happened? How long did it take the waste production (popcorn) to get ahead of the environment (the "cleaner-upper")? Was equilibrium reached? ★

Look at Figure 19/4 again. If the population of the world were to continue to increase by 50 to 70 million people a year, 900 years from now there would be 60,000,000,000,000,000 people on this planet. Even if we could help the environment clean up after us, that would mean 100 people per square yard of earth. But, as the popcorn model shows, the environment could fall behind in the near future. How long could it stay behind before we became as extinct as the dinosaurs? A few scientists have said only 10 years.

To put you in the right mood for what comes next, ask your teacher if it is possible to have the custodian "forget" to clean the classroom for a week, starting today. You will see the reason for this in Section 19.8.

19.5 We pollute the air.

The United States has many factories. We need them to make things like new cars and television sets. We also need them to make food products, clothing, and other things we must have to live. It has been estimated that our factories use one-third of the world's supply of natural resources to fill our needs. What's more, every year the factories like the one in Figure 19/8 pour 17.5 million tons of smoke and soot into the air we breathe. Power plants that burn coal add greatly to air pollution unless they have special equipment to clean their smoke.

Figure 19/8
A paper mill at Wallula, Washington, pouring sulfur-laden smoke into the air.

After a while things made in factories wear out and have to be thrown away. We burn a lot of rubbish in incinerators. More smoke and soot!

And then there are the cars made in factories. Once out on the street, the cars take in air and replace it with poison gases and MORE SMOKE AND SOOT. If the smoke and soot mix with water droplets in the air, smog is formed. Sometimes a **temperature inversion**, a layer of warm air over a layer of cool air, traps the smog under it. Figure 19/9 shows Los Angeles, California on a clear day and during a temperature inversion. Heavy smogs like this one can cause illness or even death.

Materials

aquarium or earth box, bags of hot water, cold sand bags, smoke generator or punk, funnel, and tube

Activity Time: 20 minutes.

★ You can make a temperature inversion in the aquarium or earth box. Arrange it as shown in Figure 19/10. Place the top over the box, and try to fill the box with smoke. Watch what happens. Draw the temperature inversion as you see it. What happens when you take the cover off the box? ★

Many people who work in the factories take business trips or vacations. They often travel by jet. One ton of jet fuel burned in the upper atmosphere gives off over a ton of water vapor. The water vapor condenses on the particles

Figure 19/9
Los Angeles on a clear day and during a smog-producing temperature inversion.

of smoke and soot from the burning fuel to form clouds, as you can see in Figure 19/11.

Many of the workers move out of the cities and buy houses in the suburbs. The houses have lawns and trees around them. The homeowners rake the leaves and burn them. They cook meals out of doors on their lawn barbecues. MORE SMOKE AND SOOT!

And then there are the pesticides. Everyone, from the farmer in the country to the homeowner with a lawn or garden to the camper using a can of insecticide sprays

Figure 19/10 *Making a temperature inversion. The sand bags are cold and the water bags are warm.*

Figure 19/11
The water vapor given off by jet planes condenses on particles of smoke and soot to form clouds.

Figure 19/12 *Air pollution from factories in Venice, Italy has speeded up the weathering of statues and buildings.*

If there is not enough time for the whole class to do all the activities in Sections 19.5 and 19.6, you can divide them up. You may want to start the nylon and cloth activities before reaching Chapter 19 if you cover this chapter at the end of the school year.

poisons into the air. Not until recently did scientists begin research to find out which pesticides are safe and which are harmful to both people and animals.

No area in the world is completely free of air pollution. You can check the air in your own area with a few simple experiments.

Materials
piece of nylon stocking, 35 mm slide frame, transparent tape (sticky on both sides), glass slide, square of cotton cloth, gallon can with plastic liner or gallon plastic jug, slide projector, microscope or magnifying glass

Activity Time: 10 minutes. Save a piece of nylon for the comparison. The exposed piece will have broken threads resulting from acid-producing chemicals such as sulfur dioxide (SO_2).
Process: observing

Activity Time: 5 minutes. A dissecting scope will do—up to 150×.

Activity Time: 10 minutes. If possible, have some students take containers home. Otherwise this can be a class project.

Activity Time: 5 minutes. Save a new piece for comparison. The color in the cloth will fade, and the fibers may have been weakened or broken by acids in the air.
Activity Time: 10 minutes.

Acid-producing chemicals in the air rot the nylon threads. The tape catches and holds various particles ranging from soot to pollen. The gallon can catches dust and soot that settle from the air. The melted snow contains particles that snowflakes sweep from the air as the flakes fall to the ground, or that fall on it from the air afterward.

★ Nylon is a very strong material that does not weather easily. A thin nylon rope can even pull a truck out of a ditch. Stretch a piece of nylon stocking across a 35 millimeter slide frame. Staple the edges. Take the slide frame home and hang it outside in a place where the air can get at it, but where it is protected from the sun and rain. Leave it there for at least a month. Then bring the slide frame back to school and place it in a slide projector. How does the image projected on the screen compare with the image of a new piece of nylon? Figure 19/12 shows what air pollution can do to stone.

Stick a piece of the transparent tape that is sticky on both sides to a clean glass slide. Put the slide in a sheltered place as you did with the nylon. Bring it in after a week. Look at it closely. Use a low-power microscope if you have one.

Line a gallon can with a plastic bag, or cut off the top of a plastic gallon jug. Fill the can or jug one-quarter full of water. Place it on the roof of your school building for a week. Then examine the water. Use a microscope again. If possible, place cans or jugs in other parts of your city or town, too.

Put a small square of colored cotton cloth in a sheltered place as you did the tape and nylon. Examine it after a month of exposure.

If your area has snow now, collect samples of snow. Take some from the surface and some from below the surface. Let the samples melt. Examine the water that results.

You and other people in your class will have to do each of the experiments to see how many harmful substances there are in the air near your school or home. What can you tell from all the samples? ★

19.6 We pollute the water.

You remember from Section 19.4 that we throw out enough trash each day to equal six pounds per person. What happens to the trash that isn't burned? It may end up like the trash in Figure 19/13. It may go to a dump. Does that solve the problem? No! Some of the trash doesn't weather and will stay around almost forever. The trash that does weather produces chemicals containing barium, cadmium,

Figure 19/13
What can you do with something that smells and smokes when it burns, and won't weather?

chromium, copper, manganese, nickel, selenium, silver, and zinc. Most of these chemicals can be harmful to people, animals, and plants. Others haven't even been tested yet. The chemicals can leach down through the soil and be carried by ground water into rivers, lakes, and the oceans. There they join insecticides from lawns and fields. You take it from there!

Paper mills and plastic factories dump compounds of mercury and arsenic into rivers, streams, and lakes. Some of the poisoned water may become part of the water supply of nearby cities and towns. Eventually, the poisons reach the sea. The mercury becomes part of the fish that are caught, canned, and eaten around the world.

According to the Boston Globe for May 2, 1971, a partial list of spills of pollutants into U.S. waters as of April 1971 totaled 371 in 1966, 458 in 1967, 714 in 1968, and 1188 in 1969. 90 percent of the spills involved oil, and the spills amounted to almost 100 million gallons.

Offshore oil wells can leak, and tankers sometimes spill oil. Figure 19/14 shows what can happen to the wildlife near a spill. The oil forms a film on the surface of the water. The film keeps sunlight from reaching tiny plants that put new oxygen into the air. The plants die, and fish that live

Figure 19/14
Birds caught in an oil spill will die unless they are quickly and carefully cleaned off. Even so, a small number are saved in time.

Figure 19/15 *"I smell oil. Is it the crankcase or the sea?"*

on them have less and less to eat. The result will soon be fewer fish, too. According to some scientists, an oil film only one molecule thick cuts the number of microscopic food plants in half.

Iron salts and hydroxides are formed when iron ore weathers, for example. These are poisonous to fish and plants. Sulfur compounds produce sulfuric acid. Strip mining also releases copper salts.

When mining operations strip the land of plants and topsoil, oxygen and water combine with minerals to make poisons. Surface runoff from these diggings, like the water

Figure 19/16
Strip mining exposes soft coal and shale containing sulfur to the air. The sulfur oxidizes and combines with rain water to form acids. The acids are then washed into streams.

in Figure 19/16, carries the metal compounds and large amounts of sediment to streams and lakes. Fish and plants die out. Sources of drinking water become polluted.

Materials added to laundry detergents, such as phosphates, nitrates, and a few mystery ingredients, leach through the soil or are dumped directly into streams and lakes. The chemicals in detergents, along with a lot of other things, serve as an extra food source, so that a great growth of algae takes place. Soon there are many more algae than the normal amount. They can cut out the light that other plants need, and they crowd out other forms of life. Eventually they die. As they rot, they use up oxygen needed by the fish. In some places, even fish like the perch in Figure 19/17 that don't need much oxygen die by the thousands.

Activity Time: 10 minutes to prepare, overnight to dry. Window glass will work. Don't use cleaners to polish it up. Other water samples may be added. Check the area near your school The chemistry lab can supply the distilled water, or you can distill it in class as an added activity. Be sure to explain that only water goes into the condenser—just as "only water goes into the air when it evaporates." Rainwater should not contain dissolved solids. If there are some, they came from the air on the way down. Water from a dehumidifier can be used in place of rainwater, but rainwater is best. There is no simple set of tests to determine the nature of the residue. Tell the class that public action is needed to insure that water is checked frequently.

Materials
glass, pure and impure water

★ No area of the United States is completely free of water pollution. You can check the water in your area with a simple experiment. Put a drop each of rainwater, stream or pond water, drinking water, and distilled water on a clean piece of glass. Label each drop with a marker. Let each of the samples evaporate in the sunlight. Examine the spot where each drop of water used to be. How much of the stuff left on the slide is harmful to people and animals? Which water sample left the most stuff? Should rainwater

Figure 19/17
If water becomes overgrown with algae, it does not contain enough oxygen for fish. Fish may also be killed by chemical wastes from factories.

leave anything at all? Write a class letter to your local Health Department asking if they have checked out the water in your area. ★

Compare the streams in your area with the ones in Figure 19/18. Which of the pictured streams do your streams look like? Are the ones in your area polluted? Stop and think! Ninety-one United States beaches were closed to swimming in 1971 because water flowing into them was polluted. Thirteen more were labeled NOT RECOMMENDED and another ten were labeled UNSATISFACTORY FOR USE. You would be taking a chance to swim in them.

19.7 We are finding new ways to pollute.

There are things happening in our environment that some people do not call pollution. But these conditions do upset equilibriums just like water and air pollution.

Figure 19/18 *Are the streams in your area clear and clean or murky and polluted?*

Students might like to set up an experiment to study the effects of loud music on growing plants. Be sure to include control plants in the experiments.

People living near airports and in cities often complain about noise pollution. Hearing tests taken over many years show that the young people of today hear less sound than teenagers did 30 years ago. Some sound experts say that if noise keeps rising at present rates, everyone living in a city could be stone deaf by the year 2000.

Visual pollution refers to scenes like the one in Figure 19/19. Many people find this kind of sight annoying and unnecessary. How's the scenery near your home?

All types of power plants produce thermal pollution but nuclear plants are the worst offenders and, therefore, have created the new concern about the environment.

People are finding a new type of pollution called **thermal** (THUR-mul) pollution near places where nuclear reactors and some kinds of factories are located. The word "thermal" refers to heat. The nuclear reactors that generate electrical power for our big cities don't pollute the air the way coal-burning generators do. But they make a lot of heat, and large amounts of water must be used to cool them down.

Figure 19/19 *Ugly signs and billboards pollute the landscape in many communities.*

The water for cooling is piped in from nearby lakes and rivers. When a nuclear power plant returns the water to the lakes and rivers, the cooling water is warmer than before. Plants that could not survive in cooler water begin to grow. The older forms of life in the lake or river may be crowded out or die out because of the heated water. In the mouths of some rivers, the young of important food fishes may be destroyed. The old equilibrium is upset.

Question for class discussion: There are plans for 15 nuclear reactors around Lake Ontario. What could happen?

Wouldn't it be better to use the hot water to heat homes, factories, or greenhouses?

Did You Get the Point?

The population has increased because more children live to become adults and adults live longer.

An increase in population has upset the equilibrium between the rate at which human waste builds up and the rate at which the environment can clean it up.

Air, water, noise, visual, and thermal pollution are the most common types.

Answers:

1. What happens to sewage from housing developments? supermarket garbage? old oil from gas stations? by-products and waste materials from factories?

2. Call or write any factory. Papermaking plants, dye or chemical plants, and power plants are good possibilities. Be on the lookout for equivocation.

3. These letters may be based on the answers to the preceding questions.

Check Yourself

1. Tour your community. In your notebook write down the locations of pollution sources and identify the types of pollution you believe are taking place. Compare your notes with those of your classmates. Discuss the results with your teacher. What can you and your class do about the pollution problem in your community?

2. Call or write local industries, asking them what antipollution steps they have taken. Prepare a class report on what you find.

3. In a letter to your congressman, outline one of the pollution problems that affects your community. Suggest a law that would help solve it. Ask your teacher to mail your letter and those of your classmates.

Pollution and the Future

19.8 What will happen to the earth?

The old equilibrium was: custodial time $\rightleftharpoons$ student messing up.

When you asked the custodian to "forget" to clean your classroom at the end of Section 19.4, you probably imagined what would happen. But did some things happen that you did not plan on? Did the mess bother other people? Did the mess encourage other people to add to the mess? Did the mess make new problems? Did the mess spread to other areas of the building? Could you see mixed-upness taking over? What equilibrium was upset? What would happen if the custodian "forgot" to clean your classroom for a few more weeks?

With the cooperation of the school custodian you have lived in the middle of pollution. You have seen unexpected things happen to your environment as the pollution got worse. Pollution has also caused unexpected things to happen to people in the past.

The people living in Caesar's Rome may have poisoned themselves without knowing it. They added lead salts to their cosmetics because they believed the lead made the cosmetics better. They lined their cooking pots and wine jars with lead because the lead seemed to make things taste better. They lined their rainwater collectors with lead because it was easy to bend and shape. Slowly, Roman leaders, who were rich enough to afford the expensive lead, died of lead poisoning. Some scientists say that lead poisoning may have helped to cause the fall of Rome. If the Romans had known about lead pollution and had done something about it, history might have been different.

Today, air pollution may be starting to change our environment, too. Though few scientists are sure yet, most of them feel that one of two things will happen to the earth's climate if people do not recognize and solve air pollution problems.

Part of the smoke that we are adding to the air is carbon dioxide. Scientists are afraid that adding more and more CO_2 to the air may upset the energy balance by preventing the earth from radiating back into space as much energy as it takes in. If this happens, the temperature of the earth will rise, the ice caps will melt, and coastal areas will be flooded. At last, permanent, thick clouds will completely surround the earth, like the clouds the starship captain wrote about. The cloud cover might raise the surface temperatures to 800 or 900 degrees Fahrenheit, like the temperatures of the cloud-covered planet Venus.

See Teacher's Manual for an activity that illustrates this point.

The soot we put in the air could have an opposite effect. Adding more particles to the air could keep the sun out. If the earth loses more heat by radiation than it receives in the sunlight, surface temperatures will go down. The polar ice caps will grow. If it got cold enough, the oceans might even freeze. The earth could someday become like Mars. The temperatures at the equator on Mars are as warm as 60° Fahrenheit (15° Celsius) but at its south pole, the temperature is −240° Fahrenheit (−150° Celsius).

The earth's temperature increased 0.36° Fahrenheit between 1900 and 1940, but it has been decreasing since then. The earth could become warmer, or it could become colder. If conditions for both warming and cooling are working at the same time, perhaps a new equilibrium will form. Scientists just don't know for sure what will happen. But it *is* certain that neither more carbon dioxide nor more man-made particles in the air will be good for man. If both of these increase at the same time, the results could be very bad.

19.9 Laws have already helped.

It seems as if every day we read or hear about some new way that people are adding to the disorder, changing equilibriums, or killing off animal and plant types. It is not difficult to get the feeling that all is lost.

Fortunately, there is another side to the picture. In a small way, positive things are beginning to happen. Here are some examples.

London, England. In December of 1952, 4000 people died from smog that covered London. The British government then passed laws to stop all use of high-polluting coal. Since then, there has been no pea-soup fog for six years. In a recent year London had 50 percent more sunshine than ten years ago. The smog clean-up must be partly responsible. Birds are coming back to the parks, and in 1969, fish were caught again in the Thames River.

Washington, D.C. The United States government has begun to do something about pollution. In 1970 it set up an Environmental Protection Agency to unite different federal agencies in the fight against air and water pollution and other problems of the environment. The Clean Air Act of 1970 sets up controls for automobile and industrial pollution. All new cars must have antipollution devices. The idea is to cut down the amount of "stuff" that comes out of car exhausts.

Washington, D.C. The 1970 Federal Water Pollution Control Act sets fines for oil companies if they spill oil into the seas. Other industries must get government permits to dump waste into streams. This can keep lakes and streams from becoming even more polluted than they already are.

Salem, Oregon. In 1971 the state of Oregon passed a law banning nonreturnable bottles.

Albion, New York. The city of Albion plans to build a new school that has an electrical heating system. The electrical power will come from the falling water of Niagara Falls. The city chose this type of heating because hydroelectric power does not cause air or thermal pollution.

Figure 19/20 *A modern sewage disposal plant that will keep nearby Lake Tahoe from becoming polluted.*

Denver, Colorado. The city of Denver passed a strong anti-billboard law.

Carson City, Nevada. Local communities have recently completed a waste treatment plant at Lake Tahoe, on the Nevada-California border. Figure 19/20 is a photograph of it. This plant is considered to be an outstanding example of modern pollution control. It was designed with the help of the federal government to prevent pollution of the lake, whose beauty has attracted a great number of tourists and settlers.

Alamo, California. A retired engineer has developed a method for turning used paper into bricks. He claims the bricks are as good as those made of clay and can be produced at half the cost.

Palo Alto, California. The Stanford Research Institute has suggested that solid wastes can be sent through pipes from cities and towns over relatively large areas to central treatment plants or land fill areas. The solid wastes could be shredded and carried down the pipes as slurries in water. Or, garbage could be squeezed into compact wads and carried by water through the pipes.

Pittsburgh, Pennsylvania. Two young college instructors paddled up and down the Monongahela River taking samples of the polluted water. Using this evidence, the Federal government charged four large companies with polluting the river.

Industries are also beginning to help clean up. They are printing reminders on film and food wrappers asking people not to litter. Some container manufacturers are testing replacements for plastic and glass bottles and aluminum cans. They hope to develop a container that will weather quickly or dissolve into materials that will not poison the environment. In the meantime, many can manufacturers are paying for the return of aluminum cans. Bottle plants are breaking up and reusing old glass. Construction companies are using metal junk in their building materials. Paper companies are reusing old newspapers and other waste paper.

Scientists and inventors are trying to develop cleaner engines for cars and trains. Someday we may all be driving cars that are run by electricity or steam. There is a new kind of paper that will dissolve in water and does not need to be burned. One inventor claims to have a glass bottle that will "melt" in sunlight and disappear.

Farmers are trying several new ways to control pests. Instead of using pesticides, they add natural enemies like beetles and the praying mantis to the environment. They also use sterile forms of the pests themselves. When the pests already living in the area mate with the sterile ones, no new organisms are produced.

Schools are now teaching about the pollution problem. You are probably more careful than your parents were at your age about throwing things away. Perhaps you have already been on a clean-up campaign like the students in Figure 19/21. And many people are also realizing that adding to the population will make the pollution problems worse.

Figure 19/21 *Students involved in a local clean-up campaign. Have you had one in your community?*

Some scientists still think that the pollution problem will only become worse and worse. You might call it "people pollution." People are the ones making the noise, hanging the signs, burning the garbage, dirtying the waters, building the jets, and using the electric power. But many other scientists say that the pollution problem *can* be solved. You can help settle the debate. Do you know how? Help the lawmakers by requesting laws where they are needed. Help the lawmakers by obeying the laws. Help the government to enforce the laws. The laws can be made to work. Maybe it's not too late.

How You Can Help!

Park the car
take a bus
ride a bike
Report a polluter
walk to school
recycle glass
and newspapers
Turn off a light
and tin cans
and the faucet
don't buy more
Ask for answers
use it again
ask for action
plan a meeting
Pick up a candy wrapper
make a speech
clean up a park
starve a rat
Read a book
don't give up
learn the facts
ask when and why
Talk to friends
where and how
write the editor
complain to the company
Know your rights
insist on it
change a law
vote for sewers
Think of your grandchildren
save the birds
and me
get the lead out
Take a picture
get mad
show it around
one good example
Smell a flower
and one bad
smell the air
plant a tree
Read
plant a thought
this
to
a friend

Did You Get The Point?

Federal and local laws are beginning to discourage people from polluting their environment.

Scientists and inventors are working on devices that will cut down the amount of pollution being produced.

You must do your part to help save our earth.

Answers:
1. There may not be any. (Opportunity for the class to propose some to the local government.)
2. This question should start students thinking about what they can do.
3. Answers will vary.
4. School assembly—guest speakers, slides showing local pollution problems. Poster campaigns. March to pick up trash, then weigh trash collected. Campaign to have companies take back used glass, paper, aluminum, etc., for recycling.

Check Yourself

1. List the laws your community uses to reduce pollution.

2. List the things you did to reduce pollution in the past week.

3. List the things you have seen happen to pollute the environment in the past week. Suggest some means to get rid of each pollution threat.

4. Take part in organizing a class antipollution project. Your teacher will give you some suggestions.

What's Next?

What's next? That's a good question. What might future visitors to our planet find? The answers are up to us.

Skullduggery

Directions: The questions that follow are ones you might be asked if you were considered an expert in the area of pollution. Test your awareness of the pollution problems. After your teacher has corrected your answers, find your score on the table below. Are you enough aware of the pollution problem to help do something about it?

12 correct. WOW! The world needs you.
10 correct. You can help.
8 correct. You can help others in the solution.
6 correct. Listen to those who know.
Less than 6 correct. You are one of the environment's problems.

Answers are in Teacher's manual.

1. What did the starship commander and his crew uncover when they blasted the stone slab on the mystery planet?

2. What were the fossil animals discovered by the starship crew?

3. Why is dynamic equilibrium often described as a change that is not a change?

4. What can upset an equilibrium?

5. Name three environmental equilibriums being upset in your lifetime.

6. Name as many ways as you can of causing air pollution.

7. Name as many ways as you can of stopping or reducing air pollution.

8. Name as many ways as you can of causing water pollution.

9. Name as many ways as you can of stopping or reducing water pollution.

10. Name three types of pollution that do not involve particles or poisons in the air or water.

11. What type of pollution seems to cause most other types of pollution?

12. Describe two ways in which the climate of the earth could change if our pollution problems are not solved.

For Further Reading

If possible, get a class subscription to *Audobon Magazine* published by the National Audobon Society. A student membership for one year costs $6.00 and entitles you to this informative and beautiful magazine as well as information on the activities of the Society. *Environment* is another excellent magazine.

Joffe, Joyce. *Conservation—Maintaining the Natural Balance.* Garden City, New York, Natural History Press, 1970. A book about the ecology of the earth and how man is affecting the land and wildlife. Includes a good discussion of overpopulation, too.

Herbert, Fred W. *Careers in Natural Resource Conservation.* New York, Henry Z. Walck, 1965. If you are interested in a career in conservation, this book describes what a conservationist does and the background you need to become one.

Harrison, C. William. *Conservation—The Challenge of Reclaiming our Plundered Land.* New York, Julian Messner, 1963. How our land has been damaged by man and what he is doing to restore it again. The black-and-white photos show what has actually happened.

Appendix A

Metric and English Measurements

How to Remember the Metric System

distance	
one millimeter	as wide as the wire in a paper clip
one centimeter	as wide as your index fingernail
one meter	as long as 5 of these books lying side by side
one kilometer	as long as 11 football fields end to end
volume	
one milliliter	the amount of space enclosed by the metal around a pencil eraser
one liter	the amount of space in a quart milk carton
mass	
one gram	the amount needed to balance a standard paper clip on a scale

Metric and English Systems Compared

one inch	=	25.4 millimeters
one inch	=	2.54 centimeters
one foot	=	30.5 centimeters
39.4 inches	=	one meter
3280 feet	=	one kilometer
one quart	=	943 milliliters
1.06 quarts	=	one liter
one quart	=	0.948 liters
one pound	=	454 grams

What Is the Metric System?

10 millimeters	=	one centimeter
100 centimeters	=	one meter
1000 meters	=	one kilometer
1000 grams	=	one kilogram
1000 milliliters	=	one liter

Appendix B

Nearest, Brightest, and Other Stars

Nearest stars	Visual magnitude	Distance (light-years)	Temperature (degrees Kelvin)	Luminosity (Sun = 1)
Sun	−26.7	0.00002	5,800	1.00
Alpha Centauri A*	−0.01	4.3	5,800	1.5
Alpha Centauri B*	+1.4	4.3	4,200	0.33
Alpha Centauri C*	+11.0	4.3	2,800	0.0001
Barnard's Star	+9.54	6.0	2,800	0.00045
Wolf 359*	+13.66	7.7	2,700	0.00003
Lalande 21185*	+7.47	8.1	3,200	0.0055
Sirius A*	−1.43	8.7	10,400	23.0
Sirius B	+8.5	8.7	10,700	0.0024
Luyten 726-8 A*	+12.5	8.7	2,700	0.00006
Luyten 726-8 B*	+12.9	8.7	2,700	0.00002
Ross 154*	+10.6	9.6	2,800	0.00041
Ross 248*	+12.24	10.3	2,700	0.00011
Epsilon Eridani*	+3.73	10.8	4,500	0.30
Ross 128*	+11.13	11.0	2,800	0.00054
Luyten 789-6	+12.58	11.0	2,700	0.00009
61 Cygni A*	+5.19	11.1	4,200	0.084
61 Cygni B*	+6.02	11.1	3,900	0.039
Procyon A*	+0.38	11.3	6,500	7.3
Procyon B*	+10.7	11.3	7,400	0.00055
Epsilon Indi*	+4.73	11.4	4,200	0.14

* Plot the data for the stars with asterisks. Use the other stars for a more complete picture.

Brightest stars	Visual magnitude	Distance (light-years)	Temperature (degrees Kelvin)	Luminosity (Sun = 1)
Sirius A	−1.43	8.7	10,400	23.0
Canopus*	−0.72	100.0	7,400	1,500.0
Alpha Centauri A	−0.01	4.3	5,800	1.5
Arcturus*	−0.06	36.0	4,500	110.0
Vega*	+0.04	26.0	10,700	55.0
Capella	+0.05	47.0	5,900	170.0
Rigel	+0.14	800.0	11,800	40,000.0
Procyon A	+0.38	11.3	6,500	7.3
Betelgeuse*	+0.41	500.0	3,200	17,000.0
Achernar*	+0.51	65.0	14,000	200.0
Beta Centauri*	+0.63	300.0	21,000	5,000.0
Altair*	+0.77	16.5	8,000	11.0
Alpha Crucis	+1.39	400.0	21,000	4,000.0
Aldebaran*	+0.86	53.0	4,200	100.0
Spica*	+0.91	260.0	21,000	2,800.0
Antares*	+0.92	400.0	3,400	5,000.0
Fomalhaut	+1.19	23.0	9,500	14.0
Deneb*	+1.26	1,400.0	9,900	60,000.0
Beta Crucis	+1.28	500.0	22,000	6,000.0

Other stars	Visual magnitude	Distance (light-years)	Temperature (degrees Kelvin)	Luminosity (Sun = 1)
Epsilon Andromedae	+4.37	105	5,600	1.3
Delta Aquarii*	+3.28	84	9,400	24.0
Beta Aquarii	+2.86	1030	6,000	4,300.0
Beta Auriagae	+2.8	88	11,200	69.0
Beta Cassiopeiae	+2.26	45	6,700	8.2
Grw +70° 8247	+13.19	49	9,800	0.0013
02 Eridani B*	+9.5	16	11,000	0.0028
L 879-14*	+14.10	63?	6,300	0.00068
70 Ophiuchi A*	+4.3	17	5,100	0.6
Zeta Ophiuchi	+2.56	465	26,000	3,300.0
Delta Persei*	+3.03	590	17,000	1,300.0
Zeta Persei A*	+2.83	465	24,000	16,000.0
Tau Scorpii*	+2.82	233	25,000	2,500.0
Van Maanen's Star*	+12.36	14	7,500	0.00016
W 219	+15.20	46	7,400	0.00021

Appendix C

Sky Maps

The sky maps on the following pages show the positions of the main stars, constellations, and the Milky Way in the sky throughout the year. There are four maps for each season. Two on one page show the sky above the horizon as you face north and the two on the next page are for the sky as you face south.

The top map on each page names the main stars and traces out the constellations. The bottom map on each page will help you to recognize these stars by their different magnitudes, or brightnesses. Stars with circles around them are variable stars. The Milky Way is the irregular gray strip across each map.

West and east are marked on each map by "W" and "E". The curved line at the top of each map represents an imaginary arc of the sky curving over your head from west to east. Due north is at the point halfway between east and west on the north horizon maps. South is at the middle of the east-west line on the south horizon maps.

The top maps on each page also show the line of the celestial equator across the sky with the hours of right ascension marked on it reading from west to east. The line of the ecliptic appears, too. You can locate the different visible planets along the ecliptic through the year. Note the constellations named along the ecliptic on some of these maps. These are the constellations of the zodiac, like Pisces on map 1.

From season to season, see how the stars and constellations seem to march across the sky toward the west from map to map. This is the result of the earth's moving around the sun. So the stars rise in the east about 4 minutes earlier each day, or 1 hour earlier in 15 days and 2 hours earlier each month. Then they appear farther toward the west, at the same time of night, month by month.

If you use these sky maps, or others, outdoors to help you find stars and constellations, take a flashlight to look at the maps. But use a red flashlight bulb, or cover the light with red plastic over the glass taped around the neck of the flashlight. The red light leaves your eyes adapted to the darkness. Then after looking at the maps you can turn right back to the stars without waiting for a minute.

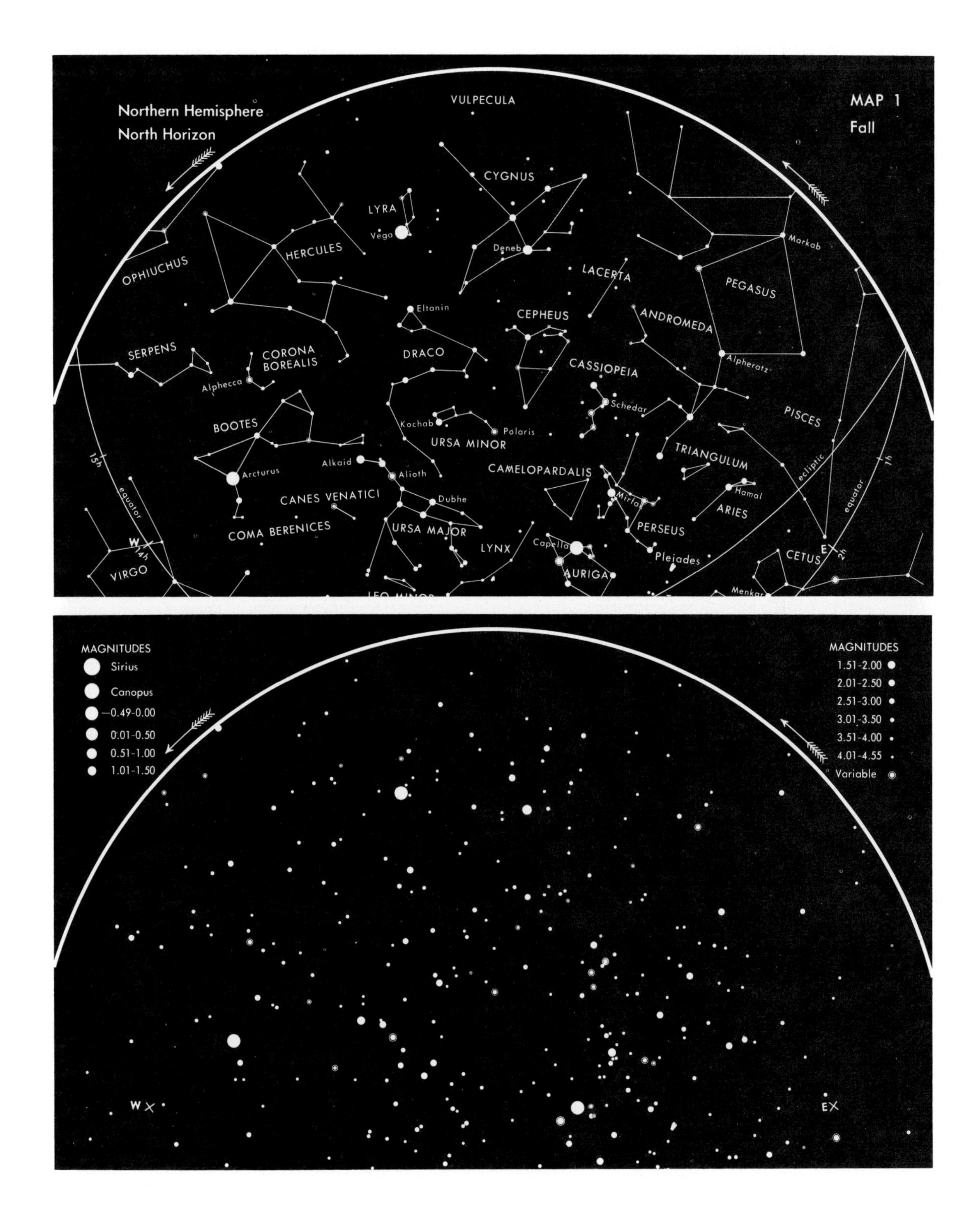

Northern Hemisphere
North Horizon
MAP 1
Fall
VULPECULA
CYGNUS
LYRA
Vega
Deneb
HERCULES
OPHIUCHUS
LACERTA
Markab
PEGASUS
Eltanin
CEPHEUS
ANDROMEDA
SERPENS
CORONA BOREALIS
DRACO
Alphecca
CASSIOPEIA
Alpheratz
Schedar
PISCES
BOOTES
Kochab
Polaris
URSA MINOR
15h
TRIANGULUM
Alkaid
Arcturus
Alioth
CAMELOPARDALIS
ecliptic
1h
equator
CANES VENATICI
Dubhe
Mirfak
Hamal
ARIES
COMA BERENICES
URSA MAJOR
PERSEUS
W
14h
LYNX
Capella
Pleiades
CETUS
E
2h
VIRGO
AURIGA
Menkar
MAGNITUDES
Sirius
Canopus
−0.49-0.00
0.01-0.50
0.51-1.00
1.01-1.50
MAGNITUDES
1.51-2.00
2.01-2.50
2.51-3.00
3.01-3.50
3.51-4.00
4.01-4.55
Variable
W
E

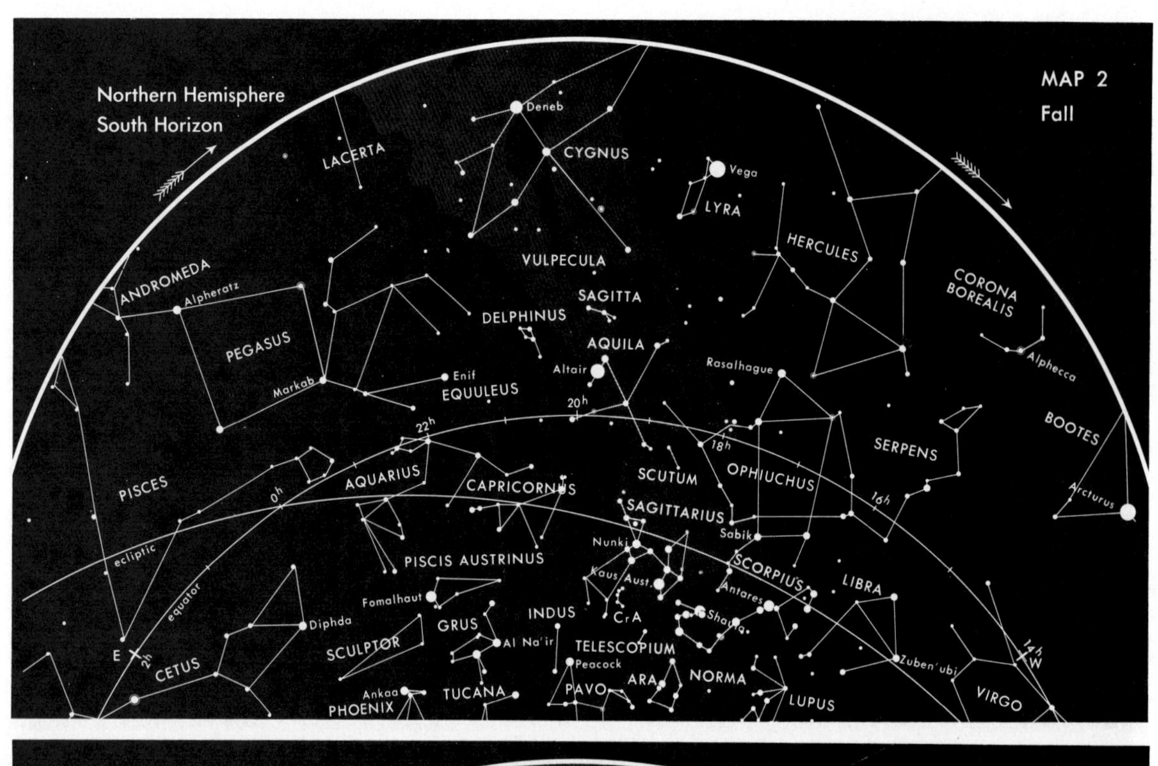
Northern Hemisphere
South Horizon
MAP 2
Fall
LACERTA
Deneb
CYGNUS
Vega
LYRA
HERCULES
CORONA BOREALIS
Alphecca
ANDROMEDA
Alpheratz
PEGASUS
Markab
VULPECULA
SAGITTA
DELPHINUS
AQUILA
Altair
Enif
EQUULEUS
Rasalhague
BOOTES
SERPENS
Arcturus
PISCES
AQUARIUS
CAPRICORNUS
SCUTUM
OPHIUCHUS
SAGITTARIUS
Sabik
Nunki
Kaus Aust.
SCORPIUS
Antares
Shaula
LIBRA
PISCIS AUSTRINUS
Fomalhaut
GRUS
INDUS
CrA
ecliptic
equator
Diphda
SCULPTOR
Al Na'ir
TELESCOPIUM
Peacock
CETUS
ARA
NORMA
Zuben'ubi
VIRGO
Ankaa
PHOENIX
TUCANA
PAVO
LUPUS
E
W

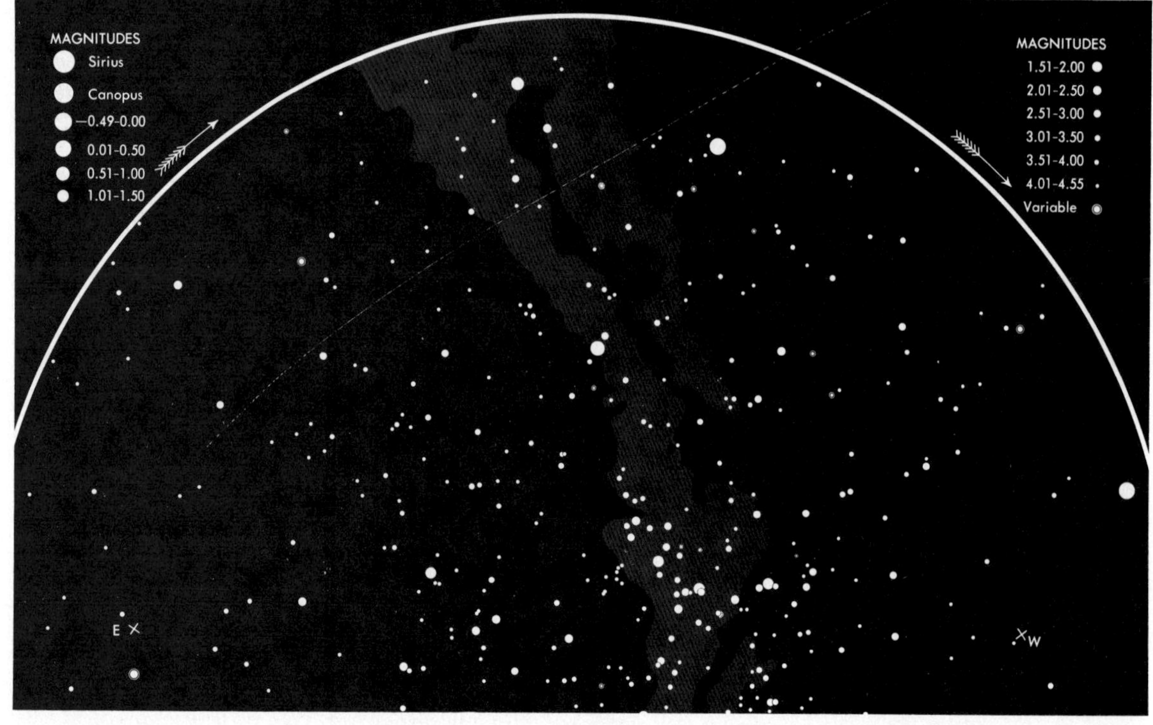
MAGNITUDES
Sirius
Canopus
−0.49-0.00
0.01-0.50
0.51-1.00
1.01-1.50
MAGNITUDES
1.51-2.00
2.01-2.50
2.51-3.00
3.01-3.50
3.51-4.00
4.01-4.55
Variable
E
W

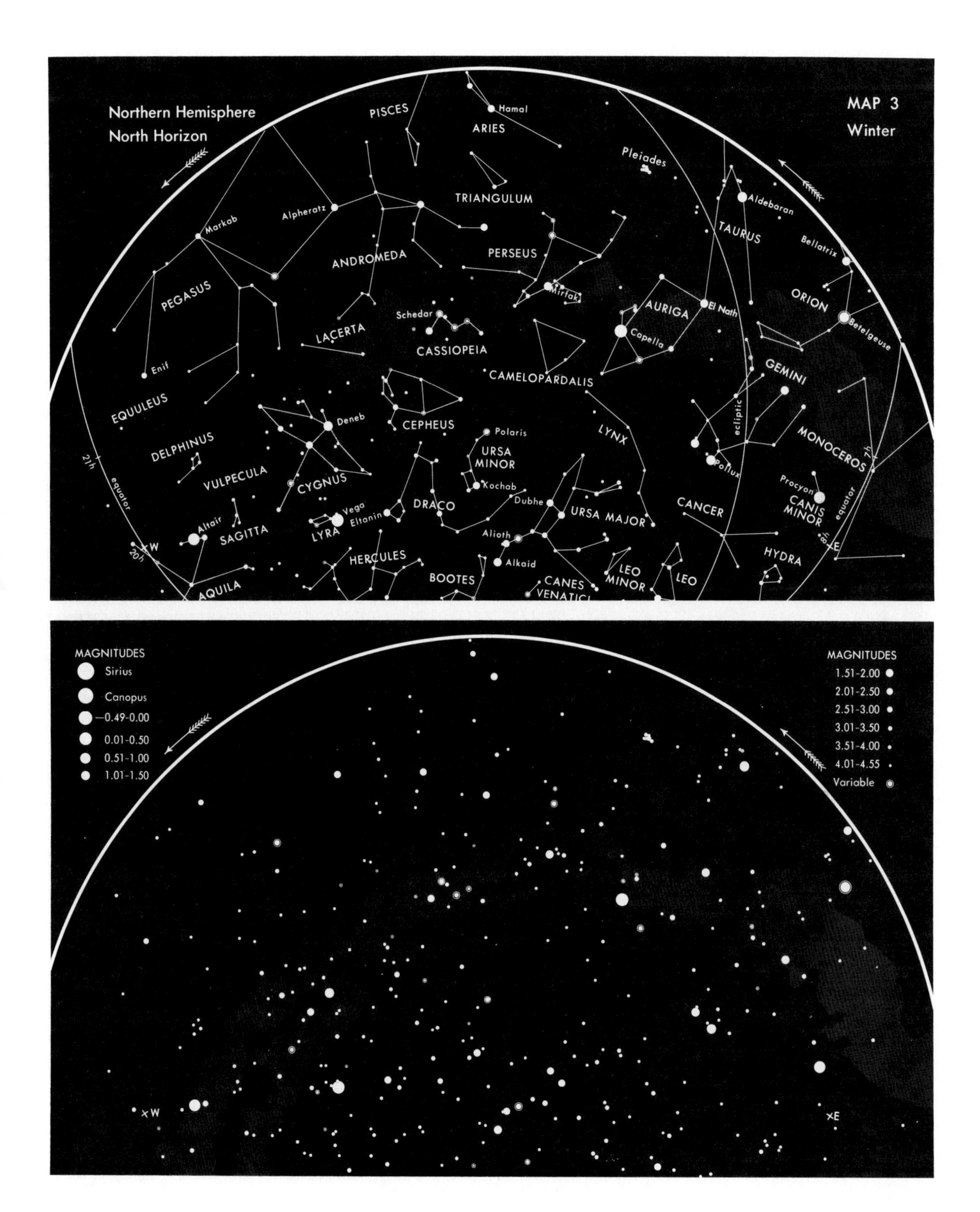
Northern Hemisphere
North Horizon
MAP 3
Winter
PISCES
Hamal
ARIES
Pleiades
TRIANGULUM
Aldebaran
TAURUS
Bellatrix
Alpheratz
Markab
ANDROMEDA
PERSEUS
PEGASUS
Mirfak
ORION
AURIGA
El Nath
Schedar
LACERTA
CASSIOPEIA
Capella
Betelgeuse
Enif
CAMELOPARDALIS
GEMINI
EQUULEUS
Deneb
CEPHEUS
Polaris
LYNX
ecliptic
DELPHINUS
URSA
MINOR
MONOCEROS
2h
Pollux
VULPECULA
CYGNUS
Kochab
equator
Procyon
CANIS
MINOR
Dubhe
DRACO
Vega
Eltanin
CANCER
Altair
SAGITTA
URSA MAJOR
LYRA
Alioth
HYDRA
W
20h
HERCULES
Alkaid
LEO
MINOR
LEO
BOOTES
CANES
VENATICI
AQUILA
MAGNITUDES
Sirius
Canopus
−0.49-0.00
0.01-0.50
0.51-1.00
1.01-1.50
MAGNITUDES
1.51-2.00
2.01-2.50
2.51-3.00
3.01-3.50
3.51-4.00
4.01-4.55
Variable
W
E

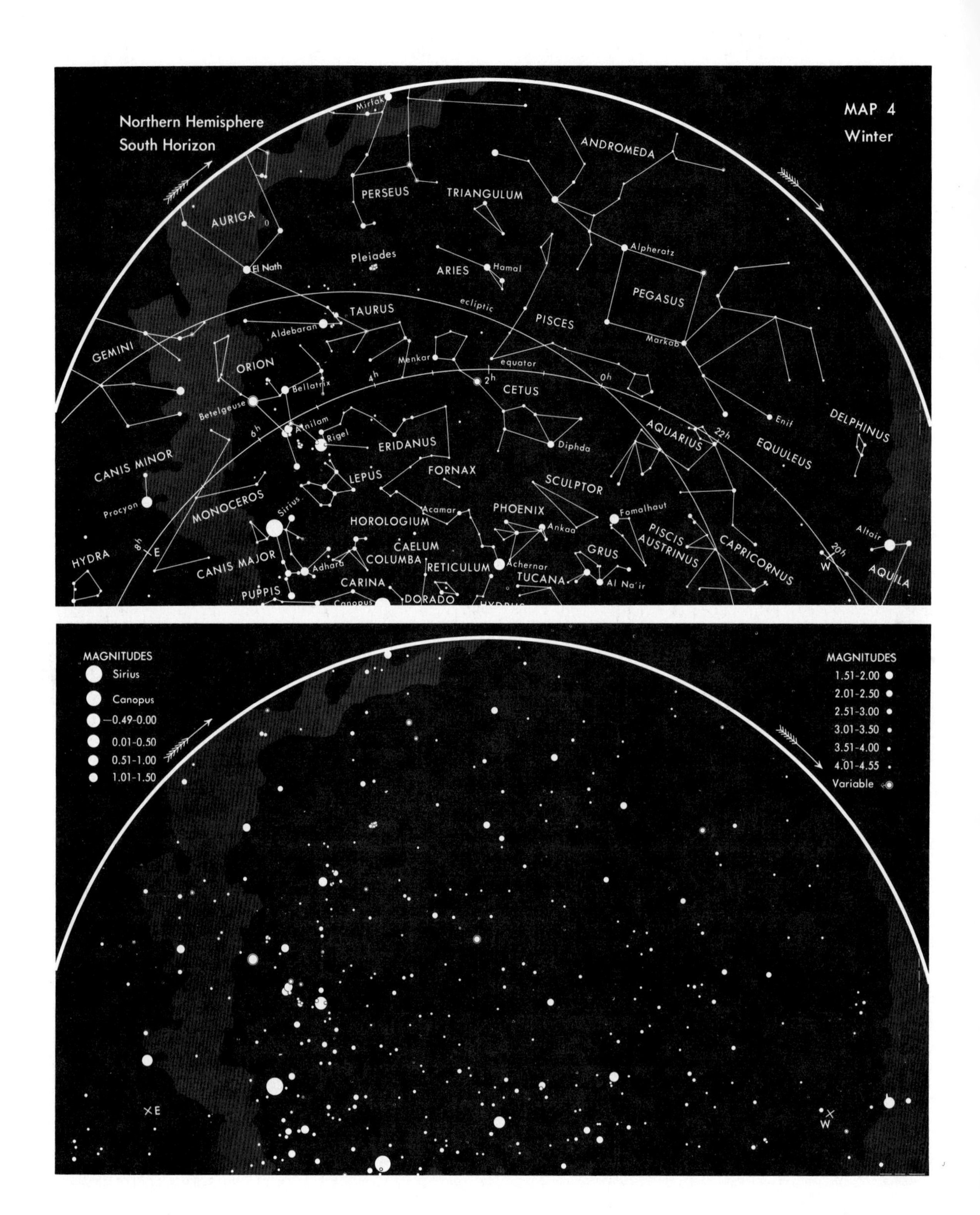
Northern Hemisphere
South Horizon
MAP 4
Winter
ANDROMEDA
PERSEUS
TRIANGULUM
AURIGA
Mirfak
El Nath
Pleiades
ARIES
Hamal
Alpheratz
PEGASUS
Markab
PISCES
ecliptic
TAURUS
Aldebaran
GEMINI
ORION
Menkar
equator
Bellatrix
Betelgeuse
CETUS
Enif
DELPHINUS
AQUARIUS
EQUULEUS
Alnilam
Rigel
ERIDANUS
Diphda
CANIS MINOR
LEPUS
FORNAX
SCULPTOR
Procyon
MONOCEROS
Acamar
PHOENIX
Fomalhaut
HOROLOGIUM
Ankaa
PISCIS AUSTRINUS
CAPRICORNUS
Altair
Sirius
HYDRA
CAELUM
GRUS
AQUILA
CANIS MAJOR
COLUMBA
RETICULUM
Achernar
Adhara
TUCANA
Al Na'ir
PUPPIS
CARINA
Canopus
DORADO
MAGNITUDES
Sirius
Canopus
−0.49-0.00
0.01-0.50
0.51-1.00
1.01-1.50
1.51-2.00
2.01-2.50
2.51-3.00
3.01-3.50
3.51-4.00
4.01-4.55
Variable

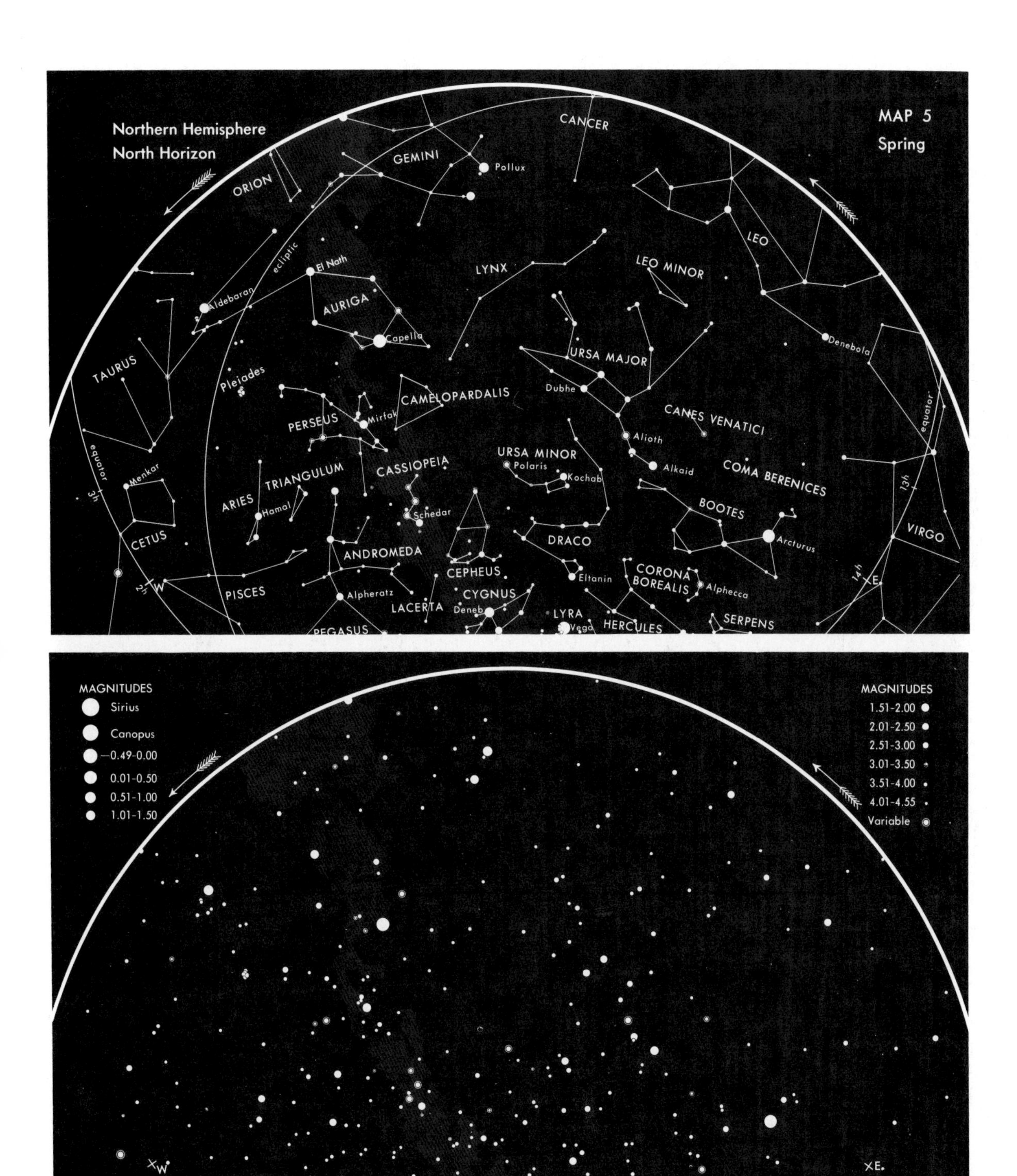

Northern Hemisphere
North Horizon
MAP 5
Spring
ORION
GEMINI
Pollux
CANCER
LEO
LEO MINOR
LYNX
ecliptic
El Nath
AURIGA
Aldebaran
Capella
TAURUS
Pleiades
URSA MAJOR
Dubhe
Denebola
CAMELOPARDALIS
PERSEUS
Mirfak
CANES VENATICI
equator
Alioth
URSA MINOR
Polaris
Kochab
Alkaid
COMA BERENICES
CASSIOPEIA
TRIANGULUM
ARIES
Menkar
Hamal
Schedar
BOOTES
13h
3h
CETUS
DRACO
Arcturus
VIRGO
ANDROMEDA
CEPHEUS
Eltanin
CORONA BOREALIS
Alphecca
2h
W
PISCES
Alpheratz
CYGNUS
LACERTA
Deneb
LYRA
Vega
HERCULES
SERPENS
PEGASUS
14h
E
MAGNITUDES
Sirius
Canopus
−0.49-0.00
0.01-0.50
0.51-1.00
1.01-1.50
MAGNITUDES
1.51-2.00
2.01-2.50
2.51-3.00
3.01-3.50
3.51-4.00
4.01-4.55
Variable
W
E

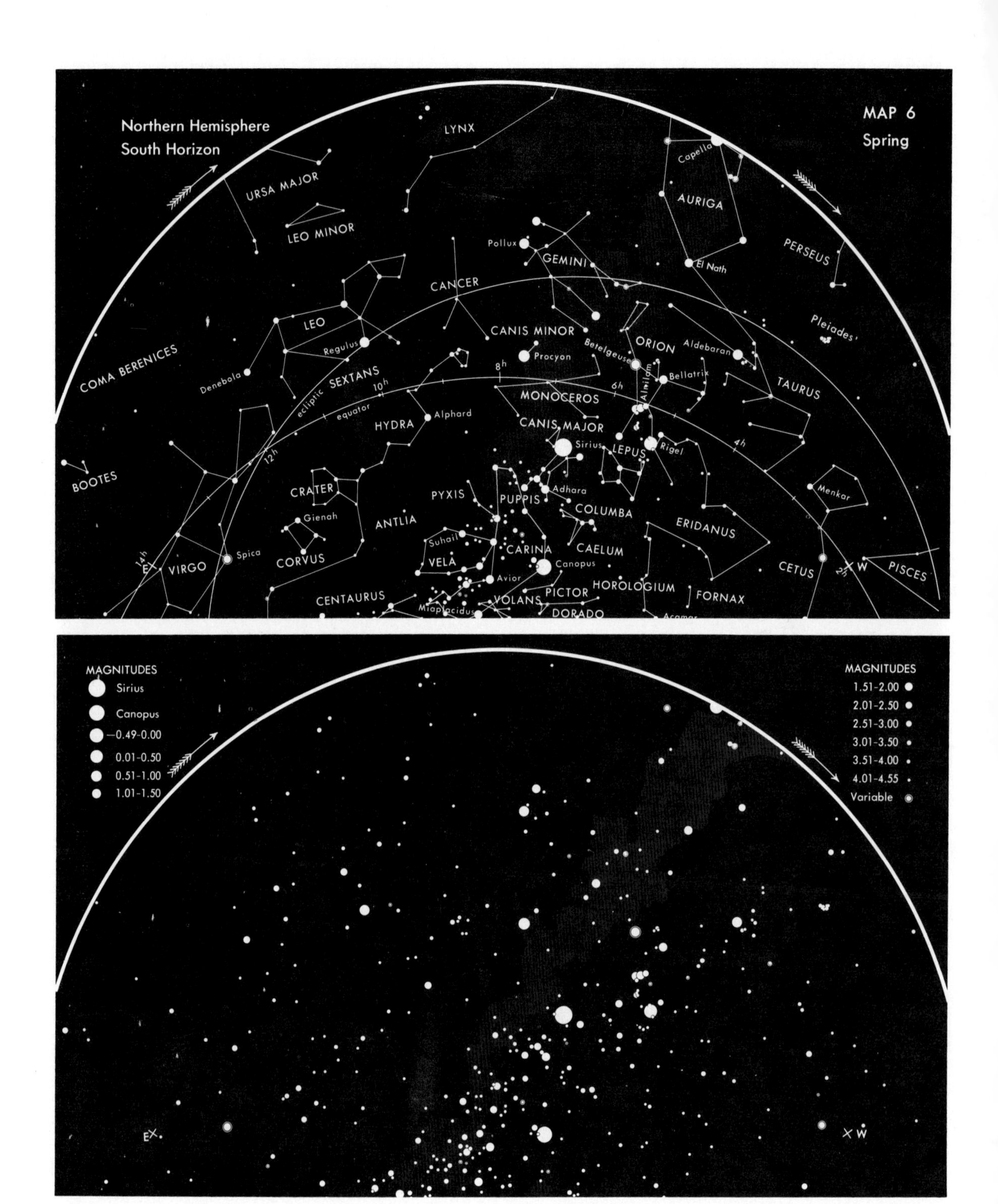
Northern Hemisphere
South Horizon
MAP 6
Spring
LYNX
URSA MAJOR
LEO MINOR
AURIGA
Capella
PERSEUS
El Nath
Pollux
GEMINI
CANCER
LEO
CANIS MINOR
Procyon
Regulus
COMA BERENICES
Denebola
SEXTANS
ecliptic
equator
ORION
Betelgeuse
Bellatrix
Alnilam
Aldebaran
Pleiades
TAURUS
MONOCEROS
HYDRA
Alphard
CANIS MAJOR
Sirius
LEPUS
Rigel
BOOTES
CRATER
PYXIS
PUPPIS
Adhara
COLUMBA
Menkar
ERIDANUS
Gienah
ANTLIA
CORVUS
Suhail
VELA
CARINA
CAELUM
Canopus
VIRGO
Spica
Avior
CETUS
PISCES
CENTAURUS
VOLANS
PICTOR
HOROLOGIUM
FORNAX
DORADO
Miaplacidus
E
W
MAGNITUDES
Sirius
Canopus
−0.49-0.00
0.01-0.50
0.51-1.00
1.01-1.50
MAGNITUDES
1.51-2.00
2.01-2.50
2.51-3.00
3.01-3.50
3.51-4.00
4.01-4.55
Variable

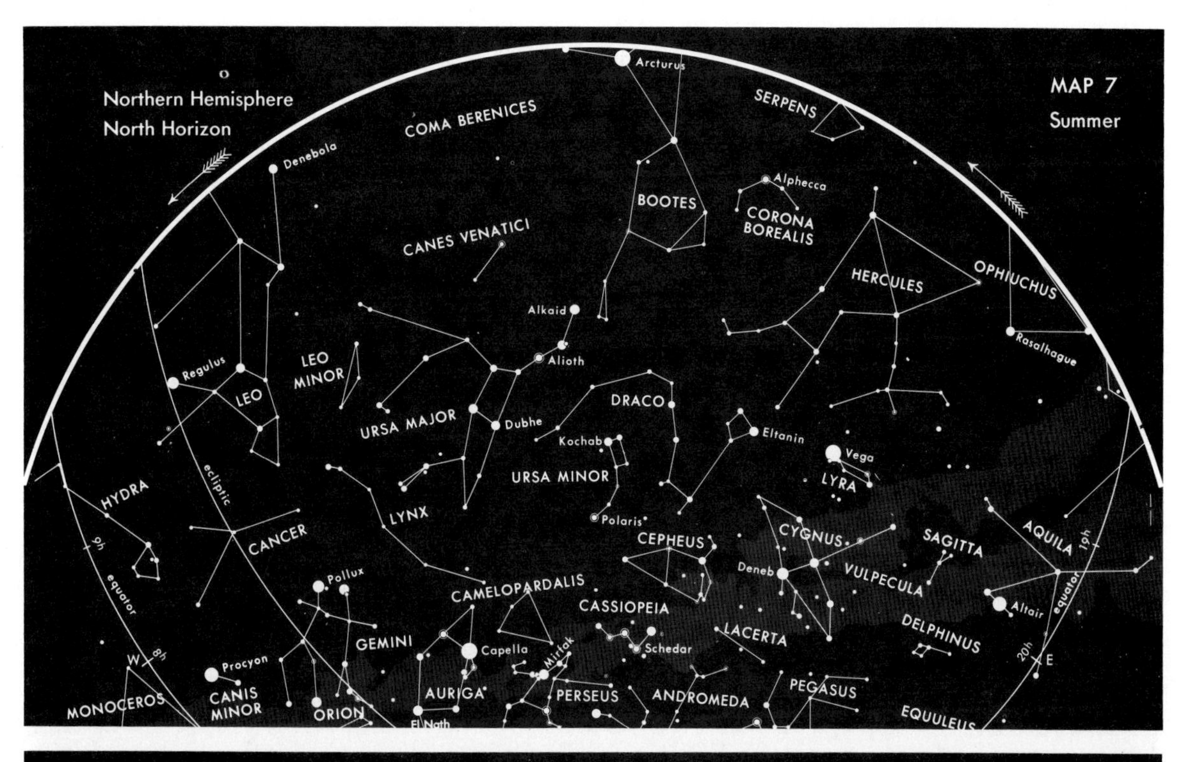
Northern Hemisphere
North Horizon
MAP 7
Summer
COMA BERENICES
Arcturus
SERPENS
Denebola
BOOTES
Alphecca
CORONA BOREALIS
CANES VENATICI
HERCULES
OPHIUCHUS
Alkaid
Rasalhague
Alioth
LEO MINOR
Regulus
LEO
DRACO
URSA MAJOR
Dubhe
Eltanin
Kochab
Vega
URSA MINOR
LYRA
HYDRA
ecliptic
LYNX
Polaris
CEPHEUS
CYGNUS
SAGITTA
AQUILA
19h
CANCER
9h
Deneb
VULPECULA
equator
Pollux
CAMELOPARDALIS
CASSIOPEIA
Altair
LACERTA
DELPHINUS
GEMINI
Capella
Mirfak
Schedar
20h
W
8h
E
Procyon
MONOCEROS
CANIS MINOR
ORION
AURIGA
PERSEUS
ANDROMEDA
PEGASUS
EQUULEUS
El Nath

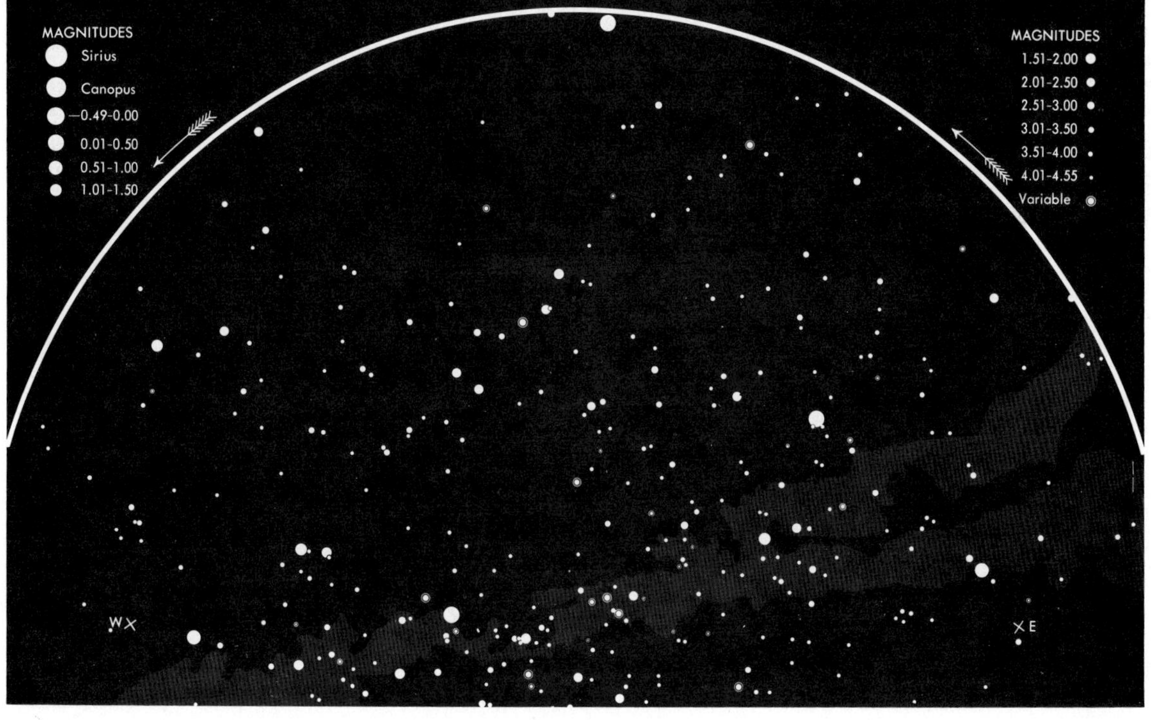
MAGNITUDES
Sirius
Canopus
−0.49-0.00
0.01-0.50
0.51-1.00
1.01-1.50
MAGNITUDES
1.51-2.00
2.01-2.50
2.51-3.00
3.01-3.50
3.51-4.00
4.01-4.55
Variable
W
E

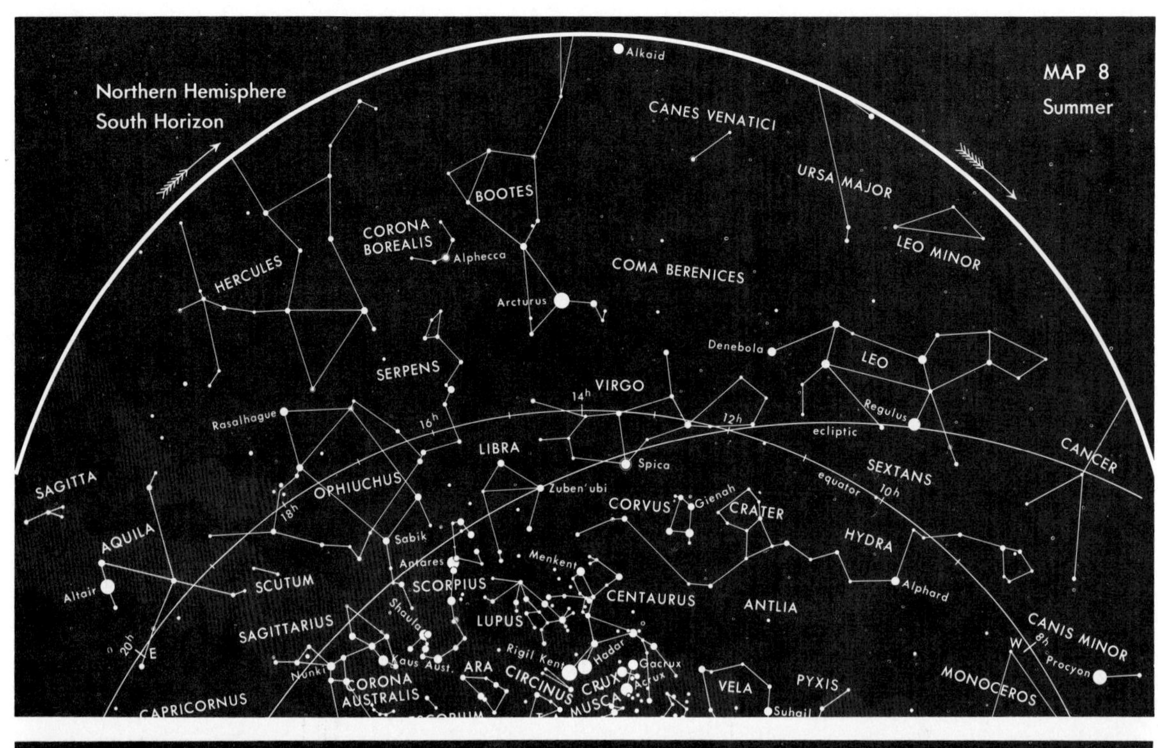

Northern Hemisphere
South Horizon
MAP 8
Summer
Alkaid
CANES VENATICI
URSA MAJOR
LEO MINOR
BOOTES
CORONA BOREALIS
Alphecca
HERCULES
COMA BERENICES
Arcturus
Denebola
LEO
SERPENS
VIRGO
14h
16h
12h
Rasalhague
Regulus
ecliptic
LIBRA
Spica
SEXTANS
CANCER
SAGITTA
OPHIUCHUS
Zuben'ubi
CORVUS
Gienah
CRATER
equator
10h
18h
AQUILA
Sabik
HYDRA
Menkent
Antares
SCUTUM
SCORPIUS
CENTAURUS
Altair
ANTLIA
Alphard
LUPUS
SAGITTARIUS
Shaula
CANIS MINOR
20h
E
Hadar
Nunki
Kaus Aust.
ARA
Rigil Kent
Gacrux
8h
W
Procyon
CIRCINUS
CRUX
Acrux
VELA
PYXIS
MONOCEROS
CORONA AUSTRALIS
MUSCA
Suhail
CAPRICORNUS

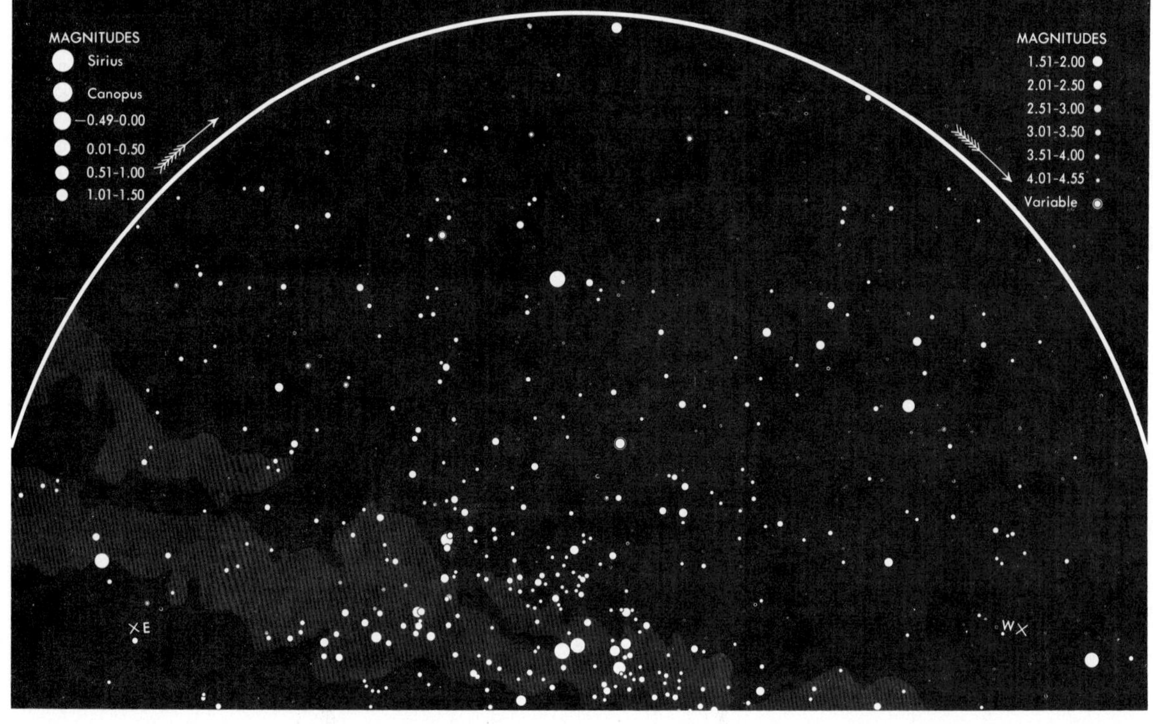

MAGNITUDES
Sirius
Canopus
−0.49-0.00
0.01-0.50
0.51-1.00
1.01-1.50
MAGNITUDES
1.51-2.00
2.01-2.50
2.51-3.00
3.01-3.50
3.51-4.00
4.01-4.55
Variable
E
W

Appendix D

Making Tables

Scientists often group records, observations, and measurements of things or events in tables. The numbers from records, observations, or measurements are dalled **data.** When there are lots of data, the numbers are often organized in tables. Tables have columns from side to side, going up and down. Tables have rows, going across. The names of the columns are given across the top of the table, the names of the rows are given up and down on the left side of the table.

For example, you can organize your estimates of the magnitudes of stars in Section 1.5 in a simple table:

Stars	Estimated magnitude	Actual magnitude
A	1	1
B	4	4
C	3	2
D	2	3
E	6	6

In making a table like this, notice that you have a name at the top of each column, and another name on the left-hand side for each of the rows. Then you put your data into the boxes for the proper columns and rows. Also, you always make a title for a table, to tell what the table is about, and put it at the top.

Appendix E

Temperature Scales

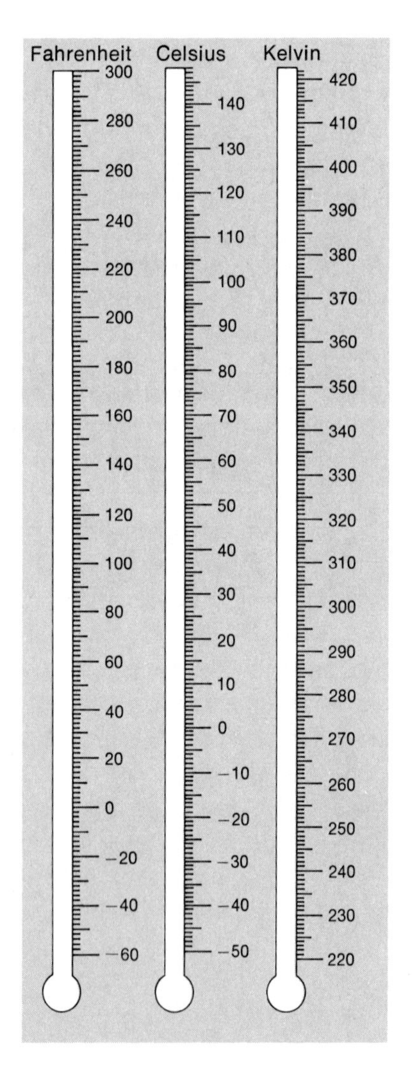

Temperature can be measured on a number of scales, just as length can be measured in either inches or centimeters. Astronomers often use the Kelvin temperature scale, which has the same size degrees as the Celsius temperature scale, and bigger degrees than the Fahrenheit scale we usually use. These three temperature scales are placed side by side in the chart. You can change a Kelvin temperature into a Fahrenheit or Celsius temperature by reading straight across the three scales. For example, you know that water freezes at 32 degrees Fahrenheit. A line straight across from 32 degrees Fahrenheit gives you 0 degrees on the Celsius table and 273 degrees on the Kelvin scale. Kelvin temperatures are always 273 degrees greater than Celsius degrees.

You can also use the following formulas to convert from one temperature scale to another:

To change from Fahrenheit(F) to Celsius(C)

$$°C = (°F - 32°) \times 5/9$$

To change from Celsius(C) to Fahrenheit(F)

$$°F = (°C \times 9/5) + 32°$$

To change from Celsius(C) to Kelvin(K)

$$°K = °C + 273°$$

Appendix F

Making Graphs

A **graph** is a picture that compares two sets of numbers. The numbers can be measurements made during an experiment. Like a picture, the graph should be constructed so that anyone looking at it will understand just what was happening during the experiment. Here are a few rules that will help you make good graphs:

1. Put your name and the date of the graph in the upper right hand corner of the page.
2. Title the graph with an explanation of the nature of the experiment.
3. Lay out the horizontal and vertical axes so that the horizontal axis is longer than the vertical axis. Measure the thing that is varying along the horizontal axis and amount of varying along the vertical axis.
4. Choose a scale that will fill the entire page of graph paper and is easy to work with. For example, 1 division = 2, 4, 5, or 10 units, not something hard to divide equally like 7. The scale need not start at zero.
5. Label each axis with the variable being plotted on it. Write the label so that the page does not have to be turned to read it. Put the label at the center of the axis it is identifying and indicate the units of measurement being used.
6. Indicate the scale being used on each axis, for example, "1 division = 10 cm."

A sample graph follows:

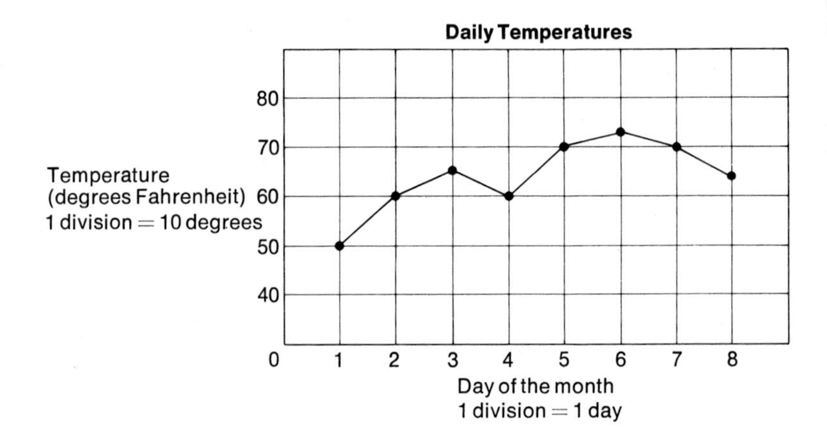
Daily Temperatures
80
70
Temperature
(degrees Fahrenheit) 60
1 division = 10 degrees
50
40
0
1
2
3
4
5
6
7
8
Day of the month
1 division = 1 day

Appendix G

Atoms, Elements, and Molecules

The sun, the earth, and people have one important thing in common. They are all made up, at least in part, of tiny particles called **atoms**. There are over 100 different kinds of atoms and everything you can see or smell consists of one or more kinds (usually more). When there is only one kind of atom in a material, that material is called an **element**.

In rocks and people, some atoms are combined into larger particles, called **molecules**. In the hotter parts of stars, the heat is so intense that molecules do not form.

How large are atoms? Think of water. Water consists of molecules. Each molecule of water has two atoms of the element hydrogen and one atom of the element oxygen. If a drop of water were as large as the earth, a hydrogen atom would be about the size of a tennis ball.

Most of the elements have names and also **symbols**, which are their short names. The symbol for oxygen, for example, is O. The symbol for hydrogen is H. Notice the list of the names and symbols for the elements, on the next page. The black lines of the spectrums from stars tell what elements are in the stars. Different combinations of these lines are made by different elements.

Atoms are made up of smaller parts. The parts of atoms are called **elementary particles**. All atoms contain the same kinds of elementary particles. The three main elementary particles are protons, neutrons, and electrons. The **protons** and **neutrons** are located in the central part of the atom and make up what is called the **atomic nucleus** (NEW-klee-us). The **electrons** move around the nucleus at great speeds. You can imagine them as a cloudlike blur many times wider than the nucleus it hides. Atoms turn out to be 99.99 percent empty space, the space in which the electrons travel.

Element	Symbol	Element	Symbol	Element	Symbol
Actinium	Ac	Hafnium	Hf	Promethium	Pm
Aluminum	Al	Helium	He	Protactinium	Pa
Americium	Am	Holmium	Ho	Radium	Ra
Antimony	Sb	Hydrogen	H	Radon	Rn
Argon	Ar	Indium	In	Rhenium	Re
Arsenic	As	Iodine	I	Rhodium	Rh
Astatine	At	Iridium	Ir	Rubidium	Rb
Barium	Ba	Iron	Fe	Ruthenium	Ru
Berkelium	Bk	Krypton	Kr	Samarium	Sm
Beryllium	Be	Lanthanum	La	Scandium	Sc
Bismuth	Bi	Lawrencium	Lr	Selenium	Se
Boron	B	Lead	Pb	Silicon	Si
Bromine	Br	Lithium	Li	Silver	Ag
Cadmium	Cd	Lutetium	Lu	Sodium	Na
Calcium	Ca	Magnesium	Mg	Strontium	Sr
Californium	Cf	Manganese	Mn	Sulfur	S
Carbon	C	Mendelevium	Md	Tantalum	Ta
Cerium	Ce	Mercury	Hg	Technetium	Tc
Cesium	Cs	Molybdenum	Mo	Tellurium	Te
Chlorine	Cl	Neodymium	Nd	Terbium	Tb
Chromium	Cr	Neon	Ne	Thallium	Tl
Cobalt	Co	Neptunium	Np	Thorium	Th
Copper	Cu	Nickel	Ni	Thulium	Tm
Curium	Cm	Niobium	Nb	Tin	Sn
Dysprosium	Dy	Nitrogen	N	Titanium	Ti
Einsteinium	Es	Nobelium	No	Tungsten	W
Erbium	Er	Osmium	Os	Uranium	U
Europium	Eu	Oxygen	O	Vanadium	V
Fermium	Fm	Palladium	Pd	Xenon	Xe
Flourine	F	Phosphorus	P	Ytterbium	Yb
Francium	Fr	Platinum	Pt	Yttrium	Y
Gadolinium	Gd	Plutonium	Pu	Zinc	Zn
Gallium	Ga	Polonium	Po	Zirconium	Zr
Germanium	Ge	Potassium	K	*	
Gold	Au	Praseodymium	Pr		

* Two more elements have been discovered but have not been named yet.

Atoms of elements differ in the numbers of elementary particles they contain. Hydrogen has 1 proton and 1 electron in each atom. Oxygen has 8 protons, 8 electrons, and 8 neutrons in each atom. An atom of uranium has 92 protons, 92 electrons, and 146 neutrons.

What holds all the particles together in whole atoms, so atoms can exist alone? Electrical forces do this. Electrical charges that are unlike, called positive and negative, attract each other. Like charges push each other apart. The attraction of unlike electrical charges holds the particles in the atoms together. Electrons are negatively charged, so they are attracted to move around the positively charged protons in the nucleus of the atom. *Neu*trons are *neu*tral and have no charge.

Atoms of elements can unite to make the stuff of the universe. A new material is formed with each union. New materials are continually being made in stars.

One way atoms unite is by sharing electrons in each other's clouds to make bigger units. The nuclei (plural of "nucleus") of the atoms are not changed. This joining process is called a **chemical reaction**. The result is the particle called the molecule. Molecules, groups of atoms that are bound together, can be named and given symbols made up from the symbols of the elements that made them. In this case, the symbols for molecules are called **formulas**. Study the list of names of some everyday materials with the numbers of each kind of atom in the formulas for their molecules.

When molecules of materials contain at least two different kinds of atoms, the materials are called **compounds**.

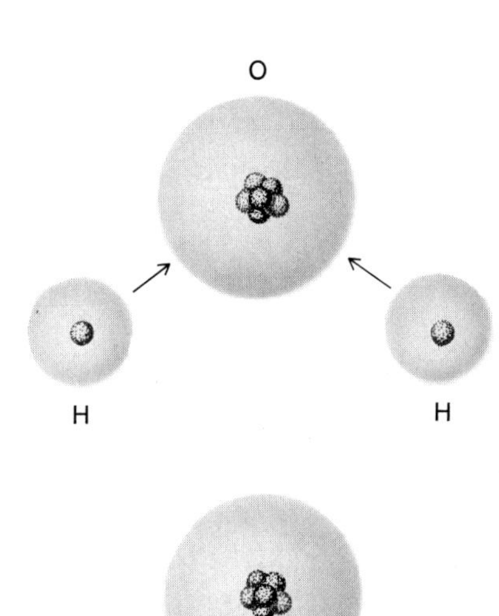

Here's a second way atoms can unite. Under special conditions, two atoms can be forced to combine their nuclei into one. This is like making one ball of clay from two. The electrons from the two original atoms form a larger cloud around the new nucleus. The process is called **fusion**.

In fusion, which happens to lighter elements, a lot of energy is given off in radiations. The fusion of two hydrogen atoms forms a helium atom. Atomic fusions that form slightly heavier elements (like lithium and carbon) from the lightest elements (like hydrogen and helium) are continually taking place within stars. The energy released accounts for what you call solar energy.

Atoms can also be split to make new materials. This is like making two balls of clay from one big one. In this way, several light atoms can be made from a heavy one, such as uranium. A lot of energy is given off. This is called **fission**, or the breaking up of atoms.

Common name	Formula
Alcohol	C_2H_5OH
Aspirin	$C_9H_8O_4$
Baking soda	$NaHCO_3$
Hydrogen peroxide	H_2O_2
Lemon juice	$C_6H_8O_7$
Marble	$CaCO_3$
Moth balls	$C_{10}H_8$
Spot remover	CCl_4
Plaster of Paris	$(CaSO_4)_2 \cdot H_2O$
Quartz	SiO_2
Starch	$(C_6H_{10}O_5)_{100}$
Sugar	$C_{12}H_{22}O_{11}$
Vinegar	$C_2H_4O_2$
Water	H_2O

Appendix H

Hubble Classification of Galaxies

Normal spiral

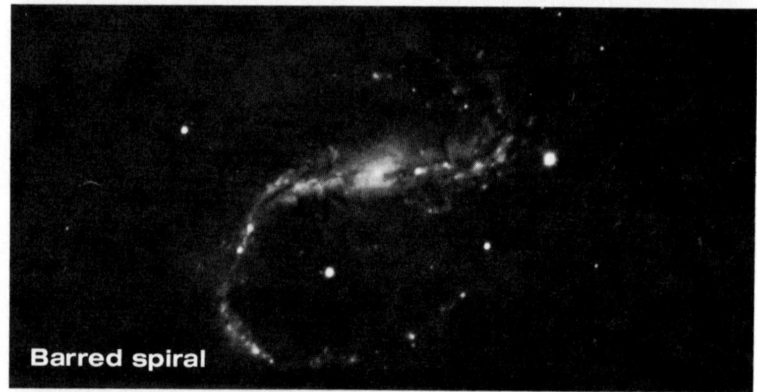
Barred spiral

Irregular

Appendix I

Earth's North Celestial Pole and Time of Rotation

In this activity, you can find out how to locate the place in the sky to which the earth's axis of rotation points. By measuring star tracks you can find out how long the earth takes for a complete rotation.

Place a small piece of tracing paper over the photograph of star tracks in Text Figure 3/3. Trace the curve of the North Star, Polaris, and the curves of a number of other stars on the tracing paper. The beginning of a tracing is shown in the appendix figure. Mark the beginning and end of each curve or track with a small dot. With a compass, find the center of the circle the curve of Polaris is a part of. Mark it with a dot.

Draw lines from the center dot out through each dot at the end of the star curves you have traced, as shown in the appendix figure. Make the lines long enough so you can use a protractor to measure the number of degrees in the angles made by each pair of lines, as shown in the appendix figure.

How many degrees did Polaris and the other stars move in the photograph? Explain why you get the results you do, no matter how far these stars are from the center.

Add up the angles you measured for the tracks of all the stars in the photograph. Divide this total by the number of angles you measured. This gives you the average angle the stars moved in each photograph.

What part of a whole circle of 360 degrees (Appendix J) is the average angle of the star tracks in the photograph? You can see from text Figure 3/3 that it took 12 hours for the star tracks to get that long. If the stars appear to have moved that far (the earth actually rotated that far) in 12

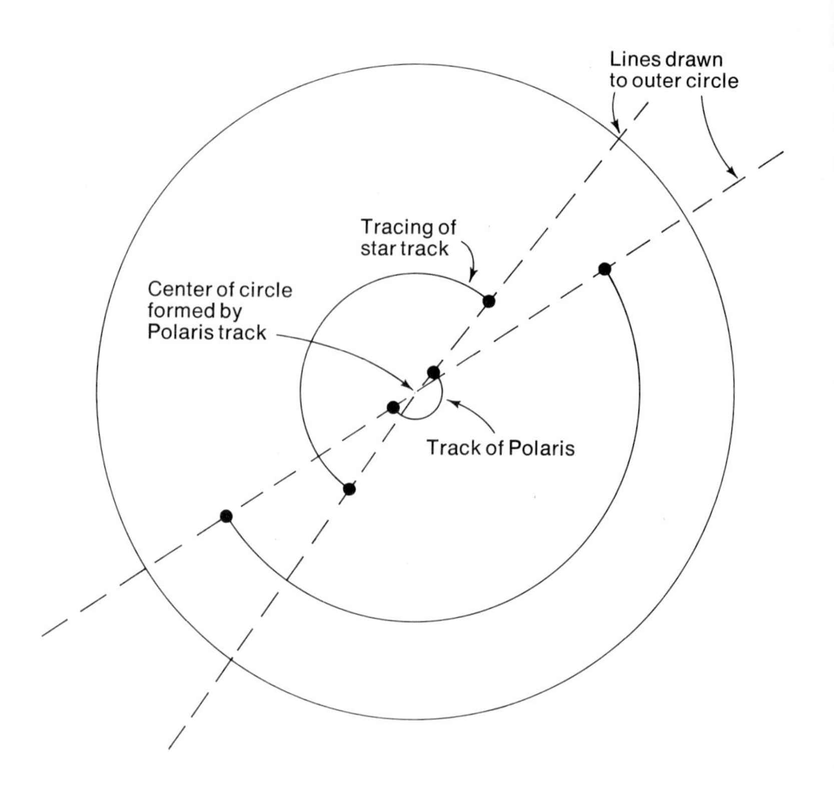

hours, how many hours would it take for the earth to rotate 360 degrees?

To find the hours necessary for a complete rotation, multiply 360 degrees by the number of hours for the tracks in the photograph. Then, divide that number by the number of degrees the stars moved.

You see now that the dot you made in the center of the circle marked out by the curve of Polaris is also the center of the tracks of all the other stars. This dot is the place where the earth's axis of rotation points.

Appendix J

Measurement of Angles

Astronomers use angles in the sky in many of their measurements. So do people who make maps of the earth, and sailors and aircraft navigators who want to find out where they are on or above the earth.

The way angles are measured is shown in the appendix figure. The unit used is a **degree**, on a scale around a full circle of 360°. The length of a curved line, called an **arc**, like those you measure in the photograph, is called **angular distance**. This is given in degrees. The distance between two things like stars in the sky is also angular distance.

For finer measurements, each degree (°) is divided up into 60 parts called **minutes** (′), and each minute into 60 **seconds** (″). Don't mix up these minutes and seconds of degrees with the minutes and seconds of time into which hours are divided. Hours are units of a different measurement scale (time) from degrees (angles).

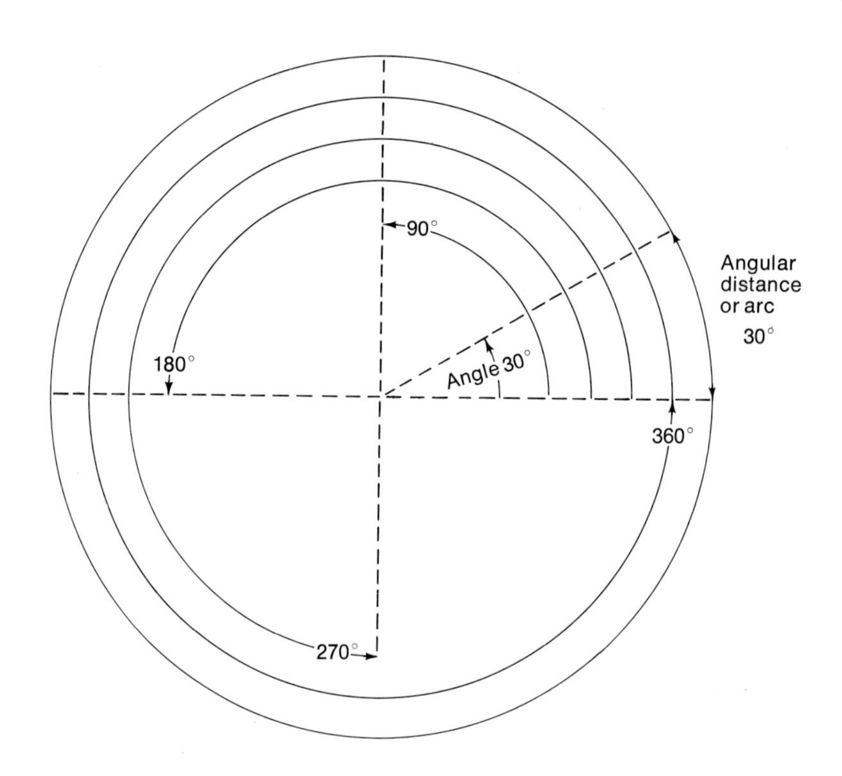
90°
180°
Angle 30°
Angular
distance
or arc
30°
360°
270°

Appendix K

Local Sun Time

Noontime (or 12:00 o'clock) on your clocks is set by the time meridian of the zone in which you are located. Noon is also the time when the sun is highest in the sky above you. This is called noon by **local sun time**. The sun is highest when it crosses your meridian, which runs directly north and south through your town. But it may not be exactly 12:00 o'clock by your watch. The appendix figure shows what makes the difference between local sun time and time-zone time. If you are near the center of your time zone, there won't be much difference between your local sun time and noon of your time zone. But if you live toward the edge of a time zone, the difference may be as much as 30 minutes.

Here is a way to predict, or to infer, how much your local sun time differs from your time-zone clock time. You can then let the sun check your prediction.

Look up your exact longitude on a large scale map of your area. Find the difference in degrees between your longitude and the longitude of the time-zone meridian that sets your regular time. Since the sun takes one hour to move 15 degrees, it will take four minutes to move one degree. Multiply the number of degrees of your distance from the time-zone meridian by four minutes. This will give the total time difference between your local sun time and the time-zone time. If you live east of your time-zone meridian, subtract this difference from 12:00 noon. If you live west of your time-zone meridian, add this difference to 12:00 noon. This is your prediction of when the sun will cross your local meridian.

You can check your prediction just before noon on any day when the sun is shining. You can do this outdoors or inside. Set your watch on time-zone time by a clock you know is right or by telephoning for the time. Use the appendix figure to help you set up this investigation.

Draw a line on a pad of paper and place the pad on a flat surface in the sunlight. Use a magnetic compass to point the line on the paper directly north and south. (Make sure the compass is not close to iron or steel that will attract its needle and give you a false reading.) The line on the paper then stands for your local meridian. Place a short pencil upright in clay or putty at the end of the paper toward the sun, and directly on the line. When you are set up, and after a minute or so, mark dots on the paper at the end of the shadow the stick casts from the sun. This will show you that the shadow is moving toward the line. When the shadow is directly along the meridian (north-south line), note the time. This is your local sun time. Does it agree with your prediction? If the sun time is not close to your prediction, what errors could you have made in your observations?

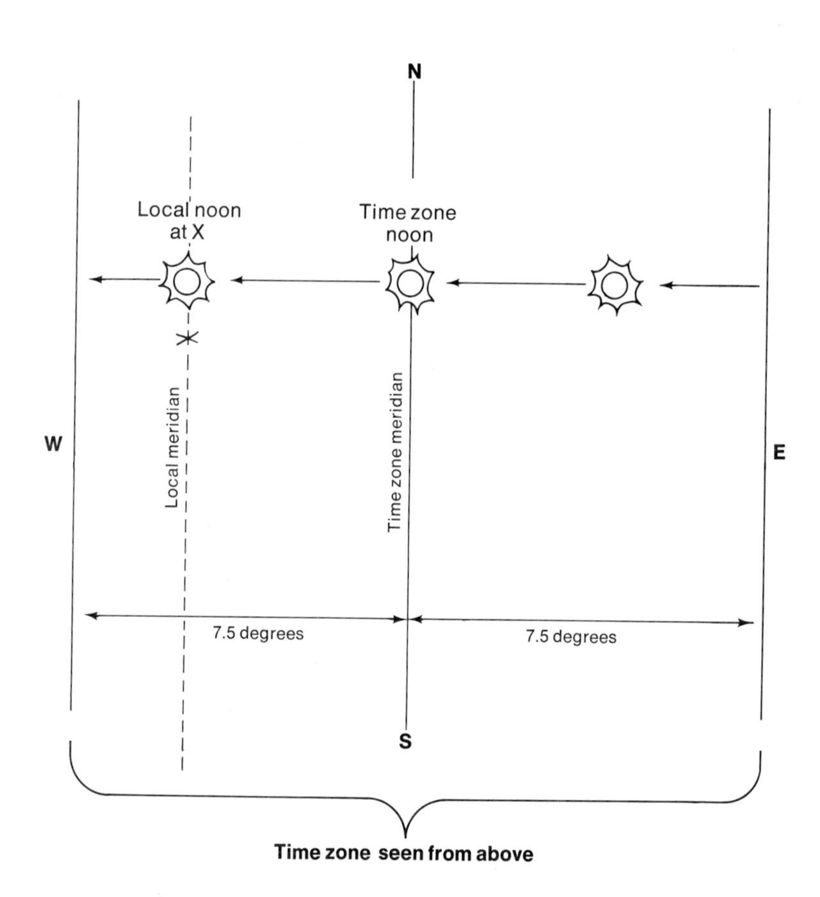

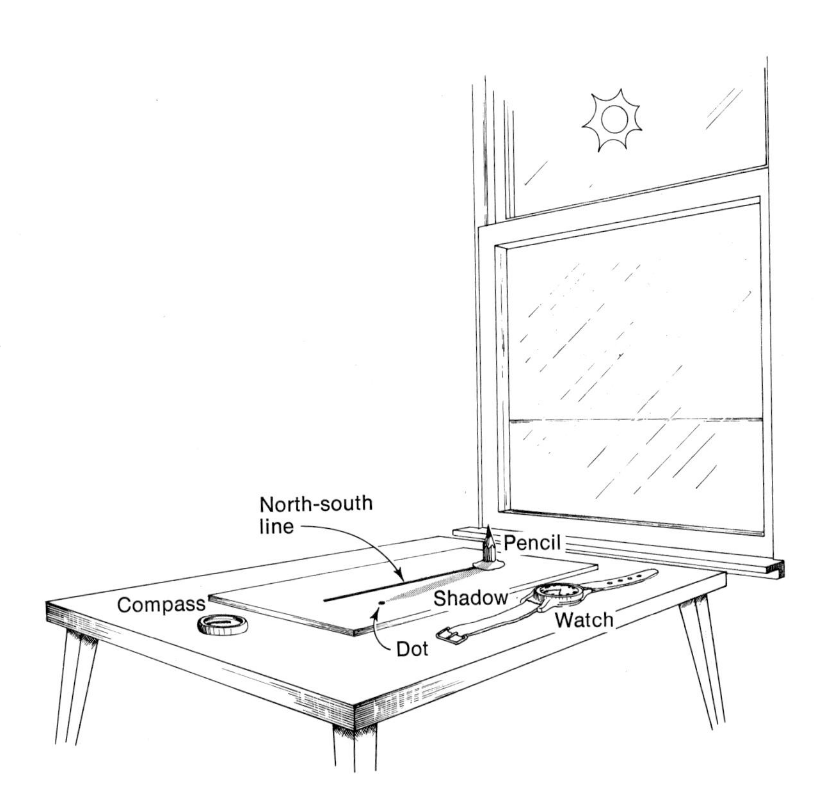
North-south line
Pencil
Compass
Shadow
Watch
Dot

Appendix L

Making and Using an Astrolabe

Men of early civilizations, like the Babylonians and Egyptians, measured angles by means of a simple instrument called an **astrolabe**. This gave them the positions of stars and other objects in the sky. Astrolabes can be made in several ways. You can make a simple one yourself with a ruler and a protractor. If you are handy with tools, you can make an astrolabe that gives you altitudes above the horizon and also direction around the horizon, which is called **azimuth** (AZ-uh-muth).

In making and using the simple astrolabe, tape or tack the protractor to a ruler, as shown in the drawing. Sight on a star or other celestial object through a plastic straw taped to the ruler. The string from which the weight hangs is fastened to a tack or through a hole in the stick just at the center of the base of the protractor.

Hold the stick or ruler so that the plumb line, with the weight pointing toward the center of the earth, does not rub against the side of the protractor. When you are sighted on the object, you can put your finger on the string to hold it against the protractor and then read the angle. The angle of altitude will be the difference between what the plumb line marks and 90°.

If you are making or using the more complete astrolabe, the direction table must be placed on its support so that the table is level with the horizon. You can use a bubble level to make the table horizontal. The post holding the sighting stick and tube must be vertical to the direction table. The direction table is moved around until the 0° mark points north, if you are working with the horizon system (Appendix M). You can obtain north in two ways, by using a compass or by sighting on Polaris at night and turning the base so the zero mark is directly under the sighting stick, aimed in the same direction.

If the protractor is mounted at right angles to the stick, the angle of altitude can be read directly.

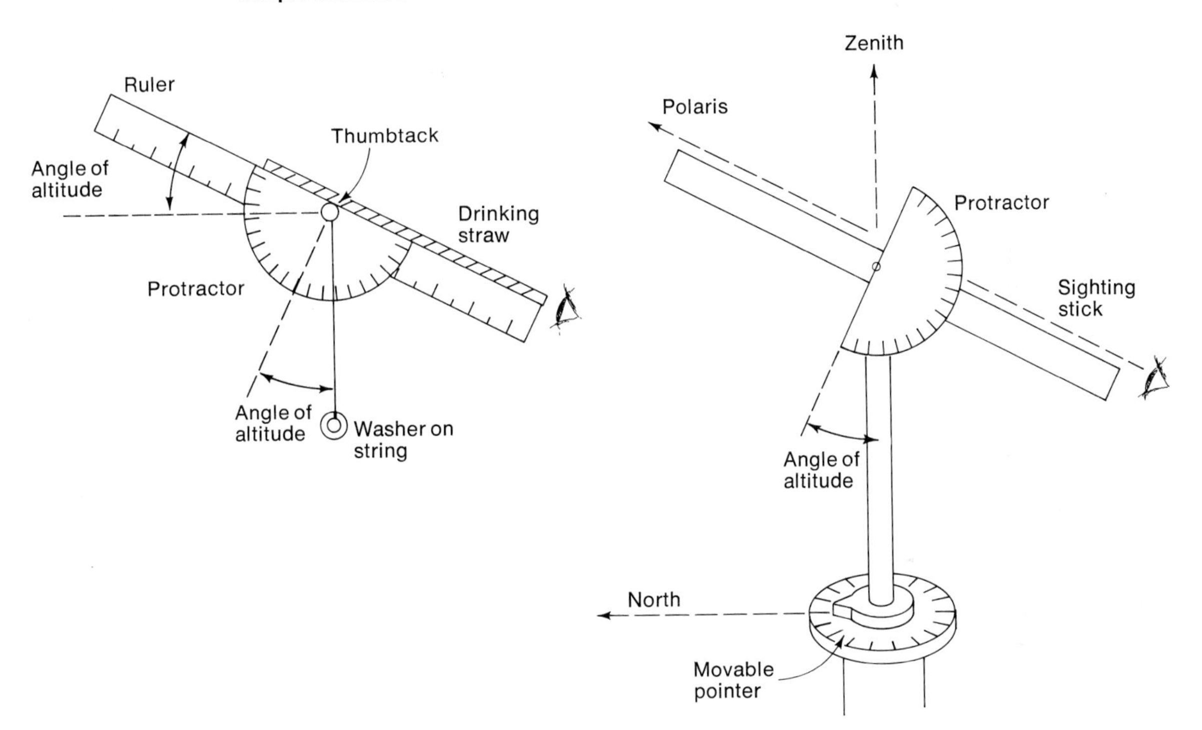

Appendix M

Locating Sky Objects

Once you have the reference point of the North Star, you can use what is called the **horizon system** for locating sky objects. The horizon system gives you local sky positions, that is, locations in the sky around your particular area.

The horizon is the reference line in the horizon system. If you have a clear view of the sky, without too many lights, use the more complete astrolabe with the direction table outdoors to work with the horizon system. Or your teacher may put up stars for you to work with in the classroom.

Adjust the direction table so that the 0° mark points north, as explained in Appendix L. The 0 degrees is your horizon reference point.

Now sight on several stars. For each star, record the number of degrees on the base of your astrolabe directly below the tube. This gives you degrees around the horizon from your 0 degree reference point, the azimuth. **Azimuth** is the angular distance of a star from north, or 0 degrees. For each star, also record the number of degrees on the protractor. This gives you the altitude of the star above the horizon. Azimuth and altitude together locate any star in the sky over a particular place.

If a star's altitude is greater than 90 degrees, it is above its zenith (ZEE-nith). The **zenith** is the point directly overhead (above the observer), at 90 degrees above any point on the horizon.

Make a horizon system sky map on a grid like that shown in the appendix figure. You might locate all of the stars you can see in the Big Dipper. Or you might like to locate all the bright stars in the constellation Pegasus (PEG-ah-suss). You can locate Pegasus and other constellations in the star charts in Appendix C.

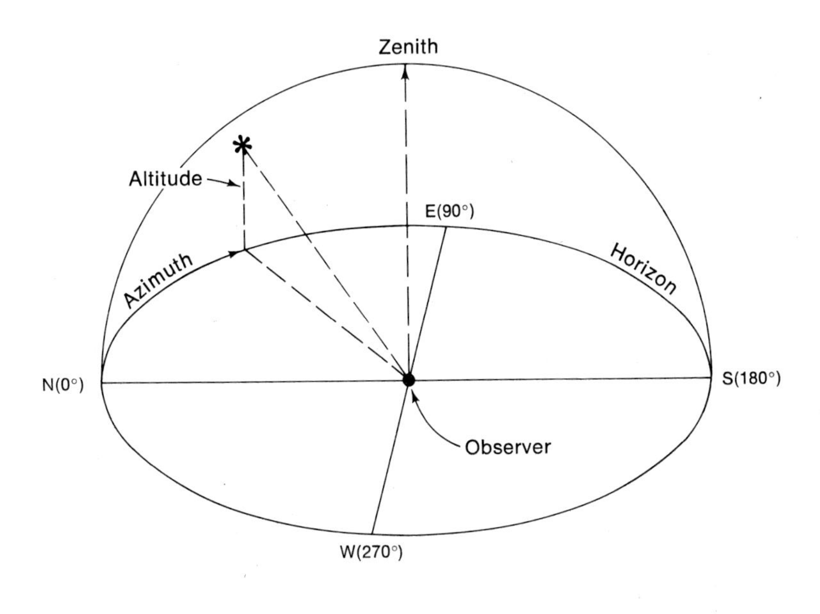
Zenith
Altitude
E(90°)
Azimuth
Horizon
N(0°)
S(180°)
Observer
W(270°)

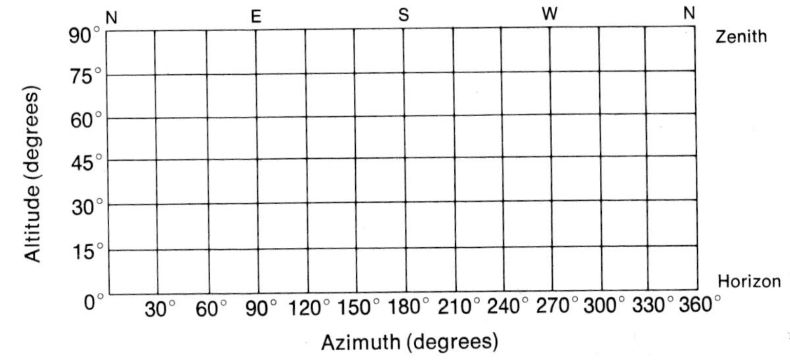
N
E
S
W
N
Zenith
90°
75°
60°
45°
30°
15°
0°
Altitude (degrees)
Horizon
30° 60° 90° 120° 150° 180° 210° 240° 270° 300° 330° 360°
Azimuth (degrees)

Appendix N

Features of the Planets

Actual celestial body	Diameter (mi)	Distance from the sun (mi)	Distance from the sun (A.U.)
Sun	870,000	—	—
Mercury	3,030	35,983,000	0.4
Venus	7,550	67,235,000	0.7
Earth	7,927	92,956,000	1.0
Mars	4,220	141,637,000	1.5
Asteroids	>500	Between Mars and Jupiter	
Jupiter	88,700	483,715,000	5.2
Saturn	75,100	890,602,000	9.5
Uranus	29,000	1,777,021,000	19.2
Neptune	28,000	2,799,435,000	30.1
Pluto	3,600?	3,654,407,000	39.4

Model celestial body	Distance from the sun (in)		Diameter (in)*			
	A	B	A	Fraction	B	Fraction**
Sun	—	—	.870	7/8	.0870	1/8
Mercury	36	4	.003	1/333	.0003	—
Venus	67	7	.007	1/143	.0007	—
Earth	93	9	.008	1/125	.0008	—
Mars	142	14	.004	1/250	.0004	—
Asteroids	—	—	Between Mars and Jupiter			
Jupiter	484	48	.089	1/8	.0089	1/64
Saturn	891	89	.075	1/16	.0075	—
Uranus	1,777	177	.029	1/32	.0029	—
Neptune	2,799	280	.028	1/32	.0028	—
Pluto	3,654	365	.003	1/333	.0003	—

*The size of Model A is one-millionth of the real size of the solar system, so 1 million miles = 1 inch. The size of Model B is one ten-millionth of the real size of the solar system, so 10 million miles = 1 inch.

**All the planets are much smaller than 1/64th of an inch with Model B, except for Jupiter, which is 1/64 inch. Small fractions are given to stress the smallness of the planets. Use smallest dots possible on the model.

Appendix O

Mars' Positions in the Sky

(March-October, 1971)

Date	Right ascension		Declination		Date	Right ascension		Declination	
	Hours	Minutes	Degrees	Minutes		Hours	Minutes	Degrees	Minutes
Mar. 22	18	25	−23	34	July 21	21	43	−20	11
Apr. 1	18	51	−23	24	July 31	21	37	−21	13
Apr. 11	19	16	−23	02	Aug. 12	21	25	−22	28
Apr. 21	19	40	−22	30	Aug. 21	21	16	−22	57
May 1	20	04	−21	50	Aug. 31	21	08	−23	05
May 11	20	26	−21	06	Aug. 5	21	05	−22	56
May 22	20	48	−20	16	Sept. 25	21	09	−21	13
June 1	21	06	−19	36	Oct. 6	21	20	−19	34
June 11	21	21	−19	06	Oct. 11	21	26	−18	49
June 21	21	33	−18	50	Oct. 21	21	41	−16	58
July 1	21	41	−18	54	Oct. 31	21	59	−14	52
July 11	21	45	−19	22					

Appendix P

Sun's Positions in the Solar System

Date	X	Y	Date	X	Y
Jan. 14, 1970	0.5	−4.8	June 19, 1986	7.3	1.1
Feb. 18, 1971	2.5	−4.0	July 24, 1987	4.9	2.7
Mar. 24, 1972	3.8	−2.2	Aug. 27, 1988	2.4	2.5
Apr. 28, 1973	3.7	−0.2	Oct. 1, 1989	0.7	1.0
June 2, 1974	2.1	2.1	Nov. 5, 1990	0.4	−1.1
July 7, 1975	−0.6	2.7	Dec. 10, 1991	1.5	−2.9
Aug. 10, 1976	−3.2	1.5	Jan. 13, 1993	3.6	−3.6
Sept. 14, 1977	−4.9	−1.1	Feb. 17, 1994	5.9	−2.9
Oct. 19, 1978	−5.0	−4.3	Mar. 24, 1995	7.7	−0.9
Nov. 23, 1979	−3.3	−7.2	Apr. 27, 1996	8.2	1.2
Dec. 27, 1980	−0.5	−8.9	June 1, 1997	7.1	5.1
Jan. 31, 1982	3.0	−9.1	July 6, 1998	4.5	7.3
Mar. 7, 1983	6.1	−7.7	Aug. 10, 1999	0.9	7.9
Apr. 10, 1984	8.1	−5.0	Sept. 13, 2000	−2.4	6.5
May 15, 1985	8.6	−1.8			

Appendix Q

Partial Eclipse of the Sun in 1971

If you wish to see just how the sun and moon move in the sky to make a solar eclipse, draw a map of their paths from the table. Set up a grid like the one shown.

The points you plot are the centers of the sun and the earth. Where the eclipse occurred, you can draw the sun and moon as they actually appear, with diameters of about 30 minutes.

	Sun				Moon			
	Right Ascension		Declination		Right Ascension		Declination	
Date	Hours	Minutes	Degrees	Minutes	Hours	Minutes	Degrees	Minutes
Aug. 16*	9	39	14	01	5	36	27	23
Aug. 17	9	43	13	42	6	34	26	23
Aug. 18	9	47	13	23	7	30	24	00
Aug. 19	9	51	13	04	8	23	20	29
Aug. 20	9	54	12	44	9	12	16	05
Aug. 20**	9	58	12	25	9	58	11	08
Aug. 21	9	58	12	24	9	58	11	06
Aug. 22	10	02	12	05	10	42	5	46

* The positions at 0:00 GMT.
** The eclipse began at 20:53 GMT on Aug. 20, reached its peak at 22:39, and was completely over by 0:25 on Aug. 21.

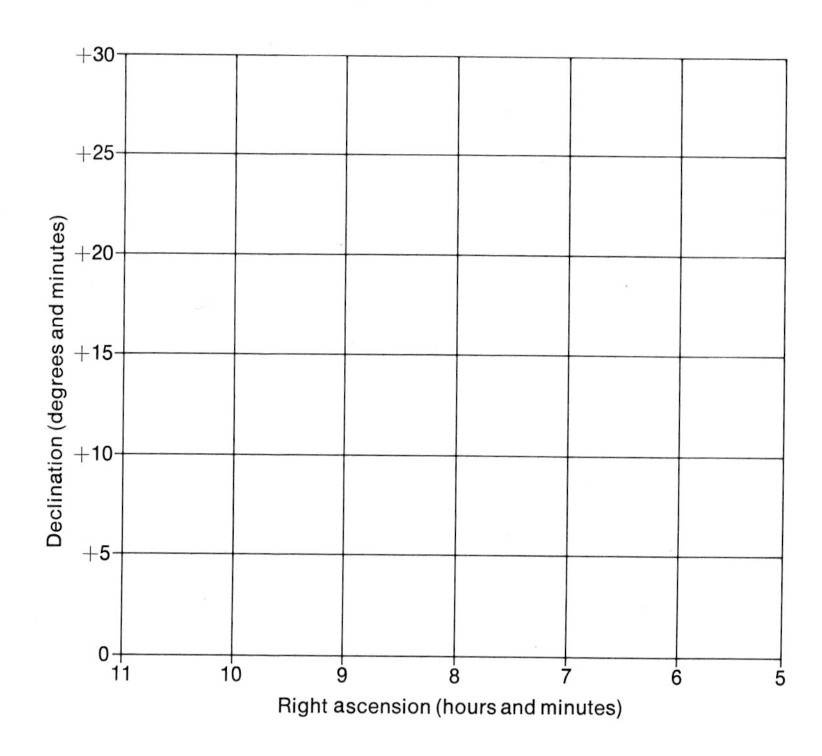
+30
+25
+20
+15
+10
+5
0
Declination (degrees and minutes)
11
10
9
8
7
6
5
Right ascension (hours and minutes)

Appendix R

Relative Humidity in Percent

Air temp. (°F)	Lowering of wet-bulb thermometer (°F)																																		
	1	2	3	4	5	6	7	8	9	10	11	12	13	14	15	16	17	18	19	20	21	22	23	24	25	26	27	28	29	30	31	32	33	34	35
0	67	33	1																																
5	73	46	20																																
10	78	56	34	13																															
15	82	64	46	29	11																														
20	85	70	55	40	26	12																													
25	87	74	62	49	37	25	13	1																											
30	89	78	67	56	46	36	26	16	6																										
35	91	81	72	63	54	45	36	27	19	10	2																								
40	92	83	75	68	60	52	45	37	29	22	15	7																							
45	93	86	78	71	64	57	51	44	38	31	25	18	12	6																					
50	93	87	80	74	67	61	55	49	43	38	32	27	21	16	10	5																			
55	94	88	82	76	70	65	59	54	49	43	38	33	28	23	19	11	9	5																	
60	94	89	83	78	73	68	63	58	53	48	43	39	34	30	26	21	17	13	9	5	1														
65	95	90	85	80	75	70	66	61	56	52	48	44	39	35	31	27	24	20	16	12	9	5	2												
70	95	90	86	81	77	72	68	64	59	55	51	48	44	40	36	33	29	25	22	19	15	12	9	6	3										
75	96	91	86	82	78	74	70	66	62	58	54	51	47	44	40	37	34	30	27	24	21	18	15	12	9	7	4	1							
80	96	91	87	83	79	75	72	68	64	61	57	54	50	47	44	41	38	35	32	29	26	23	20	18	15	12	10	7	5	3					
85	96	92	88	84	81	77	73	70	66	63	59	57	53	50	47	44	41	38	36	33	30	27	25	22	20	17	15	13	10	8	6	4	2		
90	96	92	89	85	81	78	74	71	68	65	61	58	55	52	49	47	44	41	39	36	34	31	29	26	24	22	19	17	15	13	11	9	7	5	3
95	96	93	89	86	82	79	76	73	69	66	63	61	58	55	52	50	47	44	42	39	37	34	32	30	28	25	23	21	19	17	15	13	11	10	8
100	96	93	89	86	83	80	77	73	70	68	65	62	59	56	54	51	49	46	44	41	39	37	35	33	30	28	26	24	22	21	19	17	15	13	12
105	97	93	90	87	84	81	78	75	72	69	66	64	61	58	56	53	51	49	46	44	42	40	38	36	34	32	30	28	26	24	22	21	19	17	15
110	97	93	90	87	84	81	78	75	73	70	67	65	62	60	57	55	52	50	48	46	44	42	40	38	36	34	32	30	28	26	25	23	21	20	18
115	97	94	91	88	85	82	79	76	74	71	69	66	64	61	59	57	54	52	50	48	46	44	42	40	38	36	34	33	31	29	28	26	25	23	21
120	97	94	91	88	85	82	80	77	74	72	69	76	65	62	60	58	55	53	51	49	47	45	43	41	40	38	36	34	33	31	29	28	26	25	23
125	97	94	91	88	86	83	80	78	75	73	70	68	66	64	61	59	57	55	53	51	49	47	45	44	42	40	38	37	35	33	32	30	29	27	26
130	97	94	91	89	86	83	81	78	76	73	71	69	67	64	62	60	58	56	54	52	50	48	47	45	43	41	40	38	37	35	33	32	30	29	28

Appendix S

The Layers of the Earth

Layer	Density (times heavier than water)	Pressure (times greater than at surface)	Temperature	Composition	Thickness
Crust	2.7–3.0	up to several thousand	a few hundred degrees	mostly granite and basalt	about 25 mi (40 km)
Mantle	3.3–5.5	several thousand to 1.3 million	600°–3600° F (300°–2000° C)	basalt	1800 mi (2900 km)
Outer core	9.5–11.5	1.3–3.0 million	3600° F (2000° C) or more	nickel and iron, possibly some lighter material	1300 mi (2100 km)
Inner core	about 15	up to 4 million	between 3600° and 11,700° F (2000° and 6500° C)	nickel and iron, possibly some heavier material	900 mi (1450 km) to center of earth

Appendix T

Mineral Identification

Minerals	Color	How it looks	Where to find it
Quartz	clear gray or shiny white	irregular surface	granite, sandstone, metamorphic rocks
Hematite	reddish brown on surface, darker below	irregular surface	sandstone, shale, soil
Limonite	yellowish brown on surface, darker below	irregular surface	sandstone, shale, soil
Pyrite	brass yellow	small cubic crystals, or small crystals with more than 6 sides	sedimentary and igneous rocks, usually shale
Calcite	clear	flat parallelogram surfaces	limestone
Orthoclase	white, gray, yellow, pink	flat surfaces in two planes at right angles	granite, metamorphic rocks
Biotite	black	flat surfaces in one plane, forming flakes	granite, metamorphic rocks
Muscovite	clear, or slightly brownish gray	flat surfaces in one plane, forming flakes	granite, metamorphic rocks
Hornblende	black	splintery	granite, diorite

Appendix U

Active and Extinct Volcanoes Around the World

Volcano	Latitude	Longitude	Volcano	Latitude	Longitude
Reventador	0.1° S	77.7° W	Krakatau	6.1° S	105.4° E
Ubinas	16.2° S	70.8° W	Barren I.	12.2° N	93.8° E
Maipo	34.1° S	69.9° W	Santorini	36.4° N	25.4° E
Mt. Burney	52.3° S	73.4° W	Vesuvius	40.8° N	14.4° E
São Jorge I.	38.6° N	28.0° W	Lanzarote I.	29.0° N	13.7° E
Katla	63.6° N	18.9° W	Paramushiro I.	50.3° N	155.3° E
Beerenberg	71.0° N	8.0° W	Katmai	58.3° N	155.0° W
Erta-Ale	13.6° N	40.6° E	Mt. Rainier	46.9° N	121.8° W
Niragongo	1.5° S	29.2° E	Lassen	40.5° N	121.3° W
Kartala	11.9° S	43.3° E	Tres Virgènes	27.5° N	112.7° W
Mauna Loa	19.5° N	155.9° W	Parícutin	19.5° N	102.1° W
Erebus	77.5° S	168.0° E	Poás	10.2° N	84.2° W
Pagan I.	18.1° N	145.8° E	Kanaga	51.9° N	177.2° W
Chyulu Hills	2.9° S	38.1° E	Brimstone I.	30.2° S	178.9° E
Saishuto I.	33.4° N	126.5° E	Late I.	18.8° S	174.6° W
Mayon, Luzon	13.3° N	123.7° E	Hunter I.	22.3° S	172.1° E
Makian	0.3° N	127.4° E	Tinakula	10.4° S	165.8° E
Lokon	1.4° N	124.8° E	Pago	5.6° S	150.6° E
Seroea	6.3° S	130.0 E	Fuji	35.4° N	138.2° E
Tambora	8.2° S	118.0° E			

Appendix V

Earthquakes Around the World

Date	Latitude	Longitude	Date	Latitude	Longitude
Jan. 3, 1968	51.8° N	173.3° W	July 5	7.0° S	147.3° E
Jan. 3	17.1° N	99.5° W	July 7	41.8° S	171.9° E
Jan. 7	33.5° N	141.6° E	June 21	7.4° S	155.1° E
Jan. 10	13.8° N	120.6° E	June 27	46.3° N	7.0° E
Jan. 13	31.2° S	68.4° W	July 3	34.7° N	75.1° E
May 28	56.2° N	158.0° W	July 4	36.9° N	28.6° E
May 29	62.3° N	149.1° W	July 14	20.9° S	68.8° W
May 29	3.1° N	83.7° W	July 18	8.9° N	93.9° E
May 30	15.8° S	167.6° E	June 27	8.2° S	119.7° E
June 2	41.2° N	143.4° E	July 6	58.7° S	24.9° W
June 3	36.0° N	141.2° E	July 18	9.6° N	40.2° W
June 8	87.0° N	51.3° E	July 26	22.4° S	12.6° W
June 9	39.0° N	46.0° E	June 29	20.6° S	66.2° E
June 11	41.5° S	85.4° W	June 29	19.9° S	33.6° E
June 12	24.9° N	91.9° E	July 6	1.3° S	33.3° E
June 12	13.8° N	120.7° E	July 31	80.0° N	6.3° E
May 31	36.1° N	31.0° W	Aug. 4	40.8° S	43.3° E
June 1	24.7° N	121.9° E	Aug. 4	53.0° S	9.6° E
June 5	36.1° N	66.2° E	Aug. 6	26.7° N	44.5° W
June 8	87.0° N	50.0° E	July 21	16.9° S	172.2° W

Date	Latitude	Longitude	Date	Latitude	Longitude
June 16	36.9° N	34.5° E	July 21	24.9° N	123.4° E
June 16	38.0° N	14.9° E	July 30	66.4° N	17.4° W
June 17	37.4° N	72.3° E	Aug. 10	76.0° N	8.7° E
June 17	56.0° S	27.9° W	Aug. 12	52.6° S	25.5° E
June 22	45.9° N	11.3° E	Sept. 15	17.9° S	13.0° W
May 29	35.3° N	118.5° W	Sept. 19	49.4° N	140.2° E
June 9	31.6° N	115.5° W	Oct. 7	3.2° S	146.1° E
June 24	1.6° S	15.7° W	Oct. 8	39.9° S	87.7° E
May 30	42.3° N	119.8° W	Dec. 25	26.7° S	26.9° E
June 10	18.0° S	173.1° W	Dec. 30	55.2° S	129.0° W
June 11	43.0° N	17.1° E	Jan. 2, 1969	45.0° S	167.6° E
June 23	6.0° S	103.9° E	Jan. 6	44.1° N	10.7° E
June 29	13.6° N	90.2° W	Jan. 16	27.6° N	129.2° E
June 29	11.6° S	166.4° E	Jan. 9	52.9° N	34.9° W
June 19	0.5° S	91.9° W	Jan. 10	41.6° N	32.6° E
June 9	14.3° S	167.3° E	Jan. 3	49.1° S	31.0° E
June 20	22.8° S	173.3° E	Jan. 7	36.4° N	2.8° W
June 23	5.6° S	77.3° W	Jan. 14	31.6° S	178.8° W
July 4	30.3° N	94.9° E	Jan. 22	32.2° N	70.0° E
July 4	37.8° N	23.2° E	Jan. 30	55.7° N	163.0° E

Appendix W

Intensity Scale of Earthquakes

Intensity	Effects
I	Not felt.
II	Felt by persons at rest or on upper floors.
III	Felt indoors. Hanging objects swing. Vibration feels like passing of light trucks.
IV	Hanging objects swing. Vibration like passing of heavy trucks. Standing cars rock.
V	Felt outdoors. Sleepers wakened. Liquids disturbed, some spilled. Small objects upset. Doors swing, close, open. Pictures move.
VI	Felt by all. Many people frightened and run outdoors. People walk unsteadily. Windows, dishes, glassware broken. Books fall off shelves. Pictures off walls. Furniture moved or overturned. Weak plaster cracked. Trees, bushes shaken or heard to rustle.
VII	Difficult to stand. Noticed by drivers of cars. Hanging objects shake faster. Furniture broken. Damage to plaster, loose bricks, stones, tiles. Waves on ponds; water muddy. Small slides and caving-in along sand or gravel banks.

Intensity	Effects
VIII	Steering of cars affected. Fall of some brick walls. Twisting, fall of chimneys, factory smokestacks, towers, elevated tanks. Wooden houses moved on foundations. Branches broken from trees. Changes in flow of springs. Cracks in wet ground and on steep slopes.
IX	General panic. Brick walls destroyed. General damage to foundations. Wooden houses shifted off foundations. Serious damage to reservoirs. Underground pipes broken. Large cracks in ground.
X	Most brick and wooden buildings destroyed with their foundations. Serious damage to dams. Large landslides. Water thrown on banks of canals, rivers, lakes, etc. Sand and mud shifted on beaches and flat land. Railroad tracks bent slightly.
XI	Railroad tracks bent greatly. Underground pipelines completely out of service.
XII	Damage nearly total. Large rock masses moved. Objects thrown into the air.

Glossary

agent of erosion Something that carries away weathered rock particles or soil.

air Another term for atmosphere, particularly that mixture of its gases close to the earth's surface.

air pressure The pressure created by the mass of the gases in the atmosphere above a given area.

amplitude Half of the distance between a wave crest and a wave trough.

anemometer A weather instrument for measuring the speed of the wind.

anticyclone An eddy with winds that blow clockwise in the northern hemisphere out of and around a center of high atmospheric pressure.

apogee The place in a satellite's elliptical orbit at which the satellite is farthest from the body around which it orbits.

artificial satellite A satellite in orbit by means of human action, like manned or unmanned spacecraft.

ash Fine dust blown out of a volcano during an eruption.

asteroid One of the small bodies, or minor planets, most of which orbit the sun between Mars and Jupiter.

astrolabe An instrument used to measure distances or to locate objects in the sky by their angles.

astronomer A scientist who studies the objects and events in space.

astronomical unit The average distance from the earth to the sun, about 93 million miles (150 million kilometers), used as a unit to measure distances in the solar system.

atmosphere The gases held around a planet or satellite by its gravitational attraction.

atoll A coral reef growing on top of a sunken volcanic island.

atom The smallest unit of material that can exist alone or with other similar units.

atomic nucleus The central part of an atom, containing the protons and neutrons.

atomic particles The tiny things, like protons, electrons, and neutrons, of which atoms are made.

aurora The bright sheet or fold effects in the upper atmosphere in which charged particles fall from the ends of the Van Allen belts.

axis of rotation An imaginary line around which an object spins or turns, as a ball spins or the earth turns.

barometer A weather instrument for measuring the air pressure of the atmosphere.

basalt An igneous rock, black in color, and very fine grained, that comes from volcanoes.

biosphere The zone on the earth where living things can be found.

calcite The mineral made up of molecules of calcium carbonate.

calcium carbonate The chemical name for the mineral calcite. It has the chemical formula $CaCO_3$ and is a common cement in sedimentary rocks.

caldera The crater or depression at the top of a volcano formed when rock at the top sinks or explodes away, or when the volcano collapses into the empty magma chamber below.

carbon-12 and carbon-14 Two kinds of carbon atoms. Carbon-14 is slightly heavier and is radioactive.

carbonate A chemical compound containing carbon and oxygen, such as calcium carbonate (the mineral calcite).

celestial equator The imaginary reference line in the sky on the celestial sphere that stands for the extension of the earth's equator.

celestial poles Points in the sky, directly above the earth's North and South Poles, toward which the axis of rotation points.

celestial sphere The imaginary sphere in the sky around the earth on which objects in the sky are located.

chemical weathering Weathering that changes the particle size of a material and also changes the material itself into something else.

cinder A small piece of lava blown out of a volcano during an eruption. Cinders are about the size of pebbles.

circulation The flow or motion of a fluid, like air or water, in or through a given area or volume.

climate The average state of the air, or of the weather, in a place or region over many years.

climate zone The broad bands on the earth of tropical, temperate, and polar weather conditions based on annual solar radiation.

column A narrow diagram showing a cross section of the earth's crust below a certain point on the surface.

comet A body composed largely of gases, usually in a long orbit or revolution of the sun.

compass An instrument containing a small magnet free to move in the force of a stronger magnetic field like the earth.

condensation The change from a gaseous to a liquid state, as when water vapor changes to liquid water.

conduction The transfer of heat energy in a solid by vibration from one molecule or particle to another.

cone The mound of material piled around the vent of a volcano.

constellation One of the 88 groups of stars that observers use to locate objects in the sky.

continental rise The gradual rise in the sea floor next to the continental slope.

continental shelf The edge of a continent that lies under the surface of the ocean that surrounds the continent.

continental slope The edge of the continental shelf that drops off more steeply than the shelf.

convection The transfer of heat energy through a fluid, like a gas (air) or a liquid (water) by motion of the particles in the fluid.

convection current A current in a fluid like air or water formed when the fluid is heated.

corona The sphere of intense wave and particle radiations given off by the sun.

crater A basin or depression on a moon or planet caused externally by striking of a meteorite or internally by volcanic eruption, explosion, or blowout.

crystal A piece of mineral matter that has grown into a special shape, according to the way its molecules are joined together. Most kinds of minerals have their own kinds of crystal shapes.

cluster A group of several galaxies, like the "local group" of about 20 galaxies in which we live.

current A stream of water or air moving in one direction.

cycle A series of events that keeps repeating.

cyclone An eddy with winds that blow counterclockwise in the northern hemisphere into and around a center of low atmospheric pressure.

decay, radioactive The giving off of particles by atoms, which changes the original atoms into other, non-radioactive kinds.

declination Distances north and south of the celestial equator measured in degrees, in the same way as latitude on earth.

degree A unit used in measuring circles or angles, equal to 1/360 of a complete circle.

delta A deposit of sediment found at the mouth of a stream, where it enters a lake or ocean.

deposition The laying down of sediment by water or from melting ice. Deposition can take place under water or on land.

dewpoint The temperature at which water vapor condenses from the air.

distributary A small stream that takes water away from a larger stream.

divide High ground that separates one watershed from another.

dune A mound of windblown sand.

dynamic equilibrium An equilibrium in which things are changing, but always in the same ways at the same rates of speed. In other words, the change is in equilibrium.

earthquake A shaking of the earth's crust caused by rock under pressure that suddenly breaks, or by the activity of magma under a volcano.

eclipse The blocking of light from a body by another. In a solar eclipse the moon blocks the light from the sun.

ecliptic The imaginary line across the sky standing for the path the earth follows around the sun, and close to which all the other planets move in the sky.

eddy A circular motion in air or water.

electromagnetic radiations The many kinds of wave radiations such as light, radio waves, x-rays, gamma rays, infrared waves, and microwaves.

electromagnetic spectrum The series of waves, varying in wavelength and frequency, given off by hot objects.

electron A tiny, negatively charged particle in the outer part of an atom.

element A material made up of only one kind of atom.

ellipse A flattened circle of the kind that planets and satellites follow in their orbits.

energy The ability to do work, which results in changing or moving things.

environment The total of all things around a living organism, such as air, water, plants or animals it competes with, and all else that affects its senses.

era A large subdivision of geologic time.

erosion The carrying away, usually by wind or water, of weathered rock particles or soil.

epicenter The point on the earth's surface directly over the place where an earthquake occurred.

equator The reference line around the earth halfway (90°) between the North and South Poles, used to establish parallels for measuring latitude.

equatorial bulge The bulge of the earth and other planets due to the distance around the equator being greater than the distance around the body through the poles.

equilibrium A balance between opposing forces so that there is no change.

evaporation The change from a liquid to a gaseous state, as when liquid water changes to water vapor.

evolution The development of one kind of thing into another kind of thing through a series of changes that are usually improvements or refinements.

exfoliation The kind of weathering in which flakes or layers break off a rock surface. It is usually caused by a combination of chemical and mechanical weathering. Thick exfoliation layers in granite are due to the slight expansion of the rock as rock above it is weathered away.

exosphere The outermost temperature region of the atmosphere in which gases may be escaping into space.

extinct Being a kind of organism that is no longer alive.

farside The hemisphere of the moon always turned away from the earth.

fault A break in the crust of a body like the earth or moon along which some movement of rock takes place. The Straight Wall is a fault on the moon; the San Andreas Fault is a famous one on the earth.

fault-block mountains Mountains formed by blocks of the earth's crust that have cracked away from each other and been tilted, pushed up, or dropped down.

field Any region or area throughout which an effect or a force exists and can be measured, such as the magnetic or gravitational fields of a planet.

fission The splitting of large atoms into smaller atoms, with the release of many kinds of radiations.

flood plain The flat land on either side of a stream that becomes flooded whenever the stream overflows.

focus The exact place beneath the earth's surface where an earthquake occurred.

food chain A set of animals dependent on each other for food. A food chain usually starts with plant-eating animals and ends with meat-eating animals.

fossil The remains of once living organisms found in sedimentary rock. Examples are shells, bones, wood, and footprints.

fossil cast A deposit of mineral matter that has filled in a fossil mold.

fossil mold A hollow place in a rock from which a fossil has dissolved or been weathered away.

frequency The number of waves, or other things or events, per unit of time.

front A narrow region in the atmosphere where eddies, or airflows, meet.

fusion The combination of two atoms to make a larger atom, with a great release of energy.

galaxy A group of billions of stars and nebulas in space, such as the Andromeda galaxy or the Milky Way galaxy.

gas planet One of the large planets beyond Mars, composed mostly of light materials such as gases.

Geiger counter An instrument used to detect radiation coming from a radioactive material.

geographic poles Points on the earth through which its axis of rotation passes, consisting of the North and South Poles.

geologic map A map showing where rocks of various ages are at the surface of the earth.

geologist A scientist who studies the earth and its changes.

geosyncline A portion of the ocean floor that is sinking as it is being filled in by sediment.

glacier A large body of ice, many feet thick, that flows slowly over land.

gneiss A metamorphic rock that was originally granite or a similar rock.

granite An igneous rock, relatively light in color, and coarse grained, that hardens below the earth's surface.

gravel A mixture of pebbles, larger rocks, and sand, with more pebbles than sand.

gravitation The force of attraction between any two bodies that is increased by their greater masses and decreased by their greater distance from each other.

gravity meter An instrument that measures the force of gravity.

Greenwich Mean Time (GMT) Time at the Prime Meridian, the meridian which passes through Greenwich, England.

gypsum A kind of salt different from table salt that forms from evaporating sea water.

half-life The amount of time required for half of a particular number of radioactive atoms to decay.

heat sink A place cooler than those places around it, into which heat moves.

heat source A place warmer than those places around it, from which heat moves.

high A region of high air pressure.

hillside creep A slow movement of soil downhill.

horizon The line along which the sky appears to meet the earth.

H-R diagram A graph on which stars are grouped on the basis of their surface temperatures and their luminosities.

hurricane A violent cyclonic disturbance that forms in the tropics, known as a typhoon over the Pacific.

hydration The chemical joining of water molecules to a mineral.

ice cap A large sheet of glacier ice that flows outward in all directions.

igneous rock Rock that has formed from molten material, such as lava from a volcano.

inner core The innermost zone of the earth, made up of solid iron and nickel.

intrusion A mass of magma that is forced into or against the rock above it, and then cools.

isobar A line on a weather map along which the air pressure is the same.

isostasy The balancing process in the earth's crust, in which one part of the crust rises while another part sinks.

isotherm A line on a weather map along which the temperature is the same.

latitude The distance in degrees north or south of the equator, measured by parallels.

lava Melted minerals (magma) that have reached the earth's surface. It usually contains dissolved gases.

law A general rule that describes how events will occur every time they happen.

leaching The dissolving of minerals in soil by water that moves through it.

light-year The distance light travels in one year, or about 6 trillion miles.

limestone A sedimentary rock made up mostly of the mineral calcite. The shells of sea animals contain calcite, and shells are often part of limestone.

lines of force The expression of the effects of a magnetic field on particles that can move in the forces of the field.

longitude The distance in degrees east or west of the Prime Meridian, measured by meridians.

low A region of low air pressure.

luminosity The real, or actual brightness of energy radiations of stars or other objects in the sky, usually given in comparison to the brightness of the sun.

lunar month The time (29.5 days) from one new moon to the next new moon.

magma A hot, liquid mass of minerals that lies beneath the earth's surface.

magnet A piece of iron so treated that it creates a magnetic field around it.

magnetic field A field in which effects of attraction or repulsion of unlike or like poles can be measured.

magnetite A mineral containing iron that is naturally magnetic.

magnetometer An instrument for measuring the strength and direction of a magnetic field.

magnetosphere The region around the earth, another planet, or another body in space in which its magnetic field exists.

magnitude, brightness The number which stands for the brightness of an object like a star or a planet in the sky.

magnitude, earthquake The strength, or amount of force or energy sent out by an earthquake.

marble A metamorphic rock that was originally limestone.

mare A large dark relatively smooth area on the moon covered with volcanic lava flows and soil. (plural - maria)

mass The amount of material or matter of which anything consists.

mass wasting A fast movement of soil downhill, such as a landslide or a mud flow.

mantle The layer of heavy material below the earth's crust.

meander A wide, sharp bend in a stream, formed as an old or mature stream cuts sideways in a broad valley.

mechanical weathering Weathering that changes only the particle size of a material.

mesosphere The region above the stratosphere in which many of the chemical reactions in the atmosphere take place.

metamorphic rock A rock that has been changed into another kind of rock by heat and pressure, or by heat or pressure alone.

meteor A small particle of matter from outer space that flashes when heated as it passes rapidly through the earth's atmosphere.

meteorite A meteor that reaches the earth's surface without burning up in the atmosphere.

meteorologist A scientist who studies atmospheric conditions to understand and to predict the weather.

meteor trail The bright or smoky trail left in the sky after the passage of a large meteor.

microwave One of a series of wave radiations between infrared and radio waves in the electromagnetic spectrum.

Milky Way The name of the galaxy in which we live.

mineral A non-living material, not made by man, made up of only one element or chemical compound throughout.

molecular mass The mass of a kind of molecule, the total sum of the masses of its atoms.

moon The single natural satellite of the earth; sometimes used for natural satellites of other planets.

mountain root The extension of rock downward below a mountain or mountain range.

mutation A change in a reproductive cell of an animal or plant that causes it to have offspring different from itself.

natural satellite A satellite in orbit without human action, like any moons of planets.

natural selection The natural way of selecting which kinds of animals and plants will live and where they will live. For a given area, the weakest plants and animals will die or be forced to move to another area.

nearside The hemisphere of the moon always turned toward the earth.

nebula A cloud of fine dust and gas in space, like the Great Nebula of Orion or the Crab Nebula.

neutron An uncharged, or neutral, particle in the atomic nucleus.

nova A star that explodes and brightens rapidly, then dims slowly.

obsidian An igneous rock that has hardened from magma so fast that no crystals formed. It is also called volcanic glass.

oceanic ridge An underwater mountain range.

orbit The path followed by a planet or satellite revolving around another body.

outcrop A mass of solid rock of the earth's crust that is not covered by soil.

outer core The layer inside the earth, probably of liquid iron and nickel, that lies below the mantle.

oxbow lake A body of water left standing when a meander is cut off by a new stream channel.

oxide A chemical compound formed by the joining of one or more oxygen atoms with another compound or element.

Pangaea The name given to the ancient giant continent that scientists think split up into the continents we know today.

parallel A reference line running around the earth that remains at equal distances from the equator at every point, giving latitude.

parent material Broken rock, formed by weathering, from which soil forms.

pebble A piece of rock from 4 millimeters to 64 millimeters across.

perigee The place in a satellite's elliptical orbit at which the satellite is closest to the body around which it orbits.

period The amount of time required for a particle or object to move back and forth once.

period, geologic The largest subdivision of a geologic era.

period of revolution The time a body, such as a planet or satellite, takes to complete one revolution of the body around which it orbits. The earth's period is one year, 365¼ days.

permeable Being formed with spaces through which water can flow.

phase The apparent shape of the lighted part of a body like the moon or a planet, depending on its position in relation to the sun and the observer.

plain A large flat area, on land or on the ocean floor.

pollution Change in a natural equilibrium that is harmful to one or more kinds of organisms.

planet A large body in the solar system that revolves around the sun.

plate A section of the earth's crust that moves horizontally.

pore An open space between two or more particles. Pores can be found between soil particles or the particles making up a rock.

porosity A measure of the amount of spaces between particles of a rock or soil.

precipitation Any water, liquid or solid, that falls from the atmosphere to the ground; it includes drizzle, rain, snow, sleet, and hail.

Prime Meridian The earth's meridian running through Greenwich, England, used to establish meridians for measuring longitude.

probability The likelihood of an event, often expressed as so many chances in 10 (2 in 10) that the event will happen, or so much percentage in 100 (90 percent) that the event will happen.

proton A tiny, positively charged particle in the atomic nucleus.

protractor An instrument used to measure angles around a half circle.

pulsar A star whose radiations pulse, or flash off and on, many times per second.

P wave The movement of energy from one place to another by one particle pushing against another in a more or less straight line.

quartzite A metamorphic rock that was originally sandstone.

radiation One of the kinds of energy given off by hot objects like stars.

ray A bright streak on the moon blown out as a meteorite hits the moon and makes a crater.

redshift The shift of light toward the longer, redder waves as a star or galaxy moves away from us.

reference line A line that is referred to in locating something. The equator and Prime Meridian are reference lines for locating places on earth.

reference point A point or place that is referred to in locating something. The North and South Poles are reference points for locating places on earth.

relative humidity The percentage of saturation of water vapor in the air.

revolution A motion of passing around something, like the earth moving around the sun.

right ascension Distances in the sky east from the vernal equinox measured in hours and minutes, in the same way as longitude on earth in measured in degrees.

rill A shallow or deep groove or trough on the surface of the moon that looks like a canyon on earth.

rock A piece of the hard part of the earth that was not made by man. It can be a mineral or a mixture of minerals.

rock cycle A series of changes that happen to rocks, leading from one kind to another over a long period of time.

rock salt Ordinary table salt in rock form.

rocky planet One of the four planets nearest the sun, composed mostly of minerals and rocks.

rotation A motion of turning around a center or turning on an axis, like the turning of a wheel or the turning of the earth.

sand A material made up of grains between 1/16 millimeter and 2 millimeters across that are not cemented together.

sandstone A sedimentary rock made up of sand grains that have been cemented together.

satellite Something like the moon that revolves in orbit around a planet.

saturation The state of a fluid when it can hold no more of another substance; air is saturated when it can hold no more water vapor.

seamount An underwater volcano.

season One of the four periods of the year separated from each other by different temperature, and often weather, conditions.

sediment A collection of weathered rock material that has been deposited in a body of water, or on land by a stream or wind.

sedimentary rock A rock that has been formed by sediment particles being cemented together, or solid crystals of some mineral that have been dissolved in water.

seismograph An instrument that senses and records earthquake waves.

seismometer An instrument that is sensitive to earthquake waves.

shale A sedimentary rock made up of very small particles that were originally clay and sand. The sand grains have been broken into particles smaller than sand size.

shock front A region in which a body moving through a gas or liquid creates disturbances in the gas or liquid.

slate A metamorphic rock that was originally shale.

slurry A mixture of water and a very fine material that stays suspended in the water.

soil aggregate A clump of soil particles stuck together.

solar flare An active place on the sun releasing a flood of wave and particle radiations.

solar system The group of all the things, like planets, moons, comets, dust, and particles, that are attracted by the gravitational force of the sun.

solar wind The flood of atomic particles, such as protons and electrons, cast off by our sun and other stars.

solution A mixture in which the solid material has disappeared in a liquid. The particles are too small to be seen.

source The upper end of a stream, where it begins.

spectroscope An instrument used to break up light or some other kind of radiation into the series of radiations of which it is made.

spectrum, visible The series of colors, like those of a rainbow, into which light from a radiation source can be broken.

stratosphere The region above the troposphere in the earth's atmosphere in which the temperature is relatively constant with increasing height.

stream system A large stream and all its tributaries and distributaries.

submarine canyon A deep underwater valley.

subsoil The weathered material just below the topsoil layer.

sunspot A spot of lower temperature than most of the surface of the sun. The spot looks dark because of the lower temperature.

supercluster A group of several clusters of galaxies.

supernova A star that loses almost all its mass in a giant explosion.

surface runoff Water that runs on the surface of an area, instead of soaking into soil.

suspension A mixture of very small particles in water. The particles are so small that they float freely at any depth.

S wave The movement of energy from one place to another by rows of particles moving back and forth past other rows of particles.

talus The pile of broken rock at the bottom of a cliff.

telescope An instrument used by astronomers to observe faraway objects in the sky by making them appear closer. Optical telescopes receive light waves, radio telescopes receive radio waves, and there are telescopes for all the other waves in the electromagnetic spectrum.

temperature inversion A local condition of the atmosphere in which a layer of warm air is above a layer of cool air. This condition causes pollutants to get trapped in the lower layer and form smog.

terminator The line between the lighted and dark parts of a body like the moon or a planet.

thermal pollution The upsetting of a natural equilibrium in a body of water by the addition of hot water to it.

thermometer, wet and dry bulb A weather instrument used to measure the relative humidity of the air by the difference between the wet and dry bulb readings.

thermosphere The region in the earth's atmosphere above the mesosphere, in which the temperature steadily increases with height.

tide The rise and fall of water in oceans caused by gravitational attraction of moon and sun. There are also tides in the atmosphere and the earth's crust.

till A mixture of clay, sand, pebbles, and larger rocks, with more clay and sand than pebbles and larger rocks.

time When something happens (6 o'clock) or how long something lasts (a half hour, from 6 o'clock to 6:30).

time meridian The meridians at 15° intervals east and west of the Prime Meridian on which the earth's time zones are based.

time zone The band 7.5° east and west of a time meridian in which the same time is used.

topsoil The layer of weathered rock and mineral material at the earth's surface. It usually contains dead plant material.

tornado A fast-spinning column of air below a thunder cloud.

trench An extremely deep part of the ocean, long, narrow, and deep. Trenches may be caused by the sea floor being pulled or pushed below the earth's crust.

trade winds The band of winds north and south of the equator blowing from the east.

tributary A small stream that adds water to a larger stream.

troposphere The lowest temperature region of the earth's atmosphere, in which most of the clouds occur, with temperature decreasing with height.

tsunami A large wave in the ocean caused by an earthquake.

tufa A kind of limestone formed by evaporation from spring water that contains calcite.

turbidity current A current within a body of water caused by a flow of heavier water containing suspended material.

universe Everything that exists.

Van Allen belts Regions of densely gathered charged particles trapped in the magnetic field of a body like the earth.

varve A pair of sediment layers deposited on a pond or lake bottom during a year in a cold climate. There is a coarser spring and summer layer and a finer winter layer.

velocity The speed of something moving, measured in the distance traveled over a certain length of time.

vent An opening in the crust of the earth through which magma with its dissolved gases flows, to become lava. Several vents may make up a single volcano.

vernal equinox The reference point on the celestial equator from which distances to the east are measured in the sky.

visibility A measure of the distance at which sizable objects can be clearly seen and named.

volcano An opening or series of openings in the earth's crust through which melted minerals reach the earth's surface from below.

volume The amount of space occupied by anything.

water cycle A cycle in which water moves from oceans to air to land and back to the oceans in streams.

watershed Land between streams that adds surface runoff and underground water to the streams on both sides.

waterspout A fast-spinning column of air above a body of water.

water vapor Water in its gaseous state, as opposed to its liquid or solid (ice) state.

wavelength The distance between the crests or troughs of two waves next to each other.

weather The state of the air at a given time in terms of temperature, pressure, precipitation, wind, and visibility.

weathering The natural breaking down of solid materials into small pieces, and the chemical changes that take place in solid materials as they break down.

weathering agent Something that causes weathering, such as acids, water, ice, moving air, or falling rocks.

zodiac The band across the sky along the ecliptic in which 12 constellations are found through the year.

zone A large area or region within which certain similar conditions are found.

Picture Credits

Title page photograph: R. S. Fiske, U.S. Geological Survey.

All illustrations not otherwise credited were prepared by Vantage Art, Inc., Massapequa, New York.

Teacher's Manual illustrations were prepared by Bertrick Associate Artists, Inc., Seaford, New York, except for diagram photo, page T11, from Yerkes Observatory, University of Chicago, and drawing, page T58, by Vantage Art, Inc.

The photos on the following pages were taken by John T. Urban, Arlington, Mass: pp. vi, 39, 40, 50, 72 (top), 77 (bottom), 81, 95, 100, 104, 131 (bottom), 163, 176, 188, 193 (top, right), 195 (bottom), 199, 201, 211, 231 (bottom), 232 (bottom), 245, 246, 248 (top), 251, 254, 256 (bottom, right), 257, 261 (bottom) courtesy of Karen Elitcher, 264 (top), 267 (top, left), 268, 269, 278, 292, 305, 319, 320, 341, 344 (bottom), 345, 372, 375, 377, 379, 388, 398, 409, 410, 411, 425, 460, 463, 507, 542, 555 (top).

Chapter 1

p. x, 1 Lick Observatory, University of California; **p. 2** Laurence Lowry; **p. 5** Stephen G. Maka; **p. 6** courtesy of Hale Observatories, copyright, California Institute of Technology and Carnegie Institution of Washington; **p. 7** official U.S. Navy photograph, Flagstaff Station, U.S. Naval Observatory; **p. 9** (top) Yerkes Observatory, University of Chicago; (bottom, left) G. C. Atamian, Talcott Mountain Science Center, Avon, Connecticut; (bottom, right) Adapted from Thaddaeus Hagecius, "Discourse Concerning the Appearance of a New Star of Exceptional Size and Brilliance," Frankfurt am Main, 1574; **p. 12** Donald H. Menzel's "Field Guide to the Stars and Planets," courtesy of the Harvard College Observatory; **p. 15, 16** G. C. Atamian, Talcott Mountain Science Center, Avon, Connecticut; **p. 18** Laurence Lowry; **p. 20, 21** adapted from *Investigating the Earth,* copyright © 1967 American Geological Institute; **p. 22** courtesy of Hale Observatories, copyright, California Institute of Technology and Carnegie Institution of Washington; **p. 23** adapted from *Investigating the Earth,* copyright © 1967 American Geological Institute; **p. 24** courtesy of Hale Observatories; **p. 25** Lick Observatory, University of California; **p. 27** U.S. Atomic Energy Commission.

Chapter 2

p. 30 Donald H. Menzel's "Field Guide to the Stars and Planets," courtesy of the Harvard College Observatory; **p. 32** G. C. Atamian, Talcott Mountain Science Center, Avon, Connecticut; **p. 33, 34, 35** Donald H. Menzel's "Field Guide to the Stars and Planets," courtesy of the Harvard College Observatory; **p. 36** (top) courtesy of Hale Observatories; (bottom) courtesy of Hale Observatories, copyright, California Institute of Technology and Carnegie Institution of Washington; p. 37, 38 courtesy of Hale Observatories; **p. 42, 43** photos A, B, C, D, E, G, H, I, K, M, N, O: courtesy of Hale Observatories; F, J, L, P: Lick Observatory, University of California; **p. 44** Courtesy of Hale Observatories; **p. 46, 47** The Mount Stromlo Observatory and Siding Spring Observatory, Australian National University. **p. 46** (bottom) National Radio Astronomy Observatory, operated by Associated Universities, Inc., under contract with the National Science Foundation; **p. 48** courtesy of Hale Observatories; **p. 54** (left) Leuschner Observatory, University of California, Berkeley.

Chapter 3

p. 62 (left) National Aeronautics and Space Administration; (right) David Arnold; **p. 66** Jet Propulsion Laboratory, California Institute of Technology, Pasadena; **p. 67** (top) Donald H. Menzel's "Field Guide to the Stars and Planets," courtesy of the Harvard College Observatory; **p. 76** adapted from the National Association of Watch and Clock Collectors; **p. 77** (top) National Aeronautics and Space Administration; **p. 79** adapted from "World Time Zones" map (data from U.S. Navy Oceanographic Office Chart No. 5192) in *The Earth Sciences,* 2nd Edition, by Arthur N. Strahler (Harper & Row, 1971, page 46); **p. 84** (top) National Aeronautics and Space Administration; **p. 89** adapted from map, courtesy of Jet Propulsion Laboratory, California Institute of Technology, Pasadena.

Chapter 4

p. 92 Courtesy of Hale Observatories; **p. 102** Yerkes Observatory, University of Chicago; **p. 107** Yerkes Observatory, University of Chicago; **p. 109** courtesy of Hale Observatories; **p. 110** Haleakala Observatory, University of Hawaii; **p. 111** High Altitude Observatory, National Center for Atmospheric Research; **p. 113** (bottom, left) Lowell Observatory, Flagstaff, Arizona; (bottom, right) Jet Propulsion Laboratory, California Institute of Technology, Pasadena; **p. 114, 115** Jet Propulsion Laboratory, California Institute of Technology, Pasadena; **p. 116, 117** (top) courtesy of Hale Observatories, copyright, California Institute of Technology and Carnegie Institution of Washington; **p. 117** (bottom, left) National Aeronautics and Space Administration; (bottom, right) Robert B. Leighton, California Institute of Technology, Pasadena; **p. 119** (left) Division of Radiophysics, Commonwealth Scientific and Industrial Research Organization, Australia; (right) courtesy of Hale Observatories; **p. 121** Yerkes Observatory, University of Chicago; **p. 123** Yerkes Observatory, University of Chicago.

Chapter 5

p. 128 (left) courtesy of Hale Observatories; (right) National Aeronautics and Space Administration; **p. 130** courtesy of Hale Observatories; **p. 131** (top) Jet Propulsion Laboratory, California Institute of Technology, Pasadena; **p. 132** National Oceanic and Atmospheric Administration; **p. 138** National Aeronautics and Space Administration; **p. 139** adapted from National Aeronautics and Space Administration Farside Chart; **p. 140** G. C. Atamian, Talcott Mountain Science Center, Avon, Connecticut; **p. 143** photos by G. Blouin, Information Canada; **p. 144** G. Pannella, Peabody Museum of Natural History, Yale University; **p. 149** Lick Observatory, University of California; **p. 151, 152** (top) National Aeronautics and Space Administration; **p. 152** (left, bottom) John A. Wood, Smithsonian Institution Astrophysical Observatory; **p. 154** National Aeronautics and Space Administration; **p. 158–159** National Oceanic and Atmospheric Administration.

Chapter 6

p. 160 National Aeronautics and Space Administration; **p. 164** National Center for Atmospheric Research; **p. 167** (top) Marshall Flake; (bottom) courtesy of the Boeing Company; **p. 168** Jet Propulsion Laboratory, California Institute of Technology, Pasadena; **p. 170** (left) Geophysical Institute, University of Alaska; (right) Gustav Lamprecht, Geophysical Institute, University of Alaska; **p. 179** Division of Radiophysics, Commonwealth Scientific and Industrial Research Organization, Australia.

Chapter 7:

p. 186 National Aeronautics and Space Administration; **p. 204** (left) National Environmental Satellite Service, National Oceanic and Atmospheric Administration; (right) adapted from map by P. K. Rao, *Science,* Vol. 173, pp. 529–530. Fig. 1, August 6, 1971, copyright 1971 by the American Association for the Advancement of Science, with permission of NOAA's National Environmental Satellite Service.

Chapter 8

p. 214 National Oceanic and Atmospheric Administration; **p. 218** adapted from National Oceanic and Atmospheric Administration maps; **p. 219** National Oceanic and Atmospheric Administration, National Environmental Satellite Service; **p. 223** (bottom) Mt. Washington Observatory; **p. 224, 225** courtesy of Hubbard Scientific Company, Northbrook, Illinois 60062. Used with permission. **p. 234** National Oceanic and Atmospheric Administration; **p. 235** National Oceanic and Atmospheric Administration; **p. 236** Stephen G. Maka; **p. 240, 241** Georg Gerster, from Rapho-Guillumette.

Chapter 9

p. 248 (bottom, left) Smithsonian Institution; (bottom, right) courtesy of the Buffalo Museum of Science; **p. 253** John S. Shelton; **p. 256** (top) Plate 18 "Arkose" (p. 246) in *Sedimentary Rocks* by F. J. Pettijohn, (Harper & Row, 1949); **p. 256** (bottom, left) courtesy of Hubbard Scientific Company, Northbrook, Illinois, 60062. Used with permission. **p. 258** (top) Pål-Nils Nilsson/TIO; (bottom) J. R. Stacy, U.S. Geological Survey; **p. 259** (top) Soil Conservation Service, USDA; (bottom) Diane Pitochelli; **p. 260** (top) Diane Pitochelli; (bottom) Carol Feisinger; **p. 261** (top) Bill Ratcliffe; **p. 262** George McManigle; **p. 266** courtesy of Hubbard Scientific Company, Northbrook, Illinois 60062. Used with permission. **p. 267** (top, right, bottom left and right) courtesy of Hubbard Scientific Company, Northbrook, Illinois 60062. Used with permission.

Chapter 10

p. 274 (top) Alpha Photo Associates, Inc.; (bottom) U.S. Geological Survey; **p. 277** Rochester Times-Union, Rochester, New York; **p. 279** adapted from *Volcanoes,* Science Service Program, copyright © 1968 by Nelson Doubleday, Inc. Reproduced by permission of Doubleday & Company, Inc.; **p. 280** U.S. Geological Survey; **p. 281** (right) U.S. Geological Survey, Hawaiian Volcano Observatory; **p. 282** U.S. Geological Survey; **p. 283** Camera Hawaii, Honolulu; **p. 284** (right) David C. Roberts; **p. 285** (bottom) courtesy of Japan National Tourist Organization; **p. 286** (bottom) Oregon State Highway Department, Travel Division; **p. 288** (left) Grant Heilman; (right) Jerome Wyckoff; **p. 293** (top) Prism Productions, Inc.; **p. 294** FPG; **p. 295** S. Thorarinsson, Science Institute, University of Iceland.

Chapter 11

p. 298 United Press International, Inc.; **p. 301** George McManigle; **p. 303** Redrawn after Figure 6, pg. 29 in B. Gutenberg and C. F. Richter, *Seismicity of the Earth and Associated Phenomena,* copyright 1949, 1954 by Princeton University Press. Reprinted by permission; **p. 304** adapted from L. Don Leet and Sheldon Judson, *Physical Geology,* 3rd ed., © 1965. By permission of Prentice-Hall, Inc.; **p. 306** William A. Garnett; **p. 307** courtesy of Carnegie Institution of Washington; **p. 312** adapted from L. Don Leet and Sheldon Judson, *Physical Geology,* 3rd ed., © 1965, by permission of Prentice-Hall, Inc.; **pp. 314–315** (bottom) Department of Geography and Geology, Indiana State University; **p. 324** adapted by permission from *Investigating the Earth,* experimental laboratory manual, copyright © 1967 American Geological Institute; **pp. 330, 331** William A. Garnett.

Chapter 12

pp. 332, 335 Diane Pitochelli; **p. 337** Jerome Wyckoff; **p. 338** Winston Pote, Shostal Associates, Inc.; **p. 339** George McManigle; **p. 342** Diane Pitochelli; **p. 346** (left) Diane Pitochelli; **p. 346** (right) courtesy of Habersham & Chas. Hix Photography, Hickory, N. Carolina; **p. 349** adapted by permission from *Investigating the Earth,* copyright © 1967 American Geological Institute. **p. 350** Art adapted from Soil Conservation Service, USDA, photo by Diane Pitochelli; **p. 353** portion of map taken from Olson, G. W., J. E. Witty and R. L. Marshall, 1969, "Soils and their use in the five-county area around Syracuse." Cornell Miscellaneous Bulletin 80. New York State College of Agriculture, Cornell University, Ithaca, N.Y.

Chapter 13

p. 358 Walter Dawn; **p. 361** Jerome Wyckoff; **p. 362** Bradford Washburn; **p. 363** (top) Bradford Washburn; (bottom) John S. Shelton; **p. 364** John S. Shelton; **p. 365** (top) courtesy of Sante Fe Railway; (bottom) U.S. Bureau of Reclamation; **p. 366** Arvid M. Johnson; **p. 367** John S. Shelton; **p. 370** Diane Pitochelli; **p. 376** adapted from New York State Museum Bulletin #358, "Traces of Early Man in New York", **p. 385** (top) Robert Rubic: DPI; (bottom) Jerome Wyckoff; **p. 386** Darwin Van Campen: DPI; **p. 387** National Aeronautics and Space Administration; **p. 389** National Air Photo Library, Surveys and Mapping Branch, Department of Energy, Mines, and Resources, Geological Survey of Canada, A 1814 (27); **p. 390** Carol Felsinger.

Chapter 14

p. 394 Marshall Flake; **p. 401** George McManigle; **p. 402** Ivan Massar: Black Star; **p. 403** adapted from *Science Year, The World Book Science Annual.* © 1970 Field Enterprises Education Corporation. **p. 415, 416** U.S. Naval Oceanographic Office; **p. 417** Shostal Associates, Inc.

Chapter 15

p. 426 Courtesy of Scintrex Mineral Surveys, Inc.; **p. 429** George McManigle; **p. 435, 436** adapted from S. Warren Carey, ed., "The Tectonic Approach to Continental Drift" in Continental Drift, a Symposium, University of Tasmania. **p. 436** R. W. Fairbridge, Fairbridge Associates; **p. 438** adapted from Allen Cox and R. R. Doell, "Review of Paleomagnetism," Geological Society of America Bulletin, LXXI (1960), p. 758, Fig. 33. **p. 443** Adapted from map by X. Le Pichon, *Journal of Geophysical Research,* vol. 73, page 3675, 1968. **p. 445** (top) Stephen R. Blount; **p. 458–459** William Belknap, Jr. from Rapho-Guillumette.

Chapter 16

p. 464 (left) Ernst Antevs; **p. 465** (top) Joyce Wilson: DPI; (bottom) Wide World Photos, Inc.; **p. 466** U.S. Forest Service; **p. 468** reprinted by permission of IPC Newspapers, Ltd., and Publishers-Hall Syndicate. **p. 472** (left) courtesy of Shell Development Company; (right) courtesy of Israel Museum, Jerusalem; **p. 473** (top) B. M. Shaub; (bottom) courtesy of the Center for Meteorite Studies, Arizona State University, Tempe, Arizona; **p. 476** American Museum of Natural History; **p. 482** adapted from R. S. Dietz and J. C. Holden, *Journal of Geophysical Research,* pp. 4939–4956, Vol. 75, 1970.

Chapter 17

p. 491 Jerome Wyckoff; **p. 492** Walter Dawn; **p. 498** adapted from *Nature and Science,* Vol. 5, No. 5, November 13, 1967, copyright © 1967 by the American Museum of Natural History. Reproduced by permission of Doubleday & Company, Inc. **p. 500** Adapted from *Evolution of the Earth* by R. H. Dott and R. L. Batten, copyright © 1971 by McGraw-Hill, Inc. Used with permission of McGraw-Hill Book Company. **p. 506** Redrawn from *Physical Geology Laboratory Manual,* 2nd Ed., W. K. Hamblin, J. D. Howard, 1967, Burgess Publishing Company. **p. 508** Adapted from Generalized Geologic Map of New York State with the permission of the New York State Museum and Science Service. **p. 509** Adapted from *Evolution of the Earth* by R. H. Dott and R. L. Batten, copyright © 1971 by McGraw-Hill, Inc. Used with permission of McGraw-Hill Book Company.

Chapter 18

p. 516 (top left and bottom) Pat Morris: ARDEA; **p. 516** (top right) Stephen G. Maka; **p. 517** Lamont-Doherty Geological Observatory of Columbia University; **p. 522** George McManigle; **p. 528** (top) Brookhaven National Laboratory; (bottom) copyright Life Magazine, © 1947 Time, Inc.; **p. 535** American Museum of Natural History; **p. 536** Jet Propulsion Laboratory, California Institute of Technology, Pasadena; **p. 538** American Museum of Natural History.

Chapter 19

p. 544 George McManigle; **p. 551** (top) Ernest Baxter, Black Star; (bottom) Ernst Haas, Magnum Photos; **p. 553** Cal Harbert From National Audubon Society; **p. 554** Los Angeles County Air Pollution Control District; **p. 555** (bottom) J. M. DiJoseph Jr.; **p. 556** United Press International, Inc.; **p. 558** Maurice E. Landre from National Audubon Society; **p. 559** (top) Cratie Sandlin, Van Cleve Photography; **p. 560** James N. Westwater; **p. 561** Grant Heilman; **p. 562** (left) Grant Heilman; (right) Lincoln Nutting from National Audubon Society; **p. 563** Burk Uzzle, Magnum Photos; **p. 567** by Allan J. deLay, courtesy of South Tahoe Public Utility District; **p. 569** Grant Heilman.

Appendix

p. 579–586 Maps from "Field Guide to the Stars and Planets," copyright © 1964 by Donald H. Menzel, based on designs originated by H. A. Rey. **p. 595, 596** (top) courtesy of Hale Observatories; **p. 596** (middle and bottom) Lick Observatory, University of California.

Index

M

N

O

T